To: Joyce

Liz

Sam

TABLE OF CONTENTS

Chapter VIII: Finite Dimensional Manifolds

INTRODUCTION

Compactness is frequently an annoying hypothesis in differential topology. Even when one is primarily interested in a compact manifold, associated noncompact manifolds turn up, eg. the leaf space of a foliation. Also, in technical constructions, it would be helpful to be able to dispense with compactness. For example, if f is a diffeomorphism of a compact manifold, X, then it is helpful in studying the dynamics of f to regard the integers, $\mathbb{Z}$, as a discrete manifold and look at the manifold of maps, $\mathcal{C}(\mathbb{Z},X)$. Again it is compactness that prevents one from noting that Hartman's Theorem is not only related to the structural stability theorem for Anosov diffeomorphisms but is in fact a corollary of the latter because a hyperbolic linear map is an Anosov diffeomorphism of Euclidean space.

This book describes the category of metric manifolds and metric maps to which a broad class of theorems and constructions extend from the realm of compact manifolds. The category is a broad one because all paracompact manifolds admit metric structures. Metric theorems include compact theorems because a compact manifold admits a unique metric structure and with respect to it any smooth map with compact domain is a metric map. Finally, there is a sufficient abundance of metric maps that, for example, structural stability under perturbation within the family of metric maps remains useful.

Our principal tool is the atlas. Just as most elementary constructions in p.l. topology are really simplicial constructions so are most of the elementary constructions in differential topology really atlas constructions building new atlases from old. In Chapter I we review such standard constructions as products of spaces and bundles, pull back of bundles, etc. As well as illustrating the atlas point of view, this develops notation for future use. We also make some easy definitions suggested by the emphasis on atlases. For example, if $\mathcal{a}_1 = \{U_\alpha, h_\alpha\}$ and $\mathcal{a}_2 = \{V_\beta, g_\beta\}$ are two atlases then an index preserving map $f\colon \mathcal{a}_1 \to \mathcal{a}_2$ is a continuous map, f, of the underlying manifolds and an unnamed map of the index sets $\alpha \to \beta(\alpha)$ such that $U_\alpha \subset f^{-1}(V_{\beta(\alpha)})$ for all α.

In the past this primacy of the atlas has been ignored. Where the p.l. topologist has used triangulations the differential topologist has used charts or, equivalently, local coordinates. This is because a smooth manifold has a maximal atlas consisting of all smooth charts. However, the maximal atlas lacks certain natural tools possessed by other atlases less profligate in charts. What is needed is some control over the size of the transition maps of the atlas. Size means norm in some Banach space of functions. So, in Chapter II we study various function space types.

A function space type $\mathfrak{M}$ associates to every Banach space F and every bounded open set U in a Banach space E a Banach space $\mathfrak{M}(U,F)$ of functions from U to F. For example, $\mathcal{B}(U,F)$ consists of bounded functions f with the sup norm $\|f\|_0 = \sup\{\|f(x)\|\colon x \in U\}$ and $\mathcal{C} \subset \mathcal{B}$ is the subspace of continuous functions. Defining $\|f\|_L =$ $\sup\{\|f(x) - f(y)\|/\|x - y\|\colon x \neq y$ and the segment $[x,y] \subset U\}$, we get $\mathcal{L}(U,F)$ the space of functions f with $\max(\|f\|_0, \|f\|_L) = \|f\|_{\mathcal{L}} < \infty$. This is the space of bounded, uniformly locally Lipschitz functions, or equivalently, of bounded functions, Lipschitz with respect to a natural intrinsic metric on U.

We follow Palais' "axiomatic" approach [21]. Thus, we define $f \in \mathfrak{M}^r(U,F)$ if f is r times differentiable and its r-jet, $j^r(f)$, lies in $\mathfrak{M}(U,J^r(E;F))$. $\|f\|_{\mathfrak{M}^r} = \|j^r f\|_{\mathfrak{M}}$. $\mathfrak{M}^r$ is the rth derived function space type of $\mathfrak{M}$. Then, after checking a function space property on basic examples like $\mathcal{C}$ and $\mathcal{L}$, we verify that it is inherited as we pass from

$\mathfrak{M}$ to $\mathfrak{M}^r$. For example, the <u>Gluing Property</u> states that if $\{U_\alpha\}$ is an open cover of U and for $f: U \to F$, $f|U_\alpha \in \mathfrak{M}(U_\alpha,F)$ with $\sup_\alpha \|f|U_\alpha\|_{\mathfrak{M}} < \infty$ then $f \in \mathfrak{M}(U,F)$ with $\|f\|_{\mathfrak{M}} = \sup_\alpha \|f|U_\alpha\|_{\mathfrak{M}}$.

Define the Banach space product, $\hat{\Pi}_\alpha F_\alpha$, of a family of Banach spaces to be the set of $\{x_\alpha\} \in \Pi_\alpha F_\alpha$ such that $\|\{x_\alpha\}\| = \sup\|x_\alpha\| < \infty$, with this sup norm. The <u>Strong Product Property</u> states that if $f_\alpha \in \mathfrak{M}(U,F_\alpha)$ and $\sup\|f_\alpha\|_{\mathfrak{M}} < \infty$ then the product map $\hat{\Pi} f_\alpha \in \mathfrak{M}(U,\hat{\Pi}F_\alpha)$ with $\|\hat{\Pi} f_\alpha\|_{\mathfrak{M}} = \sup\|f_\alpha\|_{\mathfrak{M}}$. The <u>Product Property</u> is this statement for finite index sets.

Smoothness of the composition map is handled as follows. Let $\mathfrak{M}$, $\mathfrak{M}_1$ and $\mathfrak{M}_2$ be function space types with $\mathfrak{M}_1 \subset \mathfrak{M}_2$ (i.e. $\mathfrak{M}_1(U,F) \subset \mathfrak{M}_2(U,F)$ as sets and $\| \ \|_{\mathfrak{M}_2} \le \| \ \|_{\mathfrak{M}_1}$ on the subset). We say that <u>$\mathfrak{M}_1$ maps $\mathfrak{M}$ to $\mathfrak{M}$ in an $\mathfrak{M}_2$ way</u> if whenever $U \subset F_1$, $G \subset F_2$ and $V \subset E_1$ are open and bounded and $H: G \to \mathfrak{M}(U,E_1)$ is an $\mathfrak{M}_2$ map with Image $H(g) \subset V$ for all $g \in G$, then the bounded linear map $\Omega_H: \mathfrak{M}_1(V,E_2) \to \mathfrak{M}_2(G,\mathfrak{M}(U,E_2))$ is well defined by $\Omega_H(f)(g) = f \circ H(g)$ and $\|\Omega_H\| \le O^*(\|H\|_{\mathfrak{M}_2})$. The latter is a typical estimate for our work and it means there exist constants K, n depending only on the function space types such that $\|\Omega_H\| \le K \max(\|H\|_{\mathfrak{M}_2},1)^n$. For example, that $\mathfrak{M}$ maps $\mathfrak{M}$ to $\mathfrak{M}$ in a $\mathfrak{M}$ way means that if $g: U \to V$ is an $\mathfrak{M}$ map then $g^*: \mathfrak{M}(V,E) \to \mathfrak{M}(U,E)$ is a bounded linear map well defined by $g^*(f) = f \circ g$ and $\|g^*\| \le O^*(\|g\|_{\mathfrak{M}})$.

The Gluing and Product Properties inherit in the obvious way. For the composition property there are two inheritance theorems: If $\mathfrak{M}_1$ maps $\mathfrak{M}$ to $\mathfrak{M}$ in an $\mathfrak{M}_2$ way then $\mathfrak{M}_1^r$ maps $\mathfrak{M}^r$ to $\mathfrak{M}^r$ in an $\mathfrak{M}_2$ way and if, in addition, $\mathfrak{M}_2 \subset \mathcal{C}$ then $\mathfrak{M}_1^r$ maps $\mathfrak{M}$ to $\mathfrak{M}$ in an $\mathfrak{M}_2^r$ way.

If $\mathfrak{M}$ satisfies a constellation of properties including $\mathfrak{M} \subset \mathcal{C}$, Gluing, Product (but not necessarily the Strong form) and $\mathfrak{M}$ maps $\mathfrak{M}$ to $\mathfrak{M}$ in a $\mathfrak{M}$ way then $\mathfrak{M}$ is called a <u>standard</u> function space type. Being standard is a heritible property. $\mathcal{C}$ and $\mathcal{L}$ are standard (it is to get the Gluing Property for $\mathcal{L}$ that we use the seminorm $\| \ \|_L$ rather than its more obvious Lipschitz constant relatives) and hence so are $\mathcal{C}^r$ and $\mathcal{L}^r$.

$\mathcal{L}^r$ also satisfies the Strong Product Property. Finally, $\mathcal{L}^{r+s}$ maps $\mathcal{C}^r$ to $\mathcal{C}^r$ in an $\mathcal{L}^s$ way (and so $\mathcal{C}^{r+s+1}$ maps $\mathcal{C}^r$ to $\mathcal{C}^r$ in a $\mathcal{C}^s$ way) and $\mathcal{L}^{r+s+1}$ maps $\mathcal{L}^r$ to $\mathcal{L}^r$ in an $\mathcal{L}^s$ way.

In Chapter III we return to atlases and describe the category of $\mathfrak{M}$ metric manifolds and maps for any standard function space type $\mathfrak{m}$. There is also an associated category of vector bundles but we will restrict discussion here to manifolds. All manifolds are assumed to be Hausdorff Banach manifolds.

An atlas $\mathcal{A} = \{U_\alpha, h_\alpha\}$ on X is a <u>bounded $\mathfrak{M}$ atlas</u> if the transition maps $h_\alpha h_\beta^{-1} \in \mathfrak{M}(h_\beta(U_\alpha \cap U_\beta), E_\alpha)$ ($h_\alpha(U_\alpha)$ is open in E_α) and $k_{\mathcal{A}} = \max(1, \sup\|h_\alpha h_\beta^{-1}\|_{\mathfrak{M}}) < \infty$. $\mathcal{A}$ is an <u>$\mathfrak{M}$ atlas</u> if each point x of X has a neighborhood U such that $\mathcal{A}|U = \{U_\alpha \cap U, h_\alpha|U_\alpha \cap U\}$ is a bounded $\mathfrak{M}$ atlas. We then define $\rho_{\mathcal{A}}\colon X \to [1,\infty)$ by $\rho_{\mathcal{A}}(x) = \inf\{k_{\mathcal{A}|U}\colon U$ is a neighborhood of $x\}$. $\rho_{\mathcal{A}}$ is clearly upper semicontinuous and it follows from the Gluing Property for $\mathfrak{M}$ that $\rho_{\mathcal{A}}$ is bounded on an open set U of X iff $\mathcal{A}|U$ is a bounded $\mathfrak{M}$ atlas and then $k_{\mathcal{A}|U} = \sup \rho_{\mathcal{A}}|U$. For example, if X is a paracompact C^r manifold and $\mathcal{A} = \{V_\alpha, h_\alpha\}$ is a locally finite C^r atlas then choosing $\{U_\alpha\}$ an open cover with $\bar{U}_\alpha \subset V_\alpha$ it is easy to check that $\mathcal{A}_1 = \{U_\alpha, h_\alpha\}$ is a $\mathcal{C}^r$ atlas.

Let $\mathcal{A}_1 = \{U_\alpha, h_\alpha\}$ and $\mathcal{A}_2 = \{V_\beta, g_\beta\}$ be $\mathfrak{M}$ atlases on X_1 and X_2. A continuous map $f\colon X_1 \to X_2$ is a bounded $\mathfrak{M}$ map $f\colon \mathcal{A}_1 \to \mathcal{A}_2$ if $\mathcal{A}_1$ is a bounded $\mathfrak{m}$ atlas and the local representatives $f_{\beta\alpha} = g_\beta f h_\alpha^{-1} \in \mathfrak{M}(h_\alpha(U_\alpha \cap f^{-1}V_\beta), E_\beta)$ and $k(f;\mathcal{A}_1,\mathcal{A}_2) = \sup\|f_{\beta\alpha}\|_{\mathfrak{m}} < \infty$. f is an $\mathfrak{M}$ map if it is locally a bounded $\mathfrak{M}$ map and we then define $\rho(f;\mathcal{A}_1,\mathcal{A}_2)(x) = \inf\{k(f|U;\mathcal{A}_1|U,\mathcal{A}_2)\colon U$ a neighborhood of $x\}$. ρ_f has properties analogous to $\rho_{\mathcal{A}}$. Again if $\mathcal{A}_1$ and $\mathcal{A}_2$ are obtained by shrinking locally finite C^r atlases as above and $f\colon X_1 \to X_2$ is a C^r map then $f\colon \mathcal{A}_1 \to \mathcal{A}_2$ is a $\mathcal{C}^r$ map. Because $\mathfrak{M}$ maps $\mathfrak{M}$ to $\mathfrak{M}$ in a $\mathfrak{s}$ way it is easy to show that if $f\colon \mathcal{A}_1 \to \mathcal{A}_2$ and $g\colon \mathcal{A}_2 \to \mathcal{A}_3$ are $\mathfrak{M}$ maps then $g{\circ}f\colon \mathcal{A}_1 \to \mathcal{A}_3$ is $\mathfrak{M}$ and

$\rho_{g\cdot f} \leq (\rho_g \cdot f) O^* (\rho_f)$.

There are two other important functions associated with an atlas $\mathfrak{a} = \{U_\alpha, h_\alpha\}$ on X. $\lambda_{\mathfrak{a}}: X \to (0,1]$ is defined by $\lambda_{\mathfrak{a}}(x) = \sup\{r \leq 1$: the ball $B(h_\alpha x, r) \subset h_\alpha(U_\alpha)$ for some U_α containing $x\}$. Clearly, $\lambda_{\mathfrak{a}}$ is lower semicontinuous. For the maximal atlas on a smooth manifold $\lambda_{\mathfrak{a}} \equiv 1$. For an $\mathfrak{M}$ atlas, $\lambda_{\mathfrak{a}}$ need not be bounded away from 0. There is a useful tension in trying to bound $\rho_{\mathfrak{a}}$ and $1/\lambda_{\mathfrak{a}}$ simultaneously.

Finally, there is a pseudometric $d_{\mathfrak{a}}$ on X. An <u>$\mathfrak{a}$-chain</u> $(x_1, \alpha_1, \ldots, \alpha_N, x_{N+1})$ is a sequence such that $x_i, x_{i+1} \in U_{\alpha_i}$ and the segment $[h_{\alpha_i} x_i, h_{\alpha_i} x_{i+1}] \subset h_{\alpha_i}(U_{\alpha_i})$ $i = 1, \ldots, N$. The length of the $\mathfrak{a}$-chain is $\Sigma_{i=1}^N \|h_{\alpha_i} x_i - h_{\alpha_i} x_{i+1}\|$. If $x, y \in X$ then $d_{\mathfrak{a}}(x,y)$ is the infinum of the lengths of all $\mathfrak{a}$-chains connecting x and y. Recall that the infinum of the empty set is ∞. Allowing ∞ as a possible value, $d_{\mathfrak{a}}$ is a pseudometric. It needn't have the topology of X. In fact, if $\mathfrak{a}$ is the maximal atlas then $d_{\mathfrak{a}}(x,y)$ is 0 or ∞ according to whether x and y do or do not lie in the same component of X. However, if $\mathfrak{M} \subset \mathcal{L}$ (eg. $\mathfrak{M} = \mathcal{C}^r$ for $r \geq 1$) and $\mathfrak{a}$ is an $\mathfrak{M}$ atlas then $d_{\mathfrak{a}}$ is a metric with topology that of X. In fact, a central result--from which the Metric Theory derives its name--is the Metric Estimate which gives an explicit local comparison between $d_{\mathfrak{a}}$ and the Banach space metric pulled back to U_α by h_α. An estimate of the size of the region on which the comparison holds and the bounds in the comparison can be computed from $\lambda_{\mathfrak{a}}$ and $\rho_{\mathfrak{a}}$. It follows that only a paracompact manifold can admit $\mathfrak{M}$ atlases with $\mathfrak{M} \subset \mathcal{L}$.

A map $\rho: X \to [1,\infty)$ is called a bound on X. We say of two bounds ρ_1 and ρ_2 on X that <u>ρ_1 dominates ρ_2</u> (written $\rho_1 \succ \rho_2$) if there exist constants K,n such that $K\rho_1^n \geq \rho_2$ on X. $\succ$ is a partial ordering with associated equivalence relation $\sim$. For example, the equivalence class containing constant functions consists of all bounds ρ with $\sup \rho < \infty$.

In essence, a metric structure on X is a choice of bound on X which dominates the growth of everything on X. In detail, an adapted $\mathfrak{M}$ atlas is a pair $(\mathcal{G},\rho)$ with $\mathcal{G}$ an $\mathfrak{M}$ atlas and ρ a bound such that $\rho > \rho_{\mathcal{G}}$. $f\colon (\mathcal{G}_1,\rho_1) \to (\mathcal{G}_2,\rho_2)$ is an $\mathfrak{M}$ map of adapted $\mathfrak{M}$ atlases if $f\colon \mathcal{G}_1 \to \mathcal{G}_2$ is an $\mathfrak{M}$ map and $\rho_1 > \max(\rho_f,\rho_2\circ f)$. Two adapted $\mathfrak{M}$ atlases are equivalent, written $(\mathcal{G}_1,\rho_1) \sim (\mathcal{G}_2,\rho_2)$ if the following equivalent conditions hold: (1) $\rho_1 \sim \rho_2$ and $(\mathcal{G}_1 \cup \mathcal{G}_2,\rho_1)$ is an adapted $\mathfrak{M}$ atlas. (2) The identity maps $1\colon (\mathcal{G}_1,\rho_1) \to (\mathcal{G}_2,\rho_2)$ and $1\colon (\mathcal{G}_2,\rho_2) \to (\mathcal{G}_1,\rho_1)$ are $\mathfrak{M}$ maps. An $\mathfrak{M}$ metric structure on X is an equivalence class of adapted atlases. The atlases and bounds appearing in the structure are called admissible atlases and bounds. The admissible bounds are a bound equivalence class. In practice, we need to relate the value of ρ near x to the value at x. So we assume as part of the definition of a metric structure that there exist continuous admissible bounds. A stronger condition, which we call regularity of a metric structure, gives a uniform estimate as follows: A metric structure is regular if $\mathfrak{M} \subset \mathcal{L}$ and for some $(\mathcal{G},\rho)$ in the metric structure there exist constants K and n such that $\rho|B^{d_{\mathcal{G}}}(x,(K\rho(x)^n)^{-1}) \leq K\rho(x)^n$. This condition then holds for all $(\mathcal{G},\rho)$ in the metric structure.

If X_1 and X_2 are $\mathfrak{M}$ manifolds, i.e. manifolds with an $\mathfrak{M}$ metric structure, then $f\colon X_1 \to X_2$ is an $\mathfrak{M}$ map if $f\colon (\mathcal{G}_1,\rho_1) \to (\mathcal{G}_2,\rho_2)$ is an $\mathfrak{M}$ map for some, and hence any, choice of $(\mathcal{G}_i,\rho_i)$ in the metric structure of X_i, $i = 1,2$. In general a vectorfield, Riemannian metric or any sort of gadget is called an $\mathfrak{M}$ vectorfield, $\mathfrak{M}$ Riemannian metric or $\mathfrak{M}$ gadget if its local representatives with respect to an admissible atlas are $\mathfrak{M}$ maps the growth of whose norm is dominated by an admissible bound. We thus obtain the category of $\mathfrak{M}$ manifolds and $\mathfrak{M}$ maps with all the accoutrements of the usual differentiable category.

A regular metric manifold has a natural uniform space structure. If $(\mathcal{G},\rho)$ is in the metric structure and $m(x,y)$ is either $\max(\rho(x),\rho(y))$ or

$\min(\rho(x),\rho(y))$ then the uniformity $\mathfrak{U}_X$ is generated by the sets $\{(x,y): d_{\mathfrak{a}}(x,y) < (Km(x,y)^n)^{-1}\}$ as K and n vary over the positive integers. This uniformity is clearly finer than the uniformity associated with the metric $d_{\mathfrak{a}}$ but the two uniformities agree on bounded sets ("bounded" always means ρ-bounded). Hence, the topology associated with $\mathfrak{U}_X$ is the original topology of X. A sequence $\{x_n\}$ is $\mathfrak{U}_X$ Cauchy iff it is $d_{\mathfrak{a}}$ Cauchy and bounded. Uniform notions like uniform continuity, uniform neighborhood of a set and uniform open cover are defined via $\mathfrak{U}_X$.

A regular $\mathfrak{M}$ manifold is called <u>semicomplete</u> if there exist adapted atlases $(\mathfrak{a},\rho)$ in the metric structure with $\rho > 1/\lambda_{\mathfrak{a}}$. Such atlases $\mathfrak{a}$ are then called s admissible. A semicomplete manifold is uniformly (i.e. $\mathfrak{U}_X$) complete. Given uniform completeness, $\rho > 1/\lambda_{\mathfrak{a}}$ for $\mathfrak{a} = \{U_\alpha, h_\alpha\}$ iff $\{U_\alpha\}$ is a uniform open cover of X. An $\mathfrak{M}$ manifold with bounded admissible bounds is called a bounded $\mathfrak{M}$ manifold. A bounded, semicomplete manifold is called <u>semicompact</u>.

Dominating $1/\lambda_{\mathfrak{a}}$ is often necessary and so semicomplete manifolds are of central importance in the theory. In the theory of several complex variables such functions are already in use (eg. [18; Chap. 7]). As the name suggests it is to semicompact manifolds that many compact results generalize in the metric category. This usually happens when compactness is used in the original proof to bound functions like $\rho_{\mathfrak{a}}$ and $1/\lambda_{\mathfrak{a}}$ and to get uniformity with respect to metrics like $d_{\mathfrak{a}}$. The resulting metric theorems are true extensions of the original theorems because compact manifolds admit a unique metric structure which is semicompact (cf. Chap. VIII. Sec. 5).

In Chapter IV we apply this procedure to section spaces and manifolds of maps. Let $\pi: E \to X$ be an $\mathfrak{M}$ vector bundle with admissible atlas $(\mathfrak{H},\mathfrak{a}) = \{U_\alpha, h_\alpha, \varphi_\alpha\}$. A section s of π is an $\mathfrak{M}$ section if the $(\mathfrak{H},\mathfrak{a})$ principal parts of s, defined by $\varphi_\alpha(s(x)) = (h_\alpha(x), s_\alpha(h_\alpha(x)))$, are locally $\mathfrak{M}$ maps the growth of whose $\mathfrak{M}$ norm is dominated by the admissible bounds on X. Thus, if π is a bounded $\mathfrak{M}$ bundle we can define the norm $\|s\|^{(\mathfrak{D},\mathfrak{a})} = \sup_\alpha \|s_\alpha\|_{\mathfrak{M}}$. This norm makes the vector space of $\mathfrak{M}$ sections of π, denoted $\mathfrak{M}(\pi)$, into a Banach space. The topology on $\mathfrak{M}(\pi)$ is independent of the atlas choice. If $(\Phi, 1_X): \pi_1 \to \pi_2$ is an $\mathfrak{M}$ linear vector bundle map of bounded $\mathfrak{M}$ bundles over X then the induced map of sections $\Phi_*: \mathfrak{M}(\pi_1) \to \mathfrak{M}(\pi_2)$ is a continuous linear map. If f: $X_0 \to X$ is an $\mathfrak{M}$ map with X_0 a bounded $\mathfrak{M}$ manifold then the pull back map $f^*: \mathfrak{M}(\pi) \to \mathfrak{M}(f^*\pi)$ is also continuous.

A <u>standard triple</u> $(\mathfrak{M}_1, \mathfrak{M}, \mathfrak{M}_2)$ is a trio of standard function space types satisfying: (1) $\mathfrak{M}_1 \subset \mathfrak{M}$ and $\mathfrak{M}_1 \subset \mathfrak{M}_2 \subset \mathcal{L}$. (2) $\mathfrak{M}_1$ maps $\mathfrak{M}$ to $\mathfrak{M}$ in an $\mathfrak{M}_2$ way. (3) $\mathfrak{M}_2$ satisfies the Strong Product Condition. $(\mathcal{L}^{r+s}, \mathcal{C}^r, \mathcal{L}^s)$ and $(\mathcal{L}^{r+s+1}, \mathcal{L}^r, \mathcal{L}^s)$ are standard triples.

Let $(\mathfrak{M}_1, \mathfrak{M}, \mathfrak{M}_2)$ be a standard triple, X_0 be a bounded $\mathfrak{M}$ manifold and X be a semicomplete $\mathfrak{M}_1^1$ manifold satisfying a mild technical condition (the existence of $\mathfrak{M}_1$ exponential maps) then the set of $\mathfrak{M}$ maps, $\mathfrak{M}(X_0, X)$, is a semicomplete $\mathfrak{M}_2$ manifold in a natural way. If F: $X \to X_1$ is an $\mathfrak{M}_1$ map of semicomplete $\mathfrak{M}_1^1$ manifolds admitting $\mathfrak{M}_1$ exponentials then $F_*: \mathfrak{M}(X_0, X) \to \mathfrak{M}(X_0, X_1)$ is an $\mathfrak{M}_2$ map. If h: $X_0 \to X_2$ is an $\mathfrak{M}$ map of bounded $\mathfrak{M}$ manifolds then $h^*: \mathfrak{M}(X_2, X) \to \mathfrak{M}(X_0, X)$ is an $\mathfrak{M}_2$ map. If X_1 is a semicomplete $\mathfrak{M}_1^2$ manifold admitting $\mathfrak{M}_1^1$ exponentials then, since $(\mathfrak{M}_1^1, \mathfrak{M}, \mathfrak{M}_2^1)$ is a standard triple, $\mathfrak{M}(X_0, X)$ is an $\mathfrak{M}_2^1$ manifold and its tangent bundle can be naturally identified with $\tau_{X*}: \mathfrak{M}(X_0, TX) \to \mathfrak{M}(X_0, X)$. Under this identification $T(F_*)$ is identified with $(TF)_*$ and $T(h^*)$ is identified with h^*.

On composition as a function of two variables, the following result is typical: Let X_1 be a semicompact $\mathcal{L}^{s+t+1}$ manifold, X_2 be a semicomplete $\mathcal{L}^{s+r+t+2}$ manifold (both admitting suitable exponential maps) and X_0 be a bounded $\mathcal{C}^t$ manifold $(r \geq s + 1)$. Let G be open and bounded in $\mathcal{C}^t(X_0, X_1)$. $\Omega_G(f) = f_*|G$ defines $\Omega_G: \mathcal{L}^{s+t}(X_1, X_2) \to \mathcal{L}^s(G, \mathcal{C}^t(X_0, X_2))$ an

$\mathscr{L}^{r-1}$ map. The composition map Comp: $\mathcal{C}^t(X_0,X_1) \times \mathscr{L}^{s+t}(X_1,X_2) \to \mathcal{C}^t(X_0,X_2)$ is an $\mathscr{L}^s$ map. Note that smoothness of Ω_G doesn't make sense until manifolds of maps are defined with noncompact domains. Ω_G is better behaved than Comp in that its smoothness increases with r, i.e. with the smoothness of X_2, while that of Comp does not.

While the definitions of metric structures on manifolds require a standard function space type, one can usually globalize nonstandard function space types on sufficiently smooth semicomplete manifolds. For example, on manifolds of finite type (see Chapter VIII) the Sobolev function space types can be globalized essentially the same way as on compact manifolds (eg. [30; Section 25]). While we don't consider the Sobolev spaces further in this work, Chapter V carries out this globalization for $\mathcal{C}_u^r$ and $\mathscr{L}_a^r$ $(0 < a \leq 1)$, the derived function space types for uniformly continuous and Hölder continuous functions. $\mathcal{C}_u^r$ and $\mathscr{L}_a^r$ bundles, sections and maps can be defined on semicomplete $\mathscr{L}^r$ manifolds. Actually for these special function space types the globalization is carried out in a more general context.

In various applications a manifold carries an auxiliary topology coarser than the manifold topology. For example, in looking at compact hyperbolic invariant subsets for a dynamical system, it is technically useful to think of the subset as a discrete space (and hence a semicompact manifold) and regard the original topology as such an auxiliary topology. Again, given a foliation of a compact manifold it is useful to consider the leaf space as a nonseparable, semicompact manifold and regard the original topology as auxiliary. These are both special cases of leaf immersions, examined in Chapter VII. In Chapter V auxiliary pseudometrics on a regular metric manifold (apm's) are defined as pseudometrics coarser than the admissible metrics. Structures are defined with the r-jet of everything uniformly continuous, or Hölder continuous with respect to the apm.

Chapter VI contains a description of immersions, submersions and transversality in the metric category. If X_0 and X_1 are regular $\mathfrak{M}^1$ metric manifolds $f: X_0 \to X_1$ is an <u>$\mathfrak{M}^1$ immersion</u> if $\rho_0 > f^*\rho_1$ and there exist admissible atlases $\mathcal{G}_0 = \{U_\alpha, h_\alpha\}$ and $\mathcal{G}_1 = \{V_\beta, g_\beta\}$ on X_0 and X_1 such that $f: \mathcal{G}_0 \to \mathcal{G}_1$ is index preserving and for each α, the principal parts of f, $g_{\beta(\alpha)} \cdot f \cdot h_\alpha^{-1}$ are restrictions of inclusions of a factor into a product of Banach spaces. f is then an $\mathfrak{M}^1$ map and $T_x f: T_x X_0 \to T_{fx} X_1$ is a split injection, i.e. f is an immersion in the usual sense. If $j: E_0 \to E_1$ is a split injection of Banach spaces we define the splitting constant $\theta(j) = \max(\|j\|, \inf\{\|P\|: P: E_1 \to E_0 \text{ with } P.j = I\})$. If $\| \ \|_i$ is a Finsler on τ_{X_i} associated with the metric structure (such Finslers are defined in Chapter III and are called admissible Finslers) then $\| \ \|_0$ and $\| \ \|_1$ make $T_x X_0$ and $T_{fx} X_1$ into Banach spaces and so we can define $\theta(f)(x) = \theta(T_x f)$. If f is an $\mathfrak{M}^1$ immersion then $\rho_0 > \theta(f)$. Conversely, if f is an $\mathfrak{M}^1$ map which satisfies $\rho_0 \sim f^*\rho_1$ (such a map is called metricly proper) and f is an immersion with $\rho_0 > \theta(f)$ then f is an $\mathfrak{M}^1$ immersion. For submersions and transversality the situation is similar. In each case an atlas definition is the appropriate one but tests are developed using the classical definition and some global condition like domination of a splitting bound like $\theta(f)$. Thus, all of these notions are global ones in the metric category rather than local as in the differentiable category. The global conditions easy enough to manipulate that, for example, openness of the proper metric notion of transversality still holds. However, the density theorems of transversality theory are lost.

While we prove that every paracompact C^r manifold admits semicompact $\mathcal{C}^r$ structures (Chapter VIII, Section 4), this result is mainly of negative interest. I suspected the existence of finite dimensional, connected manifolds which did not admit semicompact structures. Such a manifold could not occur as a leaf of foliation of a compact manifold, answering a question of Sondow [26]. In applying the metric theory to a problem, the mere existence of semicompact structures is not too useful because choosing a metric structure and going to the metric category restricts the maps, vectorfields, etc.with which one can deal to those which are metric with respect to the chosen structure. Thus, it is of more interest to look for metric structures naturally associated with the problem. This is a matter of looking for associated atlases $\mathfrak{a}$ and estimating $\rho_{\mathfrak{a}}$ and $\lambda_{\mathfrak{a}}$.

For example, if X is a Lie group then left translates (or right translates) of a chart about the identity form an atlas generating a semicompact structure called the left semicompact structure (resp. the right semicompact structure). Left invariant (resp. right invariant) gadgets are metric with respect to this structure. The left and right structures are usually not the same but are contained in a semicomplete structure with admissible bound $\rho(x) = \max(\|\mathrm{Ad}(x)\|, \|\mathrm{Ad}(x)^{-1}\|)$ where Ad is the adjoint representation and the norm is computed using any fixed norm on the Lie algebra.

If X is any Riemannian manifold then the natural atlas is the atlas of normal coordinates indexed by the points of X. $\rho_{\mathfrak{a}}^{\mathcal{C}^r}$ of this atlas can be estimated using the Jacobi equation by dominating the norm of the curvature tensor and $r - 1$ of its covariant derivatives. $\lambda_{\mathfrak{a}}(x)$ is essentially the distance to the cut locus.

Lack of space has prevented the inclusion of the proofs of the above remarks in this work. I hope to deal with the relations between the metric theory and differential geometry elsewhere. However, in Chapter VII we discuss Grassmanians and apply them to the following type of question: Let X_0 be a semicomplete $\mathfrak{M}^r$ manifold ($r \geq 1$) and $f: X \to X_0$ be a C^r immersion. When does X carry an $\mathfrak{M}^r$ structure such that f is an $\mathfrak{M}^r$ immersion? The Grassmanian $G(TX_0)$ and the associated lifting

$G(f): X \to G(TX_0)$ allow us to construct natural atlases on X associated with f which show that the $\mathfrak{M}^r$ structure, if it exists, is unique. The existence problem is handled inductively using smoothing results like the following: Let $f: X \to X_0$ be an $\mathfrak{M}^r$ immersion $(r \geq 1)$ and X_0 be an $\mathfrak{M}^{r+1}$ manifold. There exists an $\mathfrak{M}^{r+1}$ structure on X with respect to which f is an $\mathfrak{M}^{r+1}$ immersion iff the $\mathfrak{M}^{r-1}$ map $G(f): X \to G(TX_0)$ of $\mathfrak{M}^r$ manifolds is in fact an $\mathfrak{M}^r$ map.

In Chapter VII, we also consider leaf immersions. Let X_0 be a semicomplete $\mathcal{L}^r$ manifold with d_0 some admissible atlas metric. Let X be a semicomplete $\mathcal{C}^r$ manifold and $f: X \to X_0$ be a $\mathcal{C}^r$ immersion. f is called a $\mathcal{C}_u^r$ leaf immersion if f is a $\mathcal{C}_u^r$ map with respect to the apm $f^*d_0 = d_0 \cdot f \times f$ on X (in the sense of Chapter V). Inductively define $G^0(X_0) = X_0$, $G^0(f) = f$ and $G^{i+1}(X_0) = G(TG^i(X_0))$, $G^{i+1}(f) = G(G^i(f)): X \to G^{i+1}(X_0)$ and choose d_i some admissible atlas metric on $G^i(X_0)$. A $\mathcal{C}^r$ immersion f is a $\mathcal{C}_u^r$ leaf immersion iff f^*d_0 is uniformly equivalent to $G^r(f)^*d_r$. The standard example of a leaf immersion is the "inclusion" of the leaf space of a foliation into the ambient manifold. Leaf immersions were introduced by Hirsch, Pugh and Shub [12] who proved the existence of an atlas on the domain, called a plaquation atlas, which resembles a foliation atlas in some respects. We prove that a bijective $\mathcal{C}_u^r$ leaf immersion with closed image is a lamination or partial foliation in the sense of Ruelle and Sullivan [25] iff a certain weak factoring property holds.

A semicomplete, finite dimensional manifold X is said to be of <u>finite type</u> if there exist an s admissible atlas $\mathcal{G} = \{U_\alpha, h_\alpha\}$ and an integer N such that for each $x \in X$, $x \in U_\alpha$ for at most N values of α, i.e. if the nerve of the cover $\{U_\alpha\}$ has dimension $< N$. A classical result of dimension theory assures that with $N = \dim X + 1$ admissible atlases exist with this property. However, I was unable to preserve the uniformity of the open cover in getting such an atlas and so have been unable to prove the obvious conjecture that every semicomplete, finite dimensional manifold is of finite type. However, a rich stock of manifolds of finite type is provided by the theorem that if X is a semicomplete manifold which $\mathcal{L}^1$ immerses in X_0, a manifold of finite type, then X is of finite type. Conversely, any $\mathfrak{M}^r$ manifold of finite type $\mathfrak{M}^r$ immerses in some Euclidean space. This result and many other translations into the metric category of standard results of differential topology follow for manifolds of finite type because for such manifolds partitions of unity are available in the metric category. Using partitions of unity and transversality theory we are able to prove standard smoothing results in the metric category for manifolds of finite type. In particular, any $\mathfrak{M}^r$ metric structure of finite type can be smoothed to obtain a $\mathcal{C}^\infty$ structure of finite type, i.e. there exist in the $\mathfrak{M}^r$ structure adapted atlases $(\mathcal{G},\rho)$ with $\rho > 1/\lambda_{\mathcal{G}}$, $\rho > \rho^{\mathcal{C}^t}$ for $t = 1,2,\ldots$, and the nerve of $\mathcal{G}$ is finite dimensional.

<u>Notation</u>: All pseudometrics in this work are allowed to take the value ∞. $B^d(x,r)$ (or $B^d[x,r]$) is the open (resp. closed) d-ball about x with radius r. In cross references, we will drop the self-referrent part of a theorem's designation. Thus, Proposition III.7.2 occurs in Section 7 of Chapter III and in Chapter III it will be called Proposition 7.2 except in Section 7 where it will be called Proposition 2.

<u>Acknowledgements</u>: I would like to thank Ms. Kate March for her typing of this seemingly endless work. Niel Shilkret pointed out reference [3] to me. To Jon Sondow, Richard Palais and Dennis Sullivan I owe special gratitude for many helpful discussions during the development of these ideas.

CHAPTER I: ATLASES

As the title suggests, this work deals with manifolds modeled on real Banach spaces (or B-spaces). We assume of the reader a familiarity with the calculus of functions between open subsets of B-spaces (see [1],[2],[5],[14],[20]) and the resulting theory of Banach manifolds. We will also assume a knowledge of jets (see [1],[2],[19],[20],[21]). This preliminary chapter consists of material which will be, in the main, familiar to such a reader. Its purpose is, on the one hand, to review some definitions and constructions which will be of particular importance later and, on the other, to establish notation and definitions for future reference.

1. Banach Spaces and Jets: E, E', F_1, F_2, etc. will be B-spaces unless otherwise mentioned. For B-spaces $E_0, E_1, \ldots, E_n$, $L^n(E_1,\ldots,E_n;E_0)$ denotes the B-space of n-linear maps with the usual operator norm. If $E_1 = \cdots = E_n$, we denote it $L^n(E_1;E_0)$. L^n_s is the subspace of symmetric maps and L stands for L^1. If $E_1 = E_0$ then $L(E_1)$ is used for $L(E_1;E_0)$. $Lis(E_1;E_0)$ is the open subset of $L(E_1;E_0)$ consisting of isomorphisms and on this set we define θ by

$$\theta(T) = \begin{cases} \max(\|T\|, \|T^{-1}\|, 1) & T \in Lis \\ \infty & T \in L - Lis. \end{cases} \tag{1.1}$$

Since $\|T\|\,\|T^{-1}\| \geq \|I\|$ the 1 is superfluous except when the spaces are zero.

Recall the natural isometry $L^n(E_1,\ldots,E_n;L^m(F_1,\ldots,F_m;E_0)) \cong L^{n+m}(E_1,\ldots,E_n,F_1,\ldots,F_m;E_0)$. It leads to such isometries as:

$$(1.2) \qquad \kappa: L_{\epsilon_1}^{n_1}(E_1;L_{\epsilon_2}^{n_2}(E_2;E_0)) \cong L_{\epsilon_2}^{n_2}(E_2;L_{\epsilon_1}^{n_1}(E_1;E_0)),$$

where $L_{\epsilon_i}^{n_i}$ stands for L^{n_i} or $L_s^{n_i}$.

On the product $\Pi_\alpha E_\alpha$ we define $\| \|_{\hat{\Pi}}$ by $\|\{v_\alpha\}\|_{\hat{\Pi}} = \sup_\alpha \|v_\alpha\|$. The product $\hat{\Pi}_\alpha E_\alpha$ is the linear subspace $\{\| \|_{\hat{\Pi}} < \infty\}$ with norm $\| \|_{\hat{\Pi}}$. In the case of a finite index set $\hat{\Pi}_\alpha E_\alpha$ is just the product with the max norm. In the case of an infinite index set $\hat{\Pi}$ is a proper subspace (unless almost all of the factors are zero). By p_α and i_α we will denote the projection to and inclusion of the factor E_α. $T \to \{p_\alpha \cdot T\}$ defines an isometry $L^n(E_1,\dots,E_n;\hat{\Pi}_\alpha E_\alpha) \cong \hat{\Pi}_\alpha L^n(E_1,\dots,E_n;E_\alpha)$. $\{T_\alpha\} \to \hat{\Pi}_\alpha T_\alpha$ defines an injective (but not surjective) isometry $\hat{\Pi}_\alpha L^n(E_\alpha;F_\alpha) \to L^n(\hat{\Pi}_\alpha E_\alpha;\hat{\Pi}_\alpha F_\alpha)$. Note that $\{T_\alpha\}$ can be an element of $\hat{\Pi}_\alpha L(E_\alpha;F_\alpha)$ with $T_\alpha \in \mathrm{Lis}(E_\alpha;F_\alpha)$ for all α and $\hat{\Pi}_\alpha T_\alpha$ may not be surjective (it is injective). In fact, $\hat{\Pi}_\alpha T_\alpha$ is an isomorphism iff $\sup_\alpha \theta(T_\alpha) < \infty$ iff $\{T_\alpha^{-1}\} \in \hat{\Pi}_\alpha L(F_\alpha;E_\alpha)$. Using the projections, we can isometrically inject $L^n(E_1,\dots,E_n;E_0)$ into $L^n(\Pi_1^n E_i;E_0)$.

The jet space $J_\ell^k(E;F)$ is $L_s^\ell(E;F) \times \dots \times L_s^k(E;F)$ defined when $\ell \le k$. If $\ell = 0$ it is denoted $J^k(E;F)$ with the convention $L^0(E;F) = L_s^0(E;F) = F$. The obvious projections and inclusions among the jet spaces and factors are denoted P and i. If U is open in E and $f: U \to F$ is at least k times differentiable then for $u \in U$, $j_\ell^k(f)(u) = (D^\ell f(u),\dots,D^k f(u))$ (with $D^0 f(u) = f(u)$). From the isometry preceding (1.2) we obtain an injective isometry $J_\ell^{k+\ell}(E;F) \to J^k(E;L_s^\ell(E;F))$ taking $j_\ell^{k+\ell}(f)(u)$ to $j^k(D^\ell f)(u)$. Projecting $J^{k+\ell}$ to J_i^{k+i} for each i between 0 and ℓ and injecting we get an injective isometry.

$$(1.3) \qquad H: J^{k+\ell}(E;F) \longrightarrow J^k(E;J^\ell(E;F))$$

taking $j^{k+\ell}(f)(u)$ to $j^k(j^\ell(f))(u)$.

If $j = (T_0,\dots,T_k) \in J^k(E;F)$ then the associated polynomial map is

$\theta(j)(u) = \Sigma(n!)^{-1}T_n u^{(n)}$ summing on $n = 1,\ldots,k$. $j^k(\theta(j))(0) = j$. Taylor's theorem says that if f is C^k on U and $U^{(2)} = \{(u,h): [u,u+h] \subset U\}$ where $[u,v]$ is the segment between u and v, then $f(u+h) = \theta(j)(h) + R(u,h)h^{(k)}$ where $j = j^k(f)(u)$ and $R(u,h) = \int_0^1 [(1-t)^{k-1}/(k-1)!][D^k f(u+th) - D^k f(u)]dt$. The converse of Taylor's theorem is found in [2;Thm. 2.1] or preferably in [19;Thm. 3]. It says that if $g: U \to J^k(E;F)$ and $R: U^{(2)} \to L_s^k(E;F)$ are continuous, $f(u+h) = \theta(g(u))(h) + R(u,h)h^{(k)}$ for (u,h) in $U^{(2)}$ and $R(u,0) = 0$ then f is C^k and $j^k(f) = g$.

There is a natural linear map for each k: $\#$: $L^n(F_1,\ldots,F_n;F_0) \to L^n(J^k(E;F_1),\ldots,J^k(E;F_n);J^k(E;F_0))$, characterized as follows: Let $u \in U$ open in E and let $f_1,\ldots,f_n$ map U to $F_1,\ldots,F_n$. For $A \in L^n(F_1,\ldots,F_n;F_0)$, $A_\#(j^k(f_1)(u),\ldots,j_1^k(f_n)(u)) = j^k(A\cdot(f_1,\ldots,f_n))(u)$. If $n = 1$ and $j = (T_0,\ldots,T_k)$ then $A_\#(j) = (A(T_0),\ldots,A\cdot T_k)$. For $n = 2$, the computation of $A_\#$ is Leibniz' rule. While it is possible by induction to write down $A_\#$ explicitly, what is needed are a few key properties: $A_\# \cdot P \times\ldots\times P = P\cdot A_\#$ where P is the projection of J^k to J^{k-1}. In fact, for $n = 2$, $A_\# = A$ for $k = 0$ and for $k > 0$:

$$(1.4)\qquad A_\#((T_0,\ldots,T_k),(S_0,\ldots,S_k)) = (\ldots,A\cdot(T_0,S_k) + A\cdot(T_k,S_0)+\ldots)$$

where the remaining terms depend on $(T_0,\ldots,T_{k-1})$ and $(S_0,\ldots,S_{k-1})$. The analogous formula holds for larger n. Finally, for each k and n there is a universal constant bounding the norm of $\#$ independent of $E,F_0,\ldots,F_n$. This follows from the existence of the explicit formula.

Of particular importance are the applications of $\#$ to certain standard maps such as comp: $L(F_2;F_3) \times L(F_1;F_2) \to L(F_1;F_3)$. To illustrate, let ev: $L^n(F_1;F_2) \times F_1 \times\ldots\times F_1 \to F_2$ (n copies of F_1) be the $n+1$-linear evaluation map. From $ev_\#$ we obtain the map λ which fits into the following commutative diagram:

$$(1.5)\qquad \begin{array}{ccc} & & L^n_\epsilon(F_1;F_2) \\ & \swarrow^{\,i} & \downarrow \phi \\ J^k(E;L^n_\epsilon(F_1;F_2)) & \xrightarrow{\ \lambda\ } & L^n_\epsilon(J^k(E;F_1);J^k(E;F_2)) \\ & \searrow_{\,\kappa} & \downarrow L^n(i;I) \\ & & L^n_\epsilon(F_1;J^k(E;F_2)) \end{array}$$

Here L^n_ϵ stands for either L^n or L^n_s and i is the inclusion of F_1 or L^n_ϵ into J^k as zero jets. κ is the isometric isomorphism given by (1.2) on factors. For each n and k, there is a universal constant bounding the norm of λ.

For $n = 1$, let U be open in E and identify maps of U into F_i with sections of the trivial bundle $\epsilon_{F_i}: U \times F_i \to U$ (ϵ_{F_i} is the projection map). A map $\varphi: U \to L(F_1;F_2)$ can be identified with a linear bundle map $(\phi,1): \epsilon_{F_1} \to \epsilon_{F_2}$ by $\phi(u,v) = (u,\varphi(u)v)$. For $s: U \to F_1$, $\lambda(j^k(\varphi)(u))(j^k(s)(u)) = j^k(\phi_* s)(u)$ (where $\phi_* s$ is the map characterized by $1 \times \phi_* s = \phi\circ(1 \times s)$). Thus, on the jet level, λ associates to a bundle map its action on sections. There is also an injection of norm 1 given by $L^i_s(E;L(F_1;F_2)) \to L^{i+1}(E,\ldots,E,F_1;F_2)$ $\to L^{i+1}(E \times F_1;F_2) \xrightarrow{Sym} L^{i+1}_s(E \times F_1;F_2)$ where Sym is the projection map given by averaging over the action of the symmetric group S_{i+1}. The resulting map of norm 1:

$$(1.6)\qquad J^k(E;L(F_1;F_2)) \longrightarrow J^{k+1}_1(E \times F_1;F_2)$$

followed by the projection to J^k_1 takes $j^k(\varphi)(u)$ to $j^k_1(p_2\circ\phi)(u,0)$.

The chain rule can be described by saying that there is a natural polynomial map Comp: $J^k_1(E_1;E_2) \times J^k_1(E_0;E_1) \to J^k_1(E_0;E_2)$ characterized as follows: Let $u \in U_0$ open in E_0, $f: U_0 \to E_1$ and $g: U_1 \to E_2$ be sufficiently differentiable with $f(U_0) \subset U_1$, then

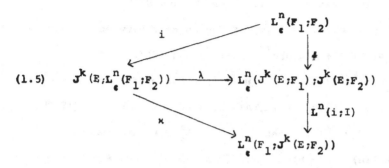

$\mathrm{Comp}(j_1^k(g)(fu), j_1^k(f)(u)) = j_1^k(g \cdot f)(u)$. Again, it is possible to write down a formula for Comp explicitly, but what is needed are the following properties: $\mathrm{Comp} \cdot P \times P = P \cdot \mathrm{Comp}$ where P is the projection from J_1^k to J_1^{k-1}. For $k = 1$, Comp = comp (linear map composition). For $k > 1$ we have

$$(1.7) \qquad \mathrm{Comp}((T_1,\ldots,T_k),(S_1,\ldots,S_k)) = (\ldots,T_1 \cdot S_k + T_k \cdot S_1^{(k)} + \ldots)$$

where the remaining terms depend on $(T_1,\ldots,T_{k-1})$ and $(S_1,\ldots,S_{k-1})$. Comp is linear in the first variable and can be regarded as an element of $L[J_1^k(E_1;E_2);J^k(J_1^k(E_0;E_1);J_1^k(E_0;E_2))]$ by identifying the polynomial map with its associated jet. Via $\varkappa$ (cf. (1.5)) we can regard this as an element of $J^k[J_1^k(E_0;E_1);L(J_1^k(E_1;E_2);J_1^k(E_0;E_2))]$ with the same norm and by (1.6) as an element of $J_1^{k+1}[J_1^k(E_1;E_2) \times J_1^k(E_0;E_1);J_1^k(E_0;E_2)]$ with no larger norm. The explicit formula yields a constant depending only on k which bounds the norm of Comp in these spaces.

Finally, it is possible to write down a polynomial map $\mathrm{Inv}: J^k(E;L(F_1;F_2)) \times L(F_2;F_1) \to J^k(E;L(F_2;F_1))$ again obtaining a uniform bound in jet norm, depending only on k and satisfying the following: Let $u \in U$ open in E, $\varphi: U \to L(F_1;F_2)$ be sufficiently differentiable with $\varphi(u) \in \mathrm{Lis}(F_1;F_2)$, then $\varphi^{-1}(\mathrm{Lis})$ is a neighborhood of u and we have $\mathrm{Inv}(j^k(\varphi)(u),\varphi(u)^{-1}) = j^k(\varphi^{-1})(u)$. In particular, if $E = F_1$, $F = F_2$ and $f: U \to F$ satisfies $Df(u) \in \mathrm{Lis}(E;F)$ then using the injection above (1.3) and Inv for $k - 1$, we have $\mathrm{Inv}(j_1^k(f)(u), Df(u)^{-1}) = j_1^k(f^{-1})(fu)$ by the inverse function theorem, providing f is at least C^1.

2. <u>Atlas Constructions</u>: All manifolds are assumed to be Hausdorff and modeled on B-spaces. An indexed family $\mathcal{G} = \{U_\alpha, h_\alpha\}$ is an atlas on a manifold X if $\{U_\alpha\}$ is an open cover of X and h_α is a homeomorphism of U_α only an open subset of a B-space E_α. The maps $h_\alpha h_\beta^{-1}: h_\beta(U_\alpha \cap U_\beta) \to E_\alpha$

are called transition maps of $\mathcal{G}$. $(\mathfrak{D},\mathcal{G}) = \{U_\alpha, h_\alpha, \varphi_\alpha\}$ is an atlas for a bundle $\pi: E \to X$ (a VB atlas) if $\mathcal{G} = \{U_\alpha, h_\alpha\}$ is an atlas on X (the atlas on the base), $\mathfrak{D} = \{\pi^{-1}(U_\alpha), \varphi_\alpha\}$ is an atlas on E (the induced total space atlas) with $\varphi_\alpha(\pi^{-1}(U_\alpha)) = h_\alpha(U_\alpha) \times F_\alpha$ for some B-space F_α, $p_1 \circ \varphi_\alpha = h_\alpha \circ \pi$, and whenever $U_\alpha \cap U_\beta \neq \emptyset$ the transition map $\varphi_\alpha \varphi_\beta^{-1}$ is given by the formula $\varphi_\alpha \varphi_\beta^{-1}(u,v) = (h_\alpha h_\beta^{-1}(u), \varphi_{\alpha\beta}(u) \cdot v)$ where $\varphi_{\alpha\beta}: h_\beta(U_\alpha \cap U_\beta) \to L(F_\beta; F_\alpha)$ is a continuous map called a transition map for $(\mathfrak{D},\mathcal{G})$. Note that $\varphi_{\alpha\beta}$ maps into $\mathrm{Lis}(F_\beta; F_\alpha)$. In fact, $\varphi_{\alpha\beta}^{-1} = \varphi_{\beta\alpha} \circ h_\alpha h_\beta^{-1}$. If $\pi_i: E_i \to X$, $i = 1, \ldots, n$ are bundles on X, an <u>n-tuple atlas</u> $(\mathfrak{D}_1, \ldots, \mathfrak{D}_n; \mathcal{G}) = \{U_\alpha, h_\alpha, \varphi_\alpha^1, \ldots, \varphi_\alpha^n\}$ is a convenient way of writing a list of VB atlases $(\mathfrak{D}_1, \mathcal{G}), \ldots, (\mathfrak{D}_n, \mathcal{G})$ with the same base space atlas (and a fortiori the same index set). In general, smoothness properties of atlases are defined in terms of the transition maps. Thus, for $1 \leq k \leq \infty$, $\mathcal{G}$ (or $(\mathfrak{D},\mathcal{G})$) is a C^k atlas if the family $\{h_\alpha h_\beta^{-1}\}$ (and the family $\{\varphi_{\alpha\beta}\}$) consists of C^k maps. Any atlas is C^0 by definition. Note that if $(\mathfrak{D},\mathcal{G})$ is C^k then the total space atlas $\mathfrak{D}$ is C^k. This kind of inheritance is the rule.

We define some operations on atlases. If $\mathcal{G} = \{U_\alpha, h_\alpha\}$ is an atlas on X (or $(\mathfrak{D},\mathcal{G}) = \{U_\alpha, h_\alpha, \varphi_\alpha\}$ is an atlas on π) and U is open in X then $\mathcal{G}|U = \{U_\alpha \cap U, h_\alpha\}$ is an atlas on U (resp. $(\mathfrak{D},\mathcal{G})|U = \{U_\alpha \cap U, h_\alpha, \varphi_\alpha\}$ is a VB atlas on $\pi_U: E_U \to U$ where $E_U = \pi^{-1}U$ and $\pi_U = \pi|E_U$) called the <u>restriction</u>. Given a collection of atlases on X or π with disjoint index sets we can take the union to obtain an atlas on X or π. Note that new transition maps arise in addition to the union of the collection of old transition maps. Thus, the union of two C^k atlases needn't be C^k. In fact, the relation between two C^k atlases that the union be C^k is an equivalence relation and an equivalence class is a C^k structure on X or π. If $\mathcal{G}_1 = \{U_\alpha, h_\alpha\}$ and $\mathcal{G}_2 = \{V_\beta, g_\beta\}$ are atlases on X then we call $\mathcal{G}_1$ a <u>refinement</u> of $\mathcal{G}_2$ if there is a map of index sets $\alpha \to \beta(\alpha)$ such that $U_\alpha \subset V_{\beta(\alpha)}$ and $h_\alpha = g_{\beta(\alpha)}|U_\alpha$. A refinement, $\mathcal{G}_1$, of $\mathcal{G}_2$ is called <u>subdivision</u> if it satisfies for all β:

$$(2.1) \qquad V_\beta = \cup\{U_\alpha : \beta(\alpha) = \beta\}.$$

For example, if $\mathfrak{W} = \{W_\gamma\}$ is an open cover of X, then $\mathcal{G}_2 \cap \mathfrak{W} = \{V_\beta \cap W_\gamma, g_\beta\}$ indexed by pairs (β,γ) is a subdivision of $\mathcal{G}_2$ with index map $(\beta,\gamma) \to \beta$. The definition of "$(\mathfrak{D}_1,\mathcal{G})$ refines $(\mathfrak{D}_2,\mathcal{G}_2)$" is analogous and in that case, $(\mathfrak{D}_1,\mathcal{G}_1)$ is a subdivision of $(\mathfrak{D}_2,\mathcal{G}_2)$ provided (2.1) holds, i.e. provided $\mathcal{G}_1$ is a subdivision of $\mathcal{G}_2$. Finally, a VB atlas construction: Let $(\mathfrak{D},\mathcal{G}) = \{U_\alpha, h_\alpha, \varphi_\alpha\}$ and $\bar{\mathcal{G}} = \{U_\alpha, \bar{h}_\alpha\}$, i.e. $\mathcal{G}$ and $\bar{\mathcal{G}}$ are atlas with the same index set and open cover. The <u>transfer of atlas construction</u> replaces $\varphi_\alpha = h_\alpha \times p_2 \cdot \varphi_\alpha : \pi^{-1}U_\alpha \to h_\alpha(U_\alpha) \times F_\alpha$ by $\bar{\varphi}_\alpha = \bar{h}_\alpha \times p_2 \cdot \varphi_\alpha : \pi^{-1}U_\alpha \to \bar{h}_\alpha(U_\alpha) \times F_\alpha$ to obtain the VB atlas $(\bar{\mathfrak{D}},\bar{\mathcal{G}}) = \{U_\alpha, \bar{h}_\alpha, \bar{\varphi}_\alpha\}$ with transition maps $\bar{\varphi}_{\alpha\beta} = \varphi_{\alpha\beta} \cdot h_\beta \bar{h}_\beta^{-1}$.

If X_0, X_1 are manifolds with atlases $\mathcal{G}_0 = \{U_\alpha, h_\alpha\}$ and $\mathcal{G}_1 = \{V_\beta, g_\beta\}$ and $f: X_0 \to X_1$ is a continuous map then by the local representatives of the atlas map $f: \mathcal{G}_0 \to \mathcal{G}_1$, or simply the local representatives of f, we mean the maps $g_\beta f h_\alpha^{-1}: h_\alpha(U_\alpha \cap f^{-1}V_\beta) \to E_\beta^1$. If $\pi_0: E_0 \to X_0$ and $\pi_1: E_1 \to X_1$ are bundles with VB atlases $(\mathfrak{D}_0,\mathcal{G}_0) = \{U_\alpha, h_\alpha, \varphi_\alpha\}$ and $(\mathfrak{D}_1,\mathcal{G}_1) = \{V_\beta, g_\beta, \psi_\beta\}$ and $(\Phi,f): \pi_0 \to \pi_1$ (i.e. $\Phi: E_0 \to E_1$ with $\pi_1 \cdot \Phi = f \cdot \pi_0$) then (Φ,f) is a VB map with local representatives $\Phi_{\beta\alpha}: h_\alpha(U_\alpha \cap f^{-1}V_\beta) \to L(F_\alpha^0;F^1)$ if the local representatives of Φ satisfy $\psi_\beta \Phi \varphi_\alpha^{-1}(u,v) = (g_\beta f h_\alpha^{-1}(u), \Phi_{\beta\alpha}(u)\cdot v)$. As a convenience we will usually write $h_{\beta\alpha}$ for $h_\beta h_\alpha^{-1}$ and $f_{\beta\alpha}$ for $g_\beta f h_\alpha^{-1}$, but the transition maps of $\mathfrak{D}$ (eg. $\varphi_\beta \varphi_\alpha^{-1}$) and local representatives of Φ (eg. $\psi_\beta \Phi \varphi_\alpha^{-1}$) should not be confused with the transition maps of $(\mathfrak{D},\mathcal{G})$ (eg. $\varphi_{\beta\alpha}$) or the local reps of (Φ,f) (eg. $\Phi_{\beta\alpha}$). As an example, note that if $X_0 = X_1 = X$ then the transition maps of the atlas $\mathcal{G}_0 \cup \mathcal{G}_1$ consist of: (1) the transition maps of $\mathcal{G}_0$ and those of $\mathcal{G}_1$ and (2) the local representatives of the identity maps $1: G_0 \to G_1$ and $1: G_1 \to G_0$ (similarly, for $\pi_0 = \pi_1 = \pi$).

An atlas map $f: G_0 \to G_1$ is called <u>index preserving</u> if there is a map of index sets $\alpha \to \beta(\alpha)$ such that $U_\alpha \subset f^{-1}V_{\beta(\alpha)}$. It is <u>strictly index preserving</u> if the index sets agree and the index map is the identity. In either case, the local representative $g_{\beta(\alpha)} f h_\alpha^{-1}$ is denoted f_α and called a <u>principal part</u> for f. $(\Phi, f): (\mathfrak{D}_0, G_0) \to (\mathfrak{D}_1, G_1)$ is index preserving or strictly index preserving if $f: G_0 \to G_1$ is, in which case Φ_α, denoting $\Phi_{\beta(\alpha)\alpha}$, is called a principal part of the VB map (Φ, f). For example, if $(\mathfrak{D}_0, \mathfrak{D}_1; G)$ is a two-tuple atlas for $\pi_0: E_0 \to X$ and $\pi_1: E_1 \to X$ and $(\Phi, 1): \pi_0 \to \pi_1$ is a VB map then $(\Phi, 1)$ is strictly index preserving. The principal parts, together with the atlases, contain all the information about the map. For example, the local representative $f_{\beta\alpha}$ is the composition $g_{\beta\beta(\alpha)} \cdot f_\alpha$ restricting to the domain $U_\alpha \cap f^{-1}V_\beta$ which maps into $V_\beta \cap V_{\beta(\alpha)}$. Similarly, $\Phi_{\beta\alpha} = \text{comp}\cdot(\varphi_{\beta\beta(\alpha)}, \Phi_\alpha)$.

If $\pi: E \to X$ is a bundle with VB atlas $(\mathfrak{D}, G) = \{U_\alpha, h_\alpha, \varphi_\alpha\}$ and $s: X \to E$ is a section, i.e. $\pi s = 1$, then $s: G \to \mathfrak{D}$ is strictly index preserving. We define the <u>principal part of the section</u> $s_\alpha: h_\alpha(U_\alpha) \to F_\alpha$ by the formula $\varphi_\alpha \cdot s \cdot h_\alpha^{-1}(u) = (u, s_\alpha(u))$. Note that the principal part of the section s is slightly different from the principal part of the map s. If O is open in E_0 and $(G, f): \pi_0|O \to \pi_1$ is a fibre preserving map with $f: G_0 \to G_1$ index preserving then we define the <u>principal part of the fibre preserving map</u> (G, f), $G_\alpha: \varphi_\alpha(O \cap \pi^{-1}U_\alpha)$ (open in $h_\alpha(U_\alpha) \times F_\alpha^0$) $\to F_\alpha^1$ by the formula $\Phi_{\beta(\alpha)} G \varphi_\alpha^{-1}(u,v) = (f_\alpha(u), G_\alpha(u,v))$. Again, if (G, f) is a VB map (Φ, f) then its principal parts <u>qua</u> VB map and <u>qua</u> fibre presering map are somewhat different. In each case the principal part represents the essential local element from which everything about the map can be computed. I hope that the utility of this jargon outweighs its abusiveness.

Various constructions on manifolds and vector bundles are really atlas constructions. We list the most important which we will call the Standard Constructions: (1) Products: If $\mathcal{G}_0 = \{U_\alpha, h_\alpha\}$ and $\mathcal{G}_1 = \{V_\beta, g_\beta\}$ are atlases on X_0, X_1 then $\mathcal{G}_0 \times \mathcal{G}_1 = \{U_\alpha \times V_\beta, h_\alpha \times g_\beta\}$ is an atlas on $X_0 \times X_1$ with $(h \times g)_{(\alpha,\beta)(\gamma,\delta)} = h_{\alpha\gamma} \times g_{\beta\delta}$. Similarly, $(\mathfrak{D}_0, \mathcal{G}_0) = \{U_\alpha, h_\alpha, \varphi_\alpha\}$, $(\mathfrak{D}_1, \mathcal{G}_1) = \{V_\beta, g_\beta, \psi_\beta\}$ on π_0, π_1 yield $(\mathfrak{D}_0, \mathcal{G}_0) \times (\mathfrak{D}_1, \mathcal{G}_1) = (\mathfrak{D}_0 \times \mathfrak{D}_1, \mathcal{G}_0 \times \mathcal{G}_1) = \{U_\alpha \times V_\beta, h_\alpha \times g_\beta, \varphi_\alpha \times \psi_\beta\}$ on $\pi_0 \times \pi_1$ with $(\varphi \times \psi)_{(\alpha,\beta)(\gamma,\delta)} = \varphi_{\alpha\gamma} \times \psi_{\beta\delta}$. (2) Trivial Bundles: $\mathcal{G} = \{U_\alpha, h_\alpha\}$ on X and a B-space F yield the atlas $(\mathcal{G} \times F, \mathcal{G}) = \{U_\alpha, h_\alpha, h_\alpha \times 1_F\}$ on $\epsilon_F : X \times F \to X$ with $(h \times 1)_{\alpha\beta}$ the constant map to the identity in $L(F;F)$. (3) Tangent Bundles: If $\mathcal{G}$ is a C^k atlas on X $(k \geq 1)$ then $(T\mathcal{G}, \mathcal{G}) = \{U_\alpha, h_\alpha, Th_\alpha\}$ is a C^{k-1} atlas on $\tau_X : TX \to X$ with $(Th)_{\alpha\beta} = Dh_{\alpha\beta}$. If $f: \mathcal{G}_0 \to \mathcal{G}_1$ then $(Tf, f): (T\mathcal{G}_0, \mathcal{G}_0) \to (T\mathcal{G}_1, \mathcal{G}_1)$ and the local transition map $(Tf)_{\alpha\beta}$ is $Df_{\alpha\beta}$. (4) Subbundles: If $(\mathfrak{D}, \mathcal{G}) = \{U_\alpha, h_\alpha, \varphi_\alpha\}$ is an atlas on $\pi: E \to X$ and E_0 is a subset of E then $(\mathfrak{D}, \mathcal{G})$ induces a subbundle atlas $(\mathfrak{D}|E_0, \mathcal{G})$ on $\pi: E_0 \to X$ if for every α there is a closed subspace F_α^0 of F_α such that $\varphi_\alpha(E_0) = h_\alpha(U_\alpha) \times F_\alpha^0$. (5) Bundle Operations: Lang [14; Sec. III.4] associates to a B-space functor an associated vector bundle functor. It is easy to check that this is an atlas construction. The examples we need are direct (Whitney) sum and linear map bundles. Let $(\mathfrak{D}_0, \mathfrak{D}_1; \mathcal{G}) = \{U_\alpha, h_\alpha, \varphi_\alpha^0, \varphi_\alpha^1\}$ be a two-tuple atlas for $\pi_0: E_0 \to X$ and $\pi_1: E_1 \to X$. We obtain $(\mathfrak{D}_0 \oplus \mathfrak{D}_1, \mathcal{G}) = \{U_\alpha, h_\alpha, (\varphi^0 \oplus \varphi^1)_\alpha\}$ on $\pi_0 \oplus \pi_1: E_0 \oplus E_1 \to X$, $(\varphi^0 \oplus \varphi^1)_{\alpha\beta} = \varphi_{\alpha\beta}^0 \times \varphi_{\alpha\beta}^1 \cdot \Delta$ where Δ is the diagonal map on $h_\beta(U_\alpha \cap U_\beta)$. On $L_\epsilon^n(\pi_0;\pi_1): L_\epsilon^n(E_0;E_1) \to X$ we obtain $(L_\epsilon^n(\mathfrak{D}_0;\mathfrak{D}_1), \mathcal{G}) = \{U_\alpha, h_\alpha, L_\epsilon^n(\varphi^0;\varphi^1)_\alpha\}$ $(L_\epsilon^n = L^n$ or $L_s^n)$ with $L_\epsilon^n(\varphi^0, \varphi^1)_{\alpha\beta} = L_\epsilon^n \cdot [(\varphi_{\alpha\beta}^0)^{-1} \times \varphi_{\alpha\beta}^1] \cdot \Delta$ where $L_\epsilon^n: L(F_\alpha^0;F_\beta^0) \times L(F_\beta^1;F_\alpha^1) \to L(L_\epsilon^n(F_\beta^0;F_\beta^1); L^n(F_\alpha^0;F_\alpha^1))$ is the map given by functoriality of L_ϵ^n. There is a bijective correspondence between sections of $L(\pi_0;\pi_1)$ and VB maps over the identity from π_0 to π_1. If $(\phi, 1): (\mathfrak{D}_0, \mathcal{G}) \to (\mathfrak{D}_1, \mathcal{G})$

and $s\colon \mathcal{a} \to L(\mathfrak{D}_0;\mathfrak{D}_1)$ are thus associated then the principal parts of the VB map $(\phi,1)$ and of the section s agree. (6) Pullbacks: Let $(\mathfrak{D},\mathcal{a}) = \{V_\beta, g_\beta, \psi_\beta\}$ be an atlas on $\pi\colon E \to X$ and $\mathcal{a}_0 = \{U_\alpha, h_\alpha\}$ be an atlas on X_0 with $f\colon \mathcal{a}_0 \to \mathcal{a}$ index preserving. We get an atlas $(f^*\mathfrak{D},\mathcal{a}_0) = \{U_\alpha, h_\alpha, (f^*\psi)_\alpha\}$ on $f^*\pi\colon f^*E \to X$ with $(f^*\psi)_{\alpha_1\alpha_2} = \psi_{\beta(\alpha_1)\beta(\alpha_2)}\cdot f_{\alpha_2}$. The natural VB map $(\phi_f, f) : f^*\pi \to \pi$ regarded as a map $(f^*\mathfrak{D},\mathcal{a}_0) \to (\mathfrak{D},\mathcal{a})$ is characterized by $\phi_{f\alpha}$ = constant map to identity in $L(F_{\beta(\alpha)};F_{\beta(\alpha)})$. Furthermore, if $(\phi,f)\colon \pi_0 \to \pi$ is a VB map with $(\mathfrak{D}_0,\mathcal{a}_0) = \{U_\alpha, h_\alpha, \varphi_\alpha\}$ on π_0, then the pullback map $(f^*\phi,1)\colon \pi_0 \to f^*\pi$ is characterized by $\phi = \phi_f\cdot f^*\phi$. The principal parts of $(\phi,f)\colon (\mathfrak{D}_0,\mathcal{a}_0) \to (\mathfrak{D},\mathcal{a})$ and $(f^*\phi,1)\colon (\mathfrak{D}_0,\mathcal{a}_0) \to (f^*\mathfrak{D},\mathcal{a}_0)$ agree. If s is a section of π, then the principal parts of the sections $s\colon \mathcal{a} \to \mathfrak{D}$ and $f^*s\colon \mathcal{a}_0 \to f^*\mathfrak{D}$ are related by $(f^*s)_\alpha = s_{\beta(\alpha)}\cdot f_\alpha$. (7) Jet Bundles: If $(\mathfrak{D},\mathcal{a}) = \{h_\alpha, U_\alpha, \varphi_\alpha\}$ is a C^k atlas on $\pi\colon E \to X$, then on $J^k(\pi)$ there is an associated atlas $(J^k(\mathfrak{D}),\mathcal{a}) = \{h_\alpha, U_\alpha, J^k(\varphi)_\alpha\}$. $J^k(\varphi)_{\alpha\beta}\colon h_\beta(U_\alpha \cap U_\beta) \to L(J^k(E_\beta;F_\beta); J^k(E_\alpha;F_\alpha))$ is characterized as follows: If $u \in h_\beta(U_\alpha \cap U_\beta) \subset E_\beta$ and $j \in J^k(E_\beta;F_\beta)$ then choosing s a C^k map from a neighborhood of u to F_β with $j^k(s)(u) = j$, then $J^k(\varphi)_{\alpha\beta}(j)$ is the jet at $h_{\alpha\beta}(u)$ of the section $ev.(\varphi_{\alpha\beta}, s)\cdot h_{\beta\alpha}$. Thus, using the maps λ, Comp and comp we can write $J^k(\varphi)_{\alpha\beta}(u)$ in terms of $j^k(\varphi_{\alpha\beta})(u)$ and $j^k(h_{\beta\alpha})(h_{\alpha\beta}(u))$. As before for future estimates the form of the formula is less important than its existence. If s is a C^k section of π and $j^k(s)$ is the associated k-jet section of $J^k(\pi)$ then the principal part $j^k(s)_\alpha$ (with respect to $(J^k(\mathfrak{D}),\mathcal{a})$) is just the k-jet $j^k(s_\alpha)$ (s_α from $(\mathfrak{D},\mathcal{a})$). If $(\mathfrak{D}_1,\mathcal{a})$ is a C^k atlas on $\pi_1\colon E_1 \to X$, $(\phi,1)\colon \pi \to \pi_1$ is a C^k VB map and $(J^k(\phi),1)\colon J^k(\pi) \to J^k(\pi_1)$ is the induced VB map then the principal part of $(J^k(\phi),1)\colon (J^k(\mathfrak{D}),\mathcal{a}) \to (J^k(\mathfrak{D}_1),\mathcal{a})$ with respect to α, $J^k(\phi)_\alpha$, is the composition $\lambda\cdot j^k(\phi_\alpha)$. The k-jet atlas is natural under restrictions to open subsets. If $f\colon X_0 \to X$ is a C^k map then there is a natural map

$(\Psi_f^k,1)\colon f^*J^k(\pi) \to J^k(f^*\pi)$ characterized by $\Psi_f^k \circ f^*j^k(s) = j^k(f^*s)$ where s is a C^k section of π. If $\mathfrak{a}_0 = \{V_\delta, g_\delta\}$ is a C^k atlas on X_0 and $f\colon \mathfrak{a}_0 \to \mathfrak{a}$ is index preserving then with respect to the two-tuple $(f^*J^k(\mathfrak{D}), J^k(f^*\mathfrak{D}); \mathfrak{a}_0)$ the principal part of $(\Psi_f^k,1)$ with respect to δ is a map $g_\delta(V_\delta) \to L[J^k(E_{\alpha(\delta)}; F_{\alpha(\delta)}); J^k(E_\delta^0; F_{\alpha(\delta)})]$ obtained by composing Comp with $j^k(f_\delta)\colon g_\delta(V_\delta) \to J^k(E_\delta^0; E_{\alpha(\delta)})$.

It is important to note that many identifications and canonical isomorphisms between various bundle constructions are in fact <u>identifications at the atlas level</u>, that is, the isomorphisms are strictly index preserving with principal parts identity maps or permutations of factors. This is a phenomenon easier to illustrate than to describe precisely: The twist isomorphism between $X_0 \times X_1$ and $X_1 \times X_0$ relates $\mathfrak{a}_0 \times \mathfrak{a}_1$ to $\mathfrak{a}_1 \times \mathfrak{a}_0$. If $f\colon X_0 \to X_1$, $g\colon X_1 \to X_2$ and $\pi\colon E \to X_2$ with $f\colon \mathfrak{a}_0 \to \mathfrak{a}_1$, $g\colon \mathfrak{a}_1 \to \mathfrak{a}_2$ index preserving and $(\mathfrak{D}, \mathfrak{a}_2)$ is an atlas on π then the canonical isomorphism $f^*g^*\pi \cong (g\circ f)^*\pi$ identifies $(f^*g^*\mathfrak{D}, \mathfrak{a}_0)$ with $((g\circ f)^*\mathfrak{D}, \mathfrak{a}_0)$. If $(\mathfrak{D}_0, \mathfrak{D}_1; \mathfrak{a})$ is a two-tuple atlas for π_1, π_2 over X then $\Delta\colon \mathfrak{a} \to \mathfrak{a} \times \mathfrak{a}$ is an index preserving map with $\alpha \to (\alpha,\alpha)$ and $(\Delta^*\mathfrak{D}_0 \times \mathfrak{D}_1, \mathfrak{a})$ is identified with $(\mathfrak{D}_0 \oplus \mathfrak{D}_1, \mathfrak{a})$ under the isomorphism $\Delta^*\pi_0 \times \pi_1 \cong \pi_0 \oplus \pi_1$. The projection map $p_1\colon E_0 \oplus E_1 \to E_0$ is a bundle with VB atlas $(\mathfrak{D}_0 \oplus \mathfrak{D}_1, \mathfrak{D}_0)$ and $\pi_0\colon \mathfrak{D}_0 \to \mathfrak{a}$ is a strictly index preserving map inducing an isomorphism $p_1 \cong \pi_0^*\pi_1$ identifying $(\mathfrak{D}_0 \oplus \mathfrak{D}_1, \mathfrak{D}_0)$ with $(\pi_0^*\mathfrak{D}_1, \mathfrak{D}_0)$. The canonical isomorphism $L(\epsilon_R; \pi) \cong \pi$ identifies $(L(\mathfrak{a}\times R; \mathfrak{D}), \mathfrak{a})$ with $(\mathfrak{D}, \mathfrak{a})$. The very naturality of isomorphisms such as these usually ensures that they are identifications at the atlas level.

Finally, we look at the tangent square of a C^1 bundle, $\pi\colon E \to X$. Define $S\colon TE \to E \oplus TX$ by $S = \tau_E \oplus T\pi$, $J\colon E \oplus E \to TE$ by $J(w_1,w_2) = \frac{d}{dt} w_1 + tw_2|_{t=0}$ (addition and differentiation in the fiber of $\pi w_1 = \pi w_2$) and $\bar{J}\colon TX \oplus E \to TE$ by $\bar{J} = (TO_\pi) + (J\circ i_2)$ where O_π is the zero section of π, and i_2 is the inclusion of E as the second factor of $E \oplus E$ (addition in the fibers of τ_E). We have the commutative diagram:

(2.2) 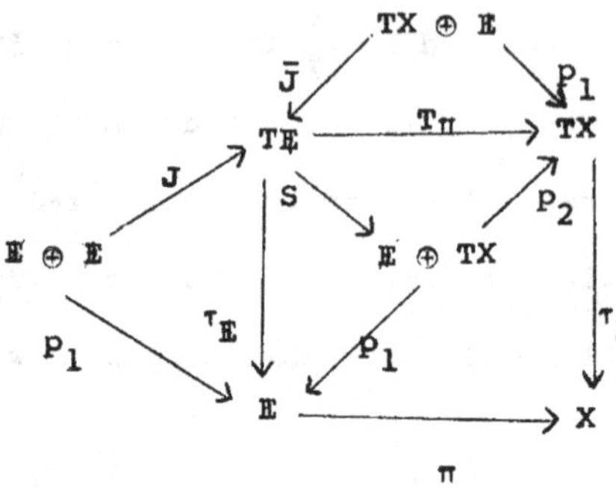

If $(\mathfrak{D},\mathfrak{A}) = \{U_\alpha, h_\alpha, \varphi_\alpha\}$ is a C^1 atlas on π then the transition maps of $T\mathfrak{D}$ on $T\mathbb{E}$ are given by:

$$(2.3)\quad T\varphi_\beta T\varphi_\alpha^{-1}(u,v,\dot{u},\dot{v}) = (h_{\beta\alpha}(u),\varphi_{\beta\alpha}(u)(v),Dh_{\beta\alpha}(u)(\dot{u}),\varphi_{\beta\alpha}(u)(\dot{v}) + D\varphi_{\beta\alpha}(u)(\dot{u})(v)),$$

and to (2.2) is associated the diagram of atlases:

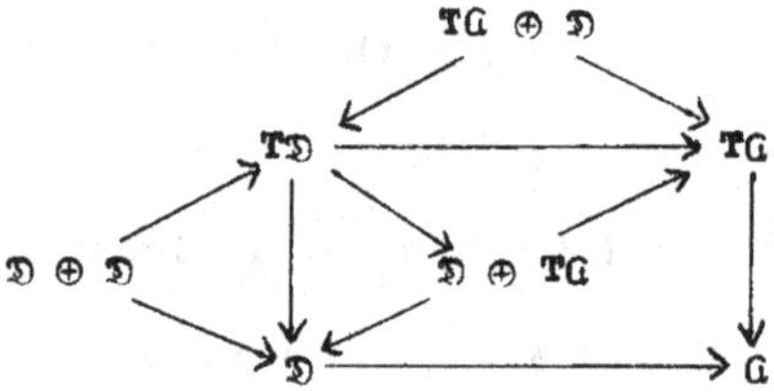

The principal parts of these maps are all inclusions of or projection to factors in various products. $0 \to \pi^*\pi \cong p_1 \xrightarrow{(J,1)} \tau_{\mathbb{E}} \xrightarrow{(S,1)} p_1 \cong \pi^*\tau_X \to 0$ is a short exact sequence of bundles over $\mathbb{E}$ with (S,1) identified with $(\pi^*T\pi,1)$. The image of (J,1) is the <u>subbundle of vertical tangents</u>, i.e. the kernel of $(T\pi,\pi)$. (2.3) shows that the atlas pair $(T\mathfrak{D},T\mathfrak{A})$ makes $T\pi$ into a bundle. Represent a vector in TX by a path in X. The corresponding fibre of $T\pi$ is represented by liftings of the path to $\mathbb{E}$. Using the bundle structure of π such lifts form a vector space and the vector space operations correspond to those in the fiber of $T\pi$. $0 \to \tau_X{}^*\pi \cong p_1 \xrightarrow{(\bar{J},1)} T\pi \xrightarrow{(S,1)} p_2 \cong \tau_X{}^*\pi \to 0$ is a short exact sequence over τ_X with (S,1) identified with $(\tau_X{}^*\tau_{\mathbb{E}},1)$. J also yields an identification at the atlas level between $\pi \oplus \pi$ and $0_X{}^*(T\pi)$. Similarly, $\bar{J}$ yields an identification at the atlas level between $\tau_X \oplus \pi$ and $0_\pi{}^*(\tau_E)$. Here 0_π is the zero section of π and $0_X = 0_{\tau_X}$ is the zero section of τ_X.

CHAPTER II: FUNCTION SPACES

In this chapter, we develop the properties of various function space types. A function space type associates to every bounded open set U in a B-space and every B-space E a B-space of functions $\mathfrak{M}(U,E)$, that is, a linear subspace of the vector space E^U of all functions from U to E together with a complete norm on the subspace. We will follow the "axiomatic" treatment of Palais [21]. This consists of associating to each type $\mathfrak{M}$ derived types $\mathfrak{M}^r$, $r = 1,2,\ldots$. For each property, we first verify that it holds for our elementary examples and then check that it is inherited by $\mathfrak{M}^r$ from $\mathfrak{M}$. Our building blocks are function spaces rather than section spaces as in [21] because we need Banach spaces rather than merely Banachable spaces.

1. <u>Elementary Properties</u>: We begin with some norms and associated function spaces: For U open and bounded in F, define the subsets of $U \times U$: $O_1 = U \times U - \Delta_U$, $O_2 = \{(u,v) \in O_1 : [u,v] \subset U\}$, $O_3^\epsilon = \{(u,v) \in O_1 : \|u - v\| < \epsilon\}$ and $O_4^\epsilon = O_3^\epsilon \cap O_2$ (with $O_i^1 = O_i$, $i = 3,4$). For $0 < a \leq 1$, $f \in E^U$ define $\Delta_a f : O_1 \to E$ by $\Delta_a f(u,v) = \|u-v\|^{-a}(f(u)-f(v))$ with $\Delta f = \Delta_1 f$. Define $\|f\|_0 = \sup_U \|f(u)\|$, $L_a(f) = \|\Delta_a f | O_3\|_0$, $\|f\|_L = \|\Delta f | O\ \|_0$, $\|f\|_{\mathcal{L}ip_a} = \max(\|f\|_0, L_a(f))$, $\|f\|_{\mathcal{L}} = \max(\|f\|_0, \|f\|_L)$. The linear subspaces of E^U on which $\|\ \|_0$, $\|\ \|_{\mathcal{L}ip_a}$ and $\|\ \|_{\mathcal{L}}$ are finite are denoted $\mathcal{B}(U,E)$, $\mathcal{L}ip_a(U,E)$ and $\mathcal{L}(U,E)$ and are assumed equipped with their defining norms. $\mathcal{C}(U,E)$, $\mathcal{C}_u(U,E)$ are the linear subspaces of $\mathcal{B}(U,E)$ consisting of continuous and uniformly continuous functions equipped with $\|\ \|_0$. In general, when $a = 1$ we delete it as a subscript.

$\mathcal{B}(U,E)$ is the set of bounded functions in E^U, $\mathcal{C}(U,E)$ the set of bounded, continuous functions, $\mathcal{L}ip(U,E)$ the set of Lipschitz functions, and $\mathcal{L}ip_a(U,E)$ the set of Holder continuous functions with exponent a. To relate these norms to the usual Lipschitz and Hölder norms we note:

(1.1) $$\|\Delta_a f\|_0 \leq \max(L_a(f), 2\|f\|_0).$$

To understand $\mathcal{L}(U,E)$ we introduce the intrinsic metric.

1 <u>PROPOSITION</u>: Let U be open in F. A chain in U between u and v is a sequence $\{u_1,\ldots,u_n\}$ with $u_1 = u$, $u_n = v$ and $[u_i,u_{i+1}] \subset U$ for $i = 1,\ldots,n-1$. Its length is $\Sigma_i\|u_{i+1} - u_i\|$. A piecewise C^1 path in U between u and v is a piecewise C^1 map u_t: $[a,b] \to U$ with $u_a = u$ and $u_b = v$. Its length is $\int_a^b \|u_t'\|dt$. $d_U(u,v)$ is equivalently defined as the infimum of the lengths of all chains between u and v or as the infimum of the lengths of all piecewise C^1 paths between u and v. By definition $\inf \emptyset = \infty \cdot d_U$ is a metric on U. $d_U(u,v) \geq \|u - v\|$ with equality if $[u,v] \subset U$.

<u>PROOF</u>: First, either definition yields $d_U(u,v) < \infty$ iff u and v lie in the same component of U. We are thus reduced to considering the case when U is connected. Let d_0 and d_1 denote the chain and C^1 definitions, respectively. To a chain $\{u_1,\ldots,u_n\}$ there is defined by linearity a piecewise C^1 (in fact, piecewise linear) path u_t: $[1,n] \to U$ between u and v with the same length. Thus, $d_0 \geq d_1$. Now given u, v and $\epsilon > 0$ choose u_t: $[a,b] \to U$ with $d_1(u,v) + \epsilon > \int_a^b \|u_t'\|dt$. Assume momentarily that u_t is C^1. By uniform continuity of u_t' there exists h_1 such that for $h < h_1$ and $t \in [a,b-h]$, $\|u_{t+h} - u_t - hu_t'\| < h\epsilon/b - a$. By compactness of $[a,b]$ we can choose h_2 such that $[u_t,u_{t+h}] \subset U$ for $h < h_2$, $t \in [a,b-h]$ (eg.pull back a convex open cover of U and pick a Lebesgue number). Finally, for $\epsilon_1 > 0$ we can choose h_3 such that if $a = t_1 <\ldots< t_n = b$ with $|t_{i+1} - t_i| < h_3$ then $\int_a^b \|u_t'\|dt + \epsilon_1 > \Sigma_i\|u_{t_i}'\|(t_{i+1} - t_i)$. For $h < \min(h_1,h_2,h_3)$,

$d_1(u,v) + \epsilon_1 + 2\epsilon > \int_a^b \|u_t'\|dt + \epsilon_1 + \epsilon > \Sigma_i(\|u_{t_i}'\| + (b-a)^{-1}\epsilon)(t_{i+1} - t_i) >$ $\Sigma_i\|u_{t_{i+1}} - u_{t_i}\| \geq d_0(u,v)$. If u_t breaks into m different C^1 pieces choose $\epsilon_1 = \epsilon/m$ for each piece. Letting $\epsilon \to 0$, $d_0 = d_1$. Clearly, $d_U(u,v) \geq \|u - v\|$ by the triangle inequality for $\| \ \|$ and equality is clear if $[u,v] \subset U$. Since symmetry and triangle inequality are clear for d_U, it follows that d_U is a metric on U agreeing with the norm metric on convex subsets of U. Q.E.D.

Now $\|f\|_L < \infty$ iff f is Lipschitz with respect to d_U on U. In fact, if V is open with $f(U) \subset V$ and $\{u_1,\ldots,u_n\}$ is a chain from u to v, then we can subdivide without changing length by inserting points on the associated p.l. path u_t and so assume that $\|u_{i+1} - u_i\| < \epsilon$ and $[f(u_{i+1}), f(u_i)] \subset V$. Then

$$(1.2) \quad \|f(u) - f(v)\| \leq d_V(f(u), f(v)) \leq \sum_i \|f(u_{i+1}) - f(u_i)\|$$
$$\leq \|f\|_L \sum_i \|u_{i+1} - u_i\| \leq \|f\|_L \, d_U(u,v).$$

This argument of <u>subdividing the segment</u> also shows that $\|f\|_L = \|\Delta f | O_2\|_0 = \|\Delta f | O_4^\epsilon\|_0$ for any $\epsilon > 0$. Thus, $\mathcal{L}(U,E)$ is the set of bounded, Lipschitz (d_U) functions in E^U. $\mathcal{L}(U,E)$ is much more important than $\mathcal{L}ip(U,E)$. $\mathcal{L}$ and its derived types are the best behaved spaces of the lot. However, we note that if $A \subset U$ with $B_\epsilon^{\| \ \|}(A) \subset U$ then

$$(1.3) \qquad \|\Delta f \mid A \times A - \Delta_A\|_0 \leq \max(\|f\|_L, \epsilon^{-1}\|f\|_0).$$

2. <u>DEFINITION</u>: (a) A function space type $\mathfrak{M}$ associates to every E and bounded open subset U of F a linear subspace $\mathfrak{M}(U,E)$ of E^U and a norm $\| \ \|_{\mathfrak{M}}$ on $\mathfrak{M}(U,E)$.

(b) For function space types $\mathfrak{M}_1$ and $\mathfrak{M}_2$, $\mathfrak{M}_1$ is included in $\mathfrak{M}_2$, written $\mathfrak{M}_1 \subset \mathfrak{M}_2$, if $\mathfrak{M}_1(U,E) \subset \mathfrak{M}_2(U,E)$ with $\| \ \|_{\mathfrak{M}_2} \leq \| \ \|_{\mathfrak{M}_1}$ on $\mathfrak{M}_1(U,E)$

for all U and E. The inclusion is <u>isometric</u> if $\| \|_{\mathfrak{M}_2} = \| \|_{\mathfrak{M}_1}$ on $\mathfrak{M}_1(U,E)$ for all U, E. The inclusion is <u>strong</u> if for $K > 0$ the closed K ball in $\mathfrak{M}_1(U,E)$ is closed in $\mathfrak{M}_2(U,E)$.

(c) For $\mathfrak{M}$ a function space type, the derived function space type $\mathfrak{M}^r$ $r = 0,1,\ldots$ is given by: $\mathfrak{M}^r(U,E)$ is the set of $f \in E^U$ r-times differentiable with $j^r f \in \mathfrak{M}(U,J^r(F;E))$ and $\|f\|_{\mathfrak{M}^r} = \|j^r f\|_{\mathfrak{M}}$, i.e. $j^r: \mathfrak{M}^r(U,E) \to \mathfrak{M}(U,J^r(F;E))$ is an injective isometry.

(d) In estimates involving function space types, we define $a \leq O^*(b_1,\ldots,b_k)$ to mean $a \leq K \max(b_1,\ldots,b_k,1)^n$ for some constants K and n. These constants are assumed to be universal, i.e. they depend only on the function space types. Where K and n depend on something else we will mention it explicitly when the estimate appears.

The definition of L_a was chosen so that $\mathcal{L}ip_a \subset \mathcal{L}ip_{a_1}$ if $a_1 \leq a$.

We now describe axioms FS1 - FS4 which we will demand of every function space type. In each case we will verify it for our examples $\mathcal{B}$, $\mathcal{C}$, $\mathcal{C}_u$, $\mathcal{L}$, $\mathcal{L}ip_a$ and then prove it is inherited from $\mathfrak{M}$ to $\mathfrak{M}^r$.

<u>FS1</u>: Let $T: E_1 \times\ldots\times E_n \to E_0$ be a bounded n-linear map. $T_*: \mathfrak{M}(U,E_1) \times\ldots\times \mathfrak{M}(U,E_n) \to \mathfrak{M}(U,E_0)$ is an n-linear map well defined by $T_*(f_1,\ldots,f_n) = T\cdot(f_1,\ldots,f_n)$. For each n, $\|T_*\| \leq O^*(1)\|T\|$ (i.e. $\|T_*\| \leq K\|T\|$ where K depends only on n and $\mathfrak{M}$) with $\|T_*\| \leq \|T\|$ for $n = 1$. If $n = 1$ and T is an isometry then T_* is an isometry with image $\{f \in \mathfrak{M}(U,E_0) : f(U) \subset \text{image } T\}$.

<u>VERIFICATION</u>: For $f_1,\ldots,f_n$ $\mathcal{B}$, $\|T_*(f_1,\ldots,f_n)\|_0 \leq \|T\|\|f_1\|_0 \cdots \|f_n\|_0$. For $u,v \in U$, $\|T_*(f_1,\ldots,f_n)(u) - T_*(f_1,\ldots,f_n)(v)\| \leq \|T\| \times \Sigma_{i=1}^n(\Pi_{j\neq i}\|f_j\|_0)\|f_i(u) - f_i(v)\|$. Thus, for $\mathcal{B}$, $\mathcal{C}$, $\mathcal{C}_u$ $\|T_*\| \leq \|T\|$ and for $\mathcal{L}$, $\mathcal{L}ip_a$ $\|T_*\| \leq n\|T\|$. The isometry result is easy.

<u>INHERITANCE</u>: Recalling $T_\#$ from Sec. I.1, we have the commutative diagram:

$$\begin{array}{ccc} \mathfrak{M}^r(U,E_1)\times\ldots\times\mathfrak{M}^r(U,E_n) & \xrightarrow{\;T_*\;} & \mathfrak{M}^r(U,E_0) \\ \Big\downarrow (j^r)^{(n)} & & \Big\downarrow j^r \\ \mathfrak{M}(U,J^r(F;E_1))\times\ldots\times\mathfrak{M}(U,J^r(F;E_n)) & \xrightarrow[(T_\#)_*]{} & \mathfrak{M}(U,J^r(F;E_0)) \end{array}$$

Thus, $j^r(T_*(f_1,\ldots,f_n)) = (T_\#)_*(j^r(f_1),\ldots,j^r(f_n)) \in \mathfrak{M}$ and so $T_*(f_1,\ldots,f_n)$ is in $\mathfrak{M}^r$. Also, $\|T_*\|_{\mathfrak{M}^r} \le \|(T_\#)_*\|_{\mathfrak{M}} \le O^*(1)\|T_\#\| \le O^*(1)\|T\|$. If $n = 1$, $\|T_\#\| = \|T\|$ and so we get $\|T_*\|_{\mathfrak{M}^r} \le \|T\|$. If $n = 1$ and T is an isometry then $T_\#$ is an isometry. Q.E.D.

The isometry result implies that the kernel of T_* is $\mathfrak{M}(U,\bar{E})$ where $\bar{E}$ is the kernel of T. However, unless T splits it is not clear that T surjective implies T_* is surjective. One can prove this for $\mathfrak{M} = \mathcal{C}$ using arguments of Michael [15], but these arguments don't inherit.

3 **PROPOSITION**: $\mathfrak{M}^0 = \mathfrak{M}, \mathfrak{M}^r \subset \mathfrak{M}^s$ if $r \le s$ and $(\mathfrak{M}^r)^s = \mathfrak{M}^{r+s}$.

PROOF: If $P: J^s(F;E) \to J^r(F;E)$ is the projection r $H: J^{r+s}(F;E) \to J^r(F;J^s(F;E))$ is the injection of I. (1.3) and f is sufficiently differentiable then $P\cdot j^s(f) = j^r(f)$ and $H\cdot j^{r+s}(f) = j^r(j^s(f))$. Thus, we have the commutative diagram:

$$\begin{array}{ccc} \mathfrak{M}^{r+s}(U,E) & = & (\mathfrak{M}^r)^s(U,E) \\ & & \Big\downarrow j^s \\ \Big\downarrow j^{r+s} & & \mathfrak{M}^r(U,J^s(F;E)) \\ & & \Big\downarrow j^r \\ \mathfrak{M}(U,J^{r+s}(F;E)) & \xrightarrow[H_*]{} & \mathfrak{M}(U,J^r(F;J^s(F;E))) \end{array} .$$

That is, the sets at the top are equal and the norms agree because all the maps are isometries. A similar diagram using P proves $\mathfrak{M}^r \subset \mathfrak{M}^s$. $\mathfrak{M}^0 = \mathfrak{M}$ is obvious. Q.E.D.

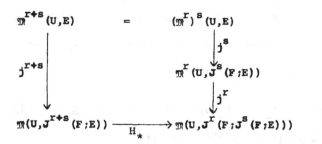

<u>FS2</u>: With U open and bounded in F, the linear maps $c: E \to \mathfrak{M}(U,E)$ and $\theta_1: L(F;E) \to \mathfrak{M}(U,E)$ are well defined by $c(v)(u) = v$ and $\theta_1(T)(u) = T(u)$. c is an isometry and $\|\theta_1\| \leq O^*(\|1_U\|_0)$.

<u>VERIFICATION</u>: $\|c(v)\|_0 = \|v\|, L(c(v)) = 0$. $\|\theta_1(T)\|_0 \leq \|1_U\|_0\|T\|$, $L(\theta_1(T)) = \|T\|$.

<u>INHERITANCE</u>: Let i_0 and i_1 be the inclusions of E and $L(F;E)$ into $J^r(F;E)$. The result for θ_1 f llows from the commutative diagram

$$\begin{array}{ccc} L(F;E) & \xrightarrow{\quad\theta_1\quad} & \mathfrak{M}^r(U,E) \\ \downarrow{\scriptstyle \theta_1 \times c} & & \downarrow{\scriptstyle j^r} \\ \mathfrak{M}(U,E) \times \mathfrak{M}(U,L(F;E)) & \xrightarrow[i_{0*}+i_{1*}]{} & \mathfrak{M}(U,J^r(F;E)). \end{array}$$

A similar diagram with $j^r \circ c_{\mathfrak{M}^r} = i_{0*} \circ c_{\mathfrak{M}}$ gives the result for c. Q.E.D.

The image of c, being complete, is closed. Note also that in FS1, $T_* \circ c_1 \times \ldots \times c_n = c_0 \circ T$ and so $\|T_*\| \geq \|T\|$. In particular, if $n = 1$, $\|T_*\| = \|T\|$.

<u>FS3</u>: $\mathfrak{M} \subset \mathfrak{B}$ and $\mathfrak{M}(U,E)$ is complete.

<u>VERIFICATION:</u> Identifying E^U with the product $\Pi_U E_u$, one copy of E for each u in U, then $\mathfrak{B}(U,E)$ is isometric to $\hat{\Pi}_U E_u$. $\mathcal{C}(U,E)$ and $\mathcal{C}_u(U,E)$ are closed subspaces. The operator $f \to \Delta f \mid O_4$ is a closed operator defined on the linear subspace $\mathcal{L}(U,E)$ to $\mathfrak{B}(O_4,E)$. The norm $\|\ \|_{\mathcal{L}}$ makes the graph of this operator an isometry. Similarly, $f \to \Delta_a f \mid O_3$ defines $\mathcal{L}ip_a(U,E)$ via the graph of a map to $\mathfrak{B}(O_3,E)$. Also, $f \to Df$ is a closed operator from $\mathfrak{B}^1(U,E)$ to $\mathfrak{B}(U,L(F;E))$ [5; Thm. VIII. 6.3] and so $\mathfrak{B}^1$ is complete.

<u>INHERITANCE</u>: If $\mathfrak{M} \subset \mathfrak{B}$, then $\mathfrak{M}^r \subset \mathfrak{M} \subset \mathfrak{B}$. We prove that $\mathfrak{M}$ complete and $\mathfrak{M} \subset \mathfrak{B}$ implies $\mathfrak{M}^1$ is complete. The result follows by induction since $(\mathfrak{M}^r)^1 = \mathfrak{M}^{r+1}$ by Prop. 3.

$$\begin{array}{ccc} \mathfrak{M}^1(U,E) & \xrightarrow{\text{inc}} & \mathfrak{B}^1(U,E) \\ \downarrow j^1 & & \downarrow j^1 \\ \mathfrak{M}(U,J^1(F;E)) & \xrightarrow[\text{inc}]{} & \mathfrak{B}(U,J^1(F,E)) \end{array}$$

$j^1(\mathfrak{M}^1) = \text{inc}^{-1}(j^1(\mathfrak{B}^1))$. By the verification step $j^1(\mathfrak{B}^1)$ is complete and so is closed. Hence $j^1(\mathfrak{M}^1)$ is closed in $\mathfrak{M}(U,J^1(F,E))$ which is complete by hypothesis. Thus, $\mathfrak{M}^1$ is complete. Q.E.D.

FS4: For U_1, U_2 open and bounded in F_1, F_2, let $g\colon F_1 \to F_2$ be an affine map, i.e. $\bar{g} = g - c(g(0))$ is linear, with $g(U_1) \subset U_2$. The bounded linear map $g^*\colon \mathfrak{M}(U_2,E) \to \mathfrak{M}(U_1,E)$ is well defined by $g^*(f) = f\circ g$. $\|g^*\| \leq O^*(\|\bar{g}\|)$ and if $\|\bar{g}\| \leq 1$ then $\|g^*\| = 1$ (unless $E = 0$).

VERIFICATION: $\|f\circ g\|_0 \leq \|f\|_0, \|f\circ g\|_L \leq \|f\|_L\|\bar{g}\|, L_a(f\circ g) \leq L_a(f)\|\bar{g}\|^a$.
INHERITANCE: Let $\bar{g}^{\#}\colon J^r(F_2;E) \to J^r(F_1;E)$ be the product of $L^i_s(\bar{g};1)$, $i = 0,\ldots,r$. $\|\bar{g}^{\#}\| \leq O^*(\|\bar{g}\|)$ and $\|\bar{g}^{\#}\| = 1$ if $\|\bar{g}\| \leq 1$. Inheritance follows from the commutative diagram:

$$\begin{array}{ccccc} \mathfrak{M}^r(U_2,E) & & \xrightarrow{\quad g^* \quad} & & \mathfrak{M}^r(U_1,E) \\ \downarrow j^r & & & & \downarrow j^r \\ \mathfrak{M}(U_2,J^r(F_2;E)) & \xrightarrow[g^*]{} & \mathfrak{M}(U_1,J^r(F_2;E)) & \xrightarrow[(\bar{g}^{\#})_*]{} & \mathfrak{M}(U_1,J^r(F_1;E)). \end{array}$$

Q.E.D.

In particular, if $U_1 \subset U_2$ and $g = \bar{g} = I$ we obtain restriction maps. We now describe some properties which follow from these axioms.

4 PROPOSITION: $\mathfrak{B}^{r+1} \subset \mathcal{L}^r$ are isometric inclusions; $\mathcal{L}^r \subset \mathcal{C}^r$ and $\mathcal{L}ip^r_a \subset \mathcal{C}^r$ are strong inclusions ($r = 0,1,\ldots$).

PROOF: If f is $\mathfrak{B}^1$ then $Df(u)v = \text{Lim}\, \Delta f(u + tv,u)$ for v a unit vector and $t > 0$ approaching 0. Hence, $\|Df\|_0 \leq \|f\|_L$. The converse inequality follows from the mean value theorem [5; Thm. VIII.5.2]. If

$f_n \to f$ (in $\mathfrak{B}$) then $\Delta_a f_n \to \Delta_a f$ pointwise. So $K \geq \Delta_a f_n | O_3$ implies $K \geq \Delta_a f | O_3$. This proves the result for $r = 0$. Inheritance is left to the reader. Q.E.D.

5 PROPOSITION: Let U be open and bounded in F. For each $r = 0,1,\ldots$ the bounded linear map φ: $J^r(F;E) \to \mathfrak{M}(U,E)$ is well defined by $j \to \varphi(j)|U$. $\|\varphi\| \leq O^*(\|1_U\|)$ constants depending on $\mathfrak{M}$ and r.

PROOF: Taking direct sums of FS1 maps $*$ for $n = 0,\ldots,r$ we obtain $*$: $J^r(F;E) \to J^r(\mathfrak{M}(U;F);\mathfrak{M}(U,E))$ with $\|*\| \leq O^*(1)$ (depending on $\mathfrak{M}$ and r). By FS2 the inclusion map 1_U: $U \to F$ is in $\mathfrak{M}(U;F)$ with $\|1_U\|_{\mathfrak{M}} \leq O^*(\|1_U\|_0)$. Finally, if $v \in F$, then ev_v: $J^r(F;E) \to E$ by $ev_v(j) = \varphi(j)(v)$ is a linear map with $\|ev_v\| \leq O^*(\|v\|)$ (depending on r). The result follows because φ is the composition $ev_{1_U} \circ *$. Q.E.D.

We turn now to properties which are not satisfied by all our examples. Preceding the statement in each case we describe those examples which work.

FS5 (Gluing: $\mathfrak{B}^r, \mathcal{C}^r, \mathcal{L}^r$): Let $\{U_\alpha\}$ be an open cover of U. Let i_α: $U_\alpha \to U$ and $i_{\alpha\beta}$: $U_{\alpha\beta} \to U_\alpha$ denote the inclusion maps where $U_{\alpha\beta} = U_\alpha \cap U_\beta$ if the latter $\neq \emptyset$ and is otherwise undefined. Define $\rho = \hat{\Pi}_\alpha i_\alpha{}^*$ and $d = \hat{\Pi}_{\alpha\beta}(i_{\alpha\beta}{}^* \circ p_\beta - i_{\beta\alpha}{}^* \circ p_\alpha)$ to get maps $0 \to \mathfrak{M}(U,E) \xrightarrow{\rho} \hat{\Pi}_\alpha \mathfrak{M}(U_\alpha,E) \xrightarrow{d} \hat{\Pi}_{\alpha\beta}\mathfrak{M}(U_{\alpha\beta},E)$. ρ is an isometry onto the kernel of d.

VERIFICATION: Note that $d \circ \rho = 0$ and ρ is injective in any case. That $\|\rho\| \leq 1$ follows from FS4. Also, $d\{f_\alpha\} = 0$ iff $\{f_\alpha\}$ fits together to define a (unique) map f in E^U with $f|U_\alpha = f_\alpha$. Clearly, $\|f\|_0 = \sup_\alpha \|f_\alpha\|_0$ and so FS5 holds for $\mathfrak{B}$. Since f is continuous iff the f_α's are it holds for $\mathcal{C}$. Finally, that $\|f\|_L \leq \sup_\alpha \|f_\alpha\|_L$ (and hence equality holds) follows from the "subdivision of the segment" argument illustrated in (1.2).

INHERITANCE: This is an easy exercise in the naturality of ρ and d with respect to j^r. Q.E.D.

Although $\mathcal{L}ip_a$ does not satisfy the Gluing Property, it is instructive to note that if the cover has Lebesgue number $\epsilon > 0$, then $\|\Delta_a f|O_3^\epsilon\|_0 \leq \sup_\alpha L_a(f_\alpha)$ and so ρ is an isomorphism onto the kernel of d with $\theta(\rho) \leq O^*(\epsilon^{-1})$. This weaker version of Gluing is also heritable and so holds for $\mathcal{L}ip_a^r$. We see that $\|f\|_L$ is the "localization" of $L(f)$, i.e. if $\{U_\alpha\}$ consists of convex open sets then $\|f\|_L = \sup_\alpha L(f_\alpha)$.

FS6 (Integral: Defined for $\mathfrak{m} \subset \mathcal{C}$): Let I be a bounded open interval in R with $a,b \in I$. The bounded linear maps $\mathcal{J}_a: \mathfrak{m}(U \times I,E) \to \mathfrak{m}(U \times I,E)$ and $\mathcal{J}_a^b: \mathfrak{m}(U \times I,E) \to \mathfrak{m}(U,E)$ are well defined by $\mathcal{J}_a(f)(x,t) = \int_a^t f(x,s)ds$ and $\mathcal{J}_a^b(f)(x) = \mathcal{J}_a(f)(x,b)$ and satisfy $\|\mathcal{J}_a\| \leq O^*(\|1_{I-a}\|_0)$ and $\|\mathcal{J}_a^b\| \leq |b-a|$.

VERIFICATION: $\|\mathcal{J}_a(f)\|_0 \leq \|1_{I-a}\|_0\|f\|_0$ and $\|\mathcal{J}_a^b(f)\|_0 \leq |b-a|\|f\|_0$. Also, $\|\mathcal{J}_a(f)(u,t) - \mathcal{J}_a(f)(v,s)\| \leq |t-s|\|f\|_0 + \|1_{J-a}\|_0 \sup_{\bar{t}\in J}|f(u,\bar{t}) - f(v,\bar{t})|$ where J is any subinterval containing t,s and a. If $t = s = b$ we can choose $J = [a,b]$. From these inequalities the results for $\mathcal{C}$, $\mathcal{C}_u$, $\mathcal{L}$ and $\mathcal{L}ip_a$ follow. To prove continuity, choose J compact and note that $\{(v,s): |f(u,s) - f(v,s)| < \epsilon\}$ is an open neighborhood of $u \times J$ in $U \times I$ and so by Wallace's Lemma [13; Thm. V.12] it contains some neighborhood $B(u,\delta)\times J$. Finally, if $f \in \mathfrak{m}^1$ (and hence $\mathcal{L}$) $\mathcal{J}_a(f)$ is in fact $\mathfrak{m}^1$ with $D\mathcal{J}_a(f)(u,t)(v,\dot{t}) = \dot{t}f(u,t) + \int_a^t D_1f(u,s)v\,ds$, computed formally and justified by the fact that $D_1f(u,s)v$ is integrable in s because, although not continuous, it is the pointwise, uniformly bounded limit of difference quotients which are. Also, $f(x+h,s) - f(x,s) - D_1f(x,s)h$ is $o(h)$ uniformly as s varies in a compact interval.

INHERITANCE ($\mathfrak{m} \subset \mathcal{C}$): Define ${}_1j^r(f): U \times I \to J^r(F;E)$ by ${}_1j^rf(x,t) = (f(x,t),D_1f(x,t),\ldots,D_1^rf(x,t))$. If f is $\mathfrak{m}^r$ then ${}_1j^rf$ is

$\mathfrak{m}$ with $\|{}_1 j^r f\|_{\mathfrak{m}} \leq \|j^r f\|_{\mathfrak{m}} = \|f\|_{\mathfrak{m}^a}$. Furthermore, by Leibniz' Rule [5; Thm. VIII. 11.2] $J^r(\mathcal{s}_a^b(f)) = \mathcal{s}_a^b({}_1 j^r(f))$. So the sharp estimate on $\|\mathcal{s}_a^b\|$ is inherited. The sloppier estimate on $\mathcal{s}_a$ is inherited by Leibniz' Rule and the Fundamental Theorem of Calculus. Q.E.D.

6 <u>PROPOSITION</u>: Assume $\mathfrak{m} \subset \mathcal{C}$ satisfies FS6.

(a) Defining $L^1(I^n)$ via Lebesgue measure on R^n, the linear map $\mathcal{s}: L^1(I^n) \to L(\mathfrak{m}(U \times I^n,E);\mathfrak{m}(U,E))$ is well defined by $q(s) \to (f(u,s) \to \int_{I^n} f(u,s)q(s)ds)$ and has norm ≤ 1.

(b) For I a bounded open interval in R, the map $\bar{\mathcal{s}}_a: I \to L(\mathfrak{m}(U \times I,E);\mathfrak{m}(U,E))$ defined by $\bar{\mathcal{s}}_a(t)(f) = \bar{\mathcal{s}}_a^t(f)$ is a $\mathcal{L}$ip map with $\|\bar{\mathcal{s}}_a\|_0 \leq \|1_{I-a}\|_0$ and $L(\bar{\mathcal{s}}_a) \leq 1$. So for $f \in \mathfrak{m}(U \times I,E)$, $\bar{\mathcal{s}}_a(f): I \to \mathfrak{m}(U,E)$ is a $\mathcal{L}$ip map with $\|\bar{\mathcal{s}}_a(f)\|_0 \leq \|1_{I-a}\|_0\|f\|$ and $L(\bar{\mathcal{s}}_a(f)) \leq \|f\|$.

(c) $\mathcal{s}_a: \mathfrak{m}(I,E) \to \mathfrak{m}^1(I,E)$ is a well defined linear map with $\|\mathcal{s}_a\| \leq 0^*(\|1_{I-a}\|_0)$.

(d) Let $\mathfrak{m}_1$ be any function space type. $\mathcal{s}_{a*}: \mathfrak{m}_1(U,\mathfrak{m}(I,E)) \to \mathfrak{m}_1(U,\mathfrak{m}^1(I,E))$ is a well defined linear map with $\|\mathcal{s}_{a*}\| \leq 0^*(\|1_{I-a}\|_0)$.

<u>PROOF</u>: (a): If q is the characteristic function of the product of subintervals of I then $\mathcal{s}(q) \in L(\mathfrak{m}(U_1 \times I^n,E);\mathfrak{m}(U,E))$ with $\|\mathcal{s}(q)\| \leq \|q\|_{L^1}$ by the estimate for $\mathcal{s}_a^b$ and induction on n. If q is a finite linear combination $\Sigma\, c_i q_i$ with the q_i's of this type, having disjoint supports, then $\|q\|_{L^1} = \Sigma|c_i|\,\|q_i\|_{L^1}$ and so the result holds for such q's. But the set of such q's is a dense linear subspace of $L^1(I^n)$ and so $\mathcal{s}$ extends to $L^1(I^n)$ with norm ≤ 1.

(b): $I \to L^1(I)$ by $t \to$ characteristic function of $[a,t]$ is a $\mathcal{L}$ip map with the given estimates. (b) is also easily proved directly as $\mathcal{s}_a^{t_1}(f) - \mathcal{s}_a^{t_2}(f) = \mathcal{s}_{t_2}^{t_1}(f)$.

(c): By FS6, $\mathcal{s}_a: \mathfrak{m} \to \mathfrak{m}$ is well defined with the right estimate. By the Fundamental Theorem of Calculus we have that $D(\mathcal{s}_a(f)): I \to L(R,E)$

$\cong E = f$. Thus, $j^1 \circ \mathscr{s}_a$ maps to $\mathfrak{M}$.

(d): From (c) and FS1 for $\mathfrak{M}_1$. Q.E.D.

From the Integral Condition for $\mathcal{C}$ follows an important tool for proving heritability:

7 **JET LEMMA**: Let $\{T_\alpha: E \to E_\alpha\}$ be a family of bounded linear maps distinguishing points of E (i.e. $u_1 \neq u_2 \in E$ implies $T_\alpha u_1 \neq T_\alpha u_2$ for some α). Let $f: U \to E$ and $g \in \mathcal{C}(U, J^r(F;E))$ satisfy: $T_\alpha \circ f$ is C^r for every α with $j^r(T_\alpha \circ f) = T_{\alpha\#} \circ g$. Then $f \in \mathcal{C}^r(U,E)$ and $j^r(f) = g$. If $g \in \mathfrak{M}(U, J^r(F;E))$ then $f \in \mathfrak{M}^r(U,E)$.

PROOF: Let $g = (A_0, \ldots, A_r)$ with $A_r: U \to L^r_s(F;E)$. As in Taylor's Theorem define $R: U^{(2)} \to L^r_s(F;E)$ by $R(u,h) = \int_0^1 [(1-t)^{r-1}/(r-1)!][A_r(u+th) - A_r(u)]dt$. I claim that for $(u,h) \in U^{(2)}$, $f(u+h) = \theta(g(u))(h) + R(u,h)h^{(r)}$. For by Taylor's theorem for $T \circ f$, $T_\alpha(f(u+h)) = \theta(T_{\alpha\#}(g(u)))(h) + T_\alpha(R(u,h)h^{(r)}) = T_\alpha(\theta(g(u))(h) + R(u,h)h^{(r)})$. Because $\{T_\alpha\}$ distinguishes points the claim holds. By FS6 for $\mathcal{C}$, R is continuous and $R(u,0) = 0$. The result then follows from the converse of Taylor's Theorem. Q.E.D.

FS7 (Product Conditions: Strong version for $\mathfrak{B}$, $\mathscr{L}^r$, $\mathscr{L}ip^r_a$): Let $\{E_\alpha\}$ be a family of B spaces. $\hat{\Pi}_\alpha p_{\alpha*}: \mathfrak{M}(U, \hat{\Pi}_\alpha E_\alpha) \to \hat{\Pi}_\alpha \mathfrak{M}(U, E_\alpha)$ is an isometry into which is onto if the index set is finite. The strong product condition (written FS7 (Strong)) holds if $\hat{\Pi}_\alpha p_{\alpha*}$ is onto for any index set.

VERIFICATION: If $f: U \to \hat{\Pi}_\alpha E_\alpha$ and $f_\alpha = p_\alpha \circ f$, we will write $f = \hat{\Pi}_\alpha f_\alpha$. Clearly, $\|\hat{\Pi}_\alpha f_\alpha\|_0 = \sup_\alpha \|f_\alpha\|_0$ and $\Delta_a \hat{\Pi}_\alpha f_\alpha = \hat{\Pi}_\alpha \Delta_a f_\alpha$. From these equations follows the strong condition for $\mathfrak{B}$, $\mathscr{L}$ and $\mathscr{L}ip_a$. Since $\mathfrak{B}^1 \subset \mathscr{L}$ is an isometric inclusion it follows that $\hat{\Pi}_\alpha p_{\alpha*}$ is isometric for $\mathfrak{B}^1$. For any function space type, $\hat{\Pi}_\alpha p_{\alpha*}$ is an isomorphism when the index set is finite. The inverse map is $\Sigma_\alpha i_{\alpha*} p_\alpha$ where $i_\alpha\colon E_\alpha \to \hat{\Pi}$ is the inclusion map. In particular, FS7 holds for $\mathcal{B}^1$, $\mathcal{C}$, $\mathcal{C}_u$. FS7 (Strong) does not hold for these.

<u>INHERITANCE</u>: Let P denote the canonical isometry $\hat{\Pi}_\alpha p_{\alpha\#}\colon J^r(F;\hat{\Pi}_\alpha E_\alpha) \to \hat{\Pi}_\alpha J^r(F;E_\alpha)$.

$$\begin{array}{ccccc} \mathfrak{M}^r(U,\hat{\Pi}_\alpha E_\alpha) & & \xrightarrow{\hat{\Pi}_\alpha p_{\alpha*}} & & \hat{\Pi}_\alpha \mathfrak{M}^r(U,E_\alpha) \\ \downarrow{\scriptstyle j^r} & & & & \downarrow{\scriptstyle \hat{\Pi}_\alpha j^r} \\ \mathfrak{M}(U,J^r(F;\hat{\Pi}_\alpha E_\alpha)) & \xrightarrow{P_*} & \mathfrak{M}(U,\hat{\Pi}_\alpha J^r(F;E_\alpha)) & \xrightarrow{\hat{\Pi}_\alpha p_{\alpha*}} & \hat{\Pi}_\alpha \mathfrak{M}(U,J^r(F;E_\alpha)). \end{array}$$

The above commutative diagram makes it clear that $\hat{\Pi}_\alpha p_{\alpha*}$ is an isometry for $\mathfrak{M}^r$. If $\{f_\alpha\} \in \hat{\Pi}_\alpha \mathfrak{M}^r(U,E_\alpha)$ with $g_\alpha = j^r f_\alpha$ and $\mathfrak{M}$ satisfies FS7 (Strong) then there exists $g \in \mathfrak{M}(U,J^r(F;\hat{\Pi}_\alpha E_\alpha))$ with $p_{\alpha\#}\cdot g = g_\alpha$. If $\mathfrak{M} \subset \mathcal{C}$ and $f = \hat{\Pi}_\alpha f_\alpha$ then the Jet Lemma applies and f is $\mathfrak{M}^r$. Q.E.D.

That $\mathfrak{M} \subset \mathcal{C}$ is necessary for inheritance is seen by the fact that $\mathcal{B}^r$ satisfies FS7 (Strong) for $r = 0$ and not for $r \geq 1$. If $\mathfrak{M}_1 \subset \mathfrak{M}_2$ and in the diagram

$$\begin{array}{ccc} \mathfrak{M}_1(U,\hat{\Pi}_\alpha E_\alpha) & \xrightarrow{\hat{\Pi}_\alpha p_{\alpha*}} & \hat{\Pi}_\alpha \mathfrak{M}_1(U,E_\alpha) \\ \downarrow{\scriptstyle i} & & \downarrow{\scriptstyle \hat{\Pi}_\alpha i_\alpha} \\ \mathfrak{M}_2(U,\hat{\Pi}_\alpha E_\alpha) & \xrightarrow[\hat{\Pi}_\alpha p_{\alpha*}]{} & \hat{\Pi}_\alpha \mathfrak{M}_2(U,E_\alpha) \end{array}$$

the image $\hat{\Pi}_\alpha p_{\alpha*}(\mathfrak{M}_2)$ contains $\hat{\Pi}_\alpha i_\alpha(\mathfrak{M}_1)$ then we say that <u>$\mathfrak{M}_1$ interchanges into $\mathfrak{M}_2$</u>. Clearly, if there exists $\mathfrak{M}$ satisfying FS7 (Strong) with $\mathfrak{M}_1 \subset \mathfrak{M} \subset \mathfrak{M}_2$, then this holds. Thus, $\mathcal{B}^{r+1}$, $\mathcal{C}^{r+1}$, $\mathcal{C}_u^{r+1}$ interchange into $\mathcal{B}^r$, $\mathcal{C}^r$, $\mathcal{C}_u^r$, and, in fact, all interchange into $\mathcal{L}^r$. If $\mathfrak{M}_1$ interchanges into $\mathfrak{M}_2$ then $\mathfrak{M}_1^r$ interchanges into $\mathfrak{M}_2^r$. The name comes from the next proposition.

Applying FS1 to the $n + 1$ linear map ev: $L_s^n(E_1;E_2) \times E_1^{(n)} \to E_2$ we obtain ev_* which induces the map λ in the following:

$$(1.4)\qquad \begin{array}{ccc} & & L^n_\epsilon(E_1;E_2) \\ & \overset{c}{\swarrow} & \downarrow * \\ \mathfrak{M}(U,L^n_\epsilon(E_1;E_2)) & \xrightarrow{\lambda} & L^n_\epsilon(\mathfrak{M}(U,E_1);\mathfrak{M}(U,E_2)) \\ & \underset{\varkappa}{\searrow} & \downarrow L^n_\epsilon(c;1) \\ & & L^n_\epsilon(E_1;\mathfrak{M}(U,E_2)) \end{array}$$

Here c and $L^n_\epsilon(c,1)$ are isometries, λ and $*$ have norms $\leq O*(1)$ (depending on n and $\mathfrak{M}$; $L^n_\epsilon = L^n$ or L^n_s). The interpretation of this diagram for $n = 1$ is similar to that of the jet diagram I.(1.5).

8 INTERCHANGE PROPOSITION: Assuming FS7, $\varkappa$ is an isometry. If $\mathfrak{M}$ satisfies FS7 (Strong) or E_1 is finite dimensional then $\varkappa$ is onto.

PROOF: Let $S = \{(v_1,\ldots,v_n) \in E_1^{(n)}: \|v_i\| = 1\}$ and let $ev_{(v_1,\ldots,v_n)}$: $L^n(E_1;E_2) \to E_2$ be the linear evaluation map. Then $T = \hat{\Pi}\, ev_{(v_1,\ldots,v_n)}: L^n(E_1;E_2) \to \hat{\Pi}_S F_2$ is an isometry and by FS1 we have the commutative diagram which must thus consist of isometries:

$$\begin{array}{ccc} \mathfrak{M}(U,L^n(E_1;E_2)) & \xrightarrow{\varkappa} & L^n(E_1;\mathfrak{M}(U,E_2)) \\ T_*\downarrow & & \downarrow T \\ \mathfrak{M}(U,\hat{\Pi}_S E_2) & \xrightarrow[\hat{\Pi}_S p_*]{} & \hat{\Pi}_S \mathfrak{M}(U,E_2) \end{array}$$

Hence $\varkappa$ is an isometry and, assuming FS7 (Strong), if $\varkappa f \in L^n(E_1;\mathfrak{M}(U,E_2))$ for $f \in L^n(E_1;E_2)^U$ then $T_*(f)$ is $\mathfrak{M}$ and so by FS1, f is $\mathfrak{M}$. If E_1 has a finite basis B of unit vectors then replacing S by the subset $S_0 = \{(v_1,\ldots,v_n) \in S: v_i \in B\}$ we can use the same argument with the finite product $\hat{\Pi}_{S_0} E_2$ and the map $T_0 = \hat{\Pi}_{S_0} ev_{(v_1,\ldots,v_n)}$ which is still an isomorphism onto a closed subspace (in fact, there is an obvious left inverse for T_0). Q.E.D.

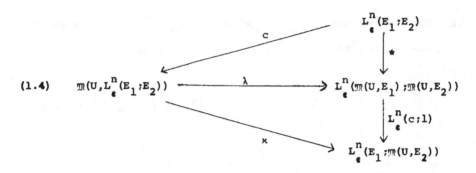

Of course, if $\mathfrak{M}_1$ interchanges into $\mathfrak{M}_2$ then the above argument shows that the image of κ for $\mathfrak{M}_2$ contains the image of $L^n_\epsilon(1;i): L^n_\epsilon(E_1;\mathfrak{M}_1(U,E_2)) \to L^n_\epsilon(E_1;\mathfrak{M}_2(U,E_2))$.

Assume $\mathfrak{M}$ satisfies FS7. For $n = 1$, if E' is the dual space of E, $j: E \to E''$ is the canonical injection and S is the unit sphere of E' then we have a sequence of isometries: $\mathfrak{M}(U,E) \xrightarrow{j_*} \mathfrak{M}(U,E'') \overset{\kappa}{\to} L(E';\mathfrak{M}(U,R)) \overset{T}{\to} \hat{\Pi}_S\, \mathfrak{M}(U,R)$, with T an isomorphism. If $f \in E^U$ then I claim $\kappa(j\cdot f) \in L(E';\mathfrak{M}(U,R))$ iff f is a <u>scalarly $\mathfrak{M}$ map</u>, i.e. for every $v' \in E'$, $v'\cdot f \in \mathfrak{M}(U,R)$. This is clearly necessary and if f is scalarly $\mathfrak{M}$ then $\{j(f(u)): E' \to R\}$ parametrized by $u \in U$ is a family in E'' and for each $v' \in E'$, $j(f(u))(v') = v'\cdot f(u)$ is $\mathfrak{M}$ and hence $\mathfrak{B}$ in u. By the resonance theorem [27; Cor. II.1.1] the family is bounded in E'', i.e. $j\cdot f \in \mathfrak{B}(U,E'')$ and so $\kappa(j\cdot f): E' \to \mathfrak{B}(U;R)$ is a continuous linear map which by hypothesis factors through the inclusion $\mathfrak{M}(U;R) \to \mathfrak{B}(U;R)$. Thus, the graph of $\kappa(j\cdot f): E' \to \mathfrak{M}(U;R)$ is closed and so $\kappa(j\cdot f)$ is bounded [27; Thm. II.6.1]. Of course, if $f \in \mathfrak{M}(U;E)$ then f is scalarly $\mathfrak{M}$. If $\mathfrak{M}$ satisfies FS7 (Strong) then the converse is true, since $j\cdot f$ is $\mathfrak{M}$ iff f is by FS1. Note that in any case the norm on $\mathfrak{M}(U,E)$ is determined by the norm on $\mathfrak{M}(U,R)$ since the map $T\cdot\kappa\cdot j_*$ is an isometry.

We now look at evaluation maps.

9 <u>LEMMA</u>: (a) The product evaluation map $e: \mathfrak{M}(U_1,L(E_1;E_2)) \times \mathfrak{M}(U_2,E_1) \to \mathfrak{M}(U_1 \times U_2,E_2)$ is a bilinear map well defined by $e(g,f)(u_1,u_2) = g(u_1)(f(u_2))$, $\|e\| \leq O^*(1)$.

(b) If V is open and bounded in E_1 then $e_V: \mathfrak{M}(U,L(E_1;E_2)) \to \mathfrak{M}(U \times V,E_2)$ is a linear map well defined by $e_V(g)(u,v) = g(u)(v)$. $\|e_V\| \leq O^*(\|1_V\|_0)$.

<u>PROOF</u>: (a): $e = ev_*\cdot(p_1^* \times p_2^*)$ where $p_i: U_1 \times U_2 \to U_i$. Result by

FS1 and FS4. (b): $e_V(g) = e(g,1_V)$. Result by (a) and FS2. Q.E.D.

10 PROPOSITION: If $u \in U$ then $Ev_u \in L(\mathfrak{M}(U,E);E)$ defined by $Ev_u(f) = f(u)$. $\|Ev_u\| = 1$ (unless $E = 0$). If $\mathfrak{M}$ satisfies FS7 (Strong) then $Ev \in \mathfrak{M}(U,L(\mathfrak{M}(U,E);E))$ with $\|Ev\|_{\mathfrak{M}} = 1$ (unless $E = 0$).

PROOF: For $u \in U$ define $u: 0 \to F$ by $u(0) = u$. Identify $\mathfrak{M}(0,E)$ with E by c (cf. FS2). Then $Ev_u = u^*$. Apply FS4. If FS7 (Strong) holds then Ev is $\mathfrak{M}$ by Proposition 8 because it is $\kappa^{-1}(I)$ with I the identity in $L(\mathfrak{M}(U,E);\mathfrak{M}(U,E))$. Q.E.D.

Note if $\mathfrak{M}_1$ interchanges into $\mathfrak{M}_2$ then $Ev \in \mathfrak{M}_2(U,L(\mathfrak{M}_1(U,E);E))$, because κEv is the inclusion in $L(\mathfrak{M}_1(U,E);\mathfrak{M}_2(U,E))$ which is the image of the identity in $L(\mathfrak{M}_1;\mathfrak{M}_1)$. From Proposition 10 and Lemma 9, FS7 (Strong) implies the following:

FS8 (Evaluation: All examples except $\mathcal{C}^r_u$): For $G \subset \mathfrak{M}(U,E)$ open and bounded, $Ev_G \in \mathfrak{M}(U \times G,E)$ is well defined by $Ev_G(u,f) = f(u)$ and $\|Ev_G\|_{\mathfrak{M}} \leq O^*(\|1_G\|_0)$.

VERIFICATION: It suffices to check that $Ev: U \times \mathcal{C}(U,E) \to E$ is continuous. This easily follows from $\|f_1(u_1) - f_2(u_2)\| \leq \|f_1(u_1) - f_1(u_2)\| + \|f_1 - f_2\|_0$.

INHERITANCE (Assuming $\mathfrak{M}^1$ interchanges to $\mathfrak{M}$ and $\mathfrak{M}^1$ interchanges to $\mathcal{C}$, or a fortiori $\mathfrak{M}^1$ interchanges to $\mathfrak{M}$ and $\mathfrak{M} \subset \mathcal{C}$): We proceed by induction proving the result for $\mathfrak{M}^1$ and then applying Proposition 3. For G open and bounded in $\mathfrak{M}^1(U,E)$ and $G^1 = \{g \in \mathfrak{M}(U,L(F;E)): \|g\| < \|1_G\|_0\}$, compute the partial derivatives. D_1Ev_G is the composition $U \times G \xrightarrow{1 \times D} U \times G^1 \xrightarrow{Ev_{G^1}} L(F,E)$ and $D_2Ev_G: U \times G \xrightarrow{p_1} U \xrightarrow{Ev} L(\mathfrak{M}^1(U,E);E)$. By hypothesis D_1Ev_G is $\mathfrak{M}$ and by the interchange results, D_2Ev_G is both $\mathfrak{M}$ and $\mathcal{C}$. By the usual proof [5; Thm. VIII. 9.1], Ev_G is differentiable and so its derivative is $\mathfrak{M}$. Thus, Ev_G is $\mathfrak{M}^1$. Q.E.D.

11 **LEMMA:** Let $U_\alpha \subset F_\alpha$ be open and bounded and $G \subset \hat{\Pi}_\alpha U_\alpha \subset \hat{\Pi}_\alpha F_\alpha$ be open. If the index set is finite or $\mathfrak{M}$ satisfies FS7 (Strong) the linear cross product map $\hat{\Pi}_\alpha \mathfrak{M}(U_\alpha, E_\alpha) \to \mathfrak{M}(G, \hat{\Pi}_\alpha E_\alpha)$ is well defined by $\{f_\alpha\} \to \hat{\Pi}_\alpha f_\alpha | G$ and has norm ≤ 1.

PROOF: Let $q_\alpha: G \to U_\alpha$ be the restriction of the projection on $\hat{\Pi}_\alpha F_\alpha$ The cross product map is the composition $[\hat{\Pi}_\alpha p_{\alpha *}]^{-1} \cdot [\hat{\Pi}_\alpha (q_\alpha{}^*)]$. Q.E.D.

12 **PROPOSITION:** For $f \in L(E_1; E_2)^U$, let $\kappa f: E_1 \to E_2{}^U$ and $f_V: U \times V \to E_2$ (V open and bounded in E_1) be the associated maps. If $f_V \in \mathfrak{M}(U \times V, E_2)$ for some V then $\kappa f \in L(E_1; \mathfrak{M}(U, E_2))$ with $\|\kappa f\| \leq 2r^{-1}\|f_V\|$ (r = sup of the radii of open balls in V). Conversely, if $\mathfrak{M}$ satisfies FS8 and $\kappa f \in L(E_1; \mathfrak{M}(U, E_2))$ then f_V is $\mathfrak{M}$ for all V with $\|f_V\| \leq O^*(\|\kappa f\|, \|1_V\|_0)$.

PROOF: For $v \in V$, $i^v: U \to U \times V$ the inclusion on $U \times v$ is affine and so $\kappa f(v) = i^{v*} f_V \in \mathfrak{M}(U, E_2)$ with $\|\kappa f(v)\| \leq \|f_V\|$ by FS4. By linearity κf maps E_1 into $\mathfrak{M}(U, E_2)$ and the neighborhood of 0, $V - V$, in E_1 is mapped into the $2\|f_V\|$ ball. Conversely, if $\kappa f \in L(E_1; \mathfrak{M}(U, E_2))$ let $G = \{g \in \mathfrak{M}(U, E_2) : \|g\| < \|\kappa f\| \|1_V\|_0\}$. $\kappa f(V) \subset G$ and $f_V = (I \times \kappa f)^* Ev_G$. Result by FS8 and FS4. Q.E.D.

2. Composition Results:

1 **DEFINITION:** Let $\mathfrak{M}_1 \subset \mathfrak{M}_2$, $\mathfrak{M}_3 \subset \mathfrak{M}_2$, $\mathfrak{M}_d \subset \mathfrak{M}_r$ be function space types. *$\mathfrak{M}_1$ maps $\mathfrak{M}_d$ to $\mathfrak{M}_r$ in an $\mathfrak{M}_2$ way rel $\mathfrak{M}_3$* if f r $U \subset F_1$, $G \subset F_2$, $V \subset E_1$ open and bounded, and $H: G \to \mathfrak{M}_d(U, E_1)$ an $\mathfrak{M}_3$ map with $H(g)(U) \subset V$ for all $g \in G$, the bounded linear map $\Omega_H: \mathfrak{M}_1(V, E_2) \to \mathfrak{M}_2(G, \mathfrak{M}_r(U, E_2))$ is well defined by $\Omega_H(f)(g) = f \cdot H(g)$ and $\|\Omega_H\| \leq O^*(\|H\|_{\mathfrak{M}_3})$. If $\mathfrak{M}_3 = \mathfrak{M}_2$ we say $\mathfrak{M}_1$ maps $\mathfrak{M}_d$ to $\mathfrak{M}_r$ in an $\mathfrak{M}_2$ way.

In many applications, H is just the inclusion of an open subset of $\mathfrak{M}_d(U, E_1)$ and the above says: if $g: U \to V$ is $\mathfrak{M}_d$ and $f: V \to E_2$

is $\mathfrak{M}_1$, then $f\circ g$ is $\mathfrak{M}_r$ and the map $g \to f\circ g$ is an $\mathfrak{M}_2$ map as g varies in G with $\mathfrak{M}_2$ bound $\leq \|f\|_{\mathfrak{M}_1} O^*(\|1_G\|_0)$. Also, since $\mathfrak{M}_3 \subset \mathfrak{M}_2 \subset \mathfrak{B}$ and for $g\colon U \to V$ as above the constant map $c(g) \in \mathfrak{M}_3(0,\mathfrak{M}_d(U,E_1))$ $\Omega_{c(g)}(f) = g^*(f) = f\circ g$ is a well defined map: $\mathfrak{M}_1(V,E_2) \to \mathfrak{M}_2(0,\mathfrak{M}_r(U,E_2)) \cong \mathfrak{M}_r(U,E_2)$ with $\|g^*\| \leq O^*(\|g\|_{\mathfrak{M}_d})$. Conversely, this condition on g^* is equivalent to saying $\mathfrak{M}_1$ maps $\mathfrak{M}_d$ to $\mathfrak{M}_r$ in a $\mathfrak{B}$ way. This yields the equivalence in the following:

<u>FS9</u>: $\mathfrak{M}$ maps $\mathfrak{M}$ to $\mathfrak{M}$ in a $\mathfrak{B}$ way, or equivalently, for $g\colon U \to V$ an $\mathfrak{M}$ map, $g^*\colon \mathfrak{M}(V,E) \to \mathfrak{M}(U,E)$ is well defined by $g^*(f) = f\circ g$ and $\|g^*\| \leq O^*(\|g\|_{\mathfrak{M}})$.

To analyze this property and the more general one in Definition 1, we will begin with a verification step, then prove two inheritance theorems.

2 <u>LEMMA:</u> (a) Let $g\colon U \to V \subset E_1$ and $f\colon V \to E_2$. $\|f\circ g\|_0 \leq \|f\|_0\|g\|_0$, $\|f\circ g\|_L \leq \|f\|_L\|g\|_L$, $L_{ab}(f\circ g) \leq 2\|f\|_{\mathcal{L}ip_a}(L_b(g))^a$.

(b) Let $g_1,g_2\colon U \to V \subset E_1$ and $f\colon V \to E_2$. If $tg_1 + (1-t)g_2(U) \subset V$ for $t \in [0,1]$ then $\|f\circ g_1 - f\circ g_2\|_0 \leq \|f\|_L\|g_1 - g_2\|_0$, $\|f\circ g_1 - f\circ g_2\|_L \leq \|Df\|_L \max(\|g_1\|_L,\|g_2\|_L)\ \|g_1 - g_2\|_0 + \|Df\|_0\|g_1 - g_2\|_L$. If $\|g_1 - g_2\|_0 < 1$ then $\|f\circ g_1 - f\circ g_2\|_0 \leq L_a(f)(\|g_1 - g_2\|_0)^a$ and if $tg_1 + (1-t)g_2(U) \subset V$ also then $L_{ab}(f\circ g_1 - f\circ g_2) \leq 2\|Df\|_{\mathcal{L}ip_a}\max(L_b(g_1),L_b(g_2))^a\|g_1 - g_2\|_0 + \|Df\|_0 L_b(g_1 - g_2)$. The estimates involving Df require f to be C^1.

<u>PROOF:</u> (a): This is easy from (1.1) and (1.2) of the previous section.

(b): The $\|\ \|_0$ estimates are easy. If f is C^1 then by Taylor's theorem: $f(v_1) - f(v_2) = \int_0^1 Df(tv_1 + (1-t)v_2)dt\cdot(v_1 - v_2)$, provided $[v_1,v_2] \subset V$. Thus, $\|(f(g_1(u_1)) - f(g_2(u_1))) - (f(g_1(u_2)) - f(g_2(u_2)))\| \leq \|\int_0^1\{Df(tg_1(u_1) + (1-t)g_2(u_1)) - Df(tg_1(u_2) + (1-t)g_2(u_2))dt\| \times \|g_1(u_1) - g_2(u_1)\| + \|\int_0^1 Df(tg_1(u_2) + (1-t)g_2(u_2))dt\|\|(g_1(u_1) - g_2(u_1)) - (g_1(u_2) - g_2(u_2))\|$. For the L_{ab} estimate, the first term is by (1.1) again bounded by $2\|Df\|_{\mathcal{L}ip_a}\int_0^1 [tL_b(g_1) + (1-t)L_b(g_2)]^a dt\|u_1-u_2\|^{ab}\|g_1-g_2\|_0$. The second term is bounded by $\|Df\|_0 L_b(g_1 - g_2)\|u_1 - u_2\|^b \le$ $\|Df\|_0 L_b(g_1 - g_2)\|u_1 - u_2\|^{ab}$ provided $\|u_1 - u_2\| < 1$. The $\| \|_L$ estimate is similar. Q.E.D.

3 <u>COROLLARY</u>: (a) FS9 holds for $\mathfrak{M} = \mathfrak{B}$, $\mathcal{C}$, $\mathcal{C}_u$, $\mathcal{L}$ and $\mathcal{L}ip$. Furthermore, $\mathcal{L}ip_a$ maps $\mathcal{L}ip_b$ to $\mathcal{L}ip_{ab}$ in a $\mathfrak{B}$ way.

(b) $\mathcal{C}_u$ maps $\mathfrak{B}$, $\mathcal{C}$ and $\mathcal{C}_u$ to $\mathfrak{B}$, $\mathcal{C}$ and $\mathcal{C}_u$ respectively in a $\mathcal{C}_u$, $\mathcal{C}$ and $\mathfrak{B}$ way. $\mathcal{L}$ maps $\mathfrak{B}$, $\mathcal{C}$ and $\mathcal{C}_u$ to $\mathfrak{B}$, $\mathcal{C}$ and $\mathcal{C}_u$ respectively in an $\mathcal{L}$ way. $\mathcal{L}ip_a$ maps $\mathfrak{B}$, $\mathcal{C}$ and $\mathcal{C}_u$ to $\mathfrak{B}$, $\mathcal{C}$ and $\mathcal{C}_u$ respectively in a $\mathcal{L}ip_{ab}$ way rel $\mathcal{L}ip_b$.

(c) $\mathcal{L}^1$ maps $\mathcal{L}$ to $\mathcal{L}$ in an $\mathcal{L}$ way. $\mathcal{L}ip_a^1$ maps $\mathcal{L}ip_b$ to $\mathcal{L}ip_{ab}$ in an $\mathcal{L}$ way.

The proof of this corollary is a good exercise in the meaning of Definition 1 and I leave it to the reader.

4 <u>THEOREM (First Inheritance Result)</u>: If $\mathfrak{M}_1$ maps $\mathfrak{M}_d$ to $\mathfrak{M}_r$ in an $\mathfrak{M}_2$ way rel $\mathfrak{M}_3$, then $\mathfrak{M}_1^s$ maps $\mathfrak{M}_d^s$ to $\mathfrak{M}_r^s$ in an $\mathfrak{M}_2$ way rel $\mathfrak{M}_3$.

<u>PROOF</u>: Let $H: G \to \mathfrak{M}_d^s(U,E_1)$ be an $\mathfrak{M}_3$ map with $H(g)(U) \subset V$ for all $g \in G$. We must show that $\Omega_H(f)$ maps G into $\mathfrak{M}_r^s(U,E_2)$, is $\mathfrak{M}_2$ on G and is bounded in f as f varies in $\mathfrak{M}_1^r(V,E_2)$. We use the following diagram:

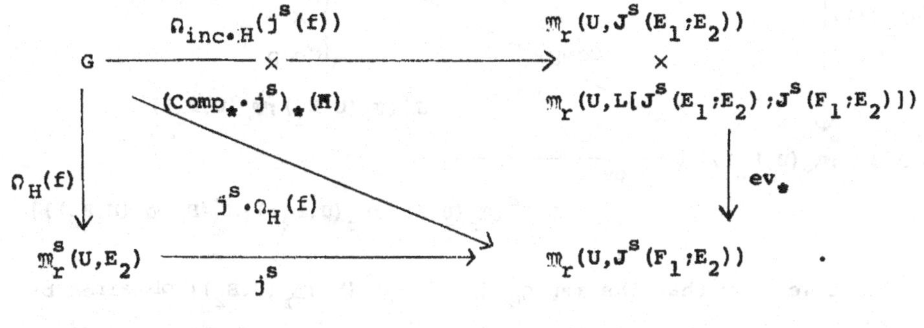

Here Comp: $J^s(F_1;E_1) \to L(J^s(E_1;E_2); J^s(F_1;E_2))$ is the polynomial map defined on page 4. By FS1 it induces a polynomial map $Comp_*$ and since

$j^s: \mathfrak{M}_d^s \to \mathfrak{M}_d$ is an isometry and $\mathfrak{M}_d \subset \mathfrak{M}_r$, $\mathfrak{M}_3 \subset \mathfrak{M}_2$ we get a polynomial map $(\mathrm{Comp}_* \circ j^s)_*: \mathfrak{M}_3(G,\mathfrak{M}_d^s(U,E_1)) \to \mathfrak{M}_2(G,\mathfrak{M}_r(U,L[J^s;J^s]))$ by FS1 again. Since the jet norm of Comp is $O^*(1)$ (depending only on s) the jet norm of $(\mathrm{Comp}_* \circ j^s)_*$ is $O^*(1)$ (depending on s and the function space types) and so $\|(\mathrm{Comp}_* \circ j^s)_*(H)\|_{\mathfrak{M}_2} \leq O^*(\|H\|_{\mathfrak{M}_3})$. $\Omega_{\mathrm{inc} \circ H}(j^s f)$ is an $\mathfrak{M}_2$ map with $\mathfrak{M}_2$ norm $\leq \|f\|_{\mathfrak{M}_1^s} O^*(\|H\|_{\mathfrak{M}_3})$ by hypothesis. On the other hand, $\Omega_H(f)(g) = f \circ H(g)$ is s times differentiable since f is $\mathfrak{M}_1^s$ and $H(g)$ is $\mathfrak{M}_d^s$ and $[j^s \circ \Omega_H(f)](g)(u) = \mathrm{Comp}(j^s(H(g))(u))[j^s(f)(H(g)(u))]$ by the chain rule definition of Comp. Thus, the upper triangle commutes. Since $\|\mathrm{ev}_*\| \leq O^*(1)$, we have $j^s \circ \Omega_H(f)$ is an $\mathfrak{M}_2$ map of norm $\leq \|f\|_{\mathfrak{M}_1^s} O^*(\|H\|_{\mathfrak{M}_3})$. Since j^s is an isometry $\Omega_H(f)$ maps into $\mathfrak{M}_r^s$ as shown. Q.E.D.

5 <u>THEOREM (Second Inheritance Result)</u>: If $\mathfrak{M}_1$ maps $\mathfrak{M}_d$ to $\mathfrak{M}_r$ in an $\mathfrak{M}_2$ way rel $\mathfrak{M}_3$ and $\mathfrak{M}_2 \subset \mathcal{C}$, then $\mathfrak{M}_1^s$ maps $\mathfrak{M}_d$ to $\mathfrak{M}_r$ in an $\mathfrak{M}_2^s$ way rel $\mathfrak{M}_3^s$. (If $\mathfrak{M}_2^s \subset \mathfrak{M}_3$ then we can delete "rel $\mathfrak{M}_3^s$").

<u>PROOF</u>: Assume that $H: G \to \mathfrak{M}_d(U,E_1)$ is $\mathfrak{M}_3^s$. For $f \in \mathfrak{M}_1^s(V,E_2)$ we compute $j^s(\Omega_H(f))$ by the following diagram:

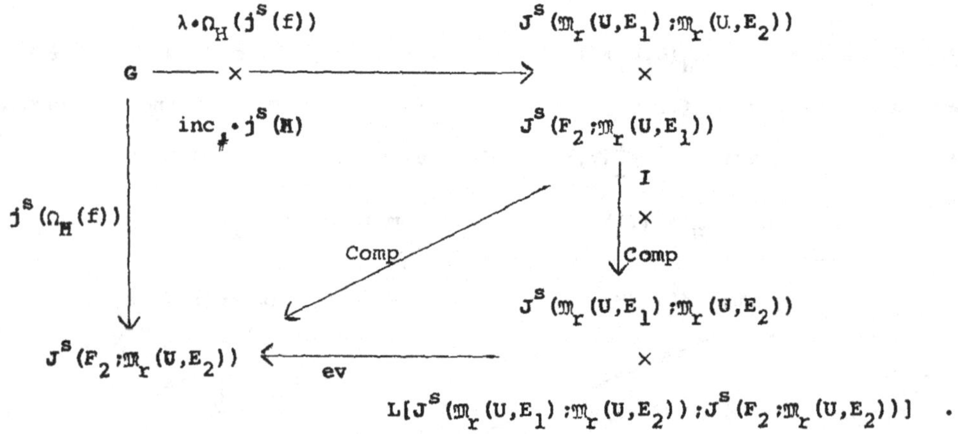

First we show that the map $q_H(f): G \to J^s(F_2;\mathfrak{M}_r(U,E_2))$ obtained by going around the right side is $\mathfrak{M}_2$ and estimate its norm. Note that the lower triangle just expresses two different definitions of Comp (cf. p. 5). inc: $\mathfrak{M}_d \to \mathfrak{M}_r$ is the inclusion. λ is the linear map of norm $\le O^*(1)$ in diagram (1.4) and so by FS1 and hypothesis $\lambda\circ\Omega_H(j^s(f))$ is $\mathfrak{M}_2$ with norm $\le \|f\|_{\mathfrak{M}_1^s} O^*(\|H\|_{\mathfrak{M}_3})$. As in the proof of Theorem 4, $\mathrm{Comp}\circ\mathrm{inc}_\#\circ j^s(H) = (\mathrm{Comp}\circ\mathrm{inc}_\#)_*(j^s(H))$ is an $\mathfrak{M}_2$ map with $\mathfrak{M}_2$ norm $\le O^*(\|H\|_{\mathfrak{M}_2^s})$. Composing with ev, $q_H(f)$ is an $\mathfrak{M}_2$ map with norm $\le \|f\|_{\mathfrak{M}_1^s} \times O^*(\|H\|_{\mathfrak{M}_2^s}, \|H\|_{\mathfrak{M}_3})$. Thus, $f \to q_H(f)$ is a bounded linear map with $\|q_H\| \le O^*(\|H\|_{\mathfrak{M}_3^s})$ (or $O^*(\|H\|_{\mathfrak{M}_2^s})$ if $\mathfrak{M}_2^s \subset \mathfrak{M}_3$ and in this case we need only assume H is $\mathfrak{M}_2^s$).

Now to show that $q_H(f) = j^s(\Omega_H(f))$ we will use the Jet Lemma with the family of linear maps $\{Ev_u: \mathfrak{M}_r(U,E_2) \to E_2\}$ indexed by U (cf. Prop. 1.10). Since $\mathfrak{M}_2 \subset \mathcal{C}$ it suffices to show $j^s(Ev_u\circ\Omega_H(f))$ is defined and equals $(Ev_u)_\#\circ q_H(f)$. $Ev_u^1: \mathfrak{M}_d(U,E_1) \to E_1$ induces $H_u = Ev_u^1\circ H: G \to E_1$ and $Ev_u\circ\Omega_H(f) = f\circ H_u$. Hence, $Ev_u\circ\Omega_H(f)$ is $\mathfrak{F}^s$ and $j^s(Ev_u\circ\Omega_H(f))$ $= \mathrm{Comp}\circ(j^s(f)\circ H_u, j^s(H_u))$. Thus, we must prove $Ev_{u\#}(\mathrm{Comp}[\lambda(j^s(f))(H(g)), j^s(H)(g)) = \mathrm{Comp}(j^s(f)(H_u(g)), j^s(H_u)(g))$. By naturality of Comp and λ we need only prove this after going from $\mathfrak{M}_r$ into $\mathfrak{F}$, which satisfies FS7 (Strong). Regarding $j^s(f)(H(g)) \in \mathfrak{F}(U,J^s(E_1;E_2))$ and $\kappa^{-1}(j^s(H)(g)) \in \mathfrak{F}(U,J^s(F_2;E_1))$ the result follows from the diagram below which commutes, again by the naturality of the explicit formula for Comp.

$$\begin{array}{ccccc}
J^s(\mathfrak{F}(U,E_1);\mathfrak{F}(U,E_2)) & & \mathfrak{F}(U,J^s(E_1;E_2)) & & J^s(E_1;E_2) \\
\times & \xleftarrow{\ \lambda\ \times\ \kappa\ } & \times & \xrightarrow{\ Ev_u\ \times\ Ev_u\ } & \times \\
J^s(F_2;\mathfrak{F}(U,E_1)) & & \mathfrak{F}(U,J^s(F_2;E_1)) & & J^s(F_2;E_1) \\
\Big\downarrow \mathrm{Comp} & & \Big\downarrow \mathrm{Comp}_* & & \Big\downarrow \mathrm{Comp} \\
J^s(F_2;\mathfrak{F}(U,E_2)) & \xleftarrow{\ \kappa\ } & \mathfrak{F}(U,J^s(F_2;E_2)) & \xrightarrow{\ Ev_u\ } & J^s(F_2;E_2)
\end{array}$$

$$J^s(F_2;\mathfrak{F}(U,E_2)) \xrightarrow{\ Ev_{u\#}\ } J^s(F_2;E_2)$$

Q.E.D.

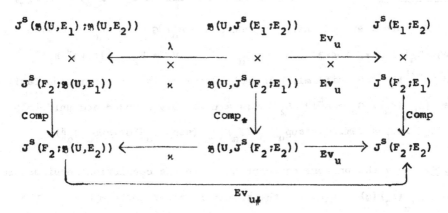

6 <u>COROLLARY (Combined Inheritance Result)</u>: If $\mathfrak{M}_1$ maps $\mathfrak{M}_d$ to $\mathfrak{M}_r$ in an $\mathfrak{M}_2$ way rel $\mathfrak{M}_3$ and $\mathfrak{M}_2 \subset \mathcal{C}$, then $\mathfrak{M}_1^{s+t}$ maps $\mathfrak{M}_d^t$ to $\mathfrak{M}_r^t$ in an $\mathfrak{M}_2^s$ way rel $\mathfrak{M}_3^s$. (If $\mathfrak{M}_2^s \subset \mathfrak{M}_3$ then delete "rel $\mathfrak{M}_3^s$".)

7 <u>COROLLARY</u>: (a) If FS9 holds for $\mathfrak{M}$ then it holds for $\mathfrak{M}^s$. In particular, $\mathfrak{B}^s$, $\mathcal{C}^s$, $\mathcal{C}_u^s$, $\mathcal{L}^s$ and $\mathcal{L}ip^s$ satisfying FS . Also, $\mathcal{L}ip_a^s$ maps $\mathcal{L}ip_b^s$ to $\mathcal{L}ip_{ab}^s$ in a $\mathfrak{B}$ way.

(b) $\mathcal{C}_u^{s+t}$ maps $\mathfrak{B}^t$, $\mathcal{C}^t$ and $\mathcal{C}_u^t$ to $\mathfrak{B}^t$, $\mathcal{C}^t$ and $\mathcal{C}_u^t$ respectively in a $\mathcal{C}^s$ and $\mathcal{C}_u^s$ way. $\mathcal{L}^{s+t}$ maps $\mathfrak{B}^t$, $\mathcal{C}^t$ and $\mathcal{C}_u^t$ to $\mathfrak{B}^t$, $\mathcal{C}^t$ and $\mathcal{C}_u^t$ respectively in an $\mathcal{L}^s$ way. $\mathcal{L}ip_a^{s+t}$ maps $\mathfrak{B}^t$, $\mathcal{C}^t$ and $\mathcal{C}_u^t$ to $\mathfrak{B}^t$, $\mathcal{C}^t$ and $\mathcal{C}_u^t$ respectively in a $\mathcal{L}ip_{ab}^s$ way rel $\mathcal{L}ip_b^s$.

(c) $\mathcal{L}^{s+t+1}$ maps $\mathcal{L}^t$ to $\mathcal{L}^t$ in an $\mathcal{L}^s$ way. $\mathcal{L}ip_a^{s+t+1}$ maps $\mathcal{L}_b^t$ to $\mathcal{L}ip_{ab}^t$ in an $\mathcal{L}^s$ way.

Note that from FS9 we cannot conclude that $\mathfrak{M}^s$ maps $\mathfrak{M}$ to $\mathfrak{M}$ in a $\mathfrak{B}^s$ way as we can't apply Theorem 5 when $\mathfrak{M}_2 = \mathfrak{B}$. In fact, with $s = 1$, $\mathfrak{M} = \mathcal{C}$ this is false, though we do have $\mathcal{C}^1$ maps $\mathcal{C}$ to $\mathcal{C}$ in an $\mathcal{L}$ way since $\mathcal{C}^1 \subset \mathcal{L}$.

For later applications we will need a "family" version of $\mathcal{C}_u^{s+t}$ maps $\mathfrak{B}^t$ to $\mathfrak{B}^t$ in a $\mathcal{C}_u^s$ way.

8 <u>PROPOSITION</u>: For α in some index set assume $U_\alpha \subset F_{1\alpha}$, $G_\alpha \subset F_{2\alpha}$, $V_\alpha \subset E_{1\alpha}$ are open and uniformly bounded. H_α: $G_\alpha \to \mathfrak{B}^t(U_\alpha, E_{1\alpha})$ with Image $H_\alpha(g) \subset V_\alpha$ for all g in G_α and f_α: $V_\alpha \to E_{2\alpha}$. If $\{j^s(H_\alpha)\}$ and $\{j^{s+t} f_\alpha\}$ are uniformly bounded and uniformly equicontinuous families, then $\{\Omega_{H_\alpha}(f_\alpha): G_\alpha \to \mathfrak{B}^t(U_\alpha, E_{2\alpha})\}$ is a uniformly bounded and uniformly equicontinuous family. $\sup_\alpha \|\Omega_{H_\alpha}(f_\alpha)\| \le (\sup_\alpha \|f_\alpha\|) O^*(\sup_\alpha \|H_\alpha\|)$.

<u>PROOF</u>: Only the uniform equicontinuity in the conclusion requires proof. Since $\Omega_{H_\alpha}(f_\alpha)(g) = f_\alpha \cdot H_\alpha(g)$ the result is clear for $s = t = 0$. The diagram in the first inheritance proof gives the results for $s = 0$ and then the diagram in the second inheritance proof gives the general result.

The key is again the uniform polynomial estimates on maps like Comp. Q.E.D.

9 PROPOSITION: Assume $\mathfrak{M}_1$ maps $\mathfrak{M}_d$ to $\mathfrak{M}_r$ in an $\mathfrak{M}_2$ way rel $\mathfrak{M}_3$ and $\mathfrak{M}_2$ satisfies FS8. Let G_0 be open and bounded in $\mathfrak{M}_1(V,E_2)$. $\text{comp}_H\colon G_0 \times G \to \mathfrak{M}_r(U,E_2)$ defined by $\text{comp}_H(f,g) = \Omega_H(g)(f)$ is an $\mathfrak{M}_2$ map with $\|\text{comp}_H\|_{\mathfrak{M}_2} \le O^*(\|H\|_{\mathfrak{M}_3}, \|1_{G_0}\|_0)$.

PROOF: Let G_1 be the open ball about 0 in $\mathfrak{M}_2(G,\mathfrak{M}_r(U,E_2))$ of radius $\|\Omega_H\|\ \|1_{G_0}\|_0$. $\Omega_H \times I(G_0 \times G) \subset G_1 \times G$ and $\text{comp}_H = (\Omega_H \times I)^*(\text{Ev}_{G_1})$. Result by FS8 and FS4 for $\mathfrak{M}_2$. Q.E.D.

10 LEMMA: Let $U \subset F$, $V \subset E_0 \times E_1$ and $g_0 \in \mathfrak{M}(U,E_0)$. g is in the interior of $\{g \in \mathfrak{M}(U,E_1)\colon g_0 \times g(U) \subset V\}$ iff $\inf_U\{\|g(u) - v\|\colon u \in U,\ v \in E_1 \text{ and } (g_0(u),v) \notin V\} > 0$. In particular, if $V \subset E_1$ then g is in the interior of $\{g \in \mathfrak{M}(U,E_1)\colon g(U) \subset V\}$ iff $\inf_U\{\|g(u) - v\|\colon u \in U,\ v \notin V\} > 0$.

PROOF: The sets described by the inf > 0 condition are in fact $\mathfrak{B}$ open and are thus the inverse of open sets under the inclusion $\mathfrak{M} \subset \mathfrak{B}$. Conversely, if $\|g(u_n) - v_n\| \to 0$ with $(g_0(u_n),v_n) \notin V$ then let $g_n = g + c(v_n - g(u_n))$. $g_n \to g$ $(\mathfrak{M})$ by FS2. But $g_0 \times g_n(U) \not\subset V$ for any n. Q.E.D.

11 PROPOSITION: Let $\mathfrak{M}_1$ map $\mathfrak{M}_d$ to $\mathfrak{M}_r$ in an $\mathfrak{M}_2$ way rel $\mathfrak{M}_3$. Let $V \subset E_0 \times E_1$, $g_0 \in \mathfrak{M}_d(U,E_0)$, G be open and bounded in $\{g \in \mathfrak{M}_d(U,E_1)\colon (g_0 \times g)(U) \subset V\}$. The map $\Omega_G^{g_0}\colon \mathfrak{M}_1(V,E_2) \to \mathfrak{M}_2(G,\mathfrak{M}_r(U,E_2))$ is a bounded linear map well defined by $\Omega_G^{g_0}(f)(g) = f\cdot(g_0 \times g)$ with $\|\Omega_G^{g_0}\| \le O^*(\|1_G\|_0, \|g_0\|_{\mathfrak{M}_d})$.

PROOF: Let $H\colon G \to \mathfrak{M}_d(U,E_0 \times E_1)$ by $H(g) = g_0 \times g = g_0 \times 0 + 0 \times g$. $\Omega_G^{g_0} = \Omega_H$ and since H is affine it is $\mathfrak{M}_3$ with $\|H\|_{\mathfrak{M}_3} \le O^*(\|1_G\|_0, \|g_0\|_{\mathfrak{M}_d})$ by FS2. Q.E.D.

If $\mathfrak{M}_2 \subset \mathcal{C}$ we can apply the second inheritance result. If $f \in \mathfrak{M}_1^s(V,E_2)$ then $\Omega_G^{g_0}(f) \in \mathfrak{M}_2^s(G,\mathfrak{M}_r(U,E_2))$. Let ${}_2j^s(f) \in \mathfrak{M}_1(V,J^s(E_1;E_2))$ be the jet maps in the second variable. Since H is affine in this case the diagram in the proof of Theorem 5 simplifies to the following diagram which is seen to be commutative by an argument using $\{Ev_u\}$ as in Theorem 5:

$$(2.1)\qquad \begin{array}{ccc} G & \xrightarrow{\quad \Omega_G^{g_0}({}_2j^s(f)) \quad} & \mathfrak{M}_r(U,J^s(E_1;E_2)) \\ & \searrow {\scriptstyle j^s(\Omega_G^{g_0}(f))} & \downarrow {\scriptstyle inc^{\#}\cdot\lambda} \\ & & J^s(\mathfrak{M}_d(U,E_1);\mathfrak{M}_r(U,E_2)). \end{array}$$

12 <u>DEFINITION</u>: A function space type $\mathfrak{M}$ is <u>standard</u> if it satisfies FS1-9, $\mathfrak{M}^1$ interchanges to $\mathfrak{M}$ and $\mathcal{C}^{r+1} \subset \mathfrak{M} \subset \mathcal{C}^r$ for some $r \geq 0$.

If $\mathfrak{M}$ is standard, then its derived types are standard. Note that the interchange condition is redundant of either $\mathcal{C}^{r+1} \subset \mathfrak{M} \subset \mathcal{L}^r$ or $\mathcal{L}^r \subset \mathfrak{M} \subset \mathcal{C}^r$. The examples of standard function space type we have are $\mathcal{C}$, $\mathcal{L}$ and $\mathfrak{B}^1$ and their derived types. Note that $\mathcal{L}$ and its deriveds are distinguished by being standard and satisfying FS7 (Strong).

13 <u>PROPOSITION</u>: (a) Let $\mathfrak{M}$ be a standard function space type. The map $C: \mathfrak{M}(U,\mathfrak{M}(V,E)) \to \mathfrak{M}(U \times V,E)$ is a linear map well defined by $C(f)(u,v) = f(u)(v)$ and $\|C\| \leq O^*(\|1_V\|_0)$.

(b) Let $\mathfrak{M}_1$ map $\mathfrak{M}$ to $\mathfrak{M}$ in an $\mathfrak{M}_2$ way rel $\mathfrak{M}_3$. The linear map $\bar{C}: \mathfrak{M}_1(U \times V,E) \to \mathfrak{M}_2(V,\mathfrak{M}(U,E))$ is well defined by $\bar{C}(f)(v)(u) = f(u,v)$ and $\|\bar{C}\| \leq O^*(\|1_U\|_0,\|1_V\|_0)$.

<u>PROOF</u>: (a) Let G be the unit ball in $\mathfrak{M}(V,E)$ and $f \in \mathfrak{M}(U,\mathfrak{M}(V,E))$ with $\|f\| < 1$. $f(U) \subset G$ and so $C(f) = (f \times 1)^* Ev_G$ where $Ev_G: V \times G \to E$. Bt FS9, $\|C(f)\| \leq O^*(\|f\|,\|1_V\|,\|Ev_G\|)$ and by FS8, $\|Ev_G\| \leq O^*(1)$. So $\|C(f)\| \leq O^*(\|1_V\|)$.

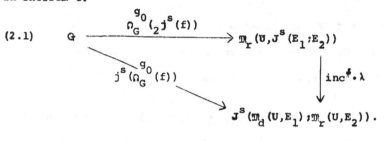

(b): Let $H: V \to \mathfrak{M}(U, E_1 \times E_2)$ (U open in E_1 and v open in E_2) be defined by $H(v) = 1_U \times c(v)$. H is an affine map and so $\|H\|_{\mathfrak{M}_3} \leq O^*(\|1_U\|_0, \|1_V\|_0)$ by FS2. $\bar{C} = \Omega_H$. Q.E.D.

In (b) if $\mathfrak{M}_1$ satisfies FS7 (Strong) we can apply $\bar{C}$ to $Ev \in \mathfrak{M}_1(U \times V, L(\mathfrak{M}_1(U \times V, E); E))$ to get an $\mathfrak{M}_2$ map $\bar{\bar{C}}$: $V \to \mathfrak{M}(U, L(\mathfrak{M}_1(U \times V, E); E))$ defined by $\bar{\bar{C}}(v)(u)(f) = f(u,v)$.

3. <u>Local Homeomorphisms and Differential Equations</u>: For $f: X \to Y$ a continuous map of Hausdorff spaces, define $S_f = ((f \times f)^{-1}\Delta_Y) - \Delta_X \subset X \times X$. Clearly, for $A \subset X$, $f|A$ is injective iff $A \times A \cap S_f = \emptyset$. In particular, f is locally injective iff S_f is closed in $X \times X$, i.e. Δ_X and S_f are disjoint closed sets with union $\overline{f \times f}^{-1}\Delta_Y$. It follows now from Wallace's Lemma [13; Thm. V. 12] that if f is locally injective, A is a compact subset of X and $f|A$ is injective then $f|U$ is injective for some open set U containing A.

1 <u>LEMMA</u>: Let $f: X \to Y$ be locally injective, and $g_1, g_2: Z \to X$ be lifts of a continuous $\bar{g}: Z \to Y$, i.e. $f \cdot g_i = \bar{g}$, $i = 1,2$. $\{y: g_1(y) = g_2(y)\}$ is open and closed in Z.

<u>PROOF</u>: The map $\bar{g} \times \bar{g} \cdot \Delta: Z \to Y \times Y$ is given by $(f \times f) \cdot (g_1 \times g_2)\Delta$. So Z is the disjoint union of closed sets $\Delta^{-1}(g_1 \times g_2)^{-1}\Delta_X$ and $\Delta^{-1}(g_1 \times g_2)^{-1}S_f$. Q.E.D.

A local homeomorphism will always mean a local homeomorphism onto an open subset of the range and so an open map. $g: V \to X$ is a local inverse for a local homeomorphism $f: X \to Y$ if V is open in Y, g is continuous and $f \cdot g = 1_V$. If $U \subset F$ is open and bounded and $f: U \to E$ is a local homeomorphism we say that f^{-1} is $\mathfrak{M}$ if there exists a constant K such that every local inverse for f is $\mathfrak{M}$ and has $\mathfrak{M}$ norm $\leq K$. The infimum of such K's is denoted $\|f^{-1}\|_{\mathfrak{M}}$.

2 **LEMMA**: Let $f\colon U \to E$ be a local homeomorphism. If f^{-1} is $\mathfrak{M}$ then for any open cover $\{U_\alpha\}$ of U with $f|U_\alpha$ injective $\{(f|U_\alpha)^{-1}\}$ $\in \hat{\Pi}_\alpha \mathfrak{M}(f(U_\alpha),F)$ with norm $\leq \|f^{-1}\|_{\mathfrak{M}}$. If $\mathfrak{M}$ satisfies FS5 and there exists an open cover $\{U_\alpha\}$ with $f|U_\alpha$ injective and $\{(f|U_\alpha)^{-1}\} \in \hat{\Pi}_\alpha \mathfrak{M}(f(U_\alpha),F)$ then f^{-1} is $\mathfrak{M}$ and $\|f^{-1}\|_{\mathfrak{M}} = \|\{(f|U_\alpha)^{-1}\}\| = \sup_\alpha \|(f|U_\alpha)^{-1}\|_{\mathfrak{M}}$.

PROOF: The first statement is obvious. Conversely, if $g\colon V \to F$ is any local inverse and $V_\alpha = g^{-1}U_\alpha \subset V \cap f(U_\alpha)$ then $g|V_\alpha = (f|U_\alpha)^{-1}|V_\alpha$. Thus, $\|g|V_\alpha\|_{\mathfrak{M}} \leq \|\{(f|U_\alpha)^{-1}\}\|$ and so by Gluing (FS5) g is $\mathfrak{M}$ with the same estimate. Q.E.D.

Next consider $Lis(E_1;E_2)$ open in $L(E_1;E_2)$ and the map $inv\colon Lis(E_1;E_2) \to Lis(E_2;E_1)$, $inv(T) = T^{-1}$. Recall [5; Thm. VIII.3.2] that inv is analytic with $D\,inv(T)(S) = -inv(T)\circ S\circ inv(T)$, i.e. the derivative of inv can be expressed in terms of inv itself and bilinear composition maps of norm 1. If for all $G \subset Lis(E_1;E_2)$ with $\sup \theta|G < \infty$ the map $inv_G\colon G \to L(E_2;E_1)$ is $\mathfrak{M}$ with $\|inv_G\|_{\mathfrak{M}} \leq O^*(\sup \theta|G)$ then we will say <u>inv is $\mathfrak{M}$</u>. The computation of the derivative implies that inv is $\mathfrak{M}^1$ if inv is $\mathfrak{M}$. Since this property is heritable and holds for $\mathcal{C}$, it holds for any standard function space type $\mathfrak{M}$. Since $\|T_1^{-1} - T_2^{-1}\| \leq \|T_1^{-1}\|\|T_2^{-1}\|\|T_1 - T_2\|$, inv is $\mathcal{L}ip$ and hence is $\mathcal{L}ip^r$ for all r.

3 **PROPOSITION**: Assume $\mathfrak{M}$ satisfies FS9 and inv is $\mathfrak{M}$, eg. $\mathfrak{M}$ is standard. Let $f \in \mathfrak{M}(U,L(E_1;E_2))$. $inv\circ f$ is a well defined member of $\mathfrak{M}(U,L(E_2;E_1))$ iff $\sup \theta\circ f < \infty$, in which case $\|inv\circ f\|_{\mathfrak{M}} \leq O^*(\|f\|_{\mathfrak{M}}, \sup \theta\circ f)$, i.e. $\|inv\circ f\|_{\mathfrak{M}} \leq O^*(\|f\|_{\mathfrak{M}}, \|inv\circ f\|_0)$.

PROOF: Clearly, $\sup \theta\circ f < \infty$ holds iff f and $inv\circ f$ are $\mathfrak{B}$. Let $G = \{T\colon \theta(T) < 2 \sup \theta\circ f\}$. $inv\circ f = f^*inv_G$. Result by FS9. Q.E.D.

4 **INVERSE FUNCTION THEOREM**: Assume $\mathfrak{M}$ is a standard function space type and $f \in \mathfrak{M}^s(U,E)$ with $s \geq 1$. f is a local homeomorphism with f^{-1}

$\mathfrak{M}^s$ iff $Df: U \to L(F;E)$ has image in $Lis(F;E)$ and $\sup \theta\cdot Df < \infty$. In that case, $\|f^{-1}\|_{\mathfrak{M}^s} \leq O^*(\|f\|_{\mathfrak{M}^s}, \|1_U\|_0, \sup \theta\cdot Df)$, i.e. $\|f^{-1}\|_{\mathfrak{M}^s} \leq O^*(\|f\|_{\mathfrak{M}^s}, \|f^{-1}\|_{\mathfrak{B}^1})$.

PROOF: Since f is C^s the ordinary inverse function theorem implies f is a local homeomorphism with f^{-1} C^s iff $Df(U) \subset Lis(F;E)$. Furthermore, it is clear that $\max(\|f\|_{\mathfrak{B}^1}, \|f^{-1}\|_{\mathfrak{B}^1}) = \max(\|f\|_0, \|1_U\|_0, \sup \theta\cdot Df)$. Thus, necessity is clear and to prove sufficiency we must show f^{-1} is $\mathfrak{M}^s$ and estimate $\|f^{-1}\|_{\mathfrak{M}^s}$. Using the polynomial function defined on page 5, $j^s((f|U_0)^{-1}) = \mathrm{Inv}\cdot(j^s(f), \mathrm{inv}\cdot Df)\cdot(f|U_0)^{-1}$ where U_0 is an open subset on which f is injective. If $(f|U_0)^{-1}$ is $\mathfrak{M}$ then it follows from FS9 and Proposition 3 that $(f|U_0)^{-1}$ is $\mathfrak{M}^s$ with $\|(f|U_0)^{-1}\|_{\mathfrak{M}^s} \leq O^*(\|f\|_{\mathfrak{M}^s}, \sup \theta\cdot Df, \|(f|U_0)^{-1}\|_{\mathfrak{M}})$. Since $\mathfrak{M}$ is standard, $\mathcal{C}^{r+1} \subset \mathfrak{M} \subset \mathcal{C}^r$ for some $r \geq 0$. Applying this result for $\mathcal{C}^{r+s}$ we have $\|(f|U_0)^{-1}\|_{\mathcal{C}^{r+s}} \leq O^*(\|f\|_{\mathcal{C}^{r+s}}, \sup \theta\cdot Df, \|1_U\|_0)$ since f^{-1} is $\mathcal{C}$ with $\|f^{-1}\|_{\mathcal{C}} = \|1_U\|_0$. Since $s \geq 1$, $\mathcal{C}^{r+s} \subset \mathfrak{M}$ and so $(f|U_0)^{-1}$ is $\mathfrak{M}$ with $\mathfrak{M}$ norm satisfying the same estimate. Since $\mathfrak{M}^s \subset \mathcal{C}^{r+s}$ we get the required estimate on $\|(f|U_0)^{-1}\|_{\mathfrak{M}^s}$ and so on $\|f^{-1}\|_{\mathfrak{M}^s}$. Q.E.D.

5 LEMMA: Let X be a connected, Hausdorff space and $f: X \to B \subset E$ a local homeomorphism onto a ball B centered at $v_0 = f(x_0)$. If for all $v \in B$, the path $v_t = tv + (1-t)v_0$, based at v_0, lifts continuously to a path in X based at x_0, then f is bijective, i.e. a homeomorphism onto B.

PROOF: Since f is injective on the path lifts it is injective on a neighborhood of each of them by compactness. Thus, for each $v \in B$, we can find U_v a convex, open set about $[v_0,v]$ and $g_v: U_v \to X$ a local inverse for f with $g_v(v_0) = x_0$. By Lemma 1, g_{v_1} and g_{v_2} agree on the convex neighborhood $U_{v_1} \cap U_{v_2}$ of v_0. The g_v's fit together to define a right inverse for f, $g: B \to X$. It suffices to show $g(B)$ is open and closed in X. If x is in the closure $\overline{g(B)}$, choose V_x open, containing x, with $f|V_x$ injective and $f(V_x)$ connected. By Lemma 1, $(f|V_x)^{-1} = g|f(V_x)$ since $V_x \cap g(B) \neq \emptyset$. Thus, $\overline{g(B)} \subset \text{Int } g(B)$. Q.E.D.

6 <u>COROLLARY</u>: Let $f: X \to E$ be a local homeomorphism and B be an open ball in E centered at $f(x_0) = v_0$. If v_t can be lifted continuously to a path based at x_0 for all $v \in B$ then there is a unique local inverse $g: B \to X$ with $g(v_0) = x_0$.

<u>PROOF</u>: Apply Lemma 5 to $f|X_1$ with X_1 the x_0 component of $f^{-1}(\ )$. Q.E.D.

The following important result comes from Hirsch and Pugh [10; Prop. 1.7].

7 <u>PROPOSITION (Size Estimate)</u>: Let $f: U \to E$ be a local homeomorphism with f^{-1} an $\mathcal{L}$ map. If the ball $B(u,r)$ contained in U, then there exists a unique local inverse, $g: B(fu,r/\|f^{-1}\|_L) \to B(u,r)$ with $g(f(u)) = u$. Thus, $B(fu,r/\|f^{-1}\|_L) \subset f(B(u,r)) \subset f(U)$. If f is also an $\mathcal{L}$ map then it is injective on $B(u,r/\|f\|_L\|f^{-1}\|_L)$ and this ball is mapped into $B(fu,r/\|f^{-1}\|_L)$.

<u>PROOF</u>: We check the hypothesis of Corollary 6 in the following form: Let $v_0 = f(u)$ and v be a unit vector in E. Let r_v be the supremum of those $s > 0$ for which the path $v_t = v_0 + tv$ has a continuous lifting on $[0,s)$ to a path based at u. It suffices to show that $r_v > \bar{r}/\|f^{-1}\|_L$ for all $\bar{r} < r$. By Lemma 1, the path lift is uniquely defined x_t for t in $[0,r_v)$ and as in the proof of Lemma 5, extends to a local inverse $\bar{g}$ defined on some neighborhood of the image of v_t. For $t_1,t_2 \in [0,r_v)$, $\|x_{t_1} - x_{t_2}\| \le \|\bar{g}\|_L\|v_{t_1} - v_{t_2}\| \le \|f^{-1}\|_L|t_1 - t_2|$. Thus, if $t_1 = 0$ and $t_2 \le \bar{r}/\|f^{-1}\|_L$ then x_{t_2} lies in the closed ball $B[u,\bar{r}]$.

Also, as t approaches r_v, x_t is Cauchy. If $r_v \leq \bar{r}/\|f^{-1}\|_L$ then as t approaches r_v, x_t has a limit x_{r_v} in $B[u,r]$ and hence in U. This extends the lift to the closed interval $[0,r_v]$ and since we can then continue the lift using f on a neighborhood of injectivity about x_r, we contradict the definition of r_v. So g exists by Corollary 6 and maps into $B(u,r)$ by the above path lift estimates. If f is $\mathcal{L}$, then we can apply this result to g with $\|g^{-1}\|_L \leq \|f\|_L$. Q.E.D.

Recall that if f is $\mathfrak{B}^1$, then $\|f\|_L = \|Df\|_0$. Clearly, if f^{-1} is $\mathfrak{B}^1$ then $\|f^{-1}\|_L = \|\text{inv.}Df\|_0$, i.e. $\max(\|f\|_L, \|f^{-1}\|_L) = \sup \theta\cdot Df$.

We turn now to differential equations.

8 <u>PROPOSITION</u>: Assume $\mathfrak{M} \subset \mathcal{C}$ and satisfies FS6. Let J be an open subinterval of $(-1,1)$ containing 0 and $A \in \mathfrak{M}(U \times J, L(F;F))$. There exists a numerical function $N(A) \geq 1$ with $N(A) \leq O^*(\|A\|_{\mathfrak{M}})$ such that the solution $B: U \times J \to L(F;F)$ of the equation

$$\frac{d}{dt} B(u,t) = \text{comp}(A(u,t), B(u,t)) \qquad B(u,0) = I \tag{3.1}$$

restricts to an element of $\mathfrak{M}(U \times J_A, L(F;F))$ with $\|B|U \times J_A\|_{\mathfrak{M}} \leq O^*(\|A\|_{\mathfrak{M}})$ and $J_A = N(A)^{-1} J$.

<u>PROOF</u>: The existence of a unique continuous B satisfying (3.1) or equivalently $B(u,t) = I + \int_0^t A(u,s)B(u,s)\,ds$ is in [14; Prop. IV.3]. For $N \geq 1$ define $B_N(u,t) = B(u, N^{-1}t)$. $B_N(u,t) = I + \frac{1}{N}\int_0^t A_N(u,s)B_N(u,s)\,ds$ on $U \times NJ$. $T_N(C)(u,t) = \frac{1}{N}\int_0^t A_N(u,s)C(u,s)\,ds$ is a linear automorphism of $\mathfrak{M}(U \times J, L(F;F))$ by FS6. Since $\|\text{comp}_*\| \leq O^*(1)$ and by FS4 $\|A_N\| \leq \|A\|$, we have $\|T_N\| \leq N^{-1}O^*(\|A\|)$. So we can define N_A so that $1 \leq N_A \leq O^*(\|A\|)$ and $\|T_N\| \leq \frac{1}{2}$ when $N = N_A$. Thus, $C \to I + T_N(C)$ has a unique fixed point of norm ≤ 2. This fixed point is B_N restricte to $U \times J$ and so by FS4 $B|U \times N^{-1}J$ has norm $\leq O^*(N)$. Q.E.D.

The need to shrink the interval in order to get a polynomial bound of norm B by norm A is seen even when $\mathfrak{M} = \mathcal{C}$, $E = \mathbb{R}$ and the equation is $dx/dt = Ax$ with A a positive constant.

9 <u>PROPOSITION</u>: Assume $\mathcal{C}^{r+1} \subset \mathfrak{M} \subset \mathcal{C}^r$ for some $r \geq 0$, $\mathfrak{M}$ maps $\mathcal{C}^{r+1}$ to $\mathfrak{M}$ in a $\mathfrak{B}$ way and $\mathfrak{M}$ satisfies FS6, eg. $\mathfrak{M}$ is standard. Let J be a subinterval of $(-1,1)$ containing 0, $1 \geq \epsilon > 0$ with $B(V,\epsilon) \subset U$, U and V open and bounded in E, and $G \in \mathfrak{M}^p(U \times J,E)$ $(p \geq 1)$. There is a numerical function $1 \leq N(G) \leq O^*(\|G\|_{\mathfrak{M}^p},\|1_U\|_0,\epsilon^{-1})$ such that the flow for G restricts to an $\mathfrak{M}^p$ map $f\colon V \times J_G \to U$ $(J_G = N(G)^{-1}J)$ with $\|f\|_{\mathfrak{M}^p} \leq O^*(\|G\|_{\mathfrak{M}^p},\|1_U\|_0,\epsilon^{-1})$.

<u>PROOF</u>: Since G is $\mathcal{C}^{p+r}$ a unique C^{p+r} flow $f\colon V \times J_{N_0} \to U$ exists with $N_0 = 3\epsilon^{-1}\|G\|_{\mathcal{C}^1}$ and $D_1f\colon V \times J_{N_0} \to L(E;E)$ is the solution of (3.1) with $A(u,t) = D_1G(f(u,t),t)$ defined on $V \times J_{N_0}$ [14; Prop. IV.1 and Thm. IV.1]. Define $N_i = N_i(G)$ inductively, $i = 1,\dots,p{+}r$ with N_i, $\|f|V \times J_{N_i}\|_{\mathcal{C}^i} \leq O^*(M)$ with $M = \max(\|G\|_{\mathfrak{M}^p},\epsilon^{-1},\|1_U\|_0)$. Note that N_0 works since $\|f|V \times J_{N_0}\|_0 \leq \|1_U\|_0$. Assume N_i has been defined for $i < p{+}r$. Since f is the flow for G, $D_2f = G\circ(f,\pi_J)\colon V \times J_{N_0} \to E$ identifying $L(\mathbb{R};E)$ with E as usual. By FS9 for $\mathcal{C}^i$, $\|D_2f|V \times J_{N_i}\|_{\mathcal{C}^i}$ $\leq O^*(\|G\|_{\mathfrak{M}^p},\|f|V \times J_{N_i}\|_{\mathcal{C}^i}) \leq O^*(M)$ by inductive hypothesis. By FS4 this inequality is preserved when we restrict to $V \times J_N$ with $N \geq N_i$. Since $i < p{+}r$, D_1G is $\mathcal{C}^i$ with $\|D_1G\|_{\mathcal{C}^i} \leq \|G\|_{\mathfrak{M}^p}$ and so $A = D_1G\circ(f,\pi_J)$ is $\mathcal{C}^i$ with $\|A|V \times J_{N_i}\|_{\mathcal{C}^i} \leq O^*(M)$. Applying Proposition 8 we obtain a factor N_A^i so that if $N_{i+1} = N_A^i\cdot N_i$, $\|D_1f|V \times J_{N_{i+1}}\|_{\mathcal{C}^i}$ $\leq O^*(\|A|V \times J_{N_i}\|_{\mathcal{C}^i})$ and $N_{i+1} \leq N_i\, O^*(\|A|V \times J_{N_i}\|_{\mathcal{C}^i})$. Thus, N_i is defined inductively through $i = p + r$. The last step, to jump from $\mathcal{C}^{p+r}$ to $\mathfrak{M}^p$ is analogous to the jumps from $\mathcal{C}^i$ to $\mathcal{C}^{i+1}$. Note that $\mathcal{C}^{p+r} \subset \mathfrak{M}^{p-1}$ and so $f|V \times J_{N_{p+r}}$ is $\mathfrak{M}^{p-1}$. $\|D_2f\|_{\mathfrak{M}^{p-1}} \leq \|D_2f\|_{\mathcal{C}^{p+r}}$ and $\|A\|_{\mathfrak{M}^{p-1}}$ are

estimated as before, using $\mathfrak{M}^{p-1}$ maps $\mathcal{C}^{p+r}$ to $\mathfrak{M}^{p-1}$ in a $\mathfrak{M}$ way for the latter. Then apply Proposition 8 again to get the estimate on $\|D_1 f\|_{\mathfrak{M}^{p-1}}$ after restriction. Q.E.D.

$\mathcal{C}^1 \subset \mathcal{L}ip_a \subset \mathcal{C}$ is not true but does hold when the domains U which we allow are convex open sets. Thus, the above proof applies to $\mathcal{L}ip_a^p$ provided U and V are convex.

If $\mathfrak{M}$ satisfies FS9 then the $\mathfrak{M}^p$ norm of $G_*(f,\pi_J)$ on $V \times J_{N_G}$ is again $\leq O^*(M)$. Thus, since $f(u,t) = u + \int_0^t G(f(u,s),s)ds$, Proposition 1.6 (b) implies that the function $t \to f_t$ lies in $\mathcal{L}(J_{N_G}, \mathfrak{M}^p(V,E))$ with $\mathcal{L}$ norm $\leq O^*(\|G\|, \epsilon^{-1}, \|1_U\|_0)$.

CHAPTER III: METRIC STRUCTURES

In this chapter we develop the elements of the metric theory of manifolds and bundles. $\mathfrak{M}$ is assumed to be a standard function space type (Def. II. 2.12).

1. $\mathfrak{M}$ Atlases and Maps: An atlas $\mathfrak{A} = \{U_\alpha, h_\alpha\}$ on X is called a bounded $\mathfrak{M}$-atlas if the transition maps $h_{\alpha\beta}: h_\beta(U_\alpha \cap U_\beta) \to E_\alpha$ are $\mathfrak{M}$ maps and $\{h_{\alpha\beta}\} \in \hat{\Pi}_{\alpha\beta}\mathfrak{M}(h_\beta(U_\alpha \cap U_\beta), E_\alpha)$ i.e. $\|\{h_{\alpha\beta}\}\| = \sup\|h_{\alpha\beta}\|_{\mathfrak{M}} < \infty$. $\max(\|\{h_{\alpha\beta}\}\|, 1)$ is denoted $k_{\mathfrak{A}}^{\mathfrak{M}}$ or $k_{\mathfrak{A}}$. If $\mathfrak{A}$ is not a bounded $\mathfrak{M}$ atlas, we define $k_{\mathfrak{A}} = \infty$. A VB atlas $(\mathfrak{D}, \mathfrak{A}) = \{U_\alpha, h_\alpha, \varphi_\alpha\}$ is a bounded $\mathfrak{M}$ atlas if $\mathfrak{A}$ is a bounded $\mathfrak{M}$ atlas and the transition maps $\varphi_{\alpha\beta}: h_\beta(U_\alpha \cap U_\beta) \to L(F_\beta; F_\alpha)$ are $\mathfrak{M}$ with $\{\varphi_{\alpha\beta}\} \in \hat{\Pi}_{\alpha\beta}\mathfrak{M}(h_\beta(U_\alpha \cap U_\beta), L(F_\beta; F_\alpha))$. $\max(k_{\mathfrak{A}}, \|\{\varphi_{\alpha\beta}\}\|) = \sup \max(\|h_{\alpha\beta}\|_{\mathfrak{M}}, \|\varphi_{\alpha\beta}\|_{\mathfrak{M}}, 1)$ is denoted $k_{(\mathfrak{D},\mathfrak{A})}^{\mathfrak{M}}$ or $k_{(\mathfrak{D},\mathfrak{A})}$. If $(\mathfrak{D},\mathfrak{A})$ is not a bounded $\mathfrak{M}$ atlas $k_{(\mathfrak{D},\mathfrak{A})} = \infty$. $\mathfrak{A}$ (or $(\mathfrak{D},\mathfrak{A})$) is called an $\mathfrak{M}$ atlas if every point $x \in X$ has a neighborhood U such that $\mathfrak{A}|U$ on U (resp. $(\mathfrak{D},\mathfrak{A})|U$ on π_U) is bounded. We define $\rho_{\mathfrak{A}}^{\mathfrak{M}}(x)$ or $\rho_{\mathfrak{A}}(x) = \inf\{k_{\mathfrak{A}|U}: U$ a neighborhood of $x\}$ (resp. $\rho_{(\mathfrak{D},\mathfrak{A})}(x) = \inf\{k_{(\mathfrak{D},\mathfrak{A})|U}: U$ a neighborhood of $x\}$). $\rho_{\mathfrak{A}}$ and $\rho_{(\mathfrak{D},\mathfrak{A})}$ map X to $[1,\infty]$ with $\rho_{\mathfrak{A}} < \infty$ or $\rho_{(\mathfrak{D},\mathfrak{A})} < \infty$ iff $\mathfrak{A}$ or $(\mathfrak{D},\mathfrak{A})$ is an $\mathfrak{M}$ atlas. It is clear that $\rho_{\mathfrak{A}}$ is upper semicontinuous, i.e. $\{\rho_{\mathfrak{A}} < K\}$ is open in X for all K. It is an important consequence of the Gluing Property, FS5, that for U open in X, $\mathfrak{A}|U$ is a bounded $\mathfrak{M}$ atlas iff $\rho_{\mathfrak{A}}$ is bounded on U and in that case $k_{\mathfrak{A}|U} = \sup \rho_{\mathfrak{A}}|U$. For if $K > \sup \rho_{\mathfrak{A}}|U$ then every point $x \in U$ admits a neighborhood U_x with $k_{\mathfrak{A}|U_x} < K$, i.e. for all α, β $\|h_{\alpha\beta}|h_\beta(U_\alpha \cap U_\beta \cap U_x)\| < K$. By applying FS5 with the cover $\{h_\beta(U_\alpha \cap U_\beta \cap U_x): x \in U\}$ of $h_\beta(U_\alpha \cap U_\beta \cap U)$ we get $\|h_{\alpha\beta}|h_\beta(U_\alpha \cap U_\beta \cap U)\| < K$. The converse is obvious. Thus, $\{\rho_{\mathfrak{A}} < n\}$

is an increasing sequence of open sets on which $\mathfrak{G}$ is bounded and with union X. In particular, $\mathfrak{G}$ is a bounded $\mathfrak{M}$ atlas iff $\rho_{\mathfrak{G}}$ is bounded. Finally, if $\mathfrak{M}_1 \subset \mathfrak{M}$ then a (bounded) $\mathfrak{M}_1$ atlas is a (bounded) $\mathfrak{M}$ atlas with $\rho_{\mathfrak{G}}^{\mathfrak{M}} \leq \rho_{\mathfrak{G}}^{\mathfrak{M}_1}$ (resp. $k_{\mathfrak{G}}^{\mathfrak{M}} \leq k_{\mathfrak{G}}^{\mathfrak{M}_1}$). Identical results hold for $\rho_{(\mathfrak{D},\mathfrak{G})}$. Clearly, $\rho_{(\mathfrak{D},\mathfrak{G})} \geq \rho_{\mathfrak{G}}$.

Let $\mathfrak{G}_1 = \{U_\alpha, h_\alpha\}$, $\mathfrak{G}_2 = \{V_\gamma, g_\gamma\}$ be $\mathfrak{M}$ atlases on X_1, X_2 and $f\colon X_1 \to X_2$ be continuous. $f\colon \mathfrak{G}_1 \to \mathfrak{G}_2$ is called a <u>bounded $\mathfrak{M}$ map</u> if: (1) $\mathfrak{G}_1$ is bounded, i.e. $\sup \rho_{\mathfrak{G}_1} < \infty$, (2) $f^*\rho_{\mathfrak{G}_2}$ is bounded, i.e. $\sup f^*\rho_{\mathfrak{G}_2} < \infty$, (3) the local representatives $f_{\gamma\alpha}\colon h_\alpha(U_\alpha \cap f^{-1}V_\gamma) \to E_\gamma$ are $\mathfrak{M}$ maps and $\{f_{\gamma\alpha}\} \in \hat{\Pi}_{\gamma\alpha}\mathfrak{M}(h_\alpha(U_\alpha \cap f^{-1}V_\gamma), E_\gamma)$. $k(f;\mathfrak{G}_1,\mathfrak{G}_2)$ or k_f denotes $\|\{f_{\gamma\alpha}\}\|$. Let $(\mathfrak{D}_1,\mathfrak{G}_1) = \{U_\alpha, h_\alpha, \varphi_\alpha\}$, $(\mathfrak{D}_2,\mathfrak{G}_2) = \{V_\gamma, g_\gamma, \psi_\gamma\}$ be $\mathfrak{M}$ atlases on $\pi_1\colon E_1 \to X_1$, $\pi_2\colon E_2 \to X_2$ and $(\Phi, f)\colon \pi_1 \to \pi_2$ be a VB map. $(\Phi,f)\colon (\mathfrak{D}_1,\mathfrak{G}_1) \to (\mathfrak{D}_2,\mathfrak{G}_2)$ is called a <u>bounded $\mathfrak{M}$ map</u> or <u>bounded $\mathfrak{M}$ VB map</u> if (1) $(\mathfrak{D}_1,\mathfrak{G}_1)$ is bounded, (2) $f^*\rho_{(\mathfrak{D}_2,\mathfrak{G}_2)}$ is bounded, (3) $f\colon \mathfrak{G}_1 \to \mathfrak{G}_2$ is bounded, (4) the local representatives $\Phi_{\gamma\alpha}\colon h_\alpha(U_\alpha \cap f^{-1}V_\gamma) \to L(F_\gamma^1;F_\alpha^2)$ are $\mathfrak{M}$ maps and $\{\Phi_{\gamma\alpha}\} \in \hat{\Pi}_{\gamma\alpha}\mathfrak{M}(h_\alpha(U_\alpha \cap f^{-1}V_\gamma), L(F_\gamma^1;F_\alpha^2))$. $k(\Phi,f;\mathfrak{D}_1,\mathfrak{G}_1,\mathfrak{D}_2,\mathfrak{G}_2)$ or $k_{(\Phi,f)}$ denotes $\max(k_f, \|\{\Phi_{\gamma\alpha}\}\|)$. $f\colon \mathfrak{G}_1 \to \mathfrak{G}_2$ (or $(\Phi,f)\colon (\mathfrak{D}_1,\mathfrak{G}_1) \to (\mathfrak{D}_2,\mathfrak{G}_2)$) is an <u>$\mathfrak{M}$ map</u> (resp. an <u>$\mathfrak{M}$ map</u> or <u>$\mathfrak{M}$ VB map</u>) if every point $x \in X_1$ has a neighborhood U such that $f\colon \mathfrak{G}_1|U \to \mathfrak{G}_2$ is a bounded $\mathfrak{M}$ map (resp. $(\Phi,f)\colon (\mathfrak{D}_1,\mathfrak{G}_1)|U \to (\mathfrak{D}_2,\mathfrak{G}_2)$ is a bounded $\mathfrak{M}$ map). We define $\rho(f,\mathfrak{G}_1,\mathfrak{G}_2)(x) = \rho_f(x) = \inf\{k(f,\mathfrak{G}_1|U,\mathfrak{G})\colon U \text{ a neighborhood of } x\}$ (resp. $\rho(\Phi,f;\mathfrak{D}_1,\mathfrak{G}_1,\mathfrak{D}_2,\mathfrak{G}_2)(x) = \rho_{(\Phi,f)}(x) = \inf\{k(\Phi,f;(\mathfrak{D}_1,\mathfrak{G}_1)|U,\mathfrak{D}_2,\mathfrak{G}_2)\colon U \text{ a neighborhood of } x\}$). ρ_f is upper semicontinuous. For U open in X, $f\colon \mathfrak{G}_1|U \to \mathfrak{G}_2$ is bounded iff ρ_f and $f^*\rho_{\mathfrak{G}_2}$ are bounded on U in which case $k(f;\mathfrak{G}_1|U,\mathfrak{G}_2) = \sup \rho_f|U$. Similarly for $\rho_{(\Phi,f)}$.

1 <u>PROPOSITION</u>: (a) If $f\colon \mathfrak{G}_1 \to \mathfrak{G}_2$ and $g\colon \mathfrak{G}_2 \to \mathfrak{G}_3$ are $\mathfrak{M}$ maps then $g\circ f\colon \mathfrak{G}_1 \to \mathfrak{G}_3$ is an $\mathfrak{M}$ map and $\rho_{g\circ f} \leq (f^*\rho_g)O^*(\rho_f)$.

(b) If $(\Phi,f)\colon (\mathfrak{D}_1,\mathfrak{G}_1) \to (\mathfrak{D}_2,\mathfrak{G}_2)$ and $(\Psi,g)\colon (\mathfrak{D}_2,\mathfrak{G}_2) \to (\mathfrak{D}_3,\mathfrak{G}_3)$ are $\mathfrak{M}$ VB maps then $(\Psi\circ\Phi, g\circ f)\colon (\mathfrak{D}_1,\mathfrak{G}_1) \to (\mathfrak{D}_3,\mathfrak{G}_3)$ is an $\mathfrak{M}$ VB map and

$$\rho_{(\Psi\cdot\Phi,g\cdot f)} \leq (f^*\rho_{(\Psi,g)})(\rho_{(\Phi,f)})O^*(\rho_f) \leq (f^*\rho_{(\Psi,g)})O^*(\rho_{(\Phi,f)}).$$

<u>PROOF</u>: (a): For $x \in X_1$ and $\epsilon > 0$ choose U a neighborhood of x and V a neighborhood of $f(x)$ with $V \supset f(U)$, $k_{f|U} < \rho_f(x) + \epsilon$ and $k_{g|V} < \rho_g(fx) + \epsilon$. If $(U_\alpha,h_\alpha),(V_\beta,g_\beta)$ and (W_γ,h_γ) are charts of $\mathfrak{a}_1$, $\mathfrak{a}_2$ and $\mathfrak{a}_3$ with $U \cap U_\alpha \cap f^{-1}V_\beta \cap (g\cdot f)^{-1}W_\gamma \neq \emptyset$ then
$f_{\beta\alpha}\colon h_\alpha(U \cap U_\alpha \cap f^{-1}V_\beta \cap (g\cdot f)^{-1}W_\gamma) \to g_\beta(V \cap V_\beta \cap g^{-1}W_\gamma)$ and
$g_{\gamma\beta}\colon g_\beta(V \cap V_\beta \cap g^{-1}W_\gamma) \to E_\gamma$ are $\mathfrak{M}$ maps with norms bounded by $k_{f|U}$ and $k_{g|V}$ respectively. By FS9 $(g\cdot f)_{\gamma\alpha}\colon h_\alpha(U \cap U_\alpha \cap f^{-1}V_\beta \cap (g\cdot f)^{-1}W_\gamma)$ $\to E_\gamma$ is $\mathfrak{M}$ with norm $\leq k_{g|V}O^*(k_{f|U})$. Letting β vary over the index set of $\mathfrak{a}_2$, FS5 implies that $(g\cdot f)_{\gamma\alpha}\colon h_\alpha(U \cap U_\alpha \cap (g\cdot f)^{-1}W_\gamma) \to E_\gamma$ is $\mathfrak{M}$ with norm $\leq k_{g|V}O^*(k_{f|U})$, i.e. $k_{g\cdot f|U} \leq k_{g|V}O^*(k_{f|U})$. Thus, $\rho_{g\cdot f}(x) \leq (\rho_g(fx) + \epsilon)O^*(\rho_f(x) + \epsilon)$. Now let $\epsilon \to 0$.

(b): The proof is similar. The key fact is that
$(\Psi\cdot\Phi)_{\gamma\alpha}\colon h_\alpha(U \cap U_\alpha \cap f^{-1}V_\beta \cap (g\cdot f)^{-1}W_\gamma) \to L(F^1_\alpha;F^3_\gamma)$ satisfies
$(\Psi\cdot\Phi)_{\gamma\alpha} := \text{comp}_*(\Psi_{\gamma\beta}\cdot f_{\beta\alpha},\Phi_{\beta\alpha})$. The result follows from FS9 and FS5 again together with FS1 which says that $\|\text{comp}_*\| \leq O^*(1)$. Q.E.D.

The inequalities of Proposition 1, as well as the composition fact itself, are the keys to all that follows.

2 <u>LEMMA</u>: Let $\mathfrak{a}_1$ and $\mathfrak{a}_2$ be atlases on X. $\mathfrak{a}_1 \cup \mathfrak{a}_2$ is an $\mathfrak{M}$ atlas iff first, $\mathfrak{a}_1$ and $\mathfrak{a}_2$ are $\mathfrak{M}$ atlases and second, the identity maps $1\colon \mathfrak{a}_1 \to \mathfrak{a}_2$ and $1\colon \mathfrak{a}_2 \to \mathfrak{a}_1$ are $\mathfrak{M}$ maps. In fact, $\rho_{\mathfrak{a}_1\cup\mathfrak{a}_2} = \max(\rho_{\mathfrak{a}_1},\rho_{\mathfrak{a}_2},\rho(1;\mathfrak{a}_1,\mathfrak{a}_2),\rho(1;\mathfrak{a}_2,\mathfrak{a}_1))$. The result for atlases on a vector bundle π is identical.

If $\mathfrak{a}_1 = \{U_\alpha,h_\alpha\}$, $\mathfrak{a}_2 = \{V_\beta,g_\beta\}$ are $\mathfrak{M}$ atlases on X_1, X_2 and $f\colon \mathfrak{a}_1 \to \mathfrak{a}_2$ is an index preserving continuous map then we define $\tilde{k}(f;\mathfrak{a}_1,\mathfrak{a}_2) = \|\{f_\alpha\}\|$ when $\{f_\alpha\} \in \hat{\Pi}_\alpha\mathfrak{M}(h_\alpha(U_\alpha),E_{\beta(\alpha)})$ and ∞ otherwise. We can then define $\tilde{\rho}_f(x) = \tilde{\rho}(f;\mathfrak{a}_1,\mathfrak{a}_a)(x) = \inf\{\tilde{k}(f|U;\mathfrak{a}_1|U,\mathfrak{a}_2)\colon U$ a neighborhood of $x\}$. Similarly, if $(\pi_1,\mathfrak{a}_1) = \{U_\alpha,h_\alpha,\varphi_\alpha\}$,

$(\mathfrak{D}_2,\mathcal{G}_2) = \{V_\beta, g_\beta, \psi_\beta\}$ are $\mathfrak{M}$ atlases on π_1 and π_2 and $(\phi,f): (\mathfrak{D}_1,\mathcal{G}_1) \to (\mathfrak{D}_2,\mathcal{G}_2)$ is index preserving, we define $\tilde{k}(\phi,f;\mathfrak{D}_1,\mathcal{G}_1,\mathfrak{D}_2,\mathcal{G}_2) = \max(\tilde{k}(f;\mathcal{G}_1,\mathcal{G}_2),\|\{\varphi_\alpha\}\|)$ when $\{\varphi_\alpha\} \in \hat{\Pi}_\alpha \mathfrak{M}(h_\alpha(U_\alpha),L(F^1_\alpha;F^2_{\beta(\alpha)}))$ and ∞ otherwise and $\tilde{\rho}_{(\phi,f)}(x) = \tilde{\rho}(\phi,f;\mathfrak{D}_1,\mathcal{G}_1,\mathfrak{D}_2,\mathcal{G}_2)(x) = \inf\{\tilde{k}((\phi,f)|U;(\mathfrak{D}_1,\mathcal{G}_1)|U,\mathfrak{D}_2,\mathcal{G}_2): U \text{ a neighborhood of } x\}$.

3 <u>COROLLARY</u>: (a) Let $\mathcal{G}_1,\mathcal{G}_2$ be $\mathfrak{M}$ atlases and $f: \mathcal{G}_1 \to \mathcal{G}_2$ be index preserving. f is an $\mathfrak{M}$ map iff $\tilde{\rho}_f < \infty$ on X_1. $\tilde{\rho}_f \le \rho_f \le O^*(\rho_{\mathcal{G}_1}, f^*\rho_{\mathcal{G}_2}, \tilde{\rho}_f)$.

(b) Let $(\mathfrak{D}_1,\mathcal{G}_1)$, $(\mathfrak{D}_2,\mathcal{G}_2)$ be $\mathfrak{M}$ atlases and $(\phi,f): (\mathfrak{D}_1,\mathcal{G}_1) \to (\mathfrak{D}_2,\mathcal{G}_2)$ be an index preserving VB map. (ϕ,f) is an $\mathfrak{M}$ map iff $\tilde{\rho}_{(\phi,f)} < \infty$ on X_1. $\tilde{\rho}_{(\phi,f)} \le \rho_{(\phi,f)} \le \tilde{\rho}_{(\phi,f)} O^*(\rho_{(\mathfrak{D}_1,\mathcal{G}_1)}, f^*\rho_{(\mathfrak{D}_2,\mathcal{G}_2)}, \tilde{\rho}_f) \le O^*(\rho_{(\mathfrak{D}_1,\mathcal{G}_1)}, f^*\rho_{(\mathfrak{D}_2,\mathcal{G}_2)}, \tilde{\rho}_{(\phi,f)})$.

<u>PROOF</u>: That $\tilde{\rho} \le \rho$ in each case is clear. The reverse bounds follow from Proposition 1. The details are left to the reader with the hint that for U a neighborhood of x, with $U \cap U_\alpha \ne \emptyset$, the atlas map $f: \mathcal{G}_1|U \cap U_\alpha \to \mathcal{G}_2$ factors: $\mathcal{G}_1|U \cap U_\alpha \xrightarrow{1} \{U \cap U_\alpha, h_\alpha|U \cap U_\alpha\} \xrightarrow{f} \{V_{\beta(\alpha)}, g_{\beta(\alpha)}\} \xrightarrow{1} \mathcal{G}_2$. Q.E.D.

<u>EXAMPLE</u>: If $\mathcal{G} = \{V_\alpha, h_\alpha\}$ is a locally finite C^r atlas on X and $\{U_\alpha\}$ is an open cover of X with $\overline{U_\alpha} \subset V_\alpha$ then the refinement $\mathcal{G} = \{U_\alpha, h_\alpha\}$ is a $\mathcal{C}^r$ atlas on X. For if $x \in X$ we can choose U meeting only finitely many U_α's and $U \cap U_\alpha \ne \emptyset$ implies $U \subset V_\alpha$. Then with only finitely many transition maps to look at we can shrink U to get the r-jets on U close to the r-jet at x. If $\bar{\mathcal{G}}_1$ and $\bar{\mathcal{G}}_2$, locally finite C^r atlases on X_1 and X_2, are refined as above to get $\mathcal{G}_1$ and $\mathcal{G}_2$, $\mathcal{C}^r$ atlases, and $f: X_1 \to X_2$ is a C^r map, then $f: \mathcal{G}_1 \to \mathcal{G}_2$ is a $\mathcal{C}^r$ map.

4 <u>LEMMA</u>: If $\mathcal{G}_1$, $\mathcal{G}_2$ are $\mathfrak{M}$ atlases on X_1, X_2 and $f: X_1 \to X_2$ is continuous, then f is a $\mathcal{C}$ map and $\rho^{\mathcal{C}}(f;\mathcal{G}_1,\mathcal{G}_2) \le f^*\rho^{\mathcal{C}}_{\mathcal{G}_2} \le f^*\rho^{\mathfrak{M}}_{\mathcal{G}_2}$.

PROOF: $\|f_{\gamma\alpha}|h_\alpha(U \cap U_\alpha \cap f^{-1}V_\gamma)\|_0 \le \|1_{g_\gamma(V\cap V_\gamma)}\|_0$ when $f(U) \subset V$. Q.E.D.

THE STANDARD CONSTRUCTIONS: (cf. Sec. I.2) (1) Refinements: If $\mathfrak{a}_0$ is a refinement of $\mathfrak{a}$ then $\rho_{\mathfrak{a}_0} \le \rho_{\mathfrak{a}}$. If $\mathfrak{a}_0$ is a subdivision of $\mathfrak{a}$ then equation I.(2.1) and FS5 imply $\rho_{\mathfrak{a}_0} = \rho_{\mathfrak{a}}$. Similarly for VB atlases. (2) Products: For atlases $\mathfrak{a}_1, \mathfrak{a}_2$ on X_1, X_2, $\rho_{\mathfrak{a}_1 \times \mathfrak{a}_2} = \max(p_1^*\rho_{\mathfrak{a}_1}, p_2^*\rho_{\mathfrak{a}_2})$ where $p_i: X_1 \times X_2 \to X_i$ $(i = 1,2)$. Similarly for products of maps, bundles and VB maps. (3) Trivial Bundles: For atlases $\mathfrak{a}$ on X and $(\mathfrak{a} \times F, \mathfrak{a})$ on $\epsilon_F: X \times F \to X$, $\rho_{(\mathfrak{a}\times F,\mathfrak{a})} = \rho_{\mathfrak{a}}$. If $f: \mathfrak{a}_1 \to \mathfrak{a}_2$ and $T: F_1 \to F_2$ is a linear map then for $(\phi, f) = (f \times T, f): \epsilon_{F_1} \to \epsilon_{F_2}$, $\rho_{(\phi,f)} = \max(\rho_f, \|T\|)$. (4) Tangent Bundles: For a C^1 atlas $\mathfrak{a}$ on X and the atlas $(T\mathfrak{a}, \mathfrak{a})$ on τ_X, $\rho^{\mathfrak{M}}_{(T\mathfrak{a},\mathfrak{a})} = \rho^{\mathfrak{M}^1}_{\mathfrak{a}}$. If $f: X_1 \to X_2$ is C^1, $\mathfrak{a}_1, \mathfrak{a}_2$ are C^1 on X_1, X_2 then with $(T\mathfrak{a}_i, \mathfrak{a}_i)$ on τ_{X_i}, $\rho^{\mathfrak{M}}_{(Tf,f)} = \rho^{\mathfrak{M}^1}_f$. If $f: \mathfrak{a}_1 \to \mathfrak{a}_2$ is index preserving then $\tilde{\rho}^{\mathfrak{M}}_{(Tf,f)} = \tilde{\rho}^{\mathfrak{M}^1}_f$, too. Thus, $\mathfrak{a}$ is $\mathfrak{M}^1$ iff $(T\mathfrak{a}, \mathfrak{a})$ is $\mathfrak{M}$. f is $\mathfrak{M}^1$ iff (Tf, f) is $\mathfrak{M}$. (5) Subbundles: If $(\mathfrak{D}, \mathfrak{a})$ on $\pi: E \to X$ induces a subbundle atlas on $E_0 \subset E$ and $(J,1): \pi|E_0 \to \pi$ is the inclusion then $\rho_{(\mathfrak{D}|E_0,\mathfrak{a})} \le \rho_{(\mathfrak{D},\mathfrak{a})}$. $(J,1): (\mathfrak{D}|E_0, \mathfrak{a}) \to (\mathfrak{D}, \mathfrak{a})$ is index preserving and $\tilde{\rho}_{(J,1)} = 1$ (unless some fibers of $E_0 = 0$), $\rho_{(J,1)} \le \rho_{(\mathfrak{D},\mathfrak{a})}$. (6) Bundle Operations: For $(\mathfrak{D}_1, \mathfrak{D}_2; \mathfrak{a})$ a two-tuple atlas on $\pi_i: E_i \to X$ $(i = 1,2)$ and $(\mathfrak{D}_1 \oplus \mathfrak{D}_2, \mathfrak{a})$ on $\pi_1 \oplus \pi_2$, $(L^n_\epsilon(\mathfrak{D}_1; \mathfrak{D}_2), \mathfrak{a})$ on $L^n_\epsilon(\pi_1; \pi_2)$, $\rho_{(\mathfrak{D}_1\oplus\mathfrak{D}_2,\mathfrak{a})} \le \max(\rho_{(\mathfrak{D}_1,\mathfrak{a})}, \rho_{(\mathfrak{D}_2,\mathfrak{a})})$ and by FS1, $\rho_{(L^n(\mathfrak{D}_1;\mathfrak{D}_2),\mathfrak{a})} \le O^*(1)\max(\rho_{(\mathfrak{D}_1,\mathfrak{a})}, \rho_{(\mathfrak{D}_2,\mathfrak{a})})$ (constant depending on $\mathfrak{M}$ and n). Similar estimates for VB maps of the form $(\phi_1 \oplus \phi_2, 1)$ and $(L^n_\epsilon(\phi_1; \phi_2), 1)$. (7) Pullbacks: If $(\mathfrak{D}, \mathfrak{a})$ is an $\mathfrak{M}$ atlas on $\pi: E \to X$, $\mathfrak{a}_0$ is an $\mathfrak{M}$ atlas on X_0 and $f: \mathfrak{a}_0 \to \mathfrak{a}$ is an index preserving $\mathfrak{M}$ map, then for $(f^*\mathfrak{D}, \mathfrak{a}_0)$ on $f^*\pi$, $\rho_{(f^*\mathfrak{D},\mathfrak{a}_0)} \le f^*\rho_{(\mathfrak{D},\mathfrak{a})} O^*(\tilde{\rho}_f)$ by FS9. If $(\mathfrak{D}_0, \mathfrak{a}_0)$ is an $\mathfrak{M}$ atlas on $\pi_0: E_0 \to X_0$ and $(\phi, f): (\mathfrak{D}_0, \mathfrak{a}_0) \to (\mathfrak{D}, \mathfrak{a})$ is a (necessarily index-preserving) VB map, then $\tilde{\rho}_{(\phi,f)} = \max(\tilde{\rho}_{(f^*\phi,1)}, \tilde{\rho}_f)$. (8) Jet Bundles: If $(\mathfrak{D}, \mathfrak{a})$ is an $\mathfrak{M}^k$ atlas on π then $(J^k(\mathfrak{D}), \mathfrak{a})$ is an

$\mathfrak{M}$ atlas on $J^k(\pi)$ with $\rho^{\mathfrak{M}}_{(J^k(\mathfrak{D}),\mathcal{G})} \leq O^*(\rho^{\mathfrak{M}^k}_{(\mathfrak{D},\mathcal{G})})$ by FS1. If $(\mathfrak{D}_1,\mathcal{G})$ is an $\mathfrak{M}^k$ atlas on π_1 and $(\phi,1)\colon \pi \to \pi_1$ is an $\mathfrak{M}^k$ VB map then $(J^k(\phi),1)\colon (J^k(\mathfrak{D}),\mathcal{G}) \to (J^k(\mathfrak{D}_1),\mathcal{G})$ is an index preserving $\mathfrak{M}$ VB map with $\tilde{\rho}^{\mathfrak{M}}_{(J^k(\phi),1)} \leq \tilde{\rho}^{\mathfrak{M}^k}_{(\phi,1)} O^*(1)$. If $\mathcal{G}_0$ is an $\mathfrak{M}^k$ atlas on X_0 and $f\colon \mathcal{G}_0 \to \mathcal{G}$ is an index preserving $\mathfrak{M}^k$ map then for the index preserving VB map $(\Psi^k_f,1)\colon (f^*J^k(\mathfrak{D}),\mathcal{G}_0) \to (J^k(f^*\mathfrak{D}),\mathcal{G}_0)$ $\tilde{\rho}_{(\Psi^k_f,1)} \leq O^*(\tilde{\rho}_f)$. The constants depend on k and $\mathfrak{M}$ throughout.

Given an $\mathfrak{M}$ n-tuple atlas $(\mathfrak{D}_1,\ldots,\mathfrak{D}_n,\mathcal{G})$ on $\pi_i\colon E_i \to X$ $(i=1,\ldots,n)$ and $\mathfrak{M}$ atlas $(\mathfrak{D}_0,\mathcal{G}_0)$ on $\pi_0\colon E_0 \to X_0$, we can define $(\phi,f)\colon (\mathfrak{D}_1 \oplus\cdots\oplus \mathfrak{D}_n,\mathcal{G}) \to (\mathfrak{D}_0,\mathcal{G}_0)$ to be an $\mathfrak{M}$ n linear VB map in a fashion analogous to the above theory where $n = 1$. The analogy is sufficiently close that we will use the results when needed without bothering to derive or even state them for $n > 1$.

2. __Metrics__. Let $\mathcal{G} = \{U_\alpha,h_\alpha\}$ be an atlas on X. An __$\mathcal{G}$-chain__ between x and y in X is a sequence $(x_1,\alpha_1,\ldots,\alpha_n,x_{n+1})$ such that: (1) $x_1 = x$, $x_{n+1} = y$. (2) $x_i,x_{i+1} \in U_{\alpha_i}$. (3) The segment $[h_{\alpha_i}x_i,h_{\alpha_i}x_{i+1}] \subset h_{\alpha_i}(U_{\alpha_i})$, $(i = 1,\ldots,n)$. The __length of the $\mathcal{G}$-chain__, denoted $\ell(x_1,\alpha_1,\ldots,\alpha_n,x_{n+1})$, is $\Sigma_i\|h_{\alpha_i}x_i - h_{\alpha_i}x_{i+1}\|_{\alpha_i}$. A __piecewise C^1 $\mathcal{G}$-path__ between x and y is a continuous path x_t, $t \in [a,b]$ and a sequence $(t_1,\alpha_1,\ldots,\alpha_n,t_{n+1})$ such that (1) $a = t_1 <\cdots< t_{n+1} = b$, (2) $x_a = x$, $x_b = y$, (3) $x_t \in U_{\alpha_i}$ for $t \in [t_i,t_{i+1}]$, (4) $h_{\alpha_i}\circ x_t\colon [t_i,t_{i+1}] \to E_{\alpha_i}$ is piecewise C^1. If $\mathcal{G}$ is a C^1 atlas then (4) holds iff x_t is a piecewise C^1 path. The __length of the piecewise C^1 $\mathcal{G}$-path__ is $\Sigma_i \int_{t_i}^{t_{i+1}} \|(h_{\alpha_i}\circ x_t)'\|_{\alpha_i}\, dt$. To every $\mathcal{G}$-chain $(x_1,\alpha_1,\ldots,\alpha_n,x_{n+1})$ there is a canonically associated piecewise C^1, in fact, piecewise linear or p.l., $\mathcal{G}$-path x_t defined on $[1,n+1]$ with $h_{\alpha_i}(x_t)$ extending $h_{\alpha_i}(x_i)$ and $h_{\alpha_i}(x_{i+1})$ linearly on $[i,i+1]$. This p.l. $\mathcal{G}$-path has the same length

as the original G-chain. Using the p.l. path, we define two G-chain operations. If $i < t \leq i+1$ then $(x_1,\alpha_1,\ldots,x_i,\alpha_i,x_t)$ is a <u>truncation</u> of the chain, and $(x_1,\alpha_1,\ldots,x_i,\alpha_i,x_t,\alpha_i,x_{i+1},\ldots,\alpha_n,x_{n+1})$ is an <u>elementary subdivision</u> of the chain. Inserting more points along the path in this way one obtains <u>subdivisions</u> of the chain. The p.l. path of a truncation (or subdivision) is a reparametrization of part or all of the original path and the length is less than (is equal to) that of the original chain.

For $x,y \in X$ we define $d_G(x,y)$ to be the infimum of the lengths of all G-chains between x and y or, equivalently, the infimum of the lengths of all piecewise C^1 G-paths between x and y. The equivalence of the two definitions is proved like Proposition II. 1.1. d_G is a pseudometric on X. ∞ is an allowed value, in fact, $d_G(x,y) < \infty$ iff x and y lie in the same component of X. d_G need not be a metric in general. For example, if G is the maximal atlas for a smooth manifold then $d_G(x,y) = 0$ if x and y lie in the same component of X. However, for G an $\mathcal{L}$ atlas we will show that d_G is a metric with topology that of X.

1 <u>LEMMA</u>: Let G be an $\mathcal{L}$ atlas on X with (U_α,h_α) a chart of G. Let A be a closed subset of $h_\alpha(U_\alpha)$. If there exists $\epsilon > 0$ and $K < \infty$ such that $B(A,\epsilon) \subset h_\alpha(U_\alpha \cap \{\rho_G^{\mathcal{L}} < K\})$, then $h_\alpha^{-1}(A)$ is closed in X.

<u>PROOF</u>: Let $\{x_n\}$ be a sequence is $h_\alpha^{-1}A$ converging (in X) to y. It suffices to show $y \in U_\alpha$ since $h_\alpha^{-1}A$ is closed in U_α. Choose a chart (U_β,h_β) of G with $y \in U_\beta$. We can assume $x_n \in U_\beta$ for all n. Let $\rho = \rho_G^{\mathcal{L}}$.

<u>CLAIM</u>: There exists $\delta > 0$ such that for all n, $B(h_\alpha x_n,\delta) \subset h_\alpha(U_\alpha \cap U_\beta \cap \{\rho < K\})$.

Assuming the claim, we can apply the Size Estimate, Prop. II.3.7, to $h_\beta h_\alpha^{-1}$: $h_\alpha(U_\alpha \cap U_\beta \cap \{\rho < K\}) \to h_\beta(U_\alpha \cap U_\beta \cap \{\rho < K\})$ and conclude $B(h_\beta x_n, \delta/K) \subset h_\beta(U_\alpha \cap U_\beta \cap \{\rho < K\}) \subset h_\beta(U_\beta)$. Since $x_n \to y$ in U_β, $h_\beta x_n \to h_\beta y$ in $h_\beta(U_\beta)$ and so eventually $h_\beta y$ is in $B(h_\beta x_n, \delta/K)$. Thus, $y \in U_\alpha \cap U_\beta \cap \{\rho < K\}$.

<u>Proof of Claim</u>: If the claim is false then for each integer $i > 0$ we can choose some $n_i \geq i$ and $v_i \in E_\alpha$ such that: (1) $v_i^t = tv_i + (1-t)h_\alpha(x_{n_i})$ is in $h_\alpha(U_\alpha \cap U_\beta \cap \{\rho < K\})$ for $0 \leq t < 1$, (2) $v_i \notin h_\alpha(U_\alpha \cap U_\beta \cap \{\rho < K\})$, and (3) $\|v_i - h_\alpha(x_{n_i})\| < \min(\epsilon, i^{-1})$. To see this, choose a provisional $\bar{v}_i$ satisfying (2) and (3) and go out the ray from $h_\alpha(x_{n_i})$ towards $\bar{v}_i$. Choose v_i to be the first point satisfying (2). Then, (1) and (3) follow. By hypothesis and (3), we have (4) $v_i \in h_\alpha(U_\alpha \cap \{\rho < K\})$. By (1) and $\|h_\beta h_\alpha^{-1} | h_\alpha(U_\alpha \cap U_\beta \cap \{\rho < K\})\|_L < K$ we have for $0 \leq t_1, t_2 < 1$: (5) $\|h_\beta h_\alpha^{-1}(v_i^{t1}) - h_\beta h_\alpha^{-1}(v_i^{t2})\| \leq K\|v_i^{t1} - v_i^{t2}\|$. Thus, for each i as t increases to 1, $h_\beta h_\alpha^{-1}(v_i^t)$ is Cauchy and so converges to $u_i \in E_\beta$. Now if $u_i \in h_\beta(U_\beta)$ then $h_\alpha^{-1}(v_i^t)$ converges to $h_\beta^{-1}(u_i) \in U_\beta$. Since $h_\alpha^{-1}(v_i^t)$ converges to $h_\alpha^{-1} v_i$, and X is Hausdorff, $h_\alpha^{-1}(v_i) = h_\beta^{-1}(u_i)$ and so $v_i \in h_\alpha(U_\beta \cap U_\alpha \cap \{\rho < K\})$ contradicting (3). Thus, $u_i \notin h_\beta(U_\beta)$ for any i. But by (5) and (3) we have $\|u_i - h_\beta(x_{n_i})\| < i^{-1}K$. Thus, u_i converges to $h_\beta(y)$ and so eventually lies in $h_\beta(U_\beta)$. Contradiction. Q.E.D.

2 <u>THEOREM (Metric Estimate)</u>: Assume $\mathfrak{M} \subset \mathcal{L}$. Let $\mathfrak{a}$ be an $\mathfrak{M}$ atlas on X and (U_α, h_α) be a chart of $\mathfrak{a}$.

(a) If $x_1, x_2 \in U_\alpha$ then $d_\alpha(h_\alpha x_1, h_\alpha x_2) \geq d_{\mathfrak{a}}(x_1, x_2)$ where d_α is the intrinsic metric on $h_\alpha(U_\alpha)$. In particular, if $[h_\alpha x_1, h_\alpha x_2] \subset h_\alpha(U_\alpha)$ then $\|h_\alpha x_1 - h_\alpha x_2\|_\alpha \geq d_{\mathfrak{a}}(x_1, x_2)$.

(b) Let $x_0 \in U_\alpha$ and $\rho_{\mathfrak{a}}(x_0) < K < \infty$. If $B^{\|\ \|_\alpha}(h_\alpha x_0, \epsilon) \subset h_\alpha(U_\alpha \cap \{\rho_{\mathfrak{a}} < K\})$, then $h_\alpha^{-1}(B^{\|\ \|_\alpha}(h_\alpha x_0, /4K)) \subset B^{d_{\mathfrak{a}}}(x_0, \epsilon/4K) \subset h_\alpha^{-1}(B^{\|\ \|_\alpha}(h_\alpha x_0, /2)) \subset U_\alpha$ and for x_1, x_2 in $B^{d_{\mathfrak{a}}}(x_0, \epsilon/4K)$, $\|h_\alpha x_1 - h_\alpha x_2\|_\alpha \geq$

$d_{G}(x_1,x_2) \geq K^{-1}\|h_\alpha x_1 - h_\alpha x_2\|_\alpha$.

PROOF: (a): For any $h_\alpha(U_\alpha)$ chain there is an associated G chain with all α_i's $= \alpha$, the same endpoints and the same length.

(b): Let $V = B^{d_G}(x_0,\epsilon/4K)$, $W = h_\alpha^{-1}(B^{\|\ \|_\alpha}(h_\alpha x_0,\epsilon))$, and $\rho = \rho_G^{\ell}$. Now if $(x_1,\alpha_1,\ldots,\alpha_n,x_{n+1})$ is an G chain whose p.l. path x_t, $t \in [1,n+1]$, maps into W then $(x_1,\alpha,\ldots,\alpha,x_{n+1})$ is an G chain and

$$(2.1) \quad \ell(x_1,\alpha_1,\ldots,\alpha_n,x_{n+1}) \geq K^{-1}\ell(x_1,\alpha,\ldots,\alpha,x_{n+1}) \geq K^{-1}\|h_\alpha x_1 - h_\alpha x_{n+1}\|.$$

To prove (2.1) note that $W \subset U_\alpha \cap \{\rho < K\}$ and $h_\alpha W$ is convex. Thus, $(x_1,\alpha,\ldots,\alpha,x_{n+1})$ is an G chain. Since the p.l. path for the original chain maps into W, $[h_{\alpha_i}x_i,h_{\alpha_i}x_{i+1}] \subset h_{\alpha_i}(W \cap U_{\alpha_i})$. $\|h_\alpha h_{\alpha_i}^{-1}|h_{\alpha_i}(W \cap U_{\alpha_i})\|_L < K$. Hence, $\|h_\alpha x_i - h_\alpha x_{i+1}\|_\alpha \leq K\|h_{\alpha_i}x_i - h_{\alpha_i}x_{i+1}\|_{\alpha_i}$. Summing yields the left inequality in (2.1). The right comes from the triangle inequality.

CLAIM (x): (a) For all $x' \in V$, if $(x_1,\alpha_1,\ldots,\alpha_n,x_{n+1})$ is an G chain between x and x' with length $< \epsilon/2K$, then the associated p.l. path function maps into W.

(b) For all $x' \in V$, $d_G(x,x') \geq K^{-1}\|h_\alpha x - h_\alpha x'\|_\alpha$.

Note first that for any $x \in V$, Claim (x)(a) implies Claim (x)(b) because $d_G(x,x') < \epsilon/2K$ and so we need only look at G chains with length $< \epsilon/2K$. Claim (x)(b) then follows from Claim (x)(a) and (2.1). Also Claim (x) for all $x \in V$ implies Theorem 2 (b).

Proof of Claim (x_0) (a): If false, let $t_0 = \sup\{t: x_s \in W \text{ for } 1 \leq s < t\}$. $1 < t_0 \leq n+1$, $x_s \in W$ for $1 \leq s < t_0$ and $x_{t_0} \notin W$. By Lemma 1, $A = h_\alpha^{-1}(B^{\|\ \|_\alpha}[h_\alpha x_0,\epsilon/2])$ is closed in X and is a subset of W. Hence, x_s can't remain in A for all $1 \leq s < t_0$. So we can choose $1 < t_1 < t_0$ with $x_t \notin A$ and if $i < t_1 \leq i+1$ then $(x_1,\alpha_1,\ldots,\alpha_i,x_{t_1})$ is a truncation with length $< \epsilon/2K$.

By (2.1), $\ell(x_1,\alpha_1,\ldots,\alpha_i,x_{t_1}) \geq K^{-1}\|h_\alpha x_0 - h_\alpha x_{t_1}\|_\alpha$. The latter is $> \epsilon/2K$ since $h_\alpha x_{t_1} \notin A$. Contradiction. This proves Claim (x_0)(a) and so Claim (x_0)(b) which implies $V \subset h_\alpha^{-1}(B^{\|\ \|_\alpha}(h_\alpha x_0,\epsilon/2))$. In turn, this implies for $x \in V$ that $B^{\|\ \|_\alpha}[h_\alpha x,\epsilon/2] \subset B^{\|\ \|_\alpha}[h_\alpha x_0,r]$ for some $r < \epsilon$.

<u>Proof of Claim (x)(a)</u>: Proceed just as before, but pick x_{t_1} so that $h_\alpha x_{t_1} \notin B^{\|\ \|_\alpha}[h_\alpha x,\epsilon/2]$ and use Lemma 1 to show that $h_\alpha^{-1}(B^{\|\ \|_\alpha}[h_\alpha x_0,r])$ and hence $h_\alpha^{-1}(B^{\|\ \|_\alpha}[h_\alpha x,\epsilon/2])$ is closed in X. Q.E.D.

3 <u>COROLLARY</u>: Assume $\mathfrak{M} \subset \mathcal{L}$. If $\mathfrak{a}$ is an $\mathfrak{M}$ atlas on X, then $d_\mathfrak{a}$ is a metric yielding the topology of X. In particular, X is paracompact.

<u>PROOF</u>: That $d_\mathfrak{a}$ is a metric with the right topology follows from Theorem 3 (b). Parac mpactness follows from Stone's Theorem [13; Cor. V.35]. Q.E.D.

It is from the fundamental role of the Metric Estimate that the Metric Theory gets its name. Readers surprised by the work required in Lemma 1, for example, are referred to Palais [23; Sec. 2] where some closely related false results are discussed.

4 <u>LEMMA</u>: Assume $\mathfrak{M} \subset \mathcal{L}$. If $\mathfrak{a}$ is an $\mathfrak{M}$ atlas on X and $\mathfrak{a}_0$ is a refinement of $\mathfrak{a}$, then $d_{\mathfrak{a}_0} \geq d_\mathfrak{a}$. If $\mathfrak{a}_0$ is a subdivision of $\mathfrak{a}$ then $d_{\mathfrak{a}_0} = d_\mathfrak{a}$.

<u>PROOF</u>: To every $\mathfrak{a}_0$ chain there is an obviously associated $\mathfrak{a}$ chain with the same endpoints and length. If $\mathfrak{a}_0$ is a subdivision, then for every $\mathfrak{a}$ chain we can use equation I.(2.1) to subdivide and construct an $\mathfrak{a}_0$ chain with the same endpoints and length. Q.E.D.

5 **PROPOSITION:** Assume $\mathfrak{M} \subset \mathcal{L}$, $\mathcal{G}$ is an $\mathfrak{M}$ atlas on X and U is open in X. For $x_1, x_2 \in U$, $d_{\mathcal{G}|U}(x_1,x_2) \geq d_{\mathcal{G}}(x_1,x_2) \geq \min(d_{\mathcal{G}|U}(x_1,x_2), d_{\mathcal{G}}(x_1, X-U) + d_{\mathcal{G}}(x_2, X-U))$. In particular, if $d_{\mathcal{G}}(x_1,x_2) < d_{\mathcal{G}}(x_1,X-U) + d_{\mathcal{G}}(x_2,X-U)$ then $d_{\mathcal{G}}(x_1,x_2) = d_{\mathcal{G}|U}(x_1,x_2)$. Finally, if $B^{d_{\mathcal{G}}}(A,\epsilon) \subset U$, $x_1,x_2 \in A$ with $d_{\mathcal{G}}(x_1,x_2) < 2\epsilon$ then $d_{\mathcal{G}}(x_1,x_2) = d_{\mathcal{G}|U}(x_1,x_2)$.

PROOF: Any $\mathcal{G}|U$ chain is an $\mathcal{G}$ chain of the same length. Conversely, if the associated p.l.path for an $\mathcal{G}$ chain has image in U then the chain is an $\mathcal{G}|U$ chain. If the associated p.l. path meets $X - U$ then it breaks into two $\mathcal{G}$ chains of length $\geq d_{\mathcal{G}}(x_1, X-U)$ and $\geq d_{\mathcal{G}}(x_2, X-U)$ respectively. From these remarks the inequalities follow. The rest is easy. Q.E.D.

6 **PROPOSITION:** Assume $\mathfrak{M} \subset \mathcal{L}$, $\mathcal{G}_i$ is an $\mathfrak{M}$ atlas on X_i $(i = 1,2)$ and $f\colon \mathcal{G}_1 \to \mathcal{G}_2$ is an $\mathfrak{M}$ map. If $A \subset X_1$, $K, \epsilon > 0$ and $\rho_f^{\mathcal{L}}|B^{d_{\mathcal{G}_1}}(A,\epsilon) \leq K$, then for $x_1, x_2 \in A$ with $d_{\mathcal{G}_1}(x_1,x_2) < 2\epsilon$, $d_{\mathcal{G}_2}(fx_1, fx_2) \leq K d_{\mathcal{G}_1}(x_1,x_2)$.

PROOF: Let $(x_1, \alpha_1, \ldots, \alpha_n, x_{n+1})$ be an $\mathcal{G}_1$ chain between x_1, x_2 of length $< 2\epsilon$. The associated p.l. path x_t has image in $U = B^{d_{\mathcal{G}}}(A,\epsilon)$. Define $\alpha_t = \alpha_i$ for $i \leq t < i+1$ and $\alpha_{n+1} = \alpha_n$. For each $t \in [1,n+1]$ choose G_t and β_t such that: (1) $x_t \in G_t \subset U_{\alpha_t} \cap U$. (2) $h_{\alpha_t}(G_t)$ is open and convex. (3) $f(G_t) \subset V_{\beta_t}$. (4) $g_{\beta_t}(f(G_t))$ is contained in an open, convex subset of V_{β_t}. By compactness, we can choose $1 = t_1 < \cdots < t_{N+1} = n+1$ a sequence including the integers and such that $x_t \in G_{t_i}$ for $t_i \leq t \leq t_{i+1}$, $i = 1,\ldots,N$. $(x_{t_1}, \alpha_{t_1}, \ldots, \alpha_{t_N}, x_{t_{N+1}})$ is an $\mathcal{G}_1$ chain, in fact a subdivision of the original one. So it has the same length. On the other hand, $(fx_{t_1}, \beta_{t_1}, \ldots, \beta_{t_N}, fx_{t_{N+1}})$ is an $\mathcal{G}_2$ chain. Since $\|f_{\beta_{t_i}\alpha_{t_i}}|h_{\alpha_{t_i}}(G_{t_i})\|_L \leq K$, $\|g_{\beta_{t_i}}(fx_{t_i}) - g_{\beta_{t_i}}(fx_{t_{i+1}})\| \leq K\,\|h_{\alpha_{t_i}}(x_{t_i}) - h_{\alpha_{t_i}}(x_{t_{i+1}})\|$. Summing on $i = 1,\ldots,N$, we have

$d_{G_2}(fx_1,fx_2) \le K\ell(x_1,\alpha_1,\ldots,\alpha_n,x_{n+1})$. Take the infimum over G_1 chains of length $< 2\epsilon$ between x_1,x_2. Q.E.D.

7 <u>COROLLARY</u>: Assume $\mathfrak{M} \subset \mathcal{L}$. If $f\colon G_1 \to G_2$ is a bounded $\mathfrak{M}$ map then $f^*d_{G_2} \le (\sup \rho_f^{\mathcal{L}})d_{G_1}$.

8 <u>PROPOSITION</u>: Assume $\mathfrak{M} \subset \mathcal{L}$, G_i is an $\mathfrak{M}$ atlas on X_i $(i = 1,2)$. $2\max(p_1{}^*d_{G_1},p_2{}^*d_{G_2}) \ge d_{G_1 \times G_2} \ge \max(p_1{}^*d_{G_1},p_2{}^*d_{G_2})$ where $p_i\colon X_1 \times X_2 \to X_i$ are the projections $(i = 1,2)$.

<u>PROOF</u>: Any $G_1 \times G_2$ chain projects by p_i to an G_i chain of no greater length. For an G_1 chain $(x_1,\alpha_1,\ldots,\alpha_n,x_{n+1})$ and an G_2 chain $(y_1,\beta_1,\ldots,\beta_m,y_{m+1})$ we can define a "product" $G_1 \times G_2$ chain with length equal to the sum of the lengths of the factors. The product is $((x_1,y_1),(\alpha_1,\beta_1),\ldots,(\alpha_n,\beta_1),(x_{n+1},y_1),(\alpha_n,\beta_2),(x_{n+1},y_2),\ldots,$ $(\alpha_n,\beta_m),(x_{n+1},y_{m+1}))$. Q.E.D.

3. <u>Finslers</u>: If $\pi\colon E \to X$ is a bundle then each fiber E_x is a Banachable space, i.e. a complete, normable TVS, but there is no canonical choice of norm. Instead each chart induces a norm and different charts induce equivalent but not necessarily equal norms. A choice of norm for each fiber is called a <u>Finsler</u> on π. To understand Finslers we consider the space of norms on a Banachable space (cf. [20; Sec. IV.A]).

For E a Banachable space let $\eta(E)$ denote the set of norms on E (i.e. norms with the given topology). If $n_1,n_2 \in \eta(E)$ then there exist constants $K \ge 0$ such that $e^{-K}n_1 \le n_2 \le e^{K}n_1$ on E. The infimum of such K's defines the distance $\mathrm{dist}(u_1,u_2)$. To see that dist is a complete metric on $\eta(E)$, let $\|\ \|$ be a fixed norm on E and let S be the unit sphere $\{\|\ \| = 1\}$ in E. By homogeneity, $e^{-K}n_1 \le n_2 \le e^{K}n_1$ holds on E iff it holds on S iff $|\log n_1 - \log n_2| \le K$ on S. The map $n \to \ell(n) = \log\circ n|S$ is thus an isometry of $\eta(E)$ into the B-space $\mathcal{C}(S,R)$

taking $\| \ \|$ to 0. f is in the image of ℓ iff $n_f(v) = \|v\| \times \exp[f(v/\|v\|)]$ if $v \neq 0$ and $= 0$ if $v = 0$ satisfies the triangle inequality. This is a closed condition on f and so $\mathcal{N}(E)$ is isometric to a closed subset of $\mathcal{C}(S,R)$. The map ℓ relates the obvious partial orders on $\mathcal{N}(E)$ and $\mathcal{C}(S,R)$. $\mathcal{N}(E)$ is closed under addition and multiplication by positive scalars and so under convex combinations. A subset of $\mathcal{N}(E)$ is convex if it is closed under convex combinations. For example if $n_1 \leq n_2$ then $\{n\colon n_1 \leq n \leq n_2\}$ is closed and convex in $\mathcal{N}(E)$. With $\| \ \|$ fixed we define $\rho_{\| \ \|}\colon \mathcal{N}(E) \to [1,\infty)$ by $\rho_{\| \ \|}(n) = \exp[\mathrm{dist}(n, \| \ \|)]$.

1. <u>PROPOSITION</u>: A subset A of $\mathcal{N}(E)$ has bounded dist diameter iff $\rho_{\| \ \|}$ is bounded on A iff A has upper and lower bounds in $\mathcal{N}(E)$ with respect to $\leq$. Call such a subset bounded. In particular, any finite subset is bounded. If A is bounded then it has a least upper bound n_M and greatest lower bound n_m with respect to $\leq$ defined by: $n_M(v) = \sup\{n(v)\colon n \in A\}$ and $n_m(v) = \inf\{\Sigma_i n_i(v_i)\colon \{(v_i,n_i)\}$ is a finite sequence in $A \times E$ with $\Sigma_i v_i = v\}$. The diameter of $\{n\colon n_m \leq n \leq n_M\}$ $= \mathrm{dist}(n_m, n_M) =$ the diameter of A.

<u>PROOF</u>: The three conditions of boundedness are equivalent to the existence of K such that $e^{-K}\| \ \| \leq n \leq e^{K}\| \ \|$ for all $n \in A$. n_M and n_m in general satisfy the triangle inequality and are homogeneous. If $n_1 \geq n$ for all $n \in A$ then $n_1 \geq n_M$ and so $n_M \in \mathcal{N}(E)$. If $n_1 \leq n$ for all n then $n_1 \leq n_m$ and so $n_m \in \mathcal{N}(E)$. The diameter results are left to the reader with the remark that $\max[\mathrm{dist}(n_1,n_M),\mathrm{dist}(n_1,n_m))]$ $\leq \sup\{\mathrm{dist}(n_1,n)\colon n \in A\}$ for all $n_1 \in \mathcal{N}(E)$. Q.E.D.

2 <u>PROPOSITION</u>: Let E and F be B spaces with norms denoted $\| \ \|$.

(a) On $\mathcal{N}(E)$, $\rho_{\| \ \|}(n_1) \leq \rho_{\| \ \|}(n_2)\cdot\exp(\mathrm{dist}(n_1,n_2))$.

(b) The map $\mathcal{N}(E) \times \mathcal{N}(F) \to \mathcal{N}(E \times F)$ defined by the sup norm on the product is continuous.

(c) The map $\eta(E) \times \eta(F) \to \eta(L^p_\epsilon(E;F))$ defined by the operator norm is continuous.

(d) The map $\mathrm{Lis}(E;F) \times \eta(F) \to \eta(E)$ defined by $(T,n) \to T^*n = n.T$ is continuous.

(e) The evaluation map $\eta(E) \times E \to \mathbb{R}$ is continuous.

(f) If $\Delta_N = \{(t_0,\ldots,t_N) \in \mathbb{R}^{N+1}: t_i \geq 0 \text{ and } \Sigma\, t_i = 1\}$ then $\Delta_N \times \eta(E)^{(N+1)} \to \eta(E)$ defined by taking convex combinations is continuous.

PROOF: (a) is clear. For (b), (c) and (d) we have $\mathrm{dist}(n_1 \times \bar{n}_1, n_2 \times \bar{n}_2) \leq \max(\mathrm{dist}(n_1,n_2),\mathrm{dist}(\bar{n}_1,\bar{n}_2))$. $\mathrm{dist}(L^p_\epsilon(n_1;\bar{n}_1), L^p_\epsilon(n_2;\bar{n}_2)) \leq p\,\mathrm{dist}(n_1,n_2) + \mathrm{dist}(\bar{n}_1,\bar{n}_2)$. If $T: E \to F$ is a fixed isomorphism then $\mathrm{dist}(T^*\bar{n}_1, T^*\bar{n}_2) = \mathrm{dist}(\bar{n}_1,\bar{n}_2)$. We also have:

$$\mathrm{dist}(T^*\|\ \|,\|\ \|) = \log \theta(T) \quad \text{and} \quad \rho_{\|\ \|}(T^*\bar{n}) \leq \theta(T)\rho_{\|\ \|}(\bar{n}). \tag{3.1}$$

On the other hand, by the mean value theorem, $|\log(\bar{n}(Tv)) - \log(\bar{n}(Sv))| \leq \max(\bar{n}(Tv)^{-1},\bar{n}(Sv)^{-1})\bar{n}((T-S)v)$ and by (3.1) if $\|v\| = 1$, this is $\leq \rho_{\|\ \|}(\bar{n})^2 \max(\theta(T),\theta(S))\|T - S\|$.

For (e), let $\rho_E(v) = \max(\|v\|, 1)$. We prove (with $\rho_E = \max(\|\ \|,1)$):

$$|n_1(v_1) - n_2(v_2)| \leq [\rho_{\|\ \|}(n_1)\rho_E(v_2)\exp(\mathrm{dist}(n_1,n_2))](\|v_1 - v_2\| + \mathrm{dist}(n_1,n_2)). \tag{3.2}$$

The first term because $|n_1(v_1) - n_1(v_2)| \leq n_1(v_1 - v_2) \leq \rho_{\|\ \|}(n_1)\|v_1 - v_2\|$. The second term because $|n_1(v_2) - n_2(v_2)| = |1 - (n_2(v_2)/n_1(v_2)|(n_1(v_2))$, with $|1 - (n_2(v_2)/n_1(v_2))| \leq |1 - \exp(\mathrm{dist}(n_1,n_2))| \leq \mathrm{dist}(n_1,n_2)\cdot\exp(\mathrm{dist}(n_1,n_2))$ and $|n_1(v_2)| \leq \rho_{\|\ \|}(n_1)\rho_E(v_2)$.

(f) follows from:

$$\begin{aligned} &\mathrm{dist}\left(\sum t_i n_i, \sum t_i \bar{n}_i\right) \leq \max_i(\mathrm{dist}(n_i,\bar{n}_i)) \\ &\mathrm{dist}\left(\sum t_i n_i, \sum s_i n_i\right) \leq (N+1)\max_i(\rho_{\|\ \|}(n_i))^2 \max_i(|s_i - t_i|). \end{aligned} \tag{3.3}$$

The first is easy. For the second let $\|v\| = 1$ and note that for $(t_0,\ldots,t_N) \in \Delta_N$, $\max(\Sigma\, t_i n_i(v), 1/\Sigma\, t_i n_i(v)) \le \max_i(\rho_{\|\ \|}(n_i)) = M$. Hence, $|\log \Sigma\, t_i n_i(v) - \log \Sigma\, s_i n_i(v)| \le M|\Sigma\, t_i n_i(v) - \Sigma\, s_i n_i(v)|$ $\le (N+1)M^2 \max_i |s_i - t_i|$. Take the sup as v varies over S. Q.E.D

Let $(\mathfrak{D},\mathcal{G}) = \{U_\alpha, h_\alpha, \varphi_\alpha\}$ be an $\mathfrak{M}$ atlas for $\pi: E \to X$. Define $\|\ \|^\alpha: \pi^{-1}U_\alpha \to [0,\infty)$ by $\|w\|^\alpha = \|p_2 \circ \varphi_\alpha(w)\|_{F_\alpha}$. So for $x \in X$ we have a family of norms $\{\|\ \|^\alpha : x \in U_\alpha\}$ on E_x. Since $\rho^{\mathcal{C}}_{(\mathfrak{D},\mathcal{G})}(x)$ $\ge \sup\{\theta(\varphi_{\alpha\beta}(h_\beta x)) : x \in U_\alpha \cap U_\beta\}$, it follows from (3.1) that the diameter of this set of norms in $\eta(E_x)$ is $\le \log \rho^{\mathcal{C}}_{(\mathfrak{D},\mathcal{G})}(x)$. Thus, we can define $\|\ \|^M, \|\ \|^m : E \to [0,\infty)$ by $\|w\|^M = \sup\{\|w\|^\alpha : \pi w \in U_\alpha\}$ and $\|w\|^m = \inf\{\Sigma_i \|w_i\|^{\alpha_i}\}$ taking the inf over sequences $(w_1,\alpha_1),\ldots,(w_N,\alpha_N)$ with $\pi w_1 = \ldots = \pi w_N = \pi w \in U_{\alpha_1} \cap \ldots \cap U_{\alpha_N}$ and $\Sigma_i w_i = w$. By Proposition 1 (a), $\|\ \|^M, \|\ \|^m$ are norms on each fiber E_x with $\|\ \|^m \le \|\ \|^\alpha \le \|\ \|^M$ on $\pi^{-1}U_\alpha$ for all α. Furthermore, $\|\ \|^M \le \rho_{(\mathfrak{D},\mathcal{G})}(x)\|\ \|^m$ on E_x or as functions on E:

$$\|\ \|^m \le \|\ \|^M \le (\pi^*\rho_{(\mathfrak{D},\mathcal{G})})\|\ \|^m. \tag{3.4}$$

A Finsler on π is a function on E which restricts to an element of $\eta(E_x)$ on each fiber E_x. Thus, $\|\ \|^M, \|\ \|^m$ are Finslers on π and $\|\ \|^\alpha$ is a Finsler on $\pi|U_\alpha$. For any Finsler $n: E \to [0,\infty)$ and atlas $(\mathfrak{D},\mathcal{G}) = \{U_\alpha, h_\alpha, \varphi_\alpha\}$ the <u>principal part of the Finsler</u> $n_\alpha: h_\alpha(U_\alpha) \to \eta(F_\alpha)$ is given by $n_\alpha(u)(v) = n(\varphi_\alpha^{-1}(u,v))$ for $(u,v) \in h_\alpha(U_\alpha) \times F_\alpha$. A Finsler n is called a <u>continuous Finsler</u> if $\{n_\alpha\}$ is a family of continuous functions for some, and so by Proposition 1(b) for any, VB atlas $(\mathfrak{D},\mathcal{G})$. $\|\ \|^m$ and $\|\ \|^M$ are, in general, not continuous.

3 **PROPOSITION:** (a) If n is a continuous Finsler on $\pi: E \to X$, then n is continuous on E.

(b) Let $(\mathfrak{D},\mathcal{G}) = \{U_\alpha, h_\alpha, \varphi_\alpha\}$ be an $\mathfrak{M}$ atlas on $\pi: E \to X$ and let $\mathcal{P} = \{\varphi_\gamma\}$ be a continuous partition of 1 with supports refining $\{U_\alpha\}$.

$\| \|^{\theta}$ is a continuous Finsler on π with $\| \|^{m} \leq \| \|^{\theta} \leq \| \|^{M}$, where $\|w\|^{\theta} = \Sigma_{\gamma}\varphi_{\gamma}(\pi w)\|w\|^{\alpha(\gamma)}$.

(c) If n_i is a continuous Finsler on $\pi_i: E_i \to X_i$ $(i = 1,2)$, then the induced Finsler on $\pi_1 \times \pi_2$ is continuous. If $X_1 = X_2$, then the induced Finslers on $\pi_1 \oplus \pi_2$ and $L^p_s(\pi_1;\pi_2)$ are continuous.

<u>PROOF</u>: (a): From Proposition 2(e).

(b): For each x choose a neighborhood U of x with $U \cap \text{support } \varphi_{\gamma} \neq \emptyset$ only for $\gamma = \gamma_0,\dots,\gamma_N$ and $U \subset U_{\alpha(\gamma)}$ for $\gamma = \gamma_0,\dots,\gamma_N$. Now use the chart $(U_{\alpha(\gamma_0)}, h_{\alpha(\gamma_0)}, \varphi_{\alpha(\gamma_0)})$ to reduce the result to Proposition 2(f).

(c): From Proposition 2(b) and (c). Q.E.D.

4 <u>LEMMA</u>: If $(\mathfrak{D}_0,\mathfrak{A}_0)$ is a refinement of $(\mathfrak{D},\mathfrak{A})$ an $\| \|^{M_0}, \| \|^{M}$ and the max Finslers, $\| \|^{m_0}, \| \|^{m}$ are the min Finslers defined by $(\mathfrak{D}_0,\mathfrak{A}_0)$, $(\mathfrak{D},\mathfrak{A})$, then $\| \|^{M} \geq \| \|^{M_0} \geq \| \|^{m_0} \geq \| \|^{m}$. If $(\mathfrak{D}_0,\mathfrak{A}_0)$ is a subdivision of $(\mathfrak{D},\mathfrak{A})$ then $\| \|^{M} = \| \|^{M_0}$, $\| \|^{m_0} = \| \|^{m}$.

Using Finslers, we can now study the total space atlas $\mathfrak{D}$ of the VB atlas $(\mathfrak{D},\mathfrak{A})$. First, the metric:

5 <u>PROPOSITION</u>: Let $(\mathfrak{D},\mathfrak{A})$ be an $\mathfrak{M}$ atlas on $\pi: E \to X$ with $\mathfrak{M} \subset \mathcal{L}$ and let $0: X \to E$ be the zero section. $d_{\mathfrak{D}} \geq \pi^* d_{\mathfrak{A}}$ and $d_{\mathfrak{A}} = 0^* d_{\mathfrak{D}}$. If $\pi w_1 = \pi w_2$, then $\|w_1 - w_2\|^{m} \geq d_{\mathfrak{D}}(w_1,w_2)$. If $O \subset E$ is open and for each x, O_x is convex in E_x, $w_1,w_2 \in O$ with $\pi w_1 = \pi w_2$, then $\|w_1 - w_2\|^{\theta} \geq d_{\mathfrak{D}|O}(w_1,w_2)$ for θ any refining partition of 1.

<u>PROOF</u>: Any $\mathfrak{D}$ chain projects by π to an $\mathfrak{A}$ chain and any $\mathfrak{A}$ chain maps by 0 to a $\mathfrak{D}$ chain. In each case length is not increased. If $(\bar{w}_1,\alpha_1),\dots,(\bar{w}_N,\alpha_N)$ is a sequence with $\Sigma_i \bar{w}_i = w_1 - w_2$ then $(w_2,\alpha_1,w_2 + \bar{w}_1,\alpha_2,\dots,\alpha_N,w_2 + \Sigma_i \bar{w}_i)$ is a $\mathfrak{D}$ chain of the same length. In the open convex set case, we must suffice with (w_2,α,w_1) for any

U_α containing $\pi w_1 = \pi w_2$ and average by $\mathcal{P}$. Q.E.D.

6 PROPOSITION: Let $(\mathfrak{D}_1, \mathcal{G}_1)$ be an $\mathfrak{M}$ atlas on $\pi_1: \mathbb{E}_1 \to X_1$ and $\mathcal{P}_1$ be a refining partition of 1. $\mathfrak{D}_1$ is an $\mathfrak{M}$ atlas on $\mathbb{E}_1$ with $\rho_{\mathfrak{D}_1} \leq O^*(\pi_1{}^*\rho_{(\mathfrak{D}_1,\mathcal{G}_1)}, \| \|^{\mathcal{P}_1})$. Let $(\mathfrak{D}_2, \mathcal{G}_2)$ be an $\mathfrak{M}$ atlas on $\pi_2: \mathbb{E}_2 \to X_2$ and $(\phi, f): (\mathfrak{D}_1, \mathcal{G}_1) \to (\mathfrak{D}_2, \mathcal{G}_2)$ be an $\mathfrak{M}$ VB map. $\phi: \mathfrak{D}_1 \to \mathfrak{D}_2$ is an $\mathfrak{M}$ map satisfying $\phi^*\| \|^{M_2} \leq (\pi_1{}^*\rho^{\mathcal{C}}_{(\phi,f)})\| \|^{M_1}$ and $\rho_\phi \leq O^*(\pi_1{}^*\rho_{(\mathfrak{D}_1,\mathcal{G}_1)}, \pi_1{}^*\rho_{(\phi,f)}, \| \|^{\mathcal{P}_1})$.

__PROOF__: We prove the map result as the atlas result is the same with $(\phi, f) =$ identity. The estimate on $\phi^*\| \|^{M_2}$ is easy. Now for $w \in \mathbb{E}_1$ let $K_1 > \|w\|^{\mathcal{P}_1}$, $K_2 > \max(\rho_{(\mathfrak{D}_1,\mathcal{G}_1)}, \rho_{(\phi,f)})(\pi_1 w) = \max(\pi_1 w)$. Let O be the open set $\{\| \|^{\mathcal{P}_1} < K_1\} \cap \pi_1^{-1}\{\max < K_2\}$. By (3.4) $\| \|^{M_1} < K_1 K_2$ on O. So $\varphi_\alpha(O \cap \pi_1^{-1}U_\alpha \cap \phi^{-1}\pi_2^{-1}V_\beta) \subset h_\alpha(U_\alpha \cap f^{-1}V_\beta \cap \{\max < K_2\})$ $\times B^{\| \|_\alpha}(0, K_1K_2)$ and $\psi_\beta \phi \varphi_\alpha^{-1}$ on this set is obtained from $\phi_{\beta\alpha}: h_\alpha(U_\alpha \cap f^{-1}V_\beta \cap \{\max < K_2\}) \to L(F^1_\alpha; F^2_\beta)$, an $\mathfrak{M}$ map of norm $\leq K_2$. by Lemma II.1.9(b) $\psi_\beta \phi \varphi_\alpha^{-1}$ is $\mathfrak{M}$ on $\varphi_\alpha(O)$ with norm $\leq K_2 O^*(K_1K_2)$ $\leq O^*(K_1, K_2)$. Thus, $\rho_\phi(w) \leq O^*(K_1, K_2)$. Let $K_1 \to \|w\|^{\mathcal{P}}$ and $K_2 \to \max(\pi_1 w)$. Q.E.D.

A section $s: X \to \mathbb{E}$ of π is always index preserving as a map $s: \mathcal{G} \to \mathfrak{D}$. By Corollary 1.3, s is an $\mathfrak{M}$ map iff $\tilde{\rho}_s^{\mathfrak{M}} < \infty$ on X. Furthermore,

$$
\begin{aligned}
&s^*\| \|^M \leq \tilde{\rho}_s \\
(3.5)\qquad & \\
&\tilde{\rho}_s \leq \rho_s \leq O^*(\rho_{(\mathfrak{D},\mathcal{G})}, \tilde{\rho}_s).
\end{aligned}
$$

The first inequality is easy and the second one follows from the first one, Corollary 1.3 and Proposition 5.

STANDARD CONSTRUCTIONS: (1) Linear Map Bundles: If $(\mathfrak{D}_1,\mathfrak{D}_2;\mathfrak{a})$ is an $\mathfrak{M}$ atlas two-tuple for $\pi_i: E_i \to X$ $(i = 1,2)$, $(\phi,1): \pi_1 \to \pi_2$ is a VB map with associated section s of $L(\pi_1;\pi_2)$, $(\phi,1)$ is an $\mathfrak{M}$ VB map iff s is an $\mathfrak{M}$ map. In fact, $\tilde{\rho}_{(\phi,1)} = \tilde{\rho}_s$. The analogous result holds for L^n with $n > 1$. (2) Pullbacks: If $(\mathfrak{D},\mathfrak{a})$ is an $\mathfrak{M}$ atlas on $\pi: E \to X$, $\mathfrak{a}_0$ is an $\mathfrak{M}$ atlas on X_0, $f: \mathfrak{a}_0 \to \mathfrak{a}$ is an index preserving map, and $s: \mathfrak{a} \to \mathfrak{D}$ is an $\mathfrak{M}$ section then $f^*s: \mathfrak{a}_0 \to f^*\mathfrak{D}$ is an $\mathfrak{M}$ section and by FS9, $\tilde{\rho}_{f^*s} \leq (f^*\tilde{\rho}_s)O^*(\tilde{\rho}_f)$. (3) Jet Bundles: If $(\mathfrak{D},\mathfrak{a})$ is an $\mathfrak{M}^k$ atlas on $\pi: E \to X$, $s: \mathfrak{a} \to \mathfrak{D}$ is a C^k section, then s is an $\mathfrak{M}^k$ section iff $j^k(s): \mathfrak{a} \to J^k(\mathfrak{D})$ is an $\mathfrak{M}$ section of $J^k(\pi)$. $\tilde{\rho}^{\mathfrak{M}}_{j^k(s)} = \tilde{\rho}^{\mathfrak{M}^k}_s$ because $j^k(s_\alpha) = j^k(s)_\alpha$. (5) Tangent Square: If $(\mathfrak{D},\mathfrak{a})$ is an $\mathfrak{M}^1$ atlas on π, then $(T\mathfrak{D},T\mathfrak{a})$ is an $\mathfrak{M}$ atlas on $T\pi$, $(T\mathfrak{D},\mathfrak{D})$ is an $\mathfrak{M}$ atlas on τ_E and the VB maps, $J,T\pi,S$ are $\mathfrak{M}$ maps with principal parts bounded in $\mathfrak{M}$ norm by 1 because the principal parts are constantly inclusions or projections of factors in products.

4. Metrics and Bounds: A function $\rho: X \to [1,\infty]$ is called a bound if $\rho < \infty$, i.e. $\rho: X \to [1,\infty)$.

1 **DEFINITIONS:** (a) For $\rho_1,\rho_2: X \to [1,\infty]$, $\rho_1 > \rho_2$ means $K\rho_1^n \geq \rho_2$ on X for some constants K, n. $>$ is a partial order and $\sim$ denotes the associated equivalence relation.

(b) For a bound ρ and pseudometrics d_1,d_2 on X, $d_1 >_\rho d_2$ means the following equivalent conditions, where $m = m_\rho$ is the bound $\min(p_1^*\rho,p_2^*\rho)$ on $X \times X$: (1) $Km^n d_1 \geq \min(d_2,1)$ on $X \times X$, for some constants K,n. (2) $Km^n d_1 \geq \min(d_2,\epsilon)$ on $X \times X$, for some constants $K,n,\epsilon > 0$. (3) $Km^n d_1 \geq d_2$ on $\{d_1 < \delta m^{-n_1}\}$ for some constants $K,n,n_1,\delta > 0$. $>_\rho$ is a partial order and $\sim_\rho$ denotes the associate equivalence relation.

(c) Let $\pi: E \to X$ be a map of sets. For $N_1,N_2: E \to [1,\infty]$ and

$\rho: X \to [1,\infty]$, $N_1 >_\rho N_2$ means $K(\pi^*\rho)^n N_1 \geq N_2$ for some constants K, n. $>_\rho$ is a partial order and $\sim_\rho$ denotes the associated equivalence relation.

It is clear that $\rho_1 > \rho_2$ and ρ_1 a bound implies ρ_2 is a bound. In (b), (1) $\Rightarrow$ (2). (2) $\Rightarrow$ (3) with $\delta = \epsilon/K$ and $n_1 = n$. (3) $\Rightarrow$ (1) with K, n in (1) $= \max(K,\delta^{-1})$, $\max(n,n_1)$ from (3). As an example, note that all bounded functions are equivalent to the constant 1.

2 **LEMMA:** (a) If $d_1 >_{\rho_2} d_2$ and $\rho_1 > \rho_2$ then $d_1 >_{\rho_1} d_2$.

(b) If $N_1 >_{\rho_2} N_2$ and $\rho_1 > \rho_2$ then $N_1 >_{\rho_1} N_2$.

(c) If $N_1 >_{\rho_1} N_2$ and $\rho_1 > \rho_2$ then $\max(\pi^*\rho_1, N_1) > \max(\pi^*\rho_2, N_2)$.

Let $f: X_0 \to X$ be a set map.

(d) If $\rho_1 > \rho_2$ on X then $f^*\rho_1 > f^*\rho_2$ on X_0.

(e) If $d_1 >_\rho d_2$ on X then $f^*d_1 >_{f^*\rho} f^*d_2$ on X_0.

(f) If $N_1 >_\rho N_2$ on $\pi: E \to X$ and $(\phi, f): \pi_0 \to \pi$ is a map of projections with $\pi_0: E_0 \to X_0$ then on π_0, $\phi^*N_1 >_{f^*\rho} \phi^*N_2$.

PROOF: (a): If $K_1\rho_1^{n_1} \geq \rho_2$ and $K_2 m_2^{n_2} d_1 \geq \min(d_2,1)$, then $K_2K_1^{n_2} m_1^{n_1+n_2} d_1 \geq \min(d_2,1)$ where $m_i = m_{\rho_i}$ $i = 1,2$. The rest are similar, or easier and are left to the reader. Note that (a) and (b) imply that the orderings $>_\rho$ depend only on the equivalence class of ρ. Q.E.D.

In practice we will require a way of estimating ρ near x from the value of ρ at x. The kind of regularity condition that is needed develops from the following construction: Let d be a pseudometric on X and $\lambda: X \to [0,1]$. Define $\lambda_d(x) = \inf\{\max(d(x,x'),\lambda(x')): x' \in X\}$. Equivalently, define the pseudometric $d' = \max(p_1^*d, p_2^*d_0)$ on $X \times R$ where d_0 is the standard metric on R. Let $F_\lambda = \{(x,t): t \geq \lambda(x)\} \subset X \times [0,1]$. $\lambda_d(x) = d'((x,0),F_\lambda)$. If $\rho: X \to [1,\infty]$ then $\rho^d: X \to [1,\infty]$ is defined by $1/\rho^d = (1/\rho)_d$.

$$\rho^d(x) = \sup\{\min(1/d(x,x'),\rho(x')) : x' \in X\} \tag{4.1}$$
$$= \inf\{K: \rho|B^d(x,K^{-1}) \le K\}.$$

with the convention that 0 and ∞ are reciprocal. The equation is easy to check. Note that the infimum is achieved. It is easy to check that $F_{1/\rho}$ is a closed set disjoint from $X \times 0$ iff ρ is a bound on X, upper semicontinuous with respect to d. More generally, for ρ a bound on X, ρ^d is a bound on X iff each point of X has a ρ bounded, d open neighborhood. More generally, (4.1) implies that ρ^d is bounded on $A \subset X$ iff ρ is bounded on $B^d(A,\epsilon)$ for some $\epsilon > 0$.

3 <u>LEMMA</u>: (a) For a family $\{\rho_\alpha\}$, $(\sup\{\rho_\alpha\})^d = \sup\{\rho_\alpha{}^d\}$.

(b) If $\rho_1 > \rho_2$ (or $\rho_1 \ge \rho_2$) then $\rho_1{}^d > \rho_2{}^d$ (resp. $\rho_1{}^d \ge \rho_2{}^d$). In particular, the $\sim$ class of ρ^d depends only on the $\sim$ class of ρ.

(c) If $d_1 >_\rho d_2$ (or $d_1 \ge d_2$) then $\max(\rho,\rho_1{}^{d_2}) > \rho_1{}^{d_1}$ (resp. $\rho_1{}^{d_2} \ge \rho_1{}^{d_1}$). In particular, the $\sim$ class of ρ^d depends only on the $\sim_\rho$ class of d.

(d) Define $d_\rho(x_1,x_2) = |\rho^{-1}(x_1) - \rho^{-1}(x_2)|$. If $\rho_1 = \rho^d$ then $d \ge d_{\rho_1}$, i.e. $1/\rho^d$ is d-Lipschitz with Lipschitz constant ≤ 1. If $d >_\rho d_\rho$ then $\rho^d \sim \rho$. In particular, $(\rho^d)^d \sim \rho^d$.

(e) If $f: X_0 \to X$, then $f^*(\rho^d) \ge (f^*\rho)^{f^*d}$.

<u>PROOF</u>: (a): From the identity $\min(a,\sup\{b_\alpha\}) = \sup\{\min(a,b_\alpha)\}$.

(b): Since $\rho \ge 1$ we need only consider pairs (x,x') in (4.1) with $d(x,x') < 1$. If $K,n,a \ge 1$ and $b \ge 0$, then $K\min(a,b)^n = \min(Ka^n,Kb^n) \ge \min(a,Kb^n)$. This implies (b) for $>$.

(c): Let $K,n \ge 1$ with $K\, m_\rho^n\, d_1 \ge \min(d_2,1)$. By (4.1) $\rho_1(x') \le \rho_1{}^{d_2}(x)$ for $d_2(x,x') < (\rho_1{}^{d_2}(x))^{-1}$. Since $m_\rho(x,x') \le \rho(x)$, $d_1(x,x') < (K\max[\rho(x),\rho_1{}^{d_2}(x)]^{n+1})^{-1}$ implies $d_2(x,x') < (\rho_1{}^{d_2}(x))^{-1}$ and so $\rho_1(x') \le \rho_1{}^{d_2}(x)$. So by (4.1), $\rho^{d_1}(x) \le K\max[\rho(x),\rho_1{}^{d_2}(x)]^{n+1}$.

(d): That $1/\rho^d = (1/\rho)_d$ is d Lipschitz is clear from the d' definition of λ_d. Now assume $d >_\rho d_\rho$. Let $K, n \geq 1$ such that $d_\rho(x,x_1) \leq Km(x,x_1)^n d(x,x_1) \leq K\rho(x)^n d(x,x_1)$. Hence, $\rho(x_1)^{-1} \geq \rho(x)^{-1} - K\rho(x)^n d(x,x_1)$. So if $d(x,x_1) < (2K\rho(x)^{n+1})^{-1}$, $\rho(x_1)^{-1} \geq (2\rho(x))^{-1}$, i.e. $\rho(x_1) \leq 2\rho(x) \leq 2K\rho(x)^{n+1}$. Thus, $\rho^d \leq 2K\rho^{n+1}$.

(e): Clear. Q.E.D.

We conclude by defining a __regular pseudometric space__ to be a triple (X,d,ρ) with d, ρ a pseudometric and bound on X satisfying $\rho \sim \rho^d$. In general, for $\rho: X \to [1,\infty]$ where $\rho^d \sim \rho$ we say that __ρ is adapted to d__ or (vice-versa). By Lemma 3 this condition is independent of choice of ρ in its equivalence class and d in its $\sim_\rho$ equivalence class. If d,ρ are pseudometric and bound on X then (X,d,ρ^d) is a regular pseudometric space iff ρ^d is a bound, i.e. iff $\rho^d < \infty$, which occurs iff ρ is d locally bounded.

5. __Metric Structures:__ A pair $(\mathfrak{a},\rho)$ is called an __adapted $\mathfrak{M}$ atlas on X__ (or adapted atlas, for short) if:(1) $\mathfrak{a}$ is an $\mathfrak{M}$ atlas on X. (2) $\rho \sim \rho_1$ for some continuous bound ρ_1 on X. (3) $\rho > \rho_{\mathfrak{a}}^{\mathfrak{M}}$. $(\mathfrak{a},\rho)$ is a __regular adapted atlas__ if $\mathfrak{M} \subset \mathcal{L}$ and condition (2) is replaced by:

(2') ρ is a bound on X and $\rho^{d_{\mathfrak{a}}} \sim \rho$. Note that (2) implies $\rho^{d_{\mathfrak{a}}} < \infty$ on X but not necessarily $\rho^{d_{\mathfrak{a}}} \sim \rho$.

$(\mathfrak{D},\mathfrak{a},\rho)$ is called an __adapted $\mathfrak{M}$ atlas on π: $E \to X$__ if:(1) $(\mathfrak{a},\rho)$ is an adapted atlas on X and (2) $\rho > \rho^{\mathfrak{M}}_{(\mathfrak{D},\mathfrak{a})}$. $(\mathfrak{D},\mathfrak{a},\rho)$ is a __regular adapted atlas__ on π if, in addition, $(\mathfrak{a},\rho)$ is a regular adapted atlas on X. For adapted atlases $(\mathfrak{a}_1,\rho_1)$, $(\mathfrak{a}_2,\rho_2)$ on X_1, X_2, $f:(\mathfrak{a}_1,\rho_1) \to (\mathfrak{a}_2,\rho_2)$ is an __$\mathfrak{M}$ map of adapted atlases__ if:(1) $f: \mathfrak{a}_1 \to \mathfrak{a}_2$ is an $\mathfrak{M}$ map. (2) $\rho_1 > f^*\rho_2$. (3) $\rho_1 > \rho^{\mathfrak{M}}(f;\mathfrak{a}_1,\mathfrak{a}_2)$. For adapted atlases $(\mathfrak{D}_1,\mathfrak{a}_1,\rho_1)$, $(\mathfrak{D}_2,\mathfrak{a}_2,\rho_2)$ on π_1,π_2, (ϕ,f): $(\mathfrak{D}_1,\mathfrak{a}_1,\rho_1) \to (\mathfrak{D}_2,\mathfrak{a}_2,\rho_2)$ is an __$\mathfrak{M}$ VB map of__ adapted atlases if:(1) f: $(\mathfrak{A}_1,\rho_1) \to (\mathfrak{A}_2,\rho_2)$ is an $\mathfrak{M}$ map of adapted atlases. (2) (Φ,f): $(\mathfrak{D}_1,\mathfrak{A}_1) \to (\mathfrak{D}_2,\mathfrak{A}_2)$ is an $\mathfrak{M}$ VB map. (3) $\rho_1 >$ $\rho^{\mathfrak{M}}(\Phi,f; \mathfrak{D}_1,\mathfrak{A}_1,\mathfrak{D}_2,\mathfrak{A}_2)$. It is clear that if $(\mathfrak{A},\rho_1)$ is an adapted atlas on X, and $\rho_2 \sim \rho_1$ on X, then $(\mathfrak{A},\rho_2)$ is an adapted atlas on X and the identity maps 1: $(\mathfrak{A},\rho_1) \to (\mathfrak{A},\rho_2)$, 1: $(\mathfrak{A},\rho_2) \to (\mathfrak{A},\rho_1)$ are $\mathfrak{M}$ maps. Similarly, for adapted VB atlases. From Proposition 1.1, it is easy to prove that $\mathfrak{M}$ maps of adapted atlases, on spaces or bundles, are closed under composition.

Adapted atlases $(\mathfrak{A}_1,\rho_1)$ and $(\mathfrak{A}_2,\rho_2)$ on X are equivalent, written $(\mathfrak{A}_1,\rho_1) \sim (\mathfrak{A}_2,\rho_2)$, if the following equivalent conditions hold: (1) $\rho_1 \sim \rho_2$ and $(\mathfrak{A}_1 \cup \mathfrak{A}_2,\rho_1)$ is an adapted atlas. (2) $\rho_1 \sim \rho_2$ and $\rho_1 > \rho^{\mathfrak{M}}_{\mathfrak{A}_1 \cup \mathfrak{A}_2}$. (3) The identity maps 1: $(\mathfrak{A}_1,\rho_1) \to (\mathfrak{A}_2,\rho_2)$ and 1: $(\mathfrak{A}_2,\rho_2) \to (\mathfrak{A}_1,\rho_1)$ are $\mathfrak{M}$ maps. (1), (2) and (3) are equivalent by Lemma 1.2. Equivalence of adapted atlases $(\mathfrak{D}_1,\mathfrak{A}_1,\rho_1)$ and $(\mathfrak{D}_2,\mathfrak{A}_2,\rho_2)$ on π: E $\to$ X, written $(\mathfrak{D}_1,\mathfrak{A}_1,\rho_1) \sim (\mathfrak{D}_2,\mathfrak{A}_2,\rho_2)$, is defined by three completely analogous conditions. $(\mathfrak{D}_1,\mathfrak{A}_1,\rho_1) \sim (\mathfrak{D}_2,\mathfrak{A}_2,\rho_2)$ implies $(\mathfrak{A}_1,\rho_1) \sim (\mathfrak{A}_2,\rho_2)$. In each case, (3) shows that $\sim$ is an equivalence relation. An $\mathfrak{M}$ metric structure on X is an equivalence class of adapted $\mathfrak{M}$ atlases on X. An $\mathfrak{M}$ manifold (or metric manifold) X is a space X together with an $\mathfrak{M}$ metric structure. An $\mathfrak{M}$ metric structure on π: E $\to$ X is an equivalence class of adapted $\mathfrak{M}$ atlases on π. An $\mathfrak{M}$ bundle (or metric bundle) π: E $\to$ X is a vector bundle π together with an $\mathfrak{M}$ metric structure on π. For $\mathfrak{M}$ manifolds X_1, X_2 an $\mathfrak{M}$ map (or metric map) f: $X_1 \to X_2$ is a map f such for $(\mathfrak{A}_1,\rho_1)$, $(\mathfrak{A}_2,\rho_2)$ in the metric structures on X_1, X_2, f: $(\mathfrak{A}_1,\rho_1) \to (\mathfrak{A}_2,\rho_2)$ is an $\mathfrak{M}$ metric map of adapted atlases. For $\mathfrak{M}$ bundles π_1,π_2 an $\mathfrak{M}$ VB map (or metric map) (Φ,f): $\pi_1 \to \pi_2$ is a VB map (Φ,f) such that for $(\mathfrak{D}_1,\mathfrak{A}_1,\rho_1),(\mathfrak{D}_2,\mathfrak{A}_2,\rho_2)$ in the metric structures on π_1,π_2, (Φ,f): $(\mathfrak{D}_1,\mathfrak{A}_1,\rho_1) \to (\mathfrak{D}_2,\mathfrak{A}_2,\rho_2)$ is an $\mathfrak{M}$ VB map of adapted atlases. By (3) this is independent of the choices of atlases in the metric structures.

This defines the categories of $\mathfrak{M}$ manifolds and $\mathfrak{M}$ maps and of $\mathfrak{M}$ bundles and $\mathfrak{M}$ VB maps.

The atlases occurring in a metric structure on X or π are called <u>admissible atlases</u>. The bounds are called <u>admissible bounds</u>. The set of admissible bounds is clearly a $\sim$ equivalence class of bounds on X and any pair consisting of an admissible atlas and an admissible bound is an adapted atlas in the metric structure. X or π is called a <u>bounded</u> metric manifold or bundle if constants are admissible bounds. Any adapted atlas on X, eg. $(\mathcal{G},\rho_{\mathcal{G}})$ for $\mathcal{G}$ an $\mathfrak{M}$ atlas on X, or set of equivalent adapted atlases generates a metric structure on X, namely the equivalence class containing the atlas or atlases. Similarly, for bundles. In particular, an $\mathfrak{M}$ metric structure on $\pi: E \to X$ induces the <u>base space</u> $\mathfrak{M}$ metric structure on X, namely the metric structure generated by the set of $(\mathcal{G},\rho)$'s for $(\mathfrak{D},\mathcal{G},\rho)$ in the metric structure for π. For an $\mathfrak{M}$ manifold X a bundle $\pi: E \to X$ is an <u>$\mathfrak{M}$ bundle over X</u> if π is an $\mathfrak{M}$ bundle whose base space structure on X is the given one. If $\mathfrak{M} \subset \mathfrak{M}_1$ then any $\mathfrak{M}$ metric structure generates an $\mathfrak{M}_1$ metric structure. If f or (ϕ,f) is an $\mathfrak{M}$ map, then it is an $\mathfrak{M}_1$ map with respect to the induced $\mathfrak{M}_1$ structures.

It is clear that any refinement, and a fortiori, any subdivision, of an admissible atlas is admissible. If $(\mathfrak{D},\mathcal{G}) = \{U_\alpha,h_\alpha,\varphi_\alpha\}$ is an admissible atlas on $\pi: E \to X$ and $\mathcal{G}_1 = \{U_\alpha,g_\alpha\}$ is an admissible atlas on X (induced structure, of course) then the atlas $(\mathfrak{D}_1,\mathcal{G}_1)$ on π constructed by the transfer of atlas construction (cf. p. 7) is admissible. This is easily proved using Corollary 1.3 and FS9.

1 LEMMA: Assume $\mathfrak{M} \subset \mathfrak{M}_1$. Let X be an $\mathfrak{M}$ manifold, $\pi_1,\ldots,\pi_n$ be $\mathfrak{M}_1$ bundles over X. There exist atlas n-tuples $(\mathfrak{D}_1,\ldots,\mathfrak{D}_n;\mathcal{G})$ for $\pi_1,\ldots,\pi_n$ with $(\mathfrak{D}_i,\mathcal{G})$ an admissible $\mathfrak{M}_1$ atlas on π_i $(i = 1,\ldots,n)$ and $\mathcal{G}$ an admissible $\mathfrak{M}$ atlas on X.

__PROOF__: Begin with admissible $\mathfrak{M}_1$ atlases $(\bar{\mathfrak{D}}_i, \bar{\mathcal{G}}_i)$ on π_i, $i = 1,\ldots,n$ and $\mathcal{G}$ an admissible $\mathfrak{M}$ atlas on X. By intersecting the n + 1 covers (and taking the product of the n + 1 index sets) we can subdivide the n + 1 atlases and so assume that the open covers for $\bar{\mathcal{G}}_1,\ldots,\bar{\mathcal{G}}_n,\mathcal{G}$ are the same (and the index sets the same). Then for each $i = 1,\ldots,n$ use the transfer of atlas construction to replace $(\bar{\mathfrak{D}}_i, \bar{\mathcal{G}}_i)$ by the atlas $(\mathfrak{D}_i, \mathcal{G})$ on π_i. Q.E.D.

We now return to regularity.

2 __LEMMA__: Assume $\mathfrak{M} \subset \mathcal{L}$. If $(\mathcal{G}_1,\rho_1) \sim (\mathcal{G}_2,\rho_2)$ and $(\mathcal{G}_1,\rho_1)$ is regular, then $(\mathcal{G}_2,\rho_2)$ is regular.

__PROOF__: We know that $\rho_1^{d_{\mathcal{G}_1}} \sim \rho_1 \sim \rho_2$ and want to show $\rho_2^{d_{\mathcal{G}_2}} \sim \rho_2$. By Lemma 4.3 (b) we can assume $\rho_1 = \rho_2 = \rho$. Furthermore, intersecting the open covers for $\mathcal{G}_1$ and $\mathcal{G}_2$ we obtain subdivisions of $\mathcal{G}_1$ and $\mathcal{G}_2$ with the same open cover and index set. By Lemma 2.4 going to a subdivision doesn't change the metrics. We are thus reduced to proving: If $\mathcal{G}_1 = \{U_\alpha, h_\alpha\}$, $\mathcal{G}_2 = \{U_\alpha, g_\alpha\}$ $(\mathcal{G}_1,\rho) \sim (\mathcal{G}_2,\rho)$ and $\rho^{d_{\mathcal{G}_1}} \sim \rho$ then $\rho^{d_{\mathcal{G}_2}} \sim \rho$. Since $\rho > \rho^{d_{\mathcal{G}_1}}$ and $\rho > \rho_{\mathcal{G}_1 \cup \mathcal{G}_2}$ we can choose $K, n \geq 1$ such that $d_1(x,y) < (K\rho(x)^n)^{-1}$ implies $K\rho(x)^n > \max(\rho(y), \rho_{\mathcal{G}_1\cup\mathcal{G}_2}(y))$, where $d_i = d_{\mathcal{G}_i}$ $i = 1,2$. Let $B_1 = B^{d_1}(x, (K\rho(x)^n)^{-1})$ and $B_2 = B^{d_2}(x, (2K^2\rho(x)^{2n})^{-1})$. It suffices to show that $B_2 \subset B_1$ because then $\rho|B_2 < K\rho(x)^n < 2K^2\rho(x)^{2n}$ and so $\rho^{d_2}(x) \leq 2K^2\rho(x)^{2n}$. Let $\bar{B}$ be the closed ball $B^{d_1}[x, (2K\rho(x)^n)^{-1}]$. If $B_2 \not\subset B_1$ then for some $y \in B_1$ there is an $\mathcal{G}_2$ chain $(x_1,\alpha_1,\ldots,\alpha_n,x_{n+1})$ between x and y of $\mathcal{G}_2$ length $< (2K^2\rho(x)^{2n})^{-1}$. By subdividing the $\mathcal{G}_2$ chain if necessary we can assume that for x_t the associated p.l. path and $i = 1,\ldots,n$ the set $\{h_{\alpha_i}(x_t) : i \leq t \leq i+1\}$ lies in an open convex set in $h_{\alpha_i}(U_{\alpha_i})$. Note that $(U_{\alpha_i}, h_{\alpha_i})$ are $\mathcal{G}_1$ charts. Now since $y = x_{n+1} \not\in B_1$ there exists $i < t_1 \leq i+1$ such that $x_t \in B_1$ for all

$1 \leq t \leq t_1$ but $x_{t_1} \notin \bar{B}$, because $\bar{B}$ and $X - B_1$ are disjoint closed sets. The truncated $\mathcal{G}_2$ chain $(x_1,\alpha_1,\ldots,\alpha_i,x_{t_1})$ has $\mathcal{G}_2$ length $< (2K^2\rho(x)^{2n})^{-1}$ and its p.l. $\mathcal{G}_2$ path lies in B_1 on which $\rho^{\mathcal{L}}_{\mathcal{G}_1 \cup \mathcal{G}_2} < K\rho(x)^n$. But $(x_1,\alpha_1,\ldots,\alpha_i,x_{t_1})$ is also an $\mathcal{G}_1$ path and its $\mathcal{G}_1$ length is estimated by: $\|h_{\alpha_j}(x_{j+1}) - h_{\alpha_j}(x_j)\| \leq K\rho(x)^n \|g_{\alpha_j}(x_{j+1}) - g_{\alpha_j}(x_j)\|$ since the $\mathcal{L}$ norm of $h_{\alpha_j} g^{-1}_{\alpha_j} | g_{\alpha_j}(U_{\alpha_j} \cap \{\rho^{\mathcal{L}}_{\mathcal{G}_1 \cup \mathcal{G}_2} < K\rho(x)^n\})$ is bounded by $K\rho(x)^n$. Summing, we get that the $\mathcal{G}_1$ length of the chain $(x_1,\alpha_1,\ldots,\alpha_i,x_{t_1}) \leq K\rho(x)^n$ times the $\mathcal{G}_2$ length of the chain $< (2K\rho(x)^n)^{-1}$. This contradicts $x_{t_1} \notin \bar{B}$. Q.E.D.

An $\mathfrak{M}$ manifold is called a <u>regular $\mathfrak{M}$ manifold</u> (or <u>regular metric manifold</u>) if $\mathfrak{M} \subset \mathcal{L}$ and adapted atlases in the metric structure are regular. Lemma 2 says that if one is regular, then all are. An $\mathfrak{M}$ bundle is regular if its base space is. A metric d on a regular $\mathfrak{M}$ manifold is called an <u>admissible metric</u> if for some $(\mathcal{G},\rho)$ in the metric structure $d \sim_\rho d_{\mathcal{G}}$. Independence of choice of adapted atlas is part of:

3 <u>PROPOSITION</u>: (a) Assume $f: X_1 \to X_2$ is an $\mathfrak{M}$ map of regular $\mathfrak{M}$ manifolds. If d_1,d_2 are admissible metrics on X_1,X_2 and ρ is an admissible bound on X_1, then $d_1 >_\rho f^*d_2$. In particular, the admissible metrics on a regular metric manifold are a $\sim_\rho$ equivalence class.

(b) Conversely, if $f: X_1 \to X_2$ is a map of regular $\mathfrak{M}$ manifolds and for some choices d_i,ρ_i admissible metric and bound on X_i, $i = 1,2$, $\rho_1 > f^*\rho_2$ and $d_1 >_{\rho_1} f^*d_2$, then f is an $\mathcal{L}$ map. If $\mathfrak{M} \subset \mathcal{C}^1$ and f is a C^1 map then these conditions imply f is $\mathcal{C}^1$.

<u>PROOF</u>: (a): For $\mathcal{G}_i$ an admissible atlas on X_i with $d_i \sim_{\rho_i} d_{\mathcal{G}_i}$ $(i = 1,2)$ we have $\rho_1 > \rho = \rho_f^{d_{\mathcal{G}_1}}$ and $\rho_1 > f^*\rho_2$. So to show $d_1 >_{\rho_1} f^*d_2$, it suffices by Lemma 4.2 to show $d_{\mathcal{G}_1} >_\rho f^*d_{\mathcal{G}_2}$. $\rho_f \leq \rho(x)$ on $B^{d_{\mathcal{G}_1}}(x,\rho(x)^{-1})$ and so by Proposition 2.6, $\min(d_{\mathcal{G}_2}(fx,fx_1),1) \leq \rho(x) d_{\mathcal{G}_1}(x,x_1)$. So by symmetry:

$$\min(f^*d_{\mathcal{G}_2},1) \leq m_\rho d_{\mathcal{G}_1} \quad \text{for} \quad \rho = \rho_f^{d_{\mathcal{G}_1}}. \tag{5.1}$$

(b): In general, if $\mathfrak{M} \subset \mathcal{C}^{r+1}$ then f is $\mathcal{C}^{r+1}$ iff f is C^{r+1} and $\mathcal{L}^r$. This is because $\mathcal{C}^{r+1} \subset \mathcal{L}^r$ is an isometric inclusion. Thus for $r = 0$, the $\mathcal{L}$ result implies the $\mathcal{C}^1$ result. For the $\mathcal{L}$ result note that the hypotheses on f and Lemma 4.2 imply $\rho_1 > f^*\rho_2$ and $d_1 >_{\rho_1} f^*d_2$ for any choice of admissible metrics and bounds. Also, f is clearly continuous. Let $\mathfrak{a}_1 = \{U_\alpha, h_\alpha\}$ and $\mathfrak{a}_2 = \{V_\beta, g_\beta\}$ be admissible atlases on X_1, X_2. Choose $\rho_2 \geq \rho_{\mathfrak{a}_2}^{d_{\mathfrak{a}_2}}$ and $\rho_1 \geq f^*\rho_2$. By replacing ρ_1 by some $K\rho_1^n$ if necessary we can assume $m_1 d_1 \geq \min(f^*d_2, 1)$ where $m_1 = m_{\rho_1}$ on $X_1 \times X_1$. By the Metric Estimate, Theorem 2.2, we can choose for every $y \in X_2$ a neighborhood V_y and index $\beta = \beta(y)$ such that $V_y \subset V_\beta$, $g_\beta(V_y)$ is a ball about $g_\beta(y)$ and for y_1, y_2 in V_y, $\|g_\beta y_1 - g_\beta y_2\| \geq d_{\mathfrak{a}_2}(y_1, y_2) \geq \rho_2(y)^{-1}\|g_\beta y_1 - g_\beta y_2\|$. For each $x \in X_1$ we can choose U_x and $\alpha = \alpha(x)$ such that $U_x \subset f^{-1}V_{fx}$ and $h_\alpha(U_x)$ is a ball about $h_\alpha(x)$ with radius $< (2\rho_1^{d_{\mathfrak{a}_1}}(x))^{-1}$. $\tilde{\mathfrak{a}}_1 = \{U_x, h_{\alpha(x)}\}$, $\tilde{\mathfrak{a}}_2 = \{V_y, g_{\beta(y)}\}$ are refinements of $\mathfrak{a}_1, \mathfrak{a}_2$ and $f: \tilde{\mathfrak{a}}_1 \to \tilde{\mathfrak{a}}_2$ is index preserving. We will show that $\rho_1 > \rho^{\mathcal{L}}(f; \tilde{\mathfrak{a}}_1, \tilde{\mathfrak{a}}_2)$. By Corollary 1.3 it suffices to show $\rho_1 > \tilde{\rho}_f$. By Lemma 1.4 $\rho_1 > \rho_f^{\mathcal{C}}$. Thus, we must estimate the norm $\|f_x\|_L$ where $f_x = g_{\beta fx} f h_{\alpha x}^{-1}$ is the principal part with index x. If $u_1, u_2 \in h_\alpha(U_x)$ with $\alpha = \alpha(x)$ then $\rho_1(x)^{-1} > \|u_1 - u_2\| \geq d_{\mathfrak{a}_1}(h_\alpha^{-1}u_1, h_\alpha^{-1}u_2)$ $\geq \min(\rho_1(h_\alpha^{-1}u_1), \rho_1(h_\alpha^{-1}u_2))^{-1} d_{\mathfrak{a}_2}(fh_\alpha^{-1}u_1, fh_\alpha^{-1}u_2) \geq \rho_1^{d_{\mathfrak{a}_1}}(x)^{-1} d_{\mathfrak{a}_2}(fh_\alpha^{-1}u_1, fh_\alpha^{-1}u_2)$ $\geq (\rho_1^{d_{\mathfrak{a}_1}}(x)\rho_2(fx))^{-1}\|f_x u_1 - f_x u_2\|$. Thus, $\|f_x\|_L \leq (\rho_1^{d_{\mathfrak{a}_1}}(x))^2$. Finally, for $y \in U_x$, $\rho^{d_{\mathfrak{a}_1}}(y) < 2\rho^{d_{\mathfrak{a}_1}}(x)$ since $|(\rho^{d_{\mathfrak{a}_1}}(y))^{-1} - (\rho^{d_{\mathfrak{a}_1}}(x))^{-1}|$ $\leq d_{\mathfrak{a}_1}(x,y) < (2\rho^{d_{\mathfrak{a}_1}}(x))^{-1}$. Hence, $\max(4(\rho_1^{d_{\mathfrak{a}_1}})^2, \tilde{\rho}_f^{\mathcal{C}}) \geq \tilde{\rho}_f^{\mathcal{L}}$. But $\rho_1 \sim \max$. Q.E.D.

For an $\mathfrak{M}$ bundle $\pi: E \to X$ a Finsler $\|\ \|$ on π is called an <u>admissible Finsler</u> if for some $(\mathfrak{D}, \mathfrak{a}, \rho)$ in the metric structure and $\|\ \|^M$ the associated max Finsler, $\|\ \| \sim_\rho \|\ \|^M$ as functions on π. Independence of choice of adapted atlas is part of:

4 PROPOSITION: (a) Assume $(\Phi, f): \pi_1 \to \pi_2$ is an $\mathfrak{M}$ VB map of $\mathfrak{M}$ bundles. If $\| \ \|_1, \| \ \|_2$ are admissible Finslers on π_1, π_2 and ρ is an admissible bound on X_1, then $\| \ \|_1 >_\rho \Phi^*\| \ \|_2$. In particular, the admissible Finslers on a metric bundle are a $\sim_\rho$ equivalence class. For every $(\mathfrak{D}, \mathcal{G})$ admissible on π the Finslers $\| \ \|^M, \| \ \|^{\mathcal{C}}, \| \ \|^m$ are admissible.

(b) Conversely, if $(\Phi, f): \pi_1 \to \pi_2$ is a VB map of $\mathfrak{M}$ bundles and for some choices $\| \ \|_i$ admissible Finsler on π_i and ρ_i admissible bound on X_i, $i = 1,2$, $\rho_1 > f^*\rho_2$ and $\| \ \|_1 >_{\rho_1} \Phi^*\| \ \|_2$ then (Φ, f) is a $\mathcal{C}$ metric map.

PROOF: (a): Choose $(\mathfrak{D}_i, \mathcal{G}_i, \rho_i)$ for π_i such that $\| \ \|_i \sim_{\rho_i} \| \ \|^{M_i}$, $i = 1,2$. Since $\rho_1 > \rho^{\mathcal{C}}_{(\Phi,f)}$ it follows from Proposition 3.6 that $\| \ \|^{M_1} >_{\rho_1} \Phi^*\| \ \|^{M_2}$. So $\| \ \|_1 \sim_{\rho_1} \| \ \|^{M_1} >_{\rho_1} \Phi^*\| \ \|^{M_2}$ and $\Phi^*\| \ \|^{M_2} \sim_{f^*\rho_2} \| \ \|_2$. Result by Lemma 4.2. That $\| \ \|^m$ and $\| \ \|^{\mathcal{C}}$ are admissible follows from inequality (3.4).

(b): The condition is independent of the choices of $\| \ \|_i$, ρ_i and so we can choose $(\mathfrak{D}_i, \mathcal{G}_i, \rho_i)$ on π_i, $i = 1,2$, and assume $\| \ \|_1 = \| \ \|^{m_1}$, $\| \ \|_2 = \| \ \|^{M_2}$. So there exist constants K and n such that for $v \in E_1$, $K\rho_1(\pi_1 v)^n \|v\|^{m_1} \geq \|\Phi v\|^{M_2}$ and $K\rho_1(x)^n \geq \rho_{\mathcal{G}_2}(fx)$. By Lemma 1.4, $K\rho_1^n \geq \rho_f^{\mathcal{C}}$ and the definition of the max and min norms shows that $K\rho_1^n \geq \rho^{\mathcal{C}}_{(\Phi,f)}$ too. Q.E.D.

A continuous bound ρ_1 on an $\mathfrak{M}$ manifold X is called a <u>refining bound</u> if $\rho_1 > \rho$ for ρ an admissible bound on X. If ρ_1 is a refining bound then the pairs $(\mathcal{G}, \rho_1)$ for $\mathcal{G}$ admissible on X are adapted atlases and this set of adapted atlases generates an $\mathfrak{M}$ metric structure, the <u>ρ_1 induced refinement</u> of X. If X is regular, then ρ_1 is a <u>regular refining bound</u> or <u>induces a regular refinement</u> if $\rho_1^d \sim \rho_1$ for d an admissible metric on X. The ρ_1-refinement is then regular. It follows from Lemma 4.2 that if d is an admissible metric for X then d is an admissible metric for any regular refinement.

EXAMPLE: If X is a paracompact C^r manifold ($r \geq 1$) then X admits regular $\mathcal{C}^r$ metric structures. As on page 47 we can choose a $\mathcal{C}^r$ atlas $\mathfrak{A}$ and get the metric structure generated by $(\mathfrak{A}, \rho_{\mathfrak{A}}^{d_{\mathfrak{A}}})$. If X is a $\mathcal{C}^r$ metric manifold and $\pi\colon \mathbb{E} \to X$ is a C^r bundle then π admits a $\mathcal{C}^r$ metric bundle structure over some regular refinement of X. For as in the proof of Lemma 1, we can find $(\mathfrak{D}_0, \mathfrak{A}_0)$ a C^r atlas on π with $\mathfrak{A}_0$ an admissible $\mathcal{C}^r$ atlas on X. Subdividing as before we can obtain a $\mathcal{C}^r$ atlas $(\mathfrak{D}, \mathfrak{A})$ with $\mathfrak{A}$ still admissible on X. If ρ is an admissible bound on X, the adapted atlas $(\mathfrak{D}, \mathfrak{A}, \max(\rho, \rho_{(\mathfrak{D},\mathfrak{A})}^{d_{\mathfrak{A}}}))$ generates a $\mathcal{C}^r$ metric structure over a regular refinement of X.

THE STANDARD CONSTRUCTIONS: (1) Open Sets: For U open in an $\mathfrak{M}$ manifold X the class $\{(\mathfrak{A}|U, \rho|U)\}$, with $(\mathfrak{A}, \rho)$ in the X structure, is a class of adapted atlases generating the restriction $\mathfrak{M}$ metric structure on U. If $f\colon X_1 \to X_2$ is an $\mathfrak{M}$ map, U_i is open in X_i ($i = 1,2$) and $f(U_1) \subset U_2$ then $f\colon U_1 \to U_2$ is an $\mathfrak{M}$ map of the restrictions. In particular, the inclusion $U \to X$ is an $\mathfrak{M}$ map. Similarly, $\mathfrak{M}$ bundles restrict to open subsets of the bases. If X is regular then U is regular since $\rho|U > \rho^{d_{\mathfrak{A}}}|U \geq (\rho|U)^{d_{\mathfrak{A}}|U} \geq (\rho|U)^{d_{\mathfrak{A}|U}}$ (Lemma 4.3 and Proposition 2.5). Note that the restriction of an admissible metric d on X to U is usually not an admissible metric on U since $d_{\mathfrak{A}|U} \geq d_{\mathfrak{A}}|U$. However, $\| \|^M, \| \|^m$ on $\pi_U\colon \mathbb{E}_U \to U$ is the same whether defined using $(\mathfrak{D}, \mathfrak{A})|U$ or defined using $(\mathfrak{D}, \mathfrak{A})$ and restricted. Thus, the restriction of an admissible (continuous) Finsler is admissible (and continuous). (2) Total Space: For $\pi\colon \mathbb{E} \to X$ an $\mathfrak{M}$ bundle the class $\{(\mathfrak{D}, \max(\pi^*\rho, \| \|))\}$, with $(\mathfrak{D}, \mathfrak{A}, \rho)$ in the π structure and $\| \|$ an admissible Finsler, is a class of adapted atlases generating the total space $\mathfrak{M}$ metric structure on $\mathbb{E}$. If $(\phi, f)\colon \pi_1 \to \pi_2$ is an $\mathfrak{M}$ metric VB map then $\phi\colon \mathbb{E}_1 \to \mathbb{E}_2$ is an $\mathfrak{M}$ metric map (Proposition 3.6 and Lemma 4.2(c)). From Proposition II.1.12 it is not hard to show, conversely, that if

$\mathfrak{M}$ satisfies FS7(Strong) and $\phi: \mathbb{E}_1 \to \mathbb{E}_2$, $f: X_1 \to X_2$ are $\mathfrak{M}$ maps then (ϕ,f) is an $\mathfrak{M}$ VB map. Clearly, π is an $\mathfrak{M}$ metric map from total space to base. If $s: X \to \mathbb{E}$ is a section of π and $(\mathfrak{D},G,\rho)$ is in the π structure then s is an $\mathfrak{M}$ metric map iff $\rho > \rho_s$ iff $\rho > \tilde{\rho}_s$ (Corollary 1.3 and inequality (3.5)). If $\|\ \|$ is an admissible Finsler, then s is a $\mathcal{C}$ map iff $\rho > \|\ \|\circ s$ (Lemma 1.4). Beware: Even if π is regular, $\mathbb{E}$ might not be. (3) Products: For X_1,X_2 $\mathfrak{M}$ manifolds the class $\{(G_1 \times G_2,\max(p_1{}^*\rho_1,p_2{}^*\rho_2))\}$, for (G_i,ρ_i) in the X_i structure $(i = 1,2)$, is a class of adapted atlases generating the product $\mathfrak{M}$ metric structure on $X_1 \times X_2$. The product is characterized by the property: If $f_i: X_0 \to X_i$ $(i = 1,2)$ are continuous with X_0 an $\mathfrak{M}$ manifold then f_1,f_2 are $\mathfrak{M}$ maps iff the product $f_1 \times f_2: X_0 \to X_1 \times X_2$ is an $\mathfrak{M}$ map. In particular, the projections p_1,p_2 are $\mathfrak{M}$ maps. The product of $\mathfrak{M}$ bundles is defined similarly. The product is regular if the factors are and for d_i an admissible metric on X_i $(i = 1,2)$, $\max(p_1{}^*d_1,p_2{}^*d_2)$ is an admissible metric on $X_1 \times X_2$. Similarly, for vector bundles, $\max(p_1{}^*\|\ \|_1,p_2{}^*\|\ \|_2)$ is an admissible (continuous) Finsler on $\pi_1 \times \pi_2$ when $\|\ \|_i$ is an admissible (continuous) Finsler on π_i $(i = 1,2)$.

(4) B Spaces: The norm metric structure on a B space F is generated by the regular adapted atlas $(G_F,\max(\|\ \|,1))$ where G_F has one chart $= (F,1_F)$(FS2). If U open and bounded in F, E another B space and $f: U \to E$ then $f \in \mathfrak{M}(U,E)$ iff f is an $\mathfrak{M}$ map of the norm metric structures. For U open in F, note that $d_{G_F}|U$ is precisely the intrinsic metric on U of Proposition II.1.1. (5) Trivial Bundles: For X an $\mathfrak{M}$ manifold and F a B space with norm $\|\ \|$ the class $\{(G \times F,G,\rho)\}$, with (G,ρ) in the X structure, is a class of adapted atlases generating the trivial $\mathfrak{M}$ bundle $\epsilon_F: X \times F \to X$. $p_2{}^*\|\ \|$ $(= \|\ \|^M = \|\ \|^m \text{ rel}(G \times F,G))$ is an admissible continuous Finsler for ϵ_F. The total space of ϵ_F is the product of X and F with the norm metric structure. In particular, $f: X \to F$ is metric with respect to the norm

structure on F iff $1_X \times f$ is a metric section of ϵ_F. (6) Tangent Bundle: For X an $\mathfrak{M}^1$ manifold the class $\{(T\mathfrak{A};\mathfrak{A},\rho)\}$, with $(\mathfrak{A},\rho)$ in the X structure, is a class of $\mathfrak{M}$ adapted atlases generating the $\mathfrak{M}$ metric tangent bundle structure on τ_X. If $f: X_1 \to X_2$ is a C^1 map of $\mathfrak{M}^1$ manifolds then f is an $\mathfrak{M}^1$ map iff $(Tf,f): \tau_{X_1} \to \tau_{X_2}$ is an $\mathfrak{M}$ VB map. (7) Subbundles: Let $\pi: \mathbb{E} \to X$ be an $\mathfrak{M}$ bundle and $\mathbb{E}_0 \subset \mathbb{E}$. If there exist admissible atlases $(\mathfrak{D},\mathfrak{A})$ for π inducing subbundle atlases $(\mathfrak{D}|\mathbb{E}_0,\mathfrak{A})$ on $\pi: \mathbb{E}_0 \to X$ then the resulting class $\{(\mathfrak{D}|\mathbb{E}_0,\mathfrak{A},\rho)\}$ is a class of adapted atlases inducing the $\mathfrak{M}$ metric subbundle structure on $\pi|\mathbb{E}_0$. The subbundle structure is characterized by the following property (where $(J,1): \pi|\mathbb{E}_0 \to \pi$ is the inclusion): A VB map $(\Phi,f): \pi_1 \to \pi|\mathbb{E}_0$ is an $\mathfrak{M}$ VB map iff $(J\cdot\Phi,f): \pi_1 \to \pi$ is an $\mathfrak{M}$ VB map. In particular, $(J,1)$ is an $\mathfrak{M}$ VB map. If $\|\ \|$ is an admissible (continuous) Finsler on π then the restriction is an admissible (continuous) Finsler on $\pi|\mathbb{E}_0$ because the max Finslers $\|\ \|^M$ defined by $(\mathfrak{D},\mathfrak{A})$ and $(\mathfrak{D}|\mathbb{E}_0,\mathfrak{A})$ agree on $\mathbb{E}_0$. (8) Bundle Operations: For $\pi_i: \mathbb{E}_i \to X$ $\mathfrak{M}$ bundles $(i = 1,2)$ the classes $\{(\mathfrak{D}_1 \oplus \mathfrak{D}_2,\mathfrak{A},\rho)\}$ and $\{(L^n_\epsilon(\mathfrak{D}_1;\mathfrak{D}_2),\mathfrak{A},\rho)\}$, with $(\mathfrak{D}_1,\mathfrak{D}_2;\mathfrak{A})$ admissible two-tuples for π_1,π_2, are classes of adapted atlases generating the induced $\mathfrak{M}$ metric structures on $\pi_1 \oplus \pi_2$ and $L^n_\epsilon(\pi_1;\pi_2)$. $\pi_1 \oplus \pi_2$ is characterized by the property: If $(\Phi_i,f): \pi_0 \to \pi_i$ $(i = 1,2)$ are VB maps where $\pi_0: \mathbb{E}_0 \to X_0$ is an $\mathfrak{M}$ bundle then $(\Phi_1,f),(\Phi_2,f)$ are $\mathfrak{M}$ VB maps iff $(\Phi_1 \oplus \Phi_2,f): \pi_0 \to \pi_1 \oplus \pi_2$ is. In particular, the projections $(P_i,1): \pi_1 \oplus \pi_2 \to \pi_i$ are $\mathfrak{M}$ maps $(i = 1,2)$. The functors $\_ \oplus \_$ and $L^n_\epsilon(\_;\_)$ take pairs of $\mathfrak{M}$ VB maps over 1_X to $\mathfrak{M}$ VB maps. $(\Phi,1): \pi_1 \to \pi_2$ is an $\mathfrak{M}$ VB map iff the associated section of $L(\pi_1;\pi_2)$ is an $\mathfrak{M}$ section. If $\|\ \|_1$, $\|\ \|_2$ are (continuous) admissible Finslers on π_1,π_2 then $\max(P_1{}^*\|\ \|_1,P_2{}^*\|\ \|_2)$ is a (continuous) admissible Finsler on $\pi_1 \oplus \pi_2$ and the associated operator Finsler is a (continuous) admissible Finsler on $L^n_\epsilon(\pi_1;\pi_2)$. In particular, $(\Phi,1): \pi_1 \to \pi_2$ is a $\mathcal{C}$ VB map iff $\rho > \|\ \|\cdot s$, where $\|\ \|$ is the operator Finsler on

$L(\pi_1;\pi_2)$ and s is the section associated to Φ. (9) Pullbacks: For $\pi: E \to X$ an $\mathfrak{M}$ bundle and $f: X_0 \to X$ an $\mathfrak{M}$ map the class $[(f^*\mathfrak{D},\mathfrak{G}_0,\rho_0)]$, with $(\mathfrak{D},\mathfrak{G})$ admissible for π, $(\mathfrak{G}_0,\rho_0)$ in the X_0 structure and $f: \mathfrak{G}_0 \to \mathfrak{G}$ index preserving, is a class of adapted atlases generating the pullback $\mathfrak{M}$ metric structure on $f^*\pi: f^*E \to X_0$. The pullback structure is characterized by the following property: If $(\Psi,g): \pi_1 \to f^*\pi$ is a VB map with π_1 an $\mathfrak{M}$ bundle then (Ψ,g) is an $\mathfrak{M}$ VB map iff the composition $(\Phi_f\circ\Psi,f\circ g): \pi_1 \to \pi$ is. In particular, (Φ_f,f) is an $\mathfrak{M}$ VB map. If $\| \ \|$ is an admissible Finsler on π then $\Phi_f^*\| \ \|$ is an admissible Finsler on $f^*\pi$. In fact, if $\mathfrak{G}_0 = \{V_\beta,g_\beta\}$, $\mathfrak{G} = \{U_\alpha,h_\alpha\}$ and $\{V_\beta\}$ is a subdivision of $\{f^{-1}U_\alpha\}$ instead of being merely a refinement, i.e. if $f^{-1}U_\alpha = \cup\{V_\beta: \alpha(\beta) = \alpha\}$ and if $\| \ \|^M$, $\| \ \|^{M_0}$ are defined on π, $f^*\pi$ by $(\mathfrak{D},\mathfrak{G})$, $(f^*\mathfrak{D},\mathfrak{G}_0)$, then

$$\| \ \|^{M_0} = \Phi_f^*\| \ \|^M. \tag{5.2}$$

For example, if $(\mathfrak{D},\mathfrak{G}) = \{U_\alpha,h_\alpha,\varphi_\alpha\}$ and $\bar{\mathfrak{G}}_0 = \{V_\gamma,g_\gamma\}$ are arbitrary admissible atlases on π and X_0, then $\mathfrak{G}_0 = \bar{\mathfrak{G}}_f = \{V_\gamma \cap f^{-1}U_\alpha,g_\gamma\}$ indexed by pairs (γ,α) satisfies this condition. (10) Atlas Identifications: All identifications at the atlas level (eg. p. 11) are $\mathfrak{M}$ isomorphisms by Corollary 1.3 and so are identifications in the metric category. (11) Jet Bundles: For $\pi: E \to X$ an $\mathfrak{M}^k$ bundle the class $[(J^k(\mathfrak{D}),\mathfrak{G},\rho)]$, with $(\mathfrak{D},\mathfrak{G},\rho)$ in the π structure, is a class of adapted atlases generating the $\mathfrak{M}$ metric jet bundle structure on $J^k(\pi)$. If $(\Phi,1_X): \pi_1 \to \pi$ is an $\mathfrak{M}^k$ VB map then $(J^k(\Phi),1_X): J^k(\pi_1) \to J^k(\pi)$ is an $\mathfrak{M}$ VB map. If s is a C^k section of π then s is an $\mathfrak{M}^k$ section iff $j^k(s)$ is an $\mathfrak{M}$ section of $J^k(\pi)$. If $f: X_0 \to X$ is an $\mathfrak{M}^k$ map then $(\Psi_f^k,1_{X_0}): f^*J^k(\pi) \to J^k(f^*\pi)$ is an $\mathfrak{M}$ VB map. (12) Tangent Square: For $\pi: E \to X$ an $\mathfrak{M}^1$ bundle the class $[(T\mathfrak{D},T\mathfrak{G},\max(\tau_X^*\rho,\| \ \|)]$, with $(\mathfrak{D},\mathfrak{G},\rho)$ in the π structure and $\| \ \|$ an admissible Finsler on τ_X, is a class of adapted atlases generating the $\mathfrak{M}$ VB structure on $T\pi$ with base space the total space of τ_X. If $(\Phi,f)\colon \pi_1 \to \pi$ is an $\mathfrak{M}^1$ VB map then $(T\Phi,Tf)\colon T\pi_1 \to T\pi$ is an $\mathfrak{M}$ VB map. In the other direction, the vertical tangent bundle is an $\mathfrak{M}$ subbundle of $\tau_{\mathbb{E}}$. Finally, the metric structures on $T\mathbb{E}$ as the total space of $T\pi$ and that of $\tau_{\mathbb{E}}$ agree, for if $(\mathfrak{D},\mathfrak{A}) = \{U_\alpha,h_\alpha,\varphi_\alpha\}$ and ρ are admissible $\mathfrak{M}^1$ atlas and bound on π, then the $\mathfrak{M}$ metric structure is easily seen to be generated in each case by the class $\{(T\mathfrak{D},\bar{\rho})\}$ where $\bar{\rho}(w) = \sup\{\max(\rho(x),\|v\|_\alpha,\|\dot{u}\|_\alpha,\|\dot{v}\|_\alpha)\colon x \in U_\alpha\}$ where $x = \pi\cdot\tau_E w = \tau_X T\pi w$ and $T\varphi_\alpha(w) = (h_\alpha x,v,\dot{u},\dot{v})$.

5 PROPOSITION: (a) Let X_1, X_2 be $\mathfrak{M}$ manifolds and $f\colon X_1 \to X_2$ be a homeomorphism. f is a $\mathcal{C}$ isomorphism iff $\rho_1 \sim f^*\rho_2$ for ρ_1, ρ_2 admissible bounds on X_1,X_2.

(b) Let π_1, π_2 be $\mathfrak{M}$ bundles and $(\Phi,f)\colon \pi_1 \to \pi_2$ be an $\mathfrak{M}$ VB map with Φ bijective and $f^{-1}\colon X_2 \to X_1$ an $\mathfrak{M}$ map. $(\Phi^{-1},f^{-1})\colon \pi_2 \to \pi_1$ is an $\mathfrak{M}$ map iff (Φ^{-1},f^{-1}) is a $\mathcal{C}$ VB map iff $(f^*\Phi,1)\colon \pi_1 \to f^*\pi_2$ is a $\mathcal{C}$ VB isomorphism iff the section s of $L(\pi_1;f^*\pi_2)$ satisfies $\rho_1 > \theta\cdot s$ where θ is defined on $L(\mathbb{E}_1;f^*\mathbb{E}_2)$ by admissible Finslers $\|\ \|_1,\|\ \|_2$ on π_1 and $f^*\pi_2$ (cf. equation I.(1.1)).

(c) Let X_1,X_2 be $\mathfrak{M}^1$ manifolds and $f\colon X_1 \to X_2$ be an $\mathfrak{M}^1$ map which is a homeomorphism. $f^{-1}\colon X_2 \to X_1$ is an $\mathfrak{M}^1$ map iff f^{-1} is a $\mathcal{C}^1$ metric map iff $(f^*Tf,1)\colon \tau_{X_1} \to f^*\tau_{X_2}$ is a $\mathcal{C}$ VB isomorphism.

PROOF: (a): Lemma 1.4. (b): The $\mathfrak{M}$ iff $\mathcal{C}$ result by Proposition II.3.3. The $\mathcal{C}$ equivalence from Proposition 4. (c): The $\mathfrak{M}^1$ iff $\mathcal{C}^1$ result by Proposition II.3.4. Also, $(T(f^{-1}),f^{-1}) = (Tf,f)^{-1}$ and f^{-1} is $\mathcal{C}^1$ iff (Tf^{-1},f^{-1}) is $\mathcal{C}$. Q.E.D.

6 PROPOSITION: Let X be a regular $\mathfrak{M}$ manifold with $\mathfrak{M} \subset \mathcal{C}^1$ and let $\|\ \|$ be a continuous admissible Finsler on τ_X. The pseudo metric $d(x_1,x_2) = \inf\{\int_a^b \|x_t'\|dt\}$ taken over all piecewise C^1 paths x_t with $x_a = x_1$, $x_b = x_2$ is an admissible metric on X.

PROOF: Let $(\mathcal{G},\rho)$ be in the X structure with ρ continuous and define $\| \|^M$ and $\| \|^m$ on τ_X via $(T\mathcal{G},\mathcal{G})$. Since $\| \|^M$, $\| \|^m$ and $\| \|$ are admissible for the $\mathcal{C}^1$ metric structure on τ_X there exist constants $K,n \geq 1$ such that $K\rho(\tau_X v)^n\|v\|^m \geq \|v\|$ and $K\rho(\tau_X v)^n\|v\| \geq \|v\|^M$ for all v in TX.

By definition of the admissible metric $d_{\mathcal{G}}$, $d_{\mathcal{G}}(x_1,x_2)$ $= \inf\{\Sigma_i \int_{t_i}^{t_{i+1}} \|x_t'\|^{\alpha_i} dt\}$ taken over piecewise C^1 paths x_t between x_1,x_2 with partitions $(t_1,\alpha_1,\ldots,\alpha_n,t_{n+1})$ $x_t \in U_{\alpha_i}$ for $t_i \leq t \leq t_{i+1}$. If $d_{\mathcal{G}}(x_1,x_2) < (\rho^{d_{\mathcal{G}}}(x_1))^{-1}$ then looking at piecewise C^1 $\mathcal{G}$-paths with length $< (\rho^{d_{\mathcal{G}}}(x_1))^{-1}$ the path remains in the set $\{\rho \leq \rho^{d_{\mathcal{G}}}(x_1)\}$ and so $K(\rho^{d_{\mathcal{G}}}(x_1))^n\ \Sigma_i \int_{t_i}^{t_{i+1}} \|x_t'\|^{\alpha_i} dt \geq \int_a^b \|x_t'\|dt \geq d(x_1,x_2)$. Taking the infinum over these short $\mathcal{G}$-paths we have $K\rho^{d_{\mathcal{G}}}(x_1)^n\ d_{\mathcal{G}}(x_1,x_2) \geq \min(d(x_1,x_2),1)$. Since $\rho^{d_{\mathcal{G}}} \sim \rho$, symmetry implies $d_{\mathcal{G}} >_\rho d$.

For the reverse inequality, note that if a piecewise C^1 path x_t between x_1 and x_2 remains in the set $\{\rho \leq \rho^{d_{\mathcal{G}}}(x_1)\}$ then we can partition the domain to obtain a piecewise C^1 $\mathcal{G}$-path and as above $K(\rho^{d_{\mathcal{G}}}(x_1))^n \cdot \int_a^s \|x_t'\|dt \geq d_{\mathcal{G}}(x_1,x_s)$. I claim that if $\int_a^b \|x_t'\|dt <$ $(K\rho^{d_{\mathcal{G}}}(x_1)^{n+1})^{-1}$ then the path lies in $\{\rho \leq \rho^{d_{\mathcal{G}}}(x_1)\}$. It then follows that $d(x_1,x_2)<(K\rho^{d_{\mathcal{G}}}(x_1)^{n+1})^{-1}$ implies $K\rho^{d_{\mathcal{G}}}(x_1)^n d(x_1,x_2) \geq d_{\mathcal{G}}(x_1,x_2)$, i.e. $K\rho^{d_{\mathcal{G}}}(x_1)^{n+1}d(x_1,x_2) \geq \min(d_{\mathcal{G}}(x_1,x_2),1)$ and we will be done by symmetry. If the claim is false then let t_0 be the largest $a < t < b$ such that x_s is in $\{\rho \leq \rho^{d_{\mathcal{G}}}(x_1)\}$ for all $a \leq s \leq t$. Then $\rho^{d_{\mathcal{G}}}(x_1)^{-1} >$ $K\rho^{d_{\mathcal{G}}}(x_1)^n d(x_1,x_{t_0}) \geq d_{\mathcal{G}}(x_1,x_{t_0})$. But then there exists $t_0 < t_1 < b$ with $\rho(x_{t_1}) > \rho^{d_{\mathcal{G}}}(x_1)$ but x_{t_1} is still in the open set $B^{d_{\mathcal{G}}}(x_1,\rho^{d_{\mathcal{G}}}(x_1)^{-1})$. This is impossible. Q.E.D.

We conclude by defining <u>$\mathcal{C}^\infty$ metric structures</u>. An adapted atlas $(\mathcal{G},\rho)$ or $(\mathfrak{D},\mathcal{G},\rho)$ is a $\mathcal{C}^\infty$ adapted atlas if it is $\mathcal{C}^r$ for all $r = 1,2,\ldots$. In particular, the atlas is C^∞. Equivalence of atlases is defined as before. Thus, X or π has a $\mathcal{C}^\infty$ structure if it has compatible $\mathcal{C}^r$

structures for $r = 1,\ldots$ and there are atlases which are admissible for all of the $\mathcal{C}^r$ structures at once. A $\mathcal{C}^\infty$ map is a map which is $\mathcal{C}^r$ for all r. $\mathcal{C}^\infty$ implies $\mathfrak{M}$ for any standard function space type $\mathfrak{M}$. All of the constructions described for $\mathfrak{M}$ have obvious analogues for $\mathcal{C}^\infty$. For example, the norm metric structure on a B space and its open subsets are $\mathcal{C}^\infty$ structures.

6. Regular Pseudometric Spaces: A regular pseudometric space is a triple (X,d,ρ) with d,ρ pseudometric and bound on X satisfying $\rho^d \sim \rho$. For example, if X is a regular metric manifold and d,ρ are admissible metric and bound (X,d,ρ) is a regular metric space. For a regular pseudometric space (X,d,ρ) we define the uniformity $\mathfrak{U}_X$ to consist of those subsets U of $X \times X$ satisfying the equivalent conditions: (1) There exists K,n such that $\{(x_1,x_2): d(x_1,x_2) < (K \max(\rho(x_1),\rho(x_2))^n)^{-1}\} \subset U$. (2) There exists K,n such that $\{(x_1,x_2); d(x_1,x_2) < (K \min(\rho(x_1),\rho(x_2))^n)^{-1}\} \subset U$. Clearly, (2) implies (1) with the same K and n. For the other direction, choose $K_1,n_1 \geq 1$ such that $K_1\rho^{n_1} \geq \rho^d$. Now with K,n chosen as in (1) the choice $K_2 = KK_1$ and $n_2 = n + n_1$ works for (2). For if $d(x_1,x_2) < (K_2\rho(x_1)^{n_2})^{-1}$ then $\rho(x_2) \leq K_1\rho(x_1)^{n_1}$ and so $d(x_1,x_2) < (K\rho(x_2)^{n})^{-1}$. To check that $\mathfrak{U}_X$ is really a uniformity (cf. [13; Chap. 6]) it suffices to check the composition property. Given U in $\mathfrak{U}_X$ choose K,n as in (2) and let $V = \{(x_1,x_2): d(x_1,x_2) < (2K \max(\rho(x_1),\rho(x_2))^n)^{-1}\}$. $V \cdot V \subset U$ because $\max(\rho(x_1),\rho(x_2))$ and $\max(\rho(x_2),\rho(x_3))$ are each $\geq \min(\rho(x_1),\rho(x_3))$. The uniformity $\mathfrak{U}_X$ refines the d-uniformity generated by the sets $\{d < K^{-1}\}$ $K > 0$. On any bounded set A of X (i.e. ρ bounded) the uniformity $\mathfrak{U}_X|A \times A$ agrees with the uniformity of $d|A \times A$. In particular, since $\rho \sim \rho^d$, X can be covered by bounded d open (and hence $\mathfrak{U}_X$ open) sets. Thus, the topologies associated to d and $\mathfrak{U}_X$ agree. Since $\mathfrak{U}_X$ clearly has a countable base the uniformity $\mathfrak{U}_X$ is pseudometrizable [13; Thm. 6.13], but the resulting pseudometric is different from (i.e. not uniformly equivalent to) d unless ρ is bounded and we will never use it.

If (X,d,ρ) is a regular pseudometric space and $f\colon X_1 \to X$ is any set map then the pullback $f^*(X,d,\rho) = (X_1, f^*d, f^*\rho)$ is a regular pseudometric space because $f^*\rho > f^*(\rho^d) \geq (f^*\rho)^{f^*d} \geq f^*\rho$. The uniformity $\mathfrak{U}_{f^*X}$ is generated by $f \times f^{-1}\mathfrak{U}_X$. Clearly, $f^*\, g^*(X,d,\rho) = (g\cdot f)^*(X,d,\rho)$. If f is the inclusion of a subset A then the pullback, $(A, d|A, \rho|A)$ is called the <u>restriction</u>.

If (X_i,d_i,ρ_i) $(i = 1,2)$ are regular pseudometric spaces, then the product $(X_1,d_1,\rho_1) \times (X_2,d_2,\rho_2) = (X_1 \times X_2, \max(p_1{}^*d_1, p_2{}^*d_2), \max(p_1{}^*\rho_1, p_2{}^*\rho_2))$ is a regular pseudometric space. The uniformity contains the product uniformity, i.e. if $V_i \in \mathfrak{U}_{X_i}$ then $V_1 \times V_2 = \{((x_1,y_1),(x_2,y_2))\colon (x_1,x_2) \in V_1 \text{ and } (y_1,y_2) \in V_2\} \in \mathfrak{U}_{X_1 \times X_2}$ but $\mathfrak{U}_{X_1 \times X_2}$ is generated by sets $\{((x_1,y_1),(x_2,y_2))\colon d_1(x_1,x_2), d_2(y_1,y_2) \leq (K \min(\max(\rho_1 x_1, \rho_2 y_1), \max(\rho_1 x_2, \rho_2 y_2))^n)^{-1}\}$. A subset A of $X_1 \times X_2$ is called <u>transverse bounded</u> if $\max(p_1{}^*\rho_1, p_2{}^*\rho_2) \sim \min(p_1{}^*\rho_1, p_2{}^*\rho_2)$ on A. Clearly, any bounded set is transverse bounded and any finite union of transverse bounded sets is transverse bounded. If $X_1 = X_2 = X$ and $A_1, A_2 \subset X \times X$ are transverse bounded then A_1^{-1} and $A_1 \cdot A_2$ are. To see this for $A_1 \cdot A_2$ let $K, n \geq 1$ be such that $K \min^n \geq \max$ on A_1 and A_2, i.e. on $A_1 \cup A_2$. If $(x_1,x) \in A_1$ and $(x,x_2) \in A_2$ then $\max(\rho(x_1),\rho(x_2)) \leq \max(\rho(x_1),\rho(x),\rho(x_2)) \leq K\,\rho(x)^n \leq K \max(\rho(x_1),\rho(x))^n \leq K^{n+1}\,\rho(x_1)^{n^2}$ and so $K^{n+1} \min^{n^2} \geq \max$ on $A_1 \cdot A_2$. Note that if $V_0 = \{(x_1,x_2)\colon d(x_1,x_2) < \max(\rho^d(x_1),\rho^d(x_2))^{-1}\}$ then $V_0 \in \mathfrak{U}_X$ and $\max(p_1{}^*\rho, p_2{}^*\rho) \leq \min(p_1{}^*\rho^d, p_2{}^*\rho^d) \sim \min(p_1{}^*\rho, p_2{}^*\rho)$ on V_0. So V_0 is transverse bounded. If $V \in \mathfrak{U}_X$ then $V \cap V_0$ is a transverse bounded member of $\mathfrak{U}_X$. Thus the uniformity is generated by its transverse bounded members. If $V \in \mathfrak{U}_X$ we define $\rho^V \geq \rho \geq \rho_V$ by

$$(6.1) \qquad \rho^V(x) = \sup \rho|V[x] \qquad \rho_V(x) = \inf \rho|V[x]$$

where $V[x] = \{x_1\colon (x,x_1) \in V\}$. V is transverse bounded iff $\rho^V \sim \rho_V$ in which case both are equivalent to ρ.

As an application, note that if $A \subset X$ and $V \in \mathfrak{U}_X$ then $\bigcup_{n=1}^{\infty} V^n[A]$ is open and closed, where $V[A] = \{x_1: (x,x_1) \in V \text{ and } x \in A\}$ $= \bigcup\{V[x]: x \in A\}$. If $V^{-1}[x] \cap V^n[A] \neq \emptyset$ then $x \in V^{n+1}[A]$ and $V[x] \subset V^{n+2}[A]$. If V is transverse bounded and open then each term of this increasing union is bounded and open, if A is bounded. $A \subset X$ is of <u>finite uniform diameter</u> if for $x \in A$ and all $V \in \mathfrak{U}_X$ there exists $n < \infty$ such that $A \subset V^n[x]$. This property is independent of the choice of $x \in A$, in fact, of $x \in \bigcup_n V^n[A]$. Clearly, any compact, connected set is of finite uniform diameter.

To understand completeness in regular pseudometric spaces we define, for d a pseudometric on X, $\varkappa(d): X \to [0,1]$ by $\varkappa(d)(x) =$ $\sup\{r \leq 1: B^d[x;r] \text{ is d complete}\}$.

1 <u>LEMMA</u>: (a) $\varkappa(d)$ is lower semicontinuous (d topology) and $(1/\varkappa(d))^d \sim 1/\varkappa(d)$.

(b) If ρ is a bound on X, d_1, d_2 are pseudometrics with the same topology and $d_1 >_\rho d_2$ then $\max(\rho^{d_2}, 1/\varkappa(d_2)) > \max(\rho^{d_1}, 1/\varkappa(d_1))$.

(c) If $\{x_n\}$ is a d Cauchy sequence in X and $\{\varkappa(d)(x_n)\}$ is bounded away from 0 then $\{x_n\}$ d converges. For $A \subset X$, $\varkappa(d)|A \geq \epsilon$ iff $B^d[A;r]$ is d complete for all $r < \epsilon$.

(d) Let (X,d,ρ) be a regular pseudometric space and $\{x_n\}$ be a sequence in X. $\{x_n\}$ is $\mathfrak{U}_X$ Cauchy iff it is d Cauchy and bounded (i.e. ρ bounded).

<u>PROOF</u>: (a): If $\varkappa(d)(x) - \epsilon = \delta > 0$ for $\epsilon > 0$ then $\varkappa(d) > \epsilon$ on $B^d(x,\delta)$. In particular, $\varkappa(d) > \varkappa(d)(x)/2$ on $B^d(x,\varkappa(d)(x)/2)$ and so $2/\varkappa(d) \geq (1/\varkappa(d))^d$.

(b): If $\infty > K > \max(\rho^{d_2}(x), 1/\varkappa(d_2)(x))$ and $K_1 m_\rho^{n_1} d_1 \geq \min(d_2, 1)$ then $B^{d_2}[x;K^{-1}]$ is d_2 complete and $K_1 K^{n_1} d_1 \geq d_2$ on it. Thus, $B^{d_1}[x;(K_1 K^{n_1+1})^{-1}] \subset B^{d_2}[x;K^{-1}]$ and any d_1 Cauchy sequence in the smaller ball is d_2 Cauchy and so converges. Thus, $K_1 \max_2^{n_1+1} \geq 1/\kappa(d_1)$.

(c): If $\kappa(d)(x_n) > \epsilon$ for all n and $d(x_m,x_n) < \epsilon$ for $m,n \geq N$ then $\{x_n\}$ for $n \geq N$ is Cauchy in $B^d[x_N;\epsilon]$ which is complete. This together with the proof of (a) implies the result for $A \subset X$.

(d): Since $\mathfrak{U}_X$ and the d-uniformity agree on bounded sets, it suffices to show that any $\mathfrak{U}_X$ Cauchy sequence $\{x_n\}$ is bounded. Let V be a transverse bounded element of $\mathfrak{U}_X$ and let $(x_n,x_n) \in V$ for $n,m \geq N$. Then $\infty > \rho^V(x_N) \geq \rho(x_n)$ for all $n \geq N$. Q.E.D.

Notice that since $\mathfrak{U}_X$ is metrizable, $\mathfrak{U}_X$ is complete iff every $\mathfrak{U}_X$ Cauchy sequence converges. A regular pseudometric space (X,d,ρ) is called <u>uniformly complete</u> if the following equivalent conditions hold: (1) X is $\mathfrak{U}_X$ complete. (2) Every closed, bounded set is d complete. (3) $\rho^d \geq 1/\kappa(d)$. (4) $\rho > 1/\kappa(d)$. (1) implies (2) and (3) implies (4) are clear. If (2) holds and $0 < \epsilon < 1/\rho^d(x)$ then $B^d[x;\epsilon]$ is closed and bounded and so $\kappa(d) \geq \epsilon$. So (3) holds. If (4) holds, let $\{x_n\}$ be a $\mathfrak{U}_X$ Cauchy sequence. By Lemma 1(d) $\{x_n\}$ is bounded. By (4) $\{\kappa(d)(x_n)\}$ is bounded away from 0. By Lemma 1(c) $\{x_n\}$ converges. Thus, (1) holds.

Let (X_i,d_i,ρ_i) be regular pseudometric spaces $i = 1,2$ and let $f\colon X_1 \to X_2$. f is a <u>$\mathcal{C}$ map</u> if $\rho_1 > f^*\rho_2$ and f is continuous with respect to the d_i topology on X_i, $i = 1,2$. f is a <u>$\mathcal{C}_u$ map</u> if $\rho_1 > f^*\rho_2$ and $(f \times f)^{-1}\mathfrak{U}_{X_2} \subset \mathfrak{U}_{X_1}$ i.e. for every K_2,n_2 there exists K_1,n_1 such that $d_1(x,y) < (K_1 m_1(x,y)^{n_1})^{-1}$ implies $d_2(fx,fy) < (K_2 m_2(fx,fy)^{n_2})^{-1}$ where $m_i(a,b) = m_{\rho_i}(a,b) = \min(\rho_i(a),\rho_i(b))$, $i = 1,2$. f is a <u>$\mathcal{L}_a$ map</u> for some $0 < a \leq 1$ if $\rho_1 > f^*\rho_2$ and there exists K,n such that $K m_1^n d_1^a \geq \min(f^*d_2,1)$ or equivalently, $K m_1^n \min(d_1,1)^a \geq \min(f^*d_2,1)$. We will write $\mathcal{L}$ for $\mathcal{L}_1$. f is an <u>$\mathcal{L}$ map</u> if $\rho_1 > f^*\rho_2$ and $d_1 >_{\rho_1} f^*d_2$. If f is $\mathcal{L}_a$ then f is $\mathcal{L}_b$ for $0 < b \leq a$ since $\min(d_1,1)^a \leq \min(d_1,1)^b$ and f is $\mathcal{C}_u$ since $d_1(x,y) < [(KK_1)^{1/a} m_1(x,y)^{(n+n_1)/a}]^{-1}$ implies $d_2(fx,fy) < (K_1 m_1(x,y)^{n_1})^{-1}$ and $m_1 > (f \times f)^*m_2$. The classes of

$\mathcal{C}$, $\mathcal{C}_u$ and $\mathcal{L}$ maps are closed under composition and the composition of an $\mathcal{L}_a$ map and an $\mathcal{L}_b$ map is $\mathcal{L}_{ab}$. f is $\mathcal{C}$, $\mathcal{C}_u$, or $\mathcal{L}_a$ iff the identity $1: (X_1,d_1,\rho_1) \to f^*(X_2,d_2,\rho_2)$ is $\mathcal{C}$, $\mathcal{C}_u$ or $\mathcal{L}_a$. If (X,d_1,ρ_1) and (X,d_2,ρ_2) are regular pseudometric spaces with the same underlying set then the identity maps are $\mathcal{L}$ iff $\rho_1 \sim \rho_2$ and $d_1 \sim_{\rho_1} d_2$. The identities are $\mathcal{C}_u$ iff $\rho_1 \sim \rho_2$ and the uniformities on X agree. In the latter case we will say (d_1,ρ_1) and (d_2,ρ_2) are uniformly equivalent.

A family of regular pseudometric spaces $\{X_\alpha,d_\alpha,\rho_\alpha\}$ is called a <u>regular family</u> if $\rho_\alpha > \rho_\alpha^{d_\alpha}$ uniformly in α, i.e. there exist K,n such that $K\rho_\alpha^n \geq \rho_\alpha^{d_\alpha}$. Clearly, if the index set is finite then the family is regular. Also, if $(X_\alpha,d_\alpha,\rho_\alpha) = f_\alpha^*(X,d,\rho)$ for some family of set maps $\{f_\alpha: X_\alpha \to X\}$ then $\{X_\alpha,d_\alpha,\rho_\alpha\}$ is regular. In particular, if the family consists of restrictions to subsets of a fixed space then it is regular. A family of maps $\{f_\alpha: (X_\alpha^1,d_\alpha^1,\rho_\alpha^1) \to (X_\alpha^2,d_\alpha^2,\rho_\alpha^2)\}$ is a $\mathcal{C}$, $\mathcal{C}_u$, or $\mathcal{L}_a$ family if the domain and range are regular families, $\rho_\alpha^1 > f^*\rho_\alpha^2$ uniformly in α and in the $\mathcal{C}_u$ and $\mathcal{L}_a$ cases the choices of constants can be made independent of α, i.e. for $\mathcal{C}_u$, K_1,n_1 can be chosen to depend only on K_2,n_2 not on α and for $\mathcal{L}_a$ K,n can be chosen independent of α. The composition results for maps of families are analogous to those for maps of spaces. As an example, any B space F with norm $\| \ \|$ is a regular metric space $(F,\| \ \|,\max(\| \ \|,1))$ denoted by F (where by $\| \ \|$ in the second spot we mean the norm metric) and any family of B spaces F_α is a regular family. If $f_\alpha: (X,d,\rho) \to F_\alpha$ then $\{f_\alpha\}$ is a $\mathcal{C}_u$ or $\mathcal{L}_a$ family iff the product $\hat{\Pi} f_\alpha: (X,d,\rho) \to \hat{\Pi} F_\alpha$ is a $\mathcal{C}_u$ or $\mathcal{L}_a$ map. If $f: (X_1,d_1,\rho_1) \to (X_2,d_2,\rho_2)$ is a $\mathcal{C}$, $\mathcal{C}_u$ or $\mathcal{L}_a$ map and $\{A_\alpha\}$ is a family of subsets of X_1 then $\{f|A_\alpha: (A_\alpha,d_1|A_\alpha,\rho_1|A_\alpha) \to (X_2,d_2,\rho_2)\}$ is a $\mathcal{C}$, $\mathcal{C}_u$ or $\mathcal{L}_a$ family.

If (X,d,ρ) is a regular pseudometric space then $1/\rho^d: (X,d,\rho) \to R$ is clearly $\mathcal{L}$ by Lemma 4.3(c). Thus, $\rho^d(x_1)\rho^d(x_2)d(x_1,x_2) \geq |\rho^d(x_1) - \rho^d(x_2)|$. Now if $d(x_1,x_2) < (2\, m_{\rho^d}(x_1,x_2))^{-1}$ then

$\max(\rho^d(x_1),\rho^d(x_2)) \leq 2\min(\rho^d(x_1),\rho^d(x_2)) = m_{\rho^d}(x_1,x_2)$. Thus $2\, m_{\rho^d}(x_1,x_2)^2 d(x_1,x_2) \geq \min(|\rho^d(x_1) - \rho^d(x_2)|,1)$. Since $\rho \sim \rho^d$ it follows that $\rho^d\colon (X,d,\rho) \to R$ is an $\mathscr{L}$ map and if $\{(X_\alpha,d_\alpha,\rho_\alpha)\}$ is a regular family $\{\rho_\alpha^{d_\alpha}\colon (X_\alpha,d_\alpha,\rho_\alpha) \to R\}$ is an $\mathscr{L}$ family.

We now turn to uniform neighborhoods: Define for d a pseudometric on X and $U \subset X$, $\delta^U(d)\colon X \to [0,1]$ by $\delta^U(d)(x) = \min(d(x,X-U),1) = \sup\{r \leq 1\colon B^d(x,r) \subset U\}$. $\delta^U(d) > 0$ on the d-interior of U in X and hence on U if U is d-open in X. Thus, $1/\delta^U(d)$ is a bound on the d-interior of U in X.

2 **LEMMA**: (a) If C_U is the characteristic function of U in X then $\delta^U(d) = (C_U)_d$. Hence, $\delta^U(d)$ is d-continuous and $(1/\delta^U(d))^d \sim 1/\delta^U(d)$.

(b) If d_1,d_2 are pseudometrics on X with $d_1 >_\rho d_2$ then $\max(\rho,1/\delta^U(d_2)) > \max(\rho,1/\delta^U(d_1))$.

PROOF: $0 \leq (C_U)_d(x) = \inf\{\max(d(x,x_1),C_U(x_1))\colon x_1 \in X\} \leq C_U(x) \leq 1$. Hence, $(C_U)_d(x) = C_U(x) = \delta^U(d)(x) = 0$ for $x \notin U$ and for $x \in U$, $(C_U)_d(x) = \inf\{\max(\min(d(x,x_1),1),C_U(x_1))\colon x_1 \in X\} = \min(1,\inf\{d(x,x_1)\colon x_1 \notin U\}) = \delta^U(d)(x)$. The rest follows from Lemma 4.3. Q.E.D.

Thus, if $\lambda\colon X \to [0,1]$ with $\lambda = 0$ on $X - U$ then $\lambda \leq C_U$ and so $(1/\lambda)^d \geq 1/\delta^U(d)$. If $\lambda \geq \delta^U(d)$ or even $1/\delta^U(d) > 1/\lambda$ then it follows that $(1/\lambda)^d \sim 1/\delta^U(d)$.

Let (X,d,ρ) be a regular pseudometric space and $A \subset U \subset X$. U is a <u>full neighborhood of A in X</u> denoted $A \subset\subset U$ rel X or $A \subset\subset U$ if the following equivalent conditions hold: (1) $\max(\rho,1/\delta^U(d)) \sim \rho$ on A. (2) $V[A] \subset U$ for some $V \in \mathfrak{U}_X$. (3) For some constants K,n, $B^d(x,(K\rho(x)^n)^{-1}) \subset U$ for all $x \in A$. Clearly if $A \subset\subset U$ then $A \subset\subset$ d-interior of U.

3 **LEMMA**: Let (X_i,d_i,ρ_i) be regular pseudometric spaces $(i = 1,2)$ and

$f: X_1 \to X_2$ be a $\mathcal{C}_u$ map.

(a) If A is transverse bounded in $X_1 \times X_2$ and $A \subset\subset U$ rel $X_1 \times X_2$ then there exist $V_i \in \mathfrak{U}_{X_i}$ $(i = 1,2)$ such that $V_1 \times V_2[A] \subset U$.

(b) $V \in \mathfrak{U}_{X_1}$ iff V is a full neighborhood of the diagonal Δ in $X_1 \times X_1$.

(c) If $A \subset\subset U$ then $X_1 - U \subset\subset X_1 - A$.

(d) If $A \subset\subset U_1$ and $A \subset\subset U_2$ then $A \subset\subset U_1 \cap U_2$.

(e) If $A \subset\subset U$ then there exists U_1 with $A \subset\subset U_1$ and $U_1 \subset\subset U$.

(f) If A is transverse bounded in $X_1 \times X_1$ and $A \subset\subset U$ then $V \cdot A \cdot V \subset U$ for some $V \in \mathfrak{U}_{X_1}$. For any $V \in \mathfrak{U}_{X_1}$ and any $A \subset X_1 \times X_1$, $A \subset\subset V \cdot A \cdot V$.

(g) If $A \subset\subset U$ rel X_2 then $f^{-1}A \subset\subset f^{-1}U$ rel X_1.

(h) If $A \subset X_1$, $U_2 \subset X_2$ and $f(A) \subset\subset U_2$ rel X_2 then there exists U_1 with $A \subset\subset U_1 \subset\subset f^{-1}(U_2)$ rel X_1.

PROOF: (a), (b) and (f): For some K, n $d_1(x_1,x_2), d_2(y_1,y_2) < (K \max(\rho_1(x_1),\rho_2(y_1))^n)^{-1}$ and $(x_1,y_1) \in A$ implies $(x_2,y_2) \in U$. By transverse boundedness there thus exist K_1 and n_1 such that $d_1(x_1,x_2), d_2(y_1,y_2) < (K_1 \min(\rho_1(x_1),\rho(y_1))^{n_1})^{-1}$ and $(x_1,y_1) \in A$ imply $(x_2,y_2) \in U$. Let $V_1 = \{(x_1,x_2): d_1(x_1,x_2) < (K\rho_1(x_1)^n)^{-1}\}$ and $V_2 = \{(y_1,y_2): d_2(y_1,y_2) < (K\rho_2(y_1)^n)^{-1}\}$. Clearly, $V_1 \times V_2[A] \subset U$. Note that if $X_1 = X_2$ and $V_1, V_2 \in \mathfrak{U}_{X_1}$ then $V_1 \times V_2[A] = V_1^{-1} \cdot A \cdot V_2$. Since we can shrink V_1, V_2 to get $V_1 = V_1^{-1} = V_2$ this proves (f). The diagonal is transverse bounded and $V_1 \times V_2[\Delta] = V_1^{-1} \cdot V_2$. If $V \in \mathfrak{U}_{X_1}$ then V contains $V_1 \cdot V_1$ for some $V_1 \in \mathfrak{U}_{X_1}$ with $V_1 = V_1^{-1}$. So $V_1 \times V_1[\Delta] \subset V$ and $\Delta \subset\subset V$. Conversesly, if $\Delta \subset\subset V$, (a) implies $V_1 \cdot V_1 = V_1 \times V_1[\Delta] \subset V$ for some $V_1 \in \mathfrak{U}_{X_1}$ with $V_1 = V_1^{-1}$ and so $V \in \mathfrak{U}_{X_1}$.

(c): $V[A] \subset U$ iff $V^{-1}[X_1 - U] \subset X_1 - A$.

(d): Obvious. Note, in fact:

$$\delta^{U_1 \cap U_2}(d) = \min(\delta^{U_1}(d), \delta^{U_2}(d)). \tag{6.2}$$

(e): If $V_1[A] \subset U$ and $V_2 \cdot V_2 \subset V_1$ with $V_1, V_2 \in \mathfrak{U}_{X_1}$ then $V_2[V_2[A]] = V_2 \cdot V_2[A] \subset U$. Let $V = V_2[A]$ choosing V_2 open in $X_1 \times X_1$.

(g): Follows from $f^{-1}(V[A]) \supset (f \times f)^{-1} V\,[f^{-1}A]$. Note that if f is $\mathcal{L}$ we obtain from Lemma 1 (b) and an easy check:

$$\max(\rho, 1/\delta^{f^{-1}U}(f^*d_1)) \succ 1/\delta^{f^{-1}U}(d). \tag{6.3}$$

$$\delta^{f^{-1}U}(f^*d) \geq f^*\delta^{U}(d).$$

(h): Choose W d_1-open with $f(A) \subset\subset W \subset\subset U$ rel X_1 and let $V = f^{-1}W$. Q.E.D.

4 <u>COROLLARY</u>: Let (X_i, d_i, ρ_i) be regular pseudometric spaces with A_i bounded in X_i $(i = 1,2)$. If $A_1 \times A_2 \subset\subset U$ rel $X_1 \times X_2$ then there exist U_i open in X_i with $A_i \subset\subset U_i$ rel X_i $(i = 1,2)$ and $U_1 \times U_2 \subset\subset U$ rel $X_1 \times X_2$.

<u>PROOF</u>: $A_1 \times A_2$ is bounded and hence transverse bounded in $X_1 \times X_2$. By Lemma 3(a) and (e) there exist $V_i \in \mathfrak{U}_{X_i}$ $(i = 1,2)$ such that $V_1[A_1] \times V_2[A_2] = V_1 \times V_2[A_1 \times A_2] \subset\subset U$. Let $U_i = V_i[A_i]$. Q.E.D.

If $\mathfrak{U} = \{U_\alpha\}$ is a family of subsets of X and d is a pseudometric on X define $\delta_{\mathfrak{U}}(d): X \to [0,1]$ by $\delta_{\mathfrak{U}}(d)(x) = \sup\{r \leq 1: B^d(x;r) \subset U_\alpha$ for some $U_\alpha \in \mathfrak{U}\} = \sup_\alpha \delta^{U_\alpha}(d)(x)$. Clearly, $\delta_{\mathfrak{U}}(d) > 0$ on A if the d-interiors of the U_α's cover A. If $\inf \delta_{\mathfrak{U}}(d)|A = \epsilon > 0$ then ϵ is a Lebesgue number for $\mathfrak{U}|A = \{U_\alpha \cap A\}$.

5 <u>LEMMA</u>: (a) $\delta_{\mathfrak{U}}(d)$ is d-lower semicontinuous and $(1/\delta_{\mathfrak{U}}(d))^d \sim 1/\delta_{\mathfrak{U}}(d)$.

(b) If d_1, d_2 are pseudometrics on X with $d_1 \succ_\rho d_2$ then $\max(\rho, 1/\delta_{\mathfrak{U}}(d_2)) \succ \max(\rho, 1/\delta_{\mathfrak{U}}(d_1))$.

<u>PROOF</u>: This is easily proved directly or from Lemma 2. Q.E.D.

Let (X, d, ρ) be a regular pseudometric space and $\mathfrak{U} = \{U_\alpha\}$ be a family of subsets of X. $\mathfrak{U}$ is a <u>uniform cover</u> of X if the following equivalent conditions hold: (1) There exists $V \in \mathfrak{U}_X$ such that $\{V[x]: x \in X\}$ refines $\mathfrak{U}$. (2) $\rho > 1/\delta_{\mathfrak{U}}(d)$. (3) There exist constants K,n such that $\{B^d(x,(K\rho(x)^n)^{-1}): x \in X\}$ refines $\mathfrak{U}$.

6 LEMMA: Let $(X_1,d_1,\rho_1),(X,d,\rho)$ be regular pseudometric spaces, $\mathfrak{U}_1 = \{U_\alpha\}$ and $\mathfrak{U}_2 = \{V_\beta\}$ be uniform covers of X, $f: X_1 \to X$ be $\mathcal{C}_u$ and $A \subset X$ with inclusion map i_A.

(a) There exist constants K,n such that for all A bounded in X, $[K(\sup \rho|A)^n]^{-1}$ is a Lebesgue number for $\mathfrak{U}|A$.

(b) $\mathfrak{U}_1 \cap \mathfrak{U}_2 = \{U_\alpha \cap V_\beta\}$ is a uniform cover of X.

(c) $f^{-1}\mathfrak{U}_1 = \{f^{-1}U_\alpha\}$ is a uniform cover of X_1.

(d) $V_{\mathfrak{U}} = \cup_\alpha U_\alpha \times U_\alpha$ is an element of $\mathfrak{U}_X$.

(e) If $\mathfrak{U}$ is a uniform cover of A then there exists $\mathfrak{U}_1$ a uniform cover of X with $i_A^{-1}\mathfrak{U}_1 = \mathfrak{U}$.

PROOF: (a) is obvious.

(b) is a consequence of:

$$\delta_{\mathfrak{U}_1 \cap \mathfrak{U}_2}(d) = \min(\delta_{\mathfrak{U}_1}(d), \delta_{\mathfrak{U}_2}(d)). \tag{6.4}$$

(c) because $f^{-1}V[fx] \supset (f \times f^{-1}V)[x]$. If f is $\mathcal{L}$ we have

$$\max(\rho, 1/\delta_{f^{-1}\mathfrak{U}}(f^*d)) > 1/\delta_{f^{-1}\mathfrak{U}}(d_1)$$
$$\delta_{f^{-1}\mathfrak{U}}(f^*d) \geq f^*\delta_{\mathfrak{U}}(d), \tag{6.5}$$

by Lemma 5(b) and an easy check.

(d): If $\{V[x]\}$ refines $\mathfrak{U}_1$ then $V^{-1}\circ V = \cup_x V[x] \times V[x] \subset \cup_\alpha U_\alpha \times U_\alpha$.

(e): If $\mathfrak{U} = \{A_\alpha\}$ and $V \in \mathfrak{U}_X$ with $\{(V \cap A \times A)[x]: x \in A\}$ refining $\mathfrak{U}$ then let $U_\alpha = \text{Int}\,\cup\{V[x]: V[x] \cap A \subset A_\alpha\} \cup (X - A) \cup A_\alpha$ and $\mathfrak{U}_1 = \{U_\alpha\}$. $i_A^*\mathfrak{U}_1 = \mathfrak{U}$. If $W \in \mathfrak{U}_X$ is symmetric and open with $W\circ W \subset V$ then $\{W[x]\}$ refines $\mathfrak{U}_1$. For if $W[x] \cap A = \emptyset$ then $W[x] \subset X - \bar{A} \subset U_\alpha$ for all α and if $x_1 \in W[x] \cap A$ and $V[x_1] \cap A \subset A_\alpha$ then $W[x] \subset W\circ W[x_1] \subset V[x_1] \subset U_\alpha$. Q.E.D.

Note that if $V = V_{\mathfrak{A}} = \bigcup_\alpha U_\alpha \times U_\alpha$ for $\{U_\alpha\}$ a uniform cover of X,
$\rho^V(x) = \sup\{\rho(x_1): \{x,x_1\} \subset U_\alpha$ for some $\alpha\}$ and $\rho_V(x) = \inf\{\rho(x_1): \{x,x_1\}$
$\subset U_\alpha$ for some $\alpha\}$. Thus, $V_{\mathfrak{A}}$ is transverse bounded iff $\rho^V \sim \rho_V \sim \rho$. The
set $V_{\mathfrak{A}}[x] = \bigcup\{U_\alpha: x \in U_\alpha\}$ is called the $\underline{\mathfrak{A}\text{-star of } x}$.

The following is the analogue of the Gluing Property for $\mathcal{L}_a$ and
$\mathcal{C}_u$:

7 PROPOSITION: Let (X,d,ρ) be a regular pseudometric space, $\{X_\alpha,d_\alpha,\rho_\alpha\}$
be a regular family, $\{A_\alpha\}$ a collection of subsets of X, $\{V_\beta\}$ a uniform
cover of X and $\{f_\alpha: A_\alpha \to X_\alpha\}$ a family of maps. The family $\{f_\alpha: A_\alpha \to X_\alpha\}$
is $\mathcal{C}_u$ or $\mathcal{L}_a$ iff the family indexed by pairs (α,β) $\{f_\alpha: A_\alpha \cap V_\beta \to X_\alpha\}$
is $\mathcal{C}_u$ or $\mathcal{L}_a$.

PROOF: Since $\{V_\beta\}$ covers X, $\rho|A_\alpha > f_\alpha^*\rho_\alpha$ uniformly iff $\rho|A_\alpha \cap V_\beta$
$> f_\alpha^*\rho_\alpha|A_\alpha \cap V_\beta$ uniformly. Now if $\{f_\alpha\}$ is $\mathcal{C}_u$ or $\mathcal{L}_a$ it is clear that
$\{f_\alpha|A_\alpha \cap V_\beta\}$ is $\mathcal{C}_u$ or $\mathcal{L}_a$. We prove the converse for $\mathcal{C}_u$. The proof for
$\mathcal{L}_a$ is similar. Given K_2,n_2 there exists K_1,n_1 such that $d(x_1,x_2)$
$< (K_1 m(x_1,x_2)^{n_1})^{-1}$ and $(x_1,x_2) \in V_\beta \times V_\beta$ for some β implies
$d_\alpha(f_\alpha x_1, f_\alpha x_2) < (K_2 m_\alpha(f_\alpha x_1, f_\alpha x_2)^{n_2})^{-1}$. But $V = \bigcup_\beta V_\beta \times V_\beta \in \mathfrak{U}_X$ so there
exists K_0,n_0 such that $d(x_1,x_2) < (K_0 m(x_1,x_2)^{n_0})^{-1}$ implies $(x_1,x_2) \in V$.
$\max(K_1,K_0), \max(n_1,n_0)$ works for the family $\{f_\alpha\}$.   Q.E.D.

8 COROLLARY: Let (X_i,d_i,ρ_i) be regular pseudometric spaces $(i = 1,2)$
and let $f: X_1 \to X_2$. If $\mathfrak{A} = \{U_\alpha\}$ is a uniform cover of X_1 then the
following are equivalent: (1) f is $\mathcal{C}_u$ (or $\mathcal{L}_a$). (2) $\{f|U_\alpha: U_\alpha \to X_2\}$
is $\mathcal{C}_u$ (or $\mathcal{L}_a$). (3) $\{f|U_\alpha: U_\alpha \to f(U_\alpha)\}$ is $\mathcal{C}_u$ (or $\mathcal{L}_a$).

7. Semicompleteness: The choice of admissible metric and bound d and
ρ on a regular metric manifold X gives X the structure of a regular
metric space. Different choices of d, ρ yield $\mathcal{L}$ equivalent regular
metric space structures. Thus, all of the definitions and results of

the preceding section apply to regular metric manifolds. In particular, a regular metric manifold X has an associated uniform structure $\mathfrak{U}_X$. Also, the notion of $\mathcal{C}$, $\mathcal{C}_u$ and $\mathcal{L}_a$ maps between subsets of regular metric manifolds are defined. By Lemma 1.4 and Proposition 5.3 the resulting notion of $\mathcal{C}$ or $\mathcal{L}$ maps between regular metric manifolds agrees with the Section 5 notion of $\mathcal{C}$ or $\mathcal{L}$ metric maps. One point of contrast requires care: An open subset U of X inherits an $\mathcal{L}$-equivalence class of regular metric space structures from X as does any subset. However, an open subset is also naturally a regular metric manifold in its own right with associated regular metric space structures. These two classes of regular metric space structures do <u>not</u> in general agree. The relation between them is investigated in some detail later in this section.

If $\mathfrak{a}_i$ are $\mathcal{L}$ atlases on X_i, $d_i = d_{\mathfrak{a}_i}$ $(i = 1,2)$ and $f: \mathfrak{a}_1 \to \mathfrak{a}_2$ is an $\mathcal{L}$ map then we can make (6.3) and (6.5) uniform and get:

$$(7.1)\qquad \begin{aligned} 1/\delta^{f^{-1}U}(d_1) &\le O^*(\rho_f^{d_1}, 1/f^*\delta^U(d_2)) \\ 1/\delta_{f^{-1}\mathfrak{U}}(d_1) &\le O^*(\rho_f^{d_1}, 1/f^*\delta_{\mathfrak{U}}(d_2)). \end{aligned}$$

The first follows from (5.1) and the proofs of Lemmas 4.3 (c) and 6.2 (b). The first implies the second.

In addition to $d_{\mathfrak{a}}$ and $\rho_{\mathfrak{a}}$ an atlas $\mathfrak{a}$ on X defines another function $\lambda_{\mathfrak{a}}: X \to [0,1]$. The purpose of this section is to introduce this third member of our atlas trinity and investigate it with the tools developed for regular pseudometric spaces. For $\mathfrak{a} = \{U_\alpha, h_\alpha\}$, define $\lambda_{\mathfrak{a}}(x) = \sup\{r \le 1: B^{\|\ \|_\alpha}(h_\alpha x; r) \subset h_\alpha(U_\alpha)$ for some $\mathfrak{a}$ chart (U_α, h_α) containing $x\}$. Clearly, $\lambda_{\mathfrak{a}}$ is lower semicontinuous and $\lambda_{\mathfrak{a}} > 0$ on X, i.e. $1/\lambda_{\mathfrak{a}}$ is a bound on X.

1 <u>THEOREM</u>: Assume $\mathfrak{M} \subset \mathcal{L}$. If $\mathfrak{a}$ is an $\mathfrak{m}$ atlas on X, let $\delta_{\mathfrak{a}} = \delta_{\mathfrak{a}}(d_{\mathfrak{a}})$ (identifying $\mathfrak{a}$ with its open cover), then:

$\max(\rho_G^{d_G}, 1/\lambda_G) \sim \max(\rho_G^{d_G}, 1/\delta_G, 1/\kappa(d_G))$.

<u>PROOF</u>: First, we prove the inequality:

$$(7.2) \qquad \max(1/\delta_G, 1/\kappa(d_G)) \geq 1/\lambda_G$$

or its equivalent $\lambda_G \geq \min(\delta_G, \kappa(d_G))$. Let $\min(\delta_G(x), \kappa(d_G)(x)) > \epsilon > 0$, assuming the left is positive (otherwise the inequality holds trivially). Then $B^{d_G}[x,\epsilon]$ is d_G complete and is contained in some chart (U_α, h_α) of G. We will show that $B^{\|\ \|_\alpha}(h_\alpha x, \epsilon) \subset h_\alpha(U_\alpha)$. For v a unit vector in E_α, let $t_0 = \sup\{t: sv + h_\alpha x \in h_\alpha(U_\alpha)$ for $0 \leq s \leq t\}$. It suffices to show $t_0 \geq \epsilon$. If not, let $\{t_n\}$ be an increasing sequence with $t_1 = 0$, converging to t_0 and let $u_n = t_n v + h_\alpha x$. Since $u_n \in h_\alpha(U_\alpha)$ there exists x_n such that $h_\alpha x_n = u_n$. Clearly, $\|u_n - u_m\|_\alpha \geq d_G(x_n, x_m)$. Thus, $\{x_n\}$ is a Cauchy sequence in $B^{d_G}[x,\epsilon]$ and so converges to a point x_0 in the closed ball and so in U_α. Hence, $h_\alpha x_0 = \text{Lim } u_n = t_0 v + h_\alpha x$ lies in $h_\alpha(U_\alpha)$. Since $h_\alpha(U_\alpha)$ is open this contradicts the definition of t_0 and proves (7.2). Next, we prove:

$$(7.3) \qquad \max(1/\delta_G, 1/\kappa(d_G)) \leq O^*(\rho_G^{d_G}, 1/\lambda_G).$$

Let $\infty > K > \max(\rho_G^{d_G}(x), 1/\lambda_G(x))$. $\rho_G \leq K$ on $B^{d_G}(x, K^{-1})$ and there exists a chart (U_α, h_α) of G with $B^{\|\ \|_\alpha}(h_\alpha x, K^{-1}) \subset h_\alpha(U_\alpha)$. Clearly, $h_\alpha^{-1} B^{\|\ \|_\alpha} \subset B^{d_G}$ and so $B^{\|\ \|_\alpha} \subset h_\alpha(U_\alpha \cap \{\rho_G \leq K\})$. By the Metric Estimate if $\epsilon = 1/2K$ and $\delta = 1/4K^2$, $B^{d_G}(x,\delta) \subset h_\alpha^{-1}(B^{\|\ \|_\alpha}(h_\alpha x, \epsilon)) \subset U_\alpha$ and on $B^{d_G}(x,\delta)$: $K\, d_G(x_1, x_2) \geq \|h_\alpha x_1 - h_\alpha x_2\|_\alpha$. We first see that $\delta_G(x) \geq \delta$. Next if $\{x_n\}$ is a d_G Cauchy sequence in $B^{d_G}[x, \delta_1]$ for $\delta_1 < \delta$ then $\{h_\alpha x_n\}$ is a Cauchy sequence in $B^{\|\ \|_\alpha}[h_\alpha x, \epsilon] \subset h_\alpha(U_\alpha)$. $\{h_\alpha x_n\}$ must converge to a point in the closed ϵ ball. Since h_α is a homeomorphism $\{x_n\}$ converges to a point x_0 which must lie in the closed δ_1 ball, i.e. $\kappa(d_G)(x) \geq \delta_1$. Thus, $4\max(\rho_G^{d_G}, 1/\lambda_G)^2 \geq \max(1/\delta_G, 1/\kappa(d_G))$. Q.E.D.

2 COROLLARY: Let X be a regular metric manifold with admissible atlas, metric and bound: $\mathfrak{a}$, d and ρ. $\max(\rho, 1/\lambda_{\mathfrak{a}})$ and $\max(\rho, 1/\delta_{\mathfrak{a}}(d), 1/\kappa(d))$ lie in the same bound equivalence class and it is adapted to d (cf. p. 64). If X is uniformly complete then this same class contains $\max(\rho, 1/\delta_{\mathfrak{a}}(d))$.

PROOF: For $d = d_{\mathfrak{a}}$ this follows from Theorem 1. Independence of the choice of d and ρ follows from Lemmas 6.1 and 6.5. In the uniformly complete case $\rho \sim \max(\rho, 1/\kappa(d))$. In fact, in this case we have from (7.2) and page 80 (3):

$$(7.4) \qquad \max(\rho^{d_{\mathfrak{a}}}, 1/\delta_{\mathfrak{a}}) \geq 1/\lambda_{\mathfrak{a}}. \qquad \text{Q.E.D.}$$

Note that for a regular metric manfiold X the bounds $\max(\rho, 1/\kappa(d))$, for d, ρ admissible metric and bound, are mutually equivalent and induce a regular refinement of the metric structure of X. It is in fact the minimal uniformly complete refinement of X.

Let X be a regular metric manifold. $\mathfrak{a} = \{U_\alpha, h_\alpha\}$ be an admissible atlas and ρ an admissible bound. It follows from Theorem 1 that $\rho > 1/\lambda_{\mathfrak{a}}$ iff X is uniformly complete and $\{U_\alpha\}$ is a uniform open cover of X. If admissible atlases exist satisfying these conditions then X is called semicomplete and the atlases are called s admissible. Thus, semicompleteness implies uniform completeness and for a semicomplete X, $\mathfrak{a}$ is s admissible iff $\{U_\alpha\}$ is a uniform cover. If X is semicomplete and bounded then X is called semicompact. A regular metric bundle $\pi: \mathbf{E} \to X$ is called semicomplete (or semicompact) if the base space X is semicomplete (or semicompact) and there exist $(\mathfrak{D}, \mathfrak{a})$ admissible for π with $\mathfrak{a}$ s admissible for X. Such atlases are called s admissible for π. The above definitions apply for $\mathfrak{M}$ metric structures and for $\mathcal{C}^\infty$ structures.

If X is an $\mathfrak{M}$ manifold with $\mathfrak{M} \subset \mathcal{L}$ and ρ_1 is a regular refining bound on X then ρ is a semicomplete refining bound or ρ induces a semicomplete refinement on X if there exist atlases $\mathfrak{a}$ admissible for X such that $\rho_1 > 1/\lambda_{\mathfrak{a}}$. This implies that the refined metric structure is semicomplete. If ρ_1 is a semicomplete refining bound, then the refined structure is called a semicomplete refinement. Similarly, for bundles.

THE STANDARD CONSTRUCTIONS: (1) Atlas Refinements: If X is semicomplete, $\mathfrak{a}$ is a admissible and $\mathfrak{a}_0$ is a refinement of $\mathfrak{a}$ then $\mathfrak{a}_0$ is s admissible iff its cover is still uniform. In particular if $\mathfrak{a} = \{U_\alpha, h_\alpha\}$ and $\mathfrak{U} = \{V_\beta\}$ is a uniform open cover then $\mathfrak{a} \cap \mathfrak{U} = \{U_\alpha \cap V_\beta, h_\alpha | U_\alpha \cap V_\beta\}$ is s admissible, by (6.4). In particular, if $\pi_1, \ldots, \pi_n$ is an n-tuple of semicomplete bundles on X then the analogue of Lemma 5.1 holds and there exist s admissible atlas n-tuples for $\pi_1, \ldots, \pi_n$. (2) Products: If X_i is semicomplete with s admissible atlas $\mathfrak{a}_i$ $(i = 1,2)$, then $X_1 \times X_2$ is semicomplete with s admissible atlas $\mathfrak{a}_1 \times \mathfrak{a}_2$. Similarly for bundles. (3) B-Spaces: The norm metric structure on a B-space F is a semicomplete $\mathcal{C}^\infty$ structure. It is a refinement of a semicompact $\mathcal{C}^\infty$ structure called the group metric structure on F. If U is open and bounded in F let $U_v = U + v$ and $h_v = I - c_v$. $\{U_v, h_v\}$ is an s admissible atlas for the group metric structure. (4) Tangent Bundles: If X is a semicomplete $\mathfrak{M}^1$ metric manifold then τ_X is semicomplete. (5) Operations: If π_1, π_2 are semicomplete on X then by (1), $\pi_1 \oplus \pi_2$ and $L_2^n(\pi_1; \pi_2)$ are semicomplete. (6) Pullbacks: If X_0 and $\pi\colon \mathbb{E} \to X$ are semicomplete and $f\colon X_0 \to X$ is an $\mathfrak{M}$ metric map ($\mathfrak{M} \subset \mathcal{L}$) then $f^*\pi$ is semicomplete. By (1) if $\mathfrak{a}_0$ and $(\mathfrak{D}, \mathfrak{a})$ are s admissible for X_0 and π, then $\mathfrak{a}_0 \cap f^{-1}\mathfrak{a}$ is s admissible on X_0 and $f\colon \mathfrak{a}_0 \cap f^{-1}\mathfrak{a} \to \mathfrak{a}$ is index preserving.

In verifying semicompleteness, we can usually finesse the problem of regularity by the following:

3 <u>REGULARITY LEMMA</u>: Assume $\mathfrak{M} \subset \mathcal{L}$. Let $(\mathfrak{G},\rho)$ be an $\mathfrak{M}$ adapted atlas on X. If there exist constants K,n such that for each $x \in X$ there exists (U_α,h_α) an $\mathfrak{G}$ chart containing x such that $B^{\|\ \|^\alpha}(h_\alpha x,(K\rho(x)^n)^{-1}) \subset h_\alpha(U_\alpha \cap \{\rho < K\rho(x)^n\})$, then $\rho > \max(\rho^{d_{\mathfrak{G}}}, 1/\lambda_{\mathfrak{G}})$. Thus, the metric structure containing $(\mathfrak{G},\rho)$ is semicomplete, and in particular is regular, and $\mathfrak{G}$ is admissible.

<u>PROOF</u>: Clearly, $K\rho^n \geq 1/\lambda_{\mathfrak{G}}$. We can assume K and n are large enough that $K\rho^n \geq \rho_{\mathfrak{G}}$. Then by the Metric Estimate $B^{d_{\mathfrak{G}}}(x,(4K^2\rho(x)^{2n})^{-1}) \subset h_\alpha^{-1}(B^{\|\ \|^\alpha}(h_\alpha x,(K\rho(x)^n)^{-1})) \subset \{\rho < K\rho(x)^n\}$. Hence, $4K^2\rho^{2n} \geq \rho^{d_{\mathfrak{G}}}$. Q.E.D.

4 <u>COROLLARY</u>: If $\pi: \mathbb{E} \to X$ is a semicomplete metric bundle then the total space $\mathbb{E}$ is a semicomplete metric manifold. Furthermore, if $(\mathfrak{D},\mathfrak{G})$ is s admissible for π, then $\mathfrak{D}$ is s admissible for $\mathbb{E}$. $\lambda_{\mathfrak{D}} = \pi^*\lambda_{\mathfrak{G}}$.

<u>PROOF</u>: Given $(\mathfrak{D},\mathfrak{G})$ s admissible for π, $\rho > \max(\rho_{(\mathfrak{D},\mathfrak{G})}, 1/\lambda_{\mathfrak{G}})$ an admissible bound on X and $d = d_{\mathfrak{G}}$, we apply the Regularity Lemma to $(\mathfrak{D},\max(\pi^*(\rho^d),\|\ \|^M))$. For $v \in \mathbb{E}$, let $x = \pi v$ and $K = \max(\rho^d(x),\|v\|^M)$. There exists a chart $(U_\alpha,h_\alpha,\varphi_\alpha)$ such that $B^{\|\ \|^\alpha}(h_\alpha x;K^{-1}) \subset h_\alpha(U_\alpha)$ and on $h_\alpha^{-1}B^{\|\ \|^\alpha}(h_\alpha x;K^{-1}) \subset B^d(x;K^{-1})$, $\rho \leq K$. Since $\|v\|^\alpha \leq \|v\|^M \leq K$, $B^{\|\ \|^\alpha}(\varphi_\alpha v;K^{-1}) \subset h_\alpha(U_\alpha) \times F_\alpha$ and in fact lies in $B^{\|\ \|^\alpha}(h_\alpha x;K^{-1}) \times B^{\|\ \|^\alpha}(0,K + K^{-1}) \subset B^{\|\ \|^\alpha}(h_\alpha x;K^{-1}) \times B^{\|\ \|^\alpha}(0,2K)$. Thus, on $\varphi_\alpha^{-1}B^{\|\ \|^\alpha}(\varphi_\alpha v; K^{-1})$, $\|\ \|^\alpha$ is bounded by 2K and $\pi^*\rho_{(\mathfrak{D},\mathfrak{G})}$ is bounded by K. By inequality (3.4), $\|\ \|^M$ is bounded by $2K^2$. So on $\varphi_\alpha^{-1}B^{\|\ \|^\alpha}$, $\max(\pi^*\rho,\|\ \|^M)$ is bounded by $2K^2$. Since $\max(\pi^*\rho^d,\|\ \|^M) \sim \max(\pi^*\rho,\|\ \|^M)$ the result follows from the previous Lemma. Q.E.D.

Now if U is open in a regular metric manifold X let d, ρ be admissible metric and bound on X and d_0 be an admissible metric for the restriction metric structure on U. As remarked earlier, $(U,d|U,\rho|U)$, the restricted regular metric space, the extrinsic metric on U, and $(U,d_0,\rho|U)$, the intrinsic metric, are not uniformly equivalent.

5 PROPOSITION: Let X be a regular metric manifold and U be open in X.

(a) If $\mathcal{G}$ is an admissible atlas on X and $x \in U$ then $d_{\mathcal{G}}(x,x_1) < \delta^U(d_{\mathcal{G}})(x)$ implies $x_1 \in U$ and $d_{\mathcal{G}}(x,x_1) = d_{\mathcal{G}|U}(x,x_1)$, and so for $\epsilon < \delta^U(d_{\mathcal{G}})(x)$, $B^{d_{\mathcal{G}}}(x;\epsilon) = B^{d_{\mathcal{G}|U}}(x;\epsilon)$. It follows that the bounds $\rho_U = \max(\rho, 1/\delta^U(d))$ for ρ,d admissible bounds and metrics on X induce a regular refinement called the canonical refinement of the restriction metric structure on U. If d_0 is an admissible metric for the restriction metric structure on U, then $d \sim_{\rho_U} d_0$ on U. In particular, if $A \subset\subset U$ (rel X) then $\rho_U|A \sim \rho|A$ and $d \sim_\rho d_0$ on A.

(b) Let $A \subset V \subset U$. If $A \subset\subset V$ (rel X) then $A \subset\subset V$ (rel U for both the restriction metric structure and the canonical refinement). If $V \subset\subset U$ (rel X) and $A \subset\subset V$ (rel U for either the restriction metric structure or its canonical refinement) then $A \subset\subset V$ (rel X).

PROOF: (a): The first part is a restatement of Proposition 2.5. The rest is easy, see eg. Lemma 6.2. Note that $A \subset\subset U$ (rel X) iff $\rho_U \sim \rho$ on A. (b): For any refinement of the restriction metric structure the inclusion $U \to X$ is a metric map and so $A \subset\subset V$ (rel X) implies $A \subset\subset V$ (rel U) by Lemma 6.3 (g). If $V \subset\subset U$ (rel X) and d,d_0 are admissible metrics for X and U (restriction structure) then $d \sim_\rho d_0$ on V and $\rho_U \sim \rho$ on V by (a). Thus, $\rho > 1/\delta^V(d)$ on A iff $\rho > 1/\delta^V(d_0)$ on A. Q.E.D.

In general, the condition $A \subset\subset U$ (rel X) is extrinsic. We now show that if X is uniformly complete the condition is intrinsic, depending only on A and U.

6 PROPOSITION: Let X be a regular metric manifold with U open in X. If $\mathcal{G}$ is an admissible atlas on X then $\delta^U(d_{\mathcal{G}}) \geq \kappa(d_{\mathcal{G}|U}) \geq 3^{-1}\min(\delta^U(d_{\mathcal{G}}),\kappa(d_{\mathcal{G}}))$. Thus, if d,d_0 are admissible metrics for X

and U (restriction structure), ρ is an admissible bound on X and if X is uniformly complete then $\max(\rho, 1/\kappa(d_0)) \sim \max(\rho, 1/\delta^U(d))$ on U, i.e. when X is uniformly complete the canonical refinement on U is the minimal uniformly complete refinement on U.

PROOF: If $B^{d_{\mathcal{G}|U}}[x;\epsilon]$ is $d_{\mathcal{G}|U}$ complete in U, then $B^{d_{\mathcal{G}}}(x;\epsilon) \subset U$. If not, then there is an $\mathcal{G}$ chain of length $< \epsilon$ between x and a point of X - U. If x_t, $t \in [1,n+1]$, is the associated p.l path then $x_{n+1} \notin U$ and by truncating the path, if necessary, we can assume $x_t \in U$ for $t \in [1,n+1)$. If t_n is an increasing sequence with limit n + 1, then $\{x_{t_n}\}$ is a $d_{\mathcal{G}|U}$ Cauchy sequence in $B^{d_{\mathcal{G}|U}}[x;\epsilon]$ converging to $x_{n+1} \notin U$. Contradiction. For $x \in U$, let $\epsilon > 0$ with $\delta^U(d_{\mathcal{G}})(x) > 3\epsilon$ and $\kappa(d_{\mathcal{G}})(x) > \epsilon$. Then, $B^{d_{\mathcal{G}|U}}[x;\epsilon] \subset B^{d_{\mathcal{G}}}[x;\epsilon] \subset B^{d_{\mathcal{G}}}[x;3\epsilon] \subset U$ and $B^{d_{\mathcal{G}}}[x;\epsilon]$ is $d_{\mathcal{G}}$ complete. For $x_1 \in B^{d_{\mathcal{G}}}[x;\epsilon]$, $\delta^U(d_{\mathcal{G}})(x_1) \geq 2\epsilon$ and so by Proposition 5(a), $B^{d_{\mathcal{G}}}[x;\epsilon] = B^{d_{\mathcal{G}|U}}[x;\epsilon]$ and the metrics $d_{\mathcal{G}}$ and $d_{\mathcal{G}|U}$ agree on it. Thus, $d_{\mathcal{G}}$ completeness implies $d_{\mathcal{G}|U}$ completeness and $\kappa(d_{\mathcal{G}|U})(x) \geq \epsilon$. Q.E.D.

7 PROPOSITION: Let X be a semicomplete metric manifold and U be open in X. The canonical refinement on U is a semicomplete refinement and if $\mathcal{G}$ is s admissible for X then $\mathcal{G}|U$ is s admissible and $\max(\rho, 1/\lambda_{\mathcal{G}|U})$ is an admissible bound for U with the canonical refinement.

PROOF: By Theorem 1, $\max(\rho, 1/\lambda_{\mathcal{G}|U}) \sim \max(\rho, 1/\kappa(d_{\mathcal{G}|U}), 1/\delta_{\mathcal{G}|U})$. By Proposition 6, $\max(\rho, 1/\kappa(d_{\mathcal{G}|U}))$ is an admissible bound for the canonical refinement. Also, $\delta_{\mathcal{G}|U} \geq \delta_{\mathcal{G}}$ because if $B^{d_{\mathcal{G}}}(x;\epsilon)$ lies in some chart of $\mathcal{G}$, then $B^{d_{\mathcal{G}|U}}(x;\epsilon) \subset B^{d_{\mathcal{G}}}(x;\epsilon) \cap U$ lies in some chart of $\mathcal{G}|U$. Since $\rho > 1/\delta_{\mathcal{G}}$, $\rho > 1/\delta_{\mathcal{G}|U}$ on U. Q.E.D.

8 COROLLARY: Let X be a uniformly complete regular metric manifold, U be open in X and $A \subset U$.

(a) For ρ an admissible bound on X, ρ_U an admissible bound on U for the canonical refinement, $A \subset\subset U(\text{rel } X)$ iff $\rho \sim \rho_U$ on A. The restriction metric structure on U is uniformly complete, and hence agrees with the canonical refinement, iff U is open and closed.

(b) If X_1 is a uniformly complete metric manifold, U_1 is open in X_1 and $f: U_1 \to U$ is a metric isomorphism of restriction structures then f is an isomorphism of the canonical refinements. $A \subset\subset U$ (rel X) iff $f^{-1}A \subset\subset U_1$ (rel X_1).

PROOF: (a): The first remark is clear even with no completeness. By Proposition 6 the restriction is uniformly complete iff it agrees with the canonical and so iff $U \subset\subset U$ (rel X).

(b): "Minimal uniformly complete refinement" is clearly a functorial retraction from the isomorphism category of regular metric manifolds to the full subcategory of uniformly complete regular metric manifolds. Thus, f preserves both the original ρ bounds and the canonical ρ_U bounds. The full neighborhood result then follows from (a). Q.E.D.

9 LEMMA: Let X be a regular metric manifold with admissible metric and bound d,ρ. Let $\rho_1: X \to [1,\infty]$ with $\rho_1 > \rho$ and let $U = \{\rho_1 < \infty\}$. If $\rho_1^d \sim \rho_1$ then U is open. Conversely if U is open and d_1 is an admissible metric on U (restriction metric structure) then the following are equivalent: (1) $\rho_1^d \sim \rho_1$, (2) $\rho_1|U > \rho_U$ and $(\rho_1|U)^{d|U} \sim \rho_1|U$, (3) $\rho_1|U > \rho_U$ and $(\rho_1|U)^{d_1} \sim \rho_1|U$, (4) $\rho_1|U$ is a regular refining bound for the restriction metric structure inducing a refinement more refined than the canonical refinement. In particular, if X is uniformly complete (or semicomplete) then the $\rho_1|U$ induced refinement on U is uniformly complete (resp. a semicomplete refinement).

PROOF: Clearly, $\rho_1^d \sim \rho_1$ implies U is open and since $\rho_1 \geq 1/C_U$, $\rho_1^{d_1} > \rho_U$. Also $\rho_1^d \sim \rho_1$ iff there exist constants K,n such that $x \in U$

and $d(x,x_1) < (K\rho_1(x)^n)^{-1}$ implies $x_1 \in U$ and $\rho_1(x_1) \leq K\rho_1(x)^n$. This is clearly equivalent to (2). If $\rho_1|U > \rho_U$ then $d|U \sim_{\rho_1|U} d_1$. Thus, (2) is equivalent to (3) of which (4) is just a restatement. Q.E.D.

For example, for B-spaces E and F the bound θ on the open set $Lis(E;F) \subset L(E;F)$ induces a semicomplete refinement. For if $T \in Lis(E;F)$ and $\|T - T_1\| < (2\theta(T))^{-1}$ then $T_1 \in Lis(E;F)$ and $\theta(T_1) \leq 2\theta(T)$.

There is a useful sharpening of Lemma 6.3 (e).

10 LEMMA: Let X be a regular metric manifold with ρ, ρ_1, U as in Lemma 9. Assume ρ_1 satisfies conditions (1) - (4) of Lemma 9. If $A \subset U$ such that $\rho|A \sim \rho_1|A$ then there exists V open in U with $A \subset\subset V$ (rel X) and $\rho|V \sim \rho_1|V$.

PROOF: Since $\rho^d \sim \rho_1 \sim \rho_1^d$ on A, there exist $K,n \geq 1$ such that $K(\rho_1^d)^n \geq 2\rho^d$ and $K(\rho^d)^n > \rho_1^d$ on A. Let $V_1 = \{\rho_1 \leq 2^n K(\rho^d)^n\}$. Clearly, $\rho^d \sim \rho \sim \rho_1$ on V_1. If $x \in A$ then $B^d(x,(K\rho_1^d(x)^n)^{-1}) \subset V_1$ because if y is in the ball then $\rho_1(y) \leq \rho_1^d(x) \leq K\rho^d(x)^n$ and since the ball is contained in $B^d(x,(2\rho^d(x))^{-1})$, $\rho^d(x) \leq 2\rho^d(y)$. Q.E.D.

11 LEMMA: Let $f: X_1 \to X_2$ be a metric map of semicomplete metric manifolds, $U \subset X_2$ be open, ρ_1 an admissible bound on X_1 and $\bar{\rho}$ a bound on U inducing a semicomplete refinement. The bound $\max(\rho_1, f^*\bar{\rho})$ induces a semicomplete refinement on $f^{-1}U$.

PROOF: By (6.3), $\max(\rho_1, f^*\bar{\rho}) > \max(\rho_1, 1/f^*\delta^U(d_2)) > \max(\rho_1, 1/\delta^{f^{-1}U}(d_1))$. Since $\bar{\rho}^{d_2|U} \sim \bar{\rho}$, $\rho_1^{d_1} \sim \rho_1$ and $d_1 >_{\rho_1} f^*d_2$ we have by Lemma 4.3 (c) on $f^{-1}U$: $\max(\rho_1, f^*\bar{\rho}) \sim \max(\rho_1^{d_1}, f^*(\bar{\rho}^{d_2|U})) \geq \max(\rho_1^{d_1|f^{-1}U}, (f^*\bar{\rho})^{f^*(d_2|U)})$ $> \max(\rho_1^{d_1|f^{-1}U}, (f^*\bar{\rho})^{d_1|f^{-1}U}) = \max(\rho_1, f^*\bar{\rho})^{d_1|f^{-1}U}$. Q.E.D.

For example, let $\pi: \mathbb{E} \to X$ be a semicomplete metric bundle and U be open in X. In addition to the canonical refinement on $\mathbb{E}_U = \pi^{-1}U$

there are two other ways of obtaining a semicomplete refinement on $\mathbb{E}_U$. First, we can "pull back" the canonical refinement on U by Lemma 10. Second, the canonical refinement on U makes $\pi : \mathbb{E}_U \to U$ a semicomplete bundle with a semicomplete total space. These refinements all agree. For if d_1 and d are admissible metrics on $\mathbb{E}$ and X, ρ_1 is an admissible bound on $\mathbb{E}$, and $(\mathfrak{J},\mathfrak{G})$ is s admissible on π, then:

$$(7.5)\quad \max(\rho_1, 1/\delta^{\mathbb{E}_U}(d_1)) \sim \max(\rho_1, 1/\lambda_{\mathfrak{J}|\mathbb{E}_U}) = \max(\rho_1, \pi^*(1/\lambda_{\mathfrak{G}|U}))$$
$$\sim \max(\rho_1, \pi^*(1/\delta^U(d))).$$

For $\pi: \mathbb{E} \to X$ a metric bundle with admissible Finsler and bound $\|\ \|$, ρ, $A \subset \mathbb{E}$ is called <u>vertically bounded</u> if $\pi^*\rho > \|\ \|$ on A. For example, the image of any metric section is vertically bounded because $\rho > s^*\|\ \|$. If $(\Phi, 1_X): \pi \to \pi_1$ is a $\mathcal{C}$ metric VB map and A is vertically bounded in $\mathbb{E}$, then $\Phi(A)$ is vertically bounded in $\mathbb{E}_1$. The importance of the concept is illustrated by:

12 <u>PROPOSITION</u>: Let $\pi: \mathbb{E} \to X$ be a semicomplete metric bundle, $A \subset \mathbb{E}$ and U open in X. If $\pi(A) \subset\subset U$ rel X then $A \subset\subset \pi^{-1}U$ rel $\mathbb{E}$. If $A \subset\subset \pi^{-1}U$ rel $\mathbb{E}$ and A is vertically bounded then $\pi A \subset\subset U$ rel X.

<u>PROOF</u>: The first is a special case of Lemma 6.3 (g). For the converse let d_1, d be admissible metrics on $\mathbb{E}$ and X. We have seen above that $\max(\pi^*\rho, 1/\delta^{\mathbb{E}_U}(d_1)) \sim \pi^* \max(\rho, 1/\delta^U(d))$. Since A is vertically bounded, $\pi^*\rho \sim \max(\pi^*\rho, \|\ \|)$ on A. Hence, if $A \subset\subset \pi^{-1}U$ rel $\mathbb{E}$ then $\pi^*\rho \sim \max(\pi^*\rho, \|\ \|) \sim \max(\pi^*\rho, \|\ \|, 1/\delta^{\mathbb{E}_U}(d_1)) \sim \pi^*\max(\rho, 1/\delta^U(d))$ on A. Q.E.D.

To continue studying open sets in total spaces we define vertical or fiberwise versions of regularity and $\delta^U(d)$. For a bundle, $\pi: \mathbb{E} \to X$, a <u>semi-Finsler</u> $\|\ \|$ on π is a function $\|\ \|: \mathbb{E} \to [0,\infty)$ which is a semi-norm on each fiber. Now for $N: \mathbb{E} \to [1,\infty]$ and $\|\ \|$ a semi-Finsler on π

define $N^{\|\ \|}: \mathbb{E} \to [1,\infty]$ by:

$$(7.6) \qquad N^{\|\ \|}(v) = \sup\{\min(1/\|v-v'\|, N(v')): v' \in \mathbb{E}_x\}$$
$$= \inf\{K: N|B^{\|\ \|}(v,K^{-1}) \le K\}.$$

For $O \subset \mathbb{E}$, $\delta^O(\|\ \|): \mathbb{E} \to [0,1]$ is defined by $\delta^O(\|\ \|)(v) = \min(\|v - (\mathbb{E}_x - O)\|, 1) = \sup\{r \le 1: B^{\|\ \|}(v,r) \subset O\}$ where the balls lie in the fiber over $x = \pi v$. The following is the (easy) analogue of Lemmas 4.3 and 6.2:

13 __LEMMA__: (a) For a family $\{N_\alpha\}$ $(\sup\{N_\alpha\})^{\|\ \|} = \sup\{N_\alpha^{\|\ \|}\}$.

(b) If $N_1 >_\rho N_2$ (or $N_1 \ge N_2$) then $N_1^{\|\ \|} >_\rho N_2^{\|\ \|}$ (or $N_1^{\|\ \|} \ge N_2^{\|\ \|}$).

(c) If $\|\ \|_1 >_\rho \|\ \|_2$ (or $\|\ \|_1 \ge \|\ \|_2$) then $N^{\|\ \|_2} >_\rho N^{\|\ \|_1}$ (or $N^{\|\ \|_2} \ge N^{\|\ \|_1}$).

(d) For $(\phi, f): \pi_1 \to \pi_2$ a VB map and N, $\|\ \|$ on $\mathbb{E}_2$ then $\phi^*(N^{\|\ \|}) \ge (\phi^*N)^{\phi^*\|\ \|}$.

e) $(\pi^*\rho)^{\|\ \|} = \pi^*\rho$. $2\max(\|\ \|,1) \ge \max(\|\ \|,1)^{\|\ \|}$. $2N^{\|\ \|} \ge (N^{\|\ \|})^{\|\ \|}$.

f) If C_O is the characteristic function of O then $1/\delta^O(\|\ \|) = (1/C_O)^{\|\ \|}$. Hence, $2/\delta^O(\|\ \|) \ge (1/\delta^O(\|\ \|))^{\|\ \|}$.

g) If $\|\ \|_1 >_\rho \|\ \|_2$ (or $\|\ \|_1 \ge \|\ \|_2$) then $1/\delta^O(\|\ \|_2) >_\rho 1/\delta^O(\|\ \|_1)$ (or $1/\delta^O(\|\ \|_2) \ge 1/\delta^O(\|\ \|_1)$).

14 __LEMMA__: Let $\pi: \mathbb{E} \to X$ be a semicomplete metric bundle, $\|\ \|$ be an admissible Finsler on π, d_1 be an admissible metric on $\mathbb{E}$, and ρ be an admissible bound on X. If $N: \mathbb{E} \to [1,\infty]$ then $\max(\pi^*\rho, \|\ \|, N^{d_1}) > \max(\pi^*\rho, \|\ \|, N^{\|\ \|})$. In particular, if $O \subset \mathbb{E}$ then $\max(\pi^*\rho, \|\ \|, 1/\delta^O(d_1)) > \max(\pi^*\rho, \|\ \|, 1/\delta^O(\|\ \|))$. In fact, $1/\delta^O(d_1) \sim (1/\delta^O(\|\ \|))^{d_1}$.

__PROOF__: Since $\max(\pi^*\rho, \|\ \|)$ is an admissible bound on $\mathbb{E}$ the estimate on N is independent of the choice of $\|\ \|$ and d_1. So let $(\mathfrak{m}, \mathfrak{a})$ be an s admissible atlas, $d_1 = d_{\mathfrak{m}}$ and $\|\ \| = \|\ \|^{\mathfrak{m}}$. From Proposition 3.5 it easily follows that $N^{d_1} \ge N^{\|\ \|}$. The sharp result for $\delta^O(\|\ \|)$ comes from the remarks after Lemma 6.2 as $C_0 \geq \delta^0(\| \|) \geq \delta^0(d_1)$. Q.E.D.

For $\pi: \mathbb{E} \to X$ a metric bundle $N: \mathbb{E} \to [1,\infty]$ is called <u>vertically regular</u> if it is <u>vertically adapted</u> to $\| \|, \rho$ for some (and so any) choice of admissible Finsler and bound, i.e. $N^{\| \|} \sim_\rho N$. If $\{N < \infty\} = 0 \subset \mathbb{E}$ and N is vertically regular then $N \geq 1/C_0$ and so $N >_\rho 1/\delta^0(\| \|)$. N is vertically regular iff there exist constants K and n such that if $\|v - v'\| < (K\rho(\pi v)^n N(v))^{-1}$ then $N(v') \leq K\rho(\pi v)^n N(v)$ for $\pi v' = \pi v$. For example, if $\| \|_i$ is an admissible Finsler on $\pi_i: \mathbb{E}_i \to X$ $(i = 1,2)$ and $\| \|$ is the induced Finsler on $L(\pi_1;\pi_2)$ then we can define $\theta: L(\mathbb{E}_1;\mathbb{E}_2) \to [1,\infty]$ by I.$\mathcal{Q}$.1) on the fibers with $\{\theta < \infty\} = \text{Lis}(\mathbb{E}_1;\mathbb{E}_2)$ open in $L(\mathbb{E}_1;\mathbb{E}_2)$. As on page 95, we see that θ is vertically regular.

Clearly, if $A \subset\subset 0$ rel $\mathbb{E}$ then $\max(\pi^*\rho, \| \|) \sim \max(\pi^*\rho, \| \|, 1/\delta^0(\| \|))$ on A. A recurring technical question is the <u>vertical fullness question</u>: namely, under what circumstances is the converse true, i.e. given 0, can we replace $\delta^0(d)$ by $\delta^0(\| \|)$ in testing for fullness? Our first affirmative answer is:

15 <u>PROPOSITION</u>: Let $\pi: \mathbb{E} \to X$ be a semicomplete metric bundle, 0 be open in $\mathbb{E}$, $s: X \to \mathbb{E}$ a $\mathcal{C}_u$ section with $s(X) \subset 0$. Then $s(X) \subset\subset 0$ rel $\mathbb{E}$ iff $\rho > 1/s^*\delta^0(\| \|)$ for some, and hence any, choice of admissible bound and Finsler.

<u>PROOF</u>: Recall that $\rho > s^*\| \|$ and so $\rho > 1/s^*\delta^0(\| \|)$ is equivalent to the vertical fullness condition on $s(X)$. To prove sufficiency, let $(\mathfrak{D},\mathcal{G})$ be s admissible for π, $d_1 = d_{\mathfrak{D}}$, $d = d_{\mathcal{G}}$, $\| \| = \| \|^m$ and $\rho_1 = \max(\pi^*\rho, \| \|)$. By the Metric Estimate and s admissibility there exist constants K, n such that $d_1(s(x),v) < (K\rho_1(s(x))^n)^{-1}$ implies $K\rho_1(s(x))^n d_1(s(x),v) \geq \|s(x) - v\|^\alpha \geq \|s(x) - v\|$ for $x = \pi v$ and some α. Since $\rho \sim s^*\rho_1$ we can replace $\rho_1(s(x))$ by $\rho(x)$ in these inequalities by increasing K,n if necessary. Finally, we can assume that $\|s(x) - v\| < (K\rho(x)^n)^{-1}$ implies

$v \in O$. Since s is $\mathcal{C}_u$ we can choose $K_1 \geq 2K^2$ and $n_1 \geq 2n$ such that $d(x,y) < (K_1\rho^d(x)^{n_1})^{-1}$ implies $d_1(s(x),s(y)) < (2K^2\rho^d(x)^{2n})^{-1}$. Recall that $d_1 \geq \pi^*d$. Now if $d_1(s(x),v) < (K_1\rho^d(x)^{n_1})^{-1}$ with $\pi v = y$ then $d_1(s(x),s(y)) < (2K^2\rho^d(x)^{2n})^{-1}$ and $\rho(y) \leq \rho^d(x)$. Thus, $d_1(s(y),v) < (K^2\rho(y)^{2n})^{-1}$ and so $\|s(y) - v\| < (K\rho(y)^n)^{-1}$. So $v \in O$. Q.E.D.

For $\delta^O(\| \|)$ we have the following analogue of (6.3): Let $(\Phi,f): \pi_1 \to \pi$ be a $\mathcal{C}$ metric VB map and let $\| \|_1$ and $\| \|$ be admissible Finslers, ρ_1 an admissible bound on X_1:

$$\text{(7.7)} \qquad \begin{aligned} \max(\pi_1{}^*\rho_1, 1/\delta^{\Phi^{-1}O}(\Phi^*\| \|)) &> 1/\delta^{\Phi^{-1}O}(\| \|_1) \\ \delta^{\Phi^{-1}O}(\Phi^*\| \|) &\geq \Phi^*\delta^O(\| \|). \end{aligned}$$

As with (7.1) if $(\Phi,f): (\mathfrak{D}_1,\mathcal{G}_1) \to (\mathfrak{D},\mathcal{G})$ and the Finslers are the max Finslers $\| \|^M$ determined by these atlases then

$$\text{(7.8)} \qquad 1/\delta^{\Phi^{-1}U}(\| \|_1) \leq O^*(\pi_1{}^*\rho_{(\Phi,f)}, 1/\delta^{\Phi^{-1}O}(\Phi^*\| \|)).$$

We conclude by giving atlas conditions for a map to be $\mathcal{C}_u$ or $\mathcal{L}_a$. Let $\mathcal{G} = \{U_\alpha,h_\alpha\}$ be an admissible atlas on a metric manifold X. $\mathcal{G}$ is called <u>star bounded</u> if $V_{\mathcal{G}} = \cup_\alpha U_\alpha \times U_\alpha$ is transversely bounded (see page 86). $\mathcal{G}$ is called <u>convex</u> if $h_\alpha(U_\alpha)$ is a convex open set in E_α for all α. $\mathcal{G}$ satisfies <u>condition $\mathcal{L}$</u> if for some, and hence every, admissible metric and bound d,ρ on X there exists K a bound on $V_{\mathcal{G}}$ with $m_\rho > K$ on $V_{\mathcal{G}}$ satisfying $K(x_1,x_2)^{-1}\min(d(x_1,x_2),1) \leq \min(\|h_\alpha x_1 - h_\alpha x_2\|,1) \leq K(x_1,x_2)\min(d(x_1,x_2),1)$ whenever $x_1,x_2 \in U_\alpha$. As for existence of such atlases, if $\mathcal{G}_0$ is any s admissible atlas on a semicomplete X it follows from the definition of s admissibility and the Metric Estimate that $\mathcal{G}_0$ admits refinements $\{U_x,h_x\}$ indexed by $x \in X$ satisfying:
(1) $h_x(U_x) = B^{\| \|_x}(h_x(x);\epsilon(x))$ where $\epsilon: X \to (0,1]$ with $\rho > 1/\epsilon$. (2) $V_{\{U_x\}}$ is transverse bounded. (3) For d an admissible metric on X there

exists K a bound on X with $\rho > K$ such that for $x_1, x_2 \in U_x$, $K(x)^{-1}\min(d(x_1,x_2),1) \leq \min(\|h_x(x_1) - h_x(x_2)\|,1) \leq K(x)\min(d(x_1,x_2),1)$.

In general, if ρ is a continuous admissible bound on X then $\{\{\frac{1}{2}\rho(x) < \rho < 2\rho(x)\}: x \in X\}$ is a star bounded open cover of X. Intersecting an admissible atlas with this cover yields a star bounded subdivision of the atlas.

16 PROPOSITION: Let X_1 and X_2 be semicomplete metric manifolds with atlases $\mathcal{G}_1 = \{U_\alpha, h_\alpha\}$ and $\mathcal{G}_2 = \{V_\beta, g_\beta\}$ satisfying condition $\mathcal{L}$. If U is open in X_1 and $f: U \to X_2$ is a map such that $f: \mathcal{G}_1|U \to \mathcal{G}_2$ is index preserving then f is $\mathcal{C}_u$ or $\mathcal{L}_a$ iff the family $\{f_\alpha: (h_\alpha(U_\alpha \cap U), \|\ \|_\alpha, \rho_1 \cdot h_\alpha^{-1}) \to (g_\beta(V_\beta), \|\ \|_\beta, \rho_2 \cdot g_\beta^{-1})\}$ $(\beta = \beta(\alpha))$ is $\mathcal{C}_u$ or $\mathcal{L}_a$. If $\mathcal{G}_1$ is also star bounded, then f is $\mathcal{L}_a$ iff $f_\alpha \in \mathcal{L}ip_a(h_\alpha(U_\alpha \cap U), E_{\beta(\alpha)})$ and $\rho_1 > \max(f^*\rho_2, \tilde{\rho}^{\mathcal{L}_a}(f;\mathcal{G}_1,\mathcal{G}_2))$ where $\tilde{\rho}^{\mathcal{L}_a}(f;\mathcal{G}_1,\mathcal{G}_2)(x) = \sup\{\|f_\alpha\|_{\mathcal{L}ip_a} : x \in U_\alpha\}$. Here f is a $\mathcal{C}_u$ or $\mathcal{L}_a$ map from the restriction regular metric space structure which implies $\mathcal{C}_u$ or $\mathcal{L}_a$ from the restriction metric manifold structure.

PROOF: Condition $\mathcal{L}$ says precisely that the family $\{h_\alpha: (U_\alpha, d|U_\alpha, \rho|U_\alpha) \to (h_\alpha(U_\alpha), \|\ \|_\alpha, \rho \cdot h_\alpha^{-1})\}$ and its inverse are $\mathcal{L}$ families. The results then follow from Corollary 6.8. Q.E.D.

8. Exponential Maps: The key tool in relating sets of functions with sets of sections is the exponential map.

1 DEFINITION: (a) Let X be a semicomplete $\mathfrak{M}^1$ manifold. An $\mathfrak{M}$ exponential on X is an $\mathfrak{M}$ isomorphism $e: D \to V$ where D is a full open neighborhood of the zero section in TX and V is a full open neighborhood of the diagonal in $X \times X$. D and V are given the restriction metric structures. In addition, the following diagram commutes:

$$\begin{array}{ccccc} & & X & & \\ & {}^{0_X}\swarrow & & \searrow^{\Delta_X} & \\ D & & \xrightarrow{\;\;e\;\;} & & V \\ & {}_{\tau}\searrow & & \swarrow_{p_1} & \\ & & X & & \end{array}$$

(b) Let $\pi\colon \mathbb{E} \to X$ be a semicomplete $\mathfrak{M}$ bundle with X a semicomplete $\mathfrak{M}^1$ manifold. Let $e\colon D \to V$ be an $\mathfrak{M}$ exponential on X. Restricting the bundle $p_2\colon \mathbb{E} \oplus TX \to TX \cong \tau^*\pi\colon \tau^*\mathbb{E} \to TX$ to D and $1 \times \pi\colon X \times \mathbb{E} \to X \times X \cong p_2{}^*\pi\colon p_2{}^*\mathbb{E} \to X \times X$ to V, an <u>$\mathfrak{M}$ fiber exponential</u> $(E_0,e)\colon (p_2)_D \to (1\times\pi)_V$ is an $\mathfrak{M}$ VB isomorphism such that the following commutes:

$$\begin{array}{ccccc} & & \pi & & \\ & {}^{(I_E \oplus 0_X \circ \pi,\, 0_X)}\swarrow & & \searrow^{(\pi\times I_E,\, \Delta_X)} & \\ (p_2)_D & & \xrightarrow[(E_0,e)]{} & & (1\times\pi)_V \end{array}$$

(c) Let $\pi\colon \mathbb{E} \to X$ be a semicomplete $\mathfrak{M}^1$ bundle and $e\colon D \to V$ be an $\mathfrak{M}$ exponential on X. A <u>$\mathfrak{M}$ bundle exponential</u> on π is an $\mathfrak{M}$ VB isomorphism $(E,e)\colon (T\pi)_D \to (\pi \times \pi)_V$ such that the following commutes:

$$\begin{array}{ccccc} & & \pi & & \\ & {}^{(0_{\mathbb{E}},0_X)}\swarrow & & \searrow^{(\Delta_{\mathbb{E}},\Delta_X)} & \\ (T\pi)_D & & \xrightarrow{(E,e)} & & (\pi\times\pi)_V \\ & {}_{(\tau_{\mathbb{E}},\tau_X)}\searrow & & \swarrow_{(P_1,p_1)} & \\ & & \pi & & \end{array} .$$

The definitions of $\mathcal{C}^\infty$ exponentials on $\mathcal{C}^\infty$ spaces and bundles are as above with $\mathfrak{M}$ and $\mathfrak{M}^1$ replaced by $\mathcal{C}^\infty$ throughout.

We can always shrink V to $V_0 \subset V$ a full neighborhood of Δ_X which is transversely bounded. By Corollary 7.8(b) $D_0 = e^{-1}V_0$ is still a full neighborhood of 0_X. It is easy to check that V_0 transversely bounded implies D_0 is vertically bounded.

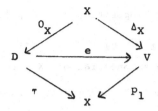

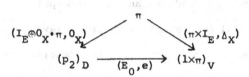

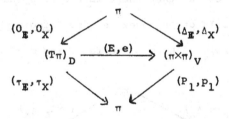

An exponential on π is the direct extension to bundles of exponentials on spaces. By Proposition 7.12, $T\pi^{-1}(D)$ and $\pi \times \pi^{-1}(V)$ are full neighborhoods of $0_{\mathbb{E}}$ and $\Delta_{\mathbb{E}}$ respectively. Thus, E is an exponential on $\mathbb{E}$ over the exponential on X. That (E,e) be a VB map is a natural condition which will be important in applications. If X is an $\mathfrak{M}^2$ semicomplete manifold with $e\colon D \to V$ an $\mathfrak{M}^1$ exponential on X then $(Te\circ\Upsilon, e)\colon (T\tau_X)_D \to (\tau_X\times\tau_X)_V$ is an $\mathfrak{M}$ exponential on the tangent bundle τ_X called the <u>Jacobi exponential</u> associated to e. Here $(\Upsilon, 1_{TX})\colon T\tau_X \to \tau_{TX}$ is the natural twist isomorphism defined locally by $\Upsilon(u,\dot{u}_1,\dot{u}_2,\ddot{u}) = (u,\dot{u}_2,\dot{u}_1,\ddot{u})$. To any $\mathfrak{M}$ bundle exponential on π there is an associated $\mathfrak{M}$ fiber exponential on π constructed as follows: The $\mathfrak{M}$ VB isomorphism (E,e) induces an $\mathfrak{M}$ isomorphism between the kernel of $(\tau_E,\tau_X)\colon (T\pi)_D \to \pi$ and the kernel of $(P_1,p_1)\colon (\pi\times\pi)_V \to \pi$. Clearly, the latter is just $(1_X \times \pi)_V$. Recall (p. 13) the identification at the atlas level between the kernel of $(\tau_{\mathbb{E}},\tau_X)$ and $p_2\colon \mathbb{E} \oplus TX \to TX$ by $(\bar{J},1)$. Modulo this isomorphism (E_0,e) is thus just the restriction of (E,e) to the kernels.

If $\pi\colon \mathbb{E} \to X$ is $\mathfrak{M}$ and X is $\mathfrak{M}^1$ (both semicomplete) and $e\colon D \to V$ is an $\mathfrak{M}$ exponential on X then there is a correspondence between $\mathfrak{M}$ metric fiber exponentials on π over e and <u>T isomorphisms</u> $(T,1_V)\colon (\pi\times 1_X)_V \to (1_X\times\pi)_V$ such that the following diagram commutes:

$$\begin{array}{ccc} & \pi & \\ {\scriptstyle (I_{\mathbb{E}}\times\pi,\Delta_X)}\swarrow & & \searrow{\scriptstyle (\pi\times I_{\mathbb{E}},\Delta_X)} \\ (\pi\times 1_X)_V & \xrightarrow[(T,1_V)]{} & (1_X\times\pi)_V \end{array} .$$

This is because, using e, we can pull back the identity on π by τ_X and p_1 to obtain an $\mathfrak{M}$ VB isomorphism $((\tau,p_1)^*I_\pi, e)\colon (p_2)_D \cong (\tau|D)^*\pi \to (p_1|D)^*\pi \cong (\pi\times 1_X)_V$. Thus, $(E_0,e)\colon (p_2)_D \to (1_X\times\pi)_V$ and $(T,1_V)$ are related by $E_0 = T\circ(\tau,p_1)^*I_\pi$ and $T = E_0\circ[(\tau,p_1)^*I_\pi]^{-1}$.

Exponentials arise naturally from sprays [14; Chap. IV, Secs. 3,4].

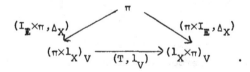

So we look at vectorfields.

2 <u>THEOREM</u>: (a) Let X be a semicomplete $\mathfrak{M}^2$ manifold and ξ be an $\mathfrak{M}^1$ vectorfield, i.e. an $\mathfrak{M}^1$ section of τ_X. There exists D an open full neighborhood of $X \times 0$ in $X \times R$ such that: (1) $D \subset X \times (-1,1)$ and $(x,t) \in D$ implies $(x,s) \in D$ for $|s| < |t|$. (2) The flow F_ξ of ξ is defined on D and is an $\mathfrak{M}^1$ metric map $F_\xi: D \to X$.

(b) Let $\pi: \mathbb{E} \to X$ be a semicomplete $\mathfrak{M}^2$ bundle and (Ξ,ξ) be an $\mathfrak{M}^1$ linear vectorfield on π, i.e. $(\Xi,\xi): \pi \to T\pi$ is an $\mathfrak{M}^1$ VB map with $\tau_{\mathbb{E}} \circ \Xi = I_{\mathbb{E}}$. There exists D an open full neighborhood of $X \times 0$ in $X \times R$ satisfying (1) and (2) of (a) and: (3) The flow F_Ξ of Ξ is defined on $(\pi \times 1_R)^{-1}D$ and $(F_\Xi,F_\xi): (\pi \times 1_R)_D \to \pi$ is an $\mathfrak{M}^1$ VB map.

(c) If X is semicompact, then ξ in (a) and Ξ in (b) are complete vectorfields and D can be chosen to be $X \times (-1,1)$.

<u>PROOF</u>: Choose $\mathfrak{a}$ or $(\mathfrak{D},\mathfrak{a})$ s admissible for X or π and operate locally in the charts. The results in (a) and (b) easily follow from Proposition II.3.9 and Proposition II.3.8. In the semicompact case $X \times 0$ is bounded in $X \times R$ and so by Corollary 6.4, D contains $X \times [-N^{-1},+N^{-1}]$ for some integer $N \geq 1$. We can extend the flow to $X \times [-2N^{-1},+2N^{-1}]$ by $F(x,t) = F(F(x,N^{-1}),t-N^{-1})$ for $N^{-1} \leq t \leq 2N^{-1}$ and similarly for $-2N^{-1} \leq t \leq -N^{-1}$. By FS9 and the Gluing Property the resulting enlarged flow is still an $\mathfrak{M}^1$ metric map. Q.E.D.

Even if X is $\mathcal{C}^\infty$ semicomplete, i.e. there exist s admissible $\mathcal{C}^\infty$ atlases, and ξ is $\mathcal{C}^\infty$, it is not clear that F_ξ is $\mathcal{C}^\infty$ on any full neighborhood D of $X \times 0$. This is because with F_ξ $\mathcal{C}^r$ on D it may be necessary to shrink D to have F_ξ be $\mathcal{C}^{r+1}$. By (c) this shrinking is not necessary if X is semicompact. Thus, if X is a $\mathcal{C}^\infty$ semicompact manifold and ξ is a $\mathcal{C}^\infty$ vectorfield then the flow $F_\xi: X \times (-1,1) \to X$ is a $\mathcal{C}^\infty$ metric map. Similarly, for linear vectorfields on $\mathcal{C}^\infty$ semicompact vector bundles.

Now we prove that sprays yield exponentials. The last few steps are most easily handled using results from Chapter VI. Rather than defer the completion of the proof, we will simply refer ahead.

3 COROLLARY: (a) Let X be a semicomplete $\mathfrak{M}^4$ manifold and let $\xi: TX \to T^2X$ be an $\mathfrak{M}^2$ spray on X. There is an $\mathfrak{M}^2$ exponential on $X, e: D \to V$ characterized by $\frac{d}{dt} p_2 \circ e(tv)|_{t=0} = v$ and $\frac{d^2}{dt^2} p_2 \circ e(tv) = \xi(\frac{d}{dt} p_2 \circ e(tv))$.

(b) Let $\pi: E \to X$ be a semicomplete $\mathfrak{M}^4$ bundle and let $(\Xi, \xi): T\pi \to T^2\pi$ be an $\mathfrak{M}^2$ linear spray on π. There is an $\mathfrak{M}^2$ bundle exponential on π $(E,e): (T\pi)_D \to (\pi \times \pi)_V$ with $e: D \to V$ and $E: T\pi^{-1}D \to (\pi \times \pi)^{-1}V$ characterized by ξ and Ξ as in (a).

PROOF: (a): By Theorem 2(a) the flow $F_\xi: D_\xi \subset TX \times \mathbb{R} \to TX$ is an $\mathfrak{M}^2$ map on some full neighborhood $\tilde{D}$ of $TX \times 0$. Let $\|\ \|$ be a continuous admissible Finsler on τ_X, d be an admissible metric on X, and ρ be a continuous admissible bound on X. There exist $K, n \geq 1$ such that $|t| \leq (K \max(\rho(\tau v), \|v\|)^n)^{-1}$ implies $(v,t) \in \tilde{D}$. Let $\epsilon: X \to (0,1]$ be defined by $\epsilon(x) = (K\rho(x)^n)^{-1}$. Because ξ is a spray, $(v,t) \in \tilde{D}$ implies $(tv,1) \in D_\xi$ and $\tau F_\xi(v,t) = \tau F_\xi(tv,1)$. Thus, if $d(\tau v, x) < 2^{-1}\epsilon_d(x)$ and $\|v\| < 2^{-1}\epsilon_d(\pi v)$ then $2^{-1}\epsilon_d(\pi v) \leq \epsilon_d(x) \leq \epsilon(v)$. So $(\epsilon_d(x)^{-1}v, \epsilon_d(x)) \in \tilde{D}$ and $\tau F_\xi(v,1) = \tau F_\xi(\epsilon_d(x)^{-1}v, \epsilon_d(x))$. Let $D_0 = \{v: \|v\| < 2^{-1}\epsilon_d(\pi v)\}$. It is a full open neighborhood of 0_X in TX by Proposition 7.15. $e_0: D_0 \to X$ defined by $e_0(v) = \tau F_\xi(v,1)$ is an $\mathfrak{M}^2$ metric map. In fact by FS4, on $D_0 \cap \tau^{-1}B^d(x, 2^{-1}\epsilon_d(x))$, $\rho^{\mathfrak{M}^2}(e_0; TG, G)$ is bounded by $O*(\rho^{\mathfrak{M}^2}(\tau F_\xi|\tilde{D} \cap \tau^{-1}B^d; TG, G), \epsilon_d(x)^{-1})$. Let $e = \tau \times e_0: D_0 \to X \times X$. Because ξ is a spray $e_0 = p_2 \circ e$ is characterized as in the statement of (a). Clearly, $e \circ 0_X = \Delta_X$. There is an identification at the atlas level between $0_X{}^*\tau_{TX}$ and $\tau_X \oplus \tau_X$ by $\bar{J}$ (cf. p. 13). $\Delta_X{}^*\tau_{X\times X} \cong \tau_X \oplus \tau_X$. By the characterization of e_0, $(0_X, \Delta_X)^*Te: \tau_X \oplus \tau_X \to \tau_X \oplus \tau_X$ is given by the matrix $\begin{pmatrix} I & I \\ 0 & I \end{pmatrix}$: $T_xX \times T_xX \to T_xX \times T_xX$. Since e is $\mathfrak{M}^2$, it follows from Chapter VI, Corollary VI.5.3 that we can shrink D_0 to D an open full neighborhood of 0_X on which e is an $\mathfrak{M}^2$ isomorphism onto V an open subset of $X \times X$. By Corollary 7.8 (b) V is a full open neighborhood of Δ_X in $X \times X$.

(b) By Theorem 2(b) there is a full open neighborhood $\tilde{D}$ of $TX \times 0$ in $TX \times R$ such that (F_{Ξ}, F_{ξ}): $(T\pi \times 1_R)_{\tilde{D}} \to T\pi$ is an $\mathfrak{M}^2$ metric VB map. As in (a) we define D_0 a full open neighborhood of 0_X in TX and (E,e): $(T\pi)_{D_0} \to \pi \times \pi$ an $\mathfrak{M}^2$ VB map. We shrink D_0 as in (a) to get e: $D \to V$ an $\mathfrak{M}^2$ metric exponential. There is an identification at the atlas level between $0_X{*}T\pi$ and $\pi \oplus \pi$ by J (cf. p. 13). It is not hard to check that $(0_X,\Delta_X){*}(E,e)$: $0_X{*}T\pi \cong \pi \oplus \pi \to \Delta_X^*\pi \times \pi \cong \pi \oplus \pi$ is given by the matrix $\begin{pmatrix} I & 0 \\ I & I \end{pmatrix}$: $E_x \times E_x \to E_x \times E_x$. Thus, we can apply Corollary VI.2.6 and shrink D, if necessary, to get $(e{*}E,1)$: $(T\pi)_D \to e{*}(\pi\times\pi)$ an $\mathfrak{M}^2$ VB isomorphism. Then, with $V = e(D)$, (E,e): $(T\pi)_D \to (\pi\times\pi)_V$ is an $\mathfrak{M}^2$ metric VB isomorphism by Proposition 5.5. Q.E.D.

If X is semicompact, then by an argument like that of Theorem 2 (c), F_ξ is defined and $\mathfrak{M}^2$ on any bounded open set D in $TX \times R$. Thus, if X is semicompact and $\mathcal{C}^\infty$ we can find e: $D \to V$ a $\mathcal{C}^\infty$ map which is a $\mathcal{C}^2$ exponential. By Proposition 5.5, e is then a $\mathcal{C}^\infty$ exponential. Similarly, for bundles.

To an exponential on X there is an associated atlas called the <u>atlas of normal coordinates</u> induced by e.

4 <u>PROPOSITION</u>: Let X be a semicomplete $\mathfrak{M}^1$ manifold with e: $D \to V$ an $\mathfrak{M}$ exponential. Define $V_x = V[x]$, $D_x = D \cap T_xX$ and $e_x = p_2 \circ e$: $D_x \to V_x$ for $x \in X$. If V is transversely bounded, and so D is vertically bounded, then $\mathcal{a}_e = \{V_x, e_x^{-1}\}$ is an s admissible $\mathfrak{M}$ atlas on X where T_xX is made a B-space for all x by choice of admissible Finsler on τ_X.

If $\pi: \mathbb{E} \to X$ is a semicomplete $\mathfrak{M}$ bundle over X and $T: (\pi\times 1)_V \to (1\times\pi)_V$ is an $\mathfrak{M}$ T isomorphism, then T restricts to $T_x: \mathbb{E}_x \times V_x \to \{x\} \times (\mathbb{E})_{V_x} \cong (\mathbb{E})_{V_x}$. $(\mathcal{T}_T, \mathcal{G}_e) = \{V_x, e_x^{-1}, T_x^{-1}\}$ is an s admissible $\mathfrak{M}$ atlas on π where $\mathbb{E}_x$ is made a B-space for all x by choice of admissible Finsler on π.

<u>PROOF</u>: Let $\mathcal{G} = \{U_\alpha, h_\alpha\}$ be an admissible $\mathfrak{M}^1$ atlas and ρ be an admissible bound on X. It suffices to show that $(\mathcal{G} \cup \mathcal{G}_e, \rho)$ is an adapted atlas on X. In turn, it suffices to show that $\rho > \rho(1;\mathcal{G},\mathcal{G}_e)$ and $\rho(1;\mathcal{G}_e,\mathcal{G})$ since then $\rho_{\mathcal{G}_e} \leq 0^*(\rho(1;\mathcal{G}_e,\mathcal{G}),\rho(1;\mathcal{G},\mathcal{G}_e)) \prec \rho$. Now for $(x,y) \in V$ choose $\mathcal{G}$ charts (U_α,h_α) and (U_β,h_β) containing x and y respectively. To estimate $\|h_\beta \cdot e_x | e_x^{-1}(U_\beta \cap V_x)\|_{\mathfrak{M}}$ and $\|e_x^{-1} \cdot h_\beta^{-1} | h_\beta(U_\beta \cap V_x)\|_{\mathfrak{M}}$ we use the following diagram, where $V_{\alpha\beta} = V \cap U_\alpha \times U_\beta$ and $D_{\alpha\beta} = e^{-1}(V_{\alpha\beta})$:

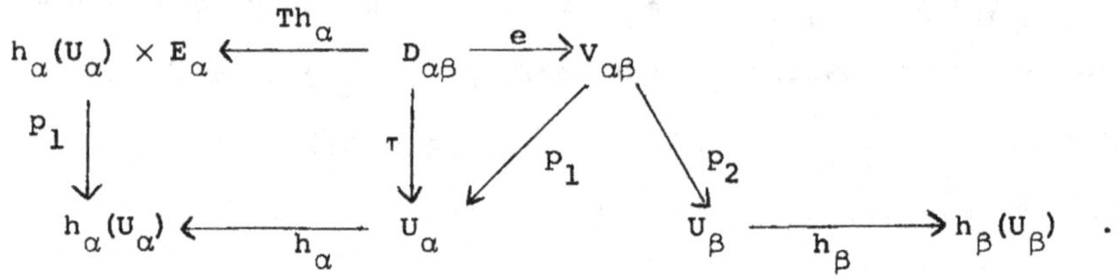

Choose $U_x \times U_y$ a neighborhood of (x,y) in $V_{\alpha\beta}$ and $h_\alpha(U_x) \times U_0$ a neighborhood of $Th_\alpha(e^{-1}(x,y))$ such that: $\|Th_\alpha e^{-1}(h_\alpha \times h_\beta)^{-1} | h_\alpha(U_x) \times h_\beta(U_y)\|_{\mathfrak{M}} < 2\rho(e^{-1};\mathcal{G}\times\mathcal{G},T\mathcal{G})(x,y)$ and $\|(h_\alpha \times h_\beta) e \cdot Th_\alpha^{-1} | h_\alpha(U_x) \times U_0\|_{\mathfrak{M}} < 2\rho(e;T\mathcal{G},\mathcal{G}\times\mathcal{G})(e^{-1}(x,y))$. By transverse and vertical boundedness, $p_1{}^*\rho > \rho_{e^{-1}}$ and $\tau^*\rho > \rho_e$. Also, $\| \ \| >_\rho \| \ \|^M$ and $\| \ \|^m >_\rho \| \ \|$ where the max and min Finslers are defined by $(T\mathcal{G},\mathcal{G})$. Let $\tilde{U}_y = e_x \cdot T_x h_\alpha^{-1}(U_0)$. It follows that for some constants K,n $\|e_x^{-1} \cdot h_\beta^{-1} | h_\beta(U_y)\|_{\mathfrak{M}} \leq K\rho(x)^n$ and $\|h_\beta \cdot e_x | e_x(\tilde{U}_y)\|_{\mathfrak{M}} \leq K\rho(x)^n$. Thus, by the Gluing Property, $\|e_x^{-1} \cdot h_\beta^{-1} | h_\beta(U_\beta \cap V_x)\|_{\mathfrak{M}}$ and $\|h_\beta \cdot e_x | e_x(U_\beta \cap V_x)\|_{\mathfrak{M}} \leq K\rho(x)^n$. Since $\rho^{V^{-1}}(y) \geq \rho(x)$ for all $y \in V_x$ we have finally, $K(\rho^{V^{-1}})^n \geq \rho_e$ and $\rho_{e^{-1}}$. Since V^{-1} is transverse bounded, $\rho^{V^{-1}} \sim \rho$. The bundle version is similar. Q.E.D.

CHAPTER IV: MANIFOLDS OF MAPPINGS

1. <u>Section Spaces</u>: Let $\pi: \mathbb{E} \to X$ be a bounded $\mathfrak{M}$ bundle with admissible atlas $(\mathfrak{D},\mathfrak{A})$. We define the norm $\| \;\|^{(\mathfrak{D},\mathfrak{A})}$ on the space of $\mathfrak{M}$ sections $\mathfrak{M}(\pi)$ by $\|s\|^{(\mathfrak{D},\mathfrak{A})} = \sup \tilde{\rho}(s;\mathfrak{A},\mathfrak{D})$ (cf. p. 46). Equivalently, if $(\mathfrak{D},\mathfrak{A}) = \{U_\alpha,h_\alpha,\varphi_\alpha\}$ and $P_\alpha: \mathfrak{M}(\pi) \to \mathfrak{M}(h_\alpha(U_\alpha),F_\alpha)$ is defined by $P_\alpha(s) = s_\alpha$, the α principal part of s, i.e. $s_\alpha = p_2 \circ \varphi_\alpha \circ s \circ h_\alpha^{-1}$, then $\hat{\Pi}\, P_\alpha: \mathfrak{M}(\pi) \to \hat{\Pi}_\alpha \mathfrak{M}(h_\alpha(U_\alpha),F_\alpha)$. By the Gluing Property, $\| \;\|^{(\mathfrak{D},\mathfrak{A})}$ is the norm on $\mathfrak{M}(\pi)$ making $\hat{\Pi} P_\alpha$ an isometry. To see that $\mathfrak{M}(\pi)$ with norm $\| \;\|^{(\mathfrak{D},\mathfrak{A})}$ is complete, or equivalently that the image of $\hat{\Pi} P_\alpha$ is closed, define for $U_\alpha \cap U_\beta \neq \emptyset$ $\delta_{\alpha\beta}: \mathfrak{M}(h_\alpha(U_\alpha),F_\alpha) \times \mathfrak{M}(h_\beta(U_\beta),F_\beta) \to \mathfrak{M}(h_\alpha(U_\alpha \cap U_\beta),F_\alpha)$ by $\delta_{\alpha\beta}(s_\alpha,s_\beta)(u) = s_\alpha(u) - \varphi_{\alpha\beta}(h_{\beta\alpha}u)[s_\beta(h_{\beta\alpha}u)]$. Define $\delta_{\alpha\beta}: \hat{\Pi}_\gamma \mathfrak{M}(h_\gamma(U_\gamma),F_\gamma) \to \mathfrak{M}(h_\alpha(U_\alpha \cap U_\beta),F_\alpha)$ by projecting to the α and β factors first. Clearly, $\delta_{\alpha\beta}$ is a continuous linear map with $\|\delta_{\alpha\beta}\| \leq O^*(k_{(\mathfrak{D},\mathfrak{A})})$. So $\delta = \hat{\Pi}_{\alpha\beta}\delta_{\alpha\beta}$ is a well defined continuous linear map and by the Gluing Property the sequence:

$$0 \longrightarrow \mathfrak{M}(\pi) \xrightarrow{\hat{\Pi}_\gamma P_\gamma} \hat{\Pi}_\gamma \mathfrak{M}(h_\gamma(U_\gamma),F_\gamma) \xrightarrow{\;\delta\;} \hat{\Pi}_{\alpha\beta}\mathfrak{M}(h_\alpha(U_\alpha \cap U_\beta),F_\alpha)$$

is an exact sequence. Thus, Image $\hat{\Pi}_\gamma P_\gamma$ = Kernel δ is closed.

1 <u>LEMMA</u>: (a) Let $(\mathfrak{D}_1,\mathfrak{A}_1)$, $(\mathfrak{D},\mathfrak{A})$ be admissible atlases on the bounded $\mathfrak{M}$ bundle $\pi: \mathbb{E} \to X$. If $(\mathfrak{D}_1,\mathfrak{A}_1)$ is a refinement of $(\mathfrak{D},\mathfrak{A})$ then $\| \;\|^{(\mathfrak{D}_1,\mathfrak{A}_1)} \leq \| \;\|^{(\mathfrak{D},\mathfrak{A})}$ with equality when $(\mathfrak{D}_1,\mathfrak{A}_1)$ is a subdivision of $(\mathfrak{D},\mathfrak{A})$. For any pair of admissible atlases, $\| \;\|^{(\mathfrak{D},\mathfrak{A})} \leq O^*(k_{(\mathfrak{D}_1 \cup \mathfrak{D},\mathfrak{A}_1 \cup \mathfrak{A})})\| \;\|^{(\mathfrak{D}_1,\mathfrak{A}_1)}$.

(b) Let $(\mathfrak{D}_i,\mathfrak{A}_i)$ be an admissible atlas on $\pi_i: \mathbb{E}_i \to X$, a bounded $\mathfrak{M}$ bundle $(i = 1,2)$. If $(\Phi,1_X): \pi_1 \to \pi_2$ is an $\mathfrak{M}$ VP map then the induced map of sections $\Phi_*: \mathfrak{M}(\pi_1) \to \mathfrak{M}(\pi_2)$ by $\Phi_*(s) = \Phi \circ s$ is a continuous linear map with $\| \ \|^{(\mathfrak{D}_2,\mathcal{G}_2)} \circ \Phi_* \leq O^*(k_{(\Phi,1)}) \| \ \|^{(\mathfrak{D}_1,\mathcal{G}_1)}$.

<u>PROOF</u>: (a): For $s\colon X \to \mathfrak{E}$ and $(\mathfrak{D}_1,\mathcal{G}_1)$ a refinement of $(\mathfrak{D},\mathcal{G})$, $\gamma(s;\mathcal{G}_1,\mathfrak{D}_1) \leq \gamma(s;\mathcal{G},\mathfrak{D})$ with equality when $(\mathfrak{D}_1,\mathcal{G}_1)$ is a subdivision. The rest follows from (b) applied to the identity map.

(b): By the subdivision result, we can intersect the covers of $\mathcal{G}_1$ and $\mathcal{G}_2$ without changing the norms. Thus, we can assume $(\mathfrak{D}_1,\mathcal{G}_1) = \{U_\alpha, h_\alpha, \varphi_\alpha\}$ and $(\mathfrak{D}_2,\mathcal{G}_2) = \{U_\alpha, g_\alpha, \psi_\alpha\}$. Define $(\Phi_\alpha)_*\colon \mathfrak{M}(h_\alpha(U_\alpha), F^1_\alpha) \to \mathfrak{M}(g_\alpha(U_\alpha), F^2_\alpha)$ by $(\Phi_\alpha)_*(s_\alpha)(u) = \Phi_\alpha(h_\alpha g_\alpha^{-1}u)[s_\alpha(h_\alpha g_\alpha^{-1}u)]$. Clearly, $(\Phi_\alpha)_*$ is a continuous linear map with $\|(\Phi_\alpha)_*\| \leq O^*(k_{(\Phi,1)})$. The result then follows from the commutative diagram:

$$\begin{array}{ccc}
\mathfrak{M}(\pi_1) & \xrightarrow{\ \hat{\Pi} P_\alpha\ } & \hat{\Pi}\mathfrak{M}(h_\alpha(U_\alpha), F^1_\alpha) \\
\Phi_* \big\downarrow & & \big\downarrow \Pi(\Phi_\alpha)_* \\
\mathfrak{M}(\pi_2) & \xrightarrow[\ \hat{\Pi} P_\alpha\ ]{} & \hat{\Pi}\mathfrak{M}(g_\alpha(U_\alpha), F^2_\alpha).
\end{array}$$

Q.E.D.

2 <u>COROLLARY</u>: Let $\pi\colon \mathfrak{E} \to X$ be a bounded $\mathfrak{M}$ bundle. The space of sections $\mathfrak{M}(\pi)$ has a Banachable space structure with norms defined by admissible atlases.

If $\mathfrak{M}_1 \subset \mathfrak{M}$ and $\pi\colon \mathfrak{E} \to X$ is a bounded $\mathfrak{M}_1$ bundle the the inclusion $\mathfrak{M}_1(\pi) \to \mathfrak{M}(\pi)$ is continuous. In fact, if $(\mathfrak{D},\mathcal{G})$ is an admissible $\mathfrak{M}_1$ atlas then $\| \ \|_{\mathfrak{M}}^{(\mathfrak{D},\mathcal{G})} \leq \| \ \|_{\mathfrak{M}_1}^{(\mathfrak{D},\mathcal{G})}$ with equality if the inclusion $\mathfrak{M}_1 \subset \mathfrak{M}$ is isometric. If $\mathfrak{M}_1 \subset \mathfrak{M}$ is a strong inclusion then $\{s\colon \|s\|_{\mathfrak{M}_1}^{(\mathfrak{D},\mathcal{G})} \leq K\}$ is closed in $\mathfrak{M}(\pi)$.

3 <u>PROPOSITION</u>: Let $(\mathfrak{D},\mathcal{G})$ be an admissible atlas on a bounded $\mathcal{C}$ bundle π and let $\| \ \|^M$ be the associated max Finsler. For $s \in \mathcal{C}(\pi)$, $\|s\|_{\mathcal{C}}^{(\mathfrak{D},\mathcal{G})} = \sup\{\|s(x)\|^M\colon x \in X\}$. For $\| \ \|$ any admissible Finsler on π, $\|s\|_0 = \sup\{\|s(x)\|\colon x \in X\}$ is a norm on $\mathcal{C}(\pi)$.

<u>PROOF</u>: That $\| \ \|_{\mathcal{C}}^{(\mathfrak{D},\mathcal{G})} = \| \ \|_0$ for $\| \ \| = \| \ \|^M$ is easy to check. For any

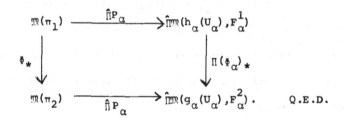

other admissible Finsler $\| \|$ on π, $\| \| \sim_1 \| \|^M$ and so the associated $\| \|_0$ norms are equivalent. Q.E.D.

THE STANDARD CONSTRUCTIONS: (1) Open Subsets: For U open in X and $(\mathfrak{D},\mathcal{G})$ admissible for π the restriction $\mathfrak{M}(\pi) \to \mathfrak{M}(\pi_U)$ is linear and norm decreasing with respect to $\| \|^{(\mathfrak{D},\mathcal{G})}$ and $\| \|^{(\mathfrak{D},\mathcal{G})|U}$. (2) Products: For $(\mathfrak{D}_i,\mathcal{G}_i)$ admissible on π_i the product map $\mathfrak{M}(\pi_1) \times \mathfrak{M}(\pi_2) \to \mathfrak{M}(\pi_1 \times \pi_2)$ by $(s_1,s_2) \to s_1 \times s_2$ is linear and norm decreasing with respect to the norms $\| \|^{(\mathfrak{D}_1,\mathcal{G}_1)\times(\mathfrak{D}_2,\mathcal{G}_2)}$ and $\| \|^{(\mathfrak{D}_i,\mathcal{G}_i)}$ $(i = 1,2)$. If neither X_1 nor X_2 is empty then the map is injective and splits. Choose $(x_1,x_2) \in X_1 \times X_2$ and map $\mathfrak{M}(\pi_1 \times \pi_2) \to \mathfrak{M}(\pi_1) \times \mathfrak{M}(\pi_2)$ by $s \to (s_1,s_2)$ where $s_1(x) = P_1 s(x,x_2)$ and similarly for s_2. (3) Trivial Bundles: For $\mathcal{G}$ admissible on a bounded manifold X and a B-space F the identification between metric maps from X to F (norm structure) and metric sections of ϵ_F: $X \times F \to X$ defines a Banachable space structure on $\mathfrak{M}(X,F)$ identified with $\mathfrak{M}(\epsilon_F)$. The norm $\| \|^{(\mathcal{G}\times F,\mathcal{G})}$ on $\mathfrak{M}(\epsilon_F)$ defines the norm $\| \|^{\mathcal{G}}$ on $\mathfrak{M}(X,F)$:
$\|f\|^{\mathcal{G}} = \|\{f_\alpha\}\|$ with $\{f_\alpha = f \cdot h_\alpha^{-1}\} \in \hat{\Pi}\, \mathfrak{M}(h_\alpha(U_\alpha),F)$. If U is open and bounded in a B-space E then $\mathcal{G}_E|U = \{U,1_U\}$ yields the original norm on $\mathfrak{M}(U,F)$. FS2 extends: The map c: $F \to \mathfrak{M}(X,F)$ taking vectors to constant functions is an isometry into with respect to $\| \|^{\mathcal{G}}$. By Proposition 3, on $\mathcal{C}(X,F)$ the topology is given by the sup norm. (4) Subbundles: For $(\mathfrak{D},\mathcal{G})$ admissible on π: $\mathbb{E} \to X$ and inducing a subbundle atlas on $\mathbb{E}_0 \subset \mathbb{E}$, the inclusion $\mathfrak{M}(\pi|\mathbb{E}_0) \to \mathfrak{M}(\pi)$ is an isometry into with respect to $\| \|^{(\mathfrak{D}|\mathbb{E}_0,\mathcal{G})}$ and $\| \|^{(\mathfrak{D},\mathcal{G})}$. (5) Operations: Let $(\mathfrak{D}_1,\mathfrak{D}_2;\mathcal{G})$ be an admissible two-tuple on π_i: $\mathbb{E}_i \to X$ $(i = 1,2)$. $P_{1*} \times P_{2*}$: $\mathfrak{M}(\pi_1 \oplus \pi_2) \to \mathfrak{M}(\pi_1) \times \mathfrak{M}(\pi_2)$ is an isometric isomorphism with respect to $\| \|^{(\mathfrak{D}_1\oplus\mathfrak{D}_2,\mathcal{G})}$ and $\| \|^{(\mathfrak{D}_i,\mathcal{G})}$ $(i = 1,2)$. For $L^n_\epsilon = L^n$ or L^n_s, the map λ: $\mathfrak{M}(L^n_\epsilon(\pi_1;\pi_2)) \to L^n_\epsilon(\mathfrak{M}(\pi_1);\mathfrak{M}(\pi_2))$ is defined by $\lambda(S)(s_1,\ldots,s_n)(x) = S(s_1(x),\ldots,s_n(x))$ where $(S,1)$ is the n linear VB map $\pi_1 \oplus \cdots \oplus \pi_1 \to \pi_2$ associated to the section S, i.e. $\lambda(S)(s_1,\ldots,s_n)(x) = ev_\circ(S,s_1,\ldots,s_n)(x)$ where

ev: $L^n_\epsilon \oplus \pi_1^{(n)} \to \pi_2$ is the n + 1 linear evaluation map. Since the map $\lambda_\alpha\colon \mathfrak{M}(h_\alpha(U_\alpha), L^n_\epsilon(F^1_\alpha;F^2_\alpha)) \to L^n_\epsilon(\mathfrak{M}(h_\alpha(U_\alpha),F^1_\alpha);\mathfrak{M}(h_\alpha(U_\alpha),F^2_\alpha))$ has norm $\leq O^*(1)$, λ is a linear map with norm $\leq O^*(1)$ with respect to $\|\ \|^{(L^n_\epsilon(\mathfrak{D}_1;\mathfrak{D}_2),\mathcal{G})}$ and $L^n_\epsilon(\|\ \|^{(\mathfrak{D}_1,\mathcal{G})}, \|\ \|^{(\mathfrak{D}_2,\mathcal{G})})$. In particular, from the trivial bundle case we generalize diagram II.(1.4):

(1.1)

$$\begin{array}{ccc}
 & & L^n_\epsilon(F_1;F_2) \\
 & \swarrow^{c} & \downarrow * \\
\mathfrak{M}(X,L^n_\epsilon(F_1;F_2)) & \xrightarrow{\ \lambda\ } & L^n_\epsilon(\mathfrak{M}(X,F_1);\mathfrak{M}(X,F_2)) \\
 & \searrow_{\kappa} & \downarrow L^n_\epsilon(c;1) \\
 & & L^n_\epsilon(F_1;\mathfrak{M}(X,F_2))
\end{array}$$

With $\mathcal{G}$ on X inducing $\|\ \|^{\mathcal{G}}$ on the mapping spaces we have $\kappa, c, L^n_\epsilon(c;1)$ are isometries and $*,\lambda$ have norm $\leq O^*(1)$ with $\|*\| \leq 1$ if n = 1. These results follow from the corresponding ones for II.(1.4) by projecting from $\mathfrak{M}(X,\_)$ to $\mathfrak{M}(h_\alpha(U_\alpha),\_)$. Similarly, it follows from Proposition II.1.8 that if $\mathfrak{M}$ satisfies FS7 (Strong) or F_1 is finite dimensional then κ is onto in (1.1). From these results various standard bilinear maps are seen to be defined and continuous: multiplication of functions: $\mathfrak{M}(X,R) \times \mathfrak{M}(X,R) \to \mathfrak{M}(X,R)$, multiplication of sections by functions: $\mathfrak{M}(X,R) \times \mathfrak{M}(\pi) \to \mathfrak{M}(\pi)$, evaluation: $\mathfrak{M}(\pi) \times \mathfrak{M}(L(\pi;\epsilon_R)) = \mathfrak{M}(\pi) \times \mathfrak{M}(\pi^*) \to \mathfrak{M}(X,R)$. With respect to $\|\ \|^{\mathcal{G}}, \|\ \|^{(\mathfrak{D},\mathcal{G})}$ and $\|\ \|^{(L(\mathfrak{D};\mathcal{G}\times R),\mathcal{G})}$ on $\mathfrak{M}(X,R)$, $\mathfrak{M}(\pi)$ and $\mathfrak{M}(\pi^*)$ these bilinear maps have norm $\leq O^*(1)$.

(6) <u>Pull Backs</u>: For $(\mathfrak{D},\mathcal{G})$ admissible on $\pi\colon \mathbb{E} \to X$, $\mathcal{G}_0$ admissible on X_0, f: $X_0 \to X$ an $\mathfrak{M}$ metric map with f: $\mathcal{G}_0 \to \mathcal{G}$ index preserving, $f^*\colon \mathfrak{M}(\pi) \to \mathfrak{M}(f^*\pi)$ is a linear map with $\|f^*\| \leq O^*(\bar{K}_f)$ by FS9 for $\|\ \|^{(\mathfrak{D},\mathcal{G})}$ and $\|\ \|^{(f^*\mathfrak{D},\mathcal{G}_0)}$ on π, $f^*\pi$. If f: $\mathcal{G}_0 \to \mathcal{G}$ is not index preserving we can apply this result to $\mathcal{G}_f = \mathcal{G}_0 \cap f^{-1}\mathcal{G}$ and get $\|f^*\| \leq O^*(k(f;\mathcal{G}_0,\mathcal{G}))$ for $\|\ \|^{(\mathfrak{D},\mathcal{G})}$ and $\|\ \|^{(f^*\mathfrak{D},\mathcal{G}_f)}$. (7) <u>Jet Bundles:</u> For $(\mathfrak{D},\mathcal{G})$ admissible on

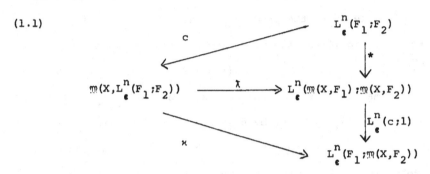

$\pi\colon \mathbb{E} \to X$ a bounded $\mathfrak{M}^k$ bundle the jet map $j^k\colon \mathfrak{M}^k(\pi) \to \mathfrak{M}(J^k(\pi))$ is an isometry with respect to $\| \|^{(\mathfrak{D},\mathcal{G})}$ and $\| \|^{(J^k(\mathfrak{D}),\mathcal{G})}$ because $j^k(s)_\alpha = j^k(s_\alpha)$. In particular, by Proposition 3 the norm $\| \|^{(\mathfrak{D},\mathcal{G})}$ on $\mathcal{C}^k(\pi)$ is given by $\|s\|^{(\mathfrak{D},\mathcal{G})} = \sup\{\|j^k(s)(x)\|^M\colon x \in X\}$ where $\| \|^M$ is induced by $(J^k(\mathfrak{D}),\mathcal{G})$ on $J^k(\pi)$. Also, if $\| \|$ is any admissible Finsler on $J^k(\pi)$ then $\|s\|_k = \|j^k(s)\|_0$ induced by $\| \|$ is a norm on $\mathcal{C}^k(\pi)$.

4 PROPOSITION: For $\pi\colon \mathbb{E} \to X$ a bounded $\mathfrak{M}$ metric bundle and $N\colon \mathbb{E} \to [1,\infty]$ define $S_N\colon \mathfrak{M}(\pi) \to [1,\infty]$ by $S_N(s) = \sup s^*N$. If N is vertically regular (cf. p. 98) then $\{S_N < \infty\}$ is open in $\mathfrak{M}(\pi)$ and $\max(\| \|_1, S_N)$ defines a semicomplete refinement of the norm metric structure on $\{S_N < \infty\}$ where $\| \|_1$ is a norm on $\mathfrak{M}(\pi)$.

PROOF: Let $\| \|$ be an admissible Finsler on π and $\| \|_0$ the associated sup norm on $\mathcal{C}(\pi)$. Since $N >_1 N^{\| \|}$ there exists K such that $\|s - s_1\|_0 < (KS_N(s))^{-1}$ implies $s_1 \le KS_N(s)$. It follows that $\{S_N < \infty\}$ is open in $\mathcal{C}(\pi)$ and $\max(\| \|_0, S_N)$ induces a semicomplete refinement. $\{S_N < \infty\}$ in $\mathfrak{M}(\pi)$ is just the inverse of the corresponding set in $\mathcal{C}(\pi)$ under the inclusion $\mathfrak{M}(\pi) \to \mathcal{C}(\pi)$. The result follows from Lemma III.7.11 or an easy direct argument. Q.E.D.

5 COROLLARY: (a) For $\| \|$ an admissible Finsler on $\pi\colon \mathbb{E} \to X$ and $O \subset \mathbb{E}$ the set $\mathfrak{M}(O) = \{s\colon \inf s^*\delta^O(\| \|) > 0\}$ is open in $\mathfrak{M}(\pi)$ and is independent of the choice of admissible Finsler.

(b) For π_1,π_2 bounded $\mathfrak{M}$ metric bundles on X the set $\mathfrak{M}(\mathrm{Lis}(\pi_1;\pi_2)) = \{S \in \mathfrak{M}(L(\pi_1;\pi_2))\colon (S,1_X)\colon \pi_1 \to \pi_2$ is an $\mathfrak{M}$ VB isomorphism$\}$, is open in $\mathfrak{M}(L(\pi_1;\pi_2))$.

PROOF: (a): Let $N = 1/\delta^O(\| \|)$. (b): For $\| \|_i$ an admissible Finsler on π_i $(i = 1,2)$ and $\theta\colon L(\mathbb{E}_1;\mathbb{E}_2) \to [1,\infty]$ the associated function, $\mathfrak{M}(\mathrm{Lis}) = \{S_\theta < \infty\}$ by Proposition III.5.5(b). Let $N = \theta$. Q.E.D.

6 LEMMA: Let $\pi: \mathbb{E} \to X$ be a semicomplete $\mathfrak{M}_1$ metric bundle with $\mathfrak{M}_1 \subset \mathfrak{M}$. Let $O \subset O_1$ be open sets in $\mathbb{E}$, X_0 be a bounded $\mathfrak{M}$ metric manifold and Z be an $\mathcal{L}$ metric manifold. Assume $f: X_0 \to X$ is an $\mathfrak{M}$ metric map and $m: Z \to \mathfrak{M}(f^*\pi)$ is an $\mathcal{L}$ metric map with $\Phi_f \circ m(z)(X_0) \subset O_1$ for all $z \in Z$. $Z_0 = \{z: \Phi_f \circ m(z)(X_0) \subset\subset O \text{ (rel } O_1)\}$ is open in Z.

PROOF: Here O_1 is given the restriction $\mathfrak{M}_1$ metric manifold structure. If $(\mathfrak{D},\mathfrak{a})$ is an s admissible atlas on π, $d_{\mathfrak{D}|O_1}$ is an admissible metric on O_1. Let $\mathfrak{a}_0$ be an admissible atlas on X_0 such that $f: \mathfrak{a}_0 \to \mathfrak{a}$ is index preserving and $\mathfrak{a}_0$ is a subdivision of $f^{-1}\mathfrak{a}$. Since the result is local we can assume Z is a bounded open set in a B-space. For $z_1 \in Z_0$ choose U open in $\mathbb{E}$ and $\varepsilon > 0$ such that: (1) $\Phi_f \circ m(z_1)(X_0) \subset\subset U \subset\subset O$ (rel O_1). (2) $d_{\mathfrak{D}|O_1}(\Phi_f \circ m(z_1)(x), v) < \varepsilon$ with $x \in X_0$, $v \in O_1$ implies $v \in U$. Let $K = \|m\|_L$ with norm induced by $(f^*\mathfrak{D},\mathfrak{a}_0)$ on $\mathfrak{M}(f^*\pi)$ and choose $0 < r < \varepsilon/K$ such that $B = B(z_1,r) \subset Z$. It suffices to show that $(x,z) \in X_0 \times B$ implies $d_{\mathfrak{D}|O}(\Phi_f \circ m(z_1)(x), \Phi_f \circ m(z)(x)) < \varepsilon$. Choose a monotone sequence $z_1, z_2, \ldots, z_N = z$ on the segment $[z_1, z]$ such that the segment $[\Phi_f \circ m(z_i)(x), \Phi_f \circ m(z_{i+1})(x)]$ lies in $O_1 \cap \mathbb{E}_{fx}$. Then
$d_{\mathfrak{D}|O_1}(\Phi_f \circ m(z_i)(x), \Phi_f \circ m(z_{i+1})(x)) \le \|\Phi_f \circ m(z_i)(x) - \Phi_f \circ m(z_{i+1})(x)\|^M$
$= \|m(z_i)(x) - m(z_{i+1})(x)\|^M \le \|m(z_i) - m(z_{i+1})\|_{\mathfrak{M}} \le K\|z_i - z_{i+1}\|$. Summing we get $d_{\mathfrak{D}|O_1}(\Phi_f \circ m(z_1)(x), \Phi_f \circ m(z)(x)) \le Kr < \varepsilon$. Q.E.D.

In particular, if $O_1 = \mathbb{E}$ and m is the identity, Lemma 6 says $\{s: \Phi_f \circ s(X_0) \subset\subset O\}$ is open in $\mathfrak{M}(f^*\pi)$.

7 PROPOSITION: Let $\pi: \mathbb{E} \to X$ be a bounded $\mathfrak{M}$ metric bundle. The evaluation map $Ev: X \times \mathfrak{M}(\pi) \to \mathbb{E}$ is an $\mathfrak{M}$ metric map. If $\mathfrak{M}$ satisfies FS7 (Strong) then $(Ev, 1_X): \varepsilon_{\mathfrak{M}(\pi)} \to \pi$ is an $\mathfrak{M}$ metric VB map.

PROOF: Let $(\mathfrak{D},\mathfrak{a}) = \{U_\alpha, h_\alpha, \varphi_\alpha\}$ be an admissible atlas on π and put the norm $\| \ \|^{(\mathfrak{D},\mathfrak{a})}$ on $\mathfrak{M}(\pi)$. If $\mathfrak{M}$ satisfies FS7(Strong) then the principal parts of the VB map are given by $Ev_\alpha = h_\alpha(U_\alpha) \xrightarrow{Ev} L(\mathfrak{M}(h_\alpha(U_\alpha),F_\alpha);F_\alpha)$ $\xrightarrow{L(P_\alpha;1)} L(\mathfrak{M}(\pi);F_\alpha)$ which by Proposition II.1.10 has norm 1 unless $F_\alpha = 0$. Thus, with respect to $(\mathcal{G} \times \mathfrak{M}(\pi),\mathcal{G})$ and $(\mathfrak{D},\mathcal{G})$, $\tilde{k}_{(Ev,1)} \leq 1$. In any case, by FS8, with respect to $\mathcal{G} \times \mathfrak{M}(\pi)$ and $\mathfrak{D}$, $\tilde{\rho}_{Ev} \leq O^*(p_2^*\| \|^{(\mathfrak{D},\mathcal{G})})$ on $X \times \mathfrak{M}(\pi)$. Also, $Ev^*\max(\| \|^{M},1) \leq \max(p_2^*\| \|^{(\mathfrak{D},\mathcal{G})},1)$. Q.E.D.

For $\pi\colon \mathbb{E} \to X$ a not necessarily bounded $\mathfrak{M}$ metric bundle there is a natural topology on $\mathfrak{M}(\pi)$. For $(\mathfrak{D},\mathcal{G},\rho)$ in the metric structure, $G \subset \mathfrak{M}(\pi)$ is a neighborhood of 0 if there exist constants K,n such that $\tilde{\rho}(s;\mathcal{G},\mathfrak{D})(x) < (K\rho(x)^n)^{-1}$ for all $x \in X$ implies $s \in G$. Neighborhoods of other sections are defined by translation. This defines a topology on $\mathfrak{M}(\pi)$ making it a topological group under addition. Also, multiplication by any nonzero scalar is a homeomorphism. But $\mathfrak{M}(\pi)$ is not a topological vector space because scalar multiplication is not continuous in the scalar variable. It is easy to show that the generalization of Lemma 1 is true, i.e. the topology is independent of the choice of $(\mathfrak{D},\mathcal{G},\rho)$ and an $\mathfrak{M}$ metric VB map $(\Phi,1_X)\colon \pi_1 \to \pi_2$ induces a continuous linear map $\Phi_*\colon \mathfrak{M}(\pi_1) \to \mathfrak{M}(\pi_2)$. The generalization of Proposition 3 states that if $\| \|$ is an admissible Finsler on π then G is a neighborhood of 0 in $\mathcal{C}(\pi)$ iff there exists K,n such that $\|s(x)\| \leq (K\rho(x)^n)^{-1}$ for all $x \in X$ implies $s \in G$. If π is semicomplete this says G is a neighborhood of 0 iff there exists O a full neighborhood of the zero section in $\mathbb{E}$ such that $s(X) \subset\subset O$ implies $s \in G$ (cf. Proposition III.7.15). The Standard Constructions generalize. The neighborhoods G of 0 define a uniformity on $\mathfrak{M}(\pi)$ by $V_G = \{(s_1,s_2)\colon s_1 - s_2 \in G\}$. This uniformity clearly has a countable base and so, by [13; Thm. 6.13] it is metrizable. If $\{s_n\}$ is a Cauchy sequence in $\mathfrak{M}(\pi)$ then for U any bounded open set in X, $\{s_n|U\}$ is Cauchy in $\mathfrak{M}(\pi|U)$ which is a Banach space. The limits on the various U's fit together to obtain a section $s\colon X \to \mathbb{E}$ with $\{s_n|U\}$ converging to $s|U$ in $\mathfrak{M}(\pi|U)$ for all bounded open U's. Clearly,

$\tilde{\rho}(s;\mathfrak{G},\mathfrak{D})$ is the pointwise limit of the $\tilde{\rho}(s_n;\mathfrak{G},\mathfrak{D})$'s. Since $\{s_n\}$ is Cauchy in $\mathfrak{M}(\pi)$ it easily follows that $s \in \mathfrak{M}(\pi)$ and $\{s_n\}$ converges to s in $\mathfrak{M}(\pi)$. Thus, $\mathfrak{M}(\pi)$ is a complete metric space and so, in particular, it is a Baire space.

A useful alternative description of the topology exists when $(\mathfrak{D},\mathfrak{G}) = \{U_\alpha, h_\alpha, \varphi_\alpha\}$ is an admissible atlas on π with $\mathfrak{G}$ star bounded (cf. p. 99). Let ρ be an admissible bound. G is a neighborhood of 0 in $\mathfrak{M}(\pi)$ iff there exist K,n such that $\|s_\alpha\| < (K\rho_\alpha^n)^{-1}$, where $\rho_\alpha = \sup \rho | U_\alpha$, imply $s \in G$.

2. Composition Results: Let $\pi_i: \mathbb{E}_i \to X$ be $\mathfrak{M}$ bundles and O be an open subset of $\mathbb{E}_1$. A fiber preserving map $(G, 1_X): \pi_1|O \to \pi_2$ can be regarded as a section of the bundle $(\pi_1|O)^*\pi_2 \cong (p_1)_O: (\mathbb{E}_1 \oplus \mathbb{E}_2)_O \to O$. If $(\mathfrak{D}_1,\mathfrak{D}_2;\mathfrak{G})$ is an admissible two-tuple for π_1,π_2 then the principal parts of the fiber preserving map $(G,1_X): (\mathfrak{D}_1|O,\mathfrak{G}) \to (\mathfrak{D}_2,\mathfrak{G})$ (cf. p. 8) are the same as those of the section G with respect to $((\pi_1|O)^*\mathfrak{D}_2,\mathfrak{D}_1)$ on $(\pi_1|O)^*\pi_2$ or the atlas identification, with respect to $(\mathfrak{D}_1 \oplus \mathfrak{D}_2,\mathfrak{D}_1)|O$ on $(p_1)_O$. In particular, $G: O \to \mathbb{E}_2$ is an $\mathfrak{M}$ map iff G is an $\mathfrak{M}$ section of $(\pi_1|O)^*\pi_2$.

1 LEMMA: Given $(G,1_X): \pi_1|O \to \pi_2$ a fiber preserving map and an admissible two-tuple $(\mathfrak{D}_1,\mathfrak{D}_2;\mathfrak{G})$ on π_1,π_2, let $\| \ \|^{M_i}$ be the max Finsler induced on π_i by $(\mathfrak{D}_i,\mathfrak{G})$ $(i = 1,2)$. On O:

$$G^*\| \ \|^{M_2} \le \tilde{\rho}(G;\mathfrak{D}_1|O,\mathfrak{D}_2) \le \rho(G;\mathfrak{D}_1|O,\mathfrak{D}_2)$$
$$\le O^*(\tilde{\rho}(G;\mathfrak{D}_1|O,\mathfrak{D}_2), \| \ \|^{M_1}, \pi_1^*\rho_{(\mathfrak{D}_1,\mathfrak{G})}, \pi_1^*\rho_{(\mathfrak{D}_2,\mathfrak{G})}).$$

In particular, if O is bounded in $\mathbb{E}_1$ then (taking suprema over O):

$$\sup G^*\| \ \|^{M_2} \le \|G\|^{(\mathfrak{D}_1 \oplus \mathfrak{D}_2,\mathfrak{D}_1)|O} \le k(G;\mathfrak{D}_1|O,\mathfrak{D}_2)$$
$$\le O^*(\|G\|^{(\mathfrak{D}_1 \oplus \mathfrak{D}_2,\mathfrak{D}_1)|O}, \sup\| \ \|^{M_1}, \sup \pi_1^*\rho_{(\mathfrak{D}_1,\mathfrak{G})}, \sup \pi_1^*\rho_{(\mathfrak{D}_2,\mathfrak{G})}).$$

PROOF: That $G^*\| \|^{M_2} \leq \rho_G^{\mathcal{C}}$ on O is easy (compare Proposition 1.3). By Proposition III.3.6 $\rho_{\mathfrak{D}_1}|O \leq O^*(\pi_1{}^*\rho_{(\mathfrak{D}_1,\mathfrak{a})}, \| \|^{M_1})$ and $G^*\rho_{\mathfrak{D}_2} \leq O^*(\pi_1{}^*\rho_{(\mathfrak{D}_2,\mathfrak{a})}, G^*\| \|^{M_2})$. If O is bounded then $\|G\| = \sup \tilde{\rho}_G$. The inequalities follow from Corollary III.1.3 (a). Q.E.D.

Notice that under the identification between $(p_1)_O\colon (\mathbb{E}_1 \oplus \mathbb{E}_2)_O \to O$ and $(\pi_1|O)^*\pi_2\colon (\pi_1|O)^*\mathbb{E}_2 \to O$ the $\mathfrak{M}$ metric VB maps $(p_2,\pi_1)\colon (p_1)_O \to \pi_2$ and $(\Phi_{\pi_1},\pi_1)\colon (\pi_1|O)^*\pi_2 \to \pi_2$ are identified. Hence, if $\| \|^M$ is the Finsler on $(p_1)_O$ induced by $(\mathfrak{D}_1 \oplus \mathfrak{D}_2,\mathfrak{D}_1)|O$, equation III (5.2) implies:

$$\| \|^M = p_2{}^*\| \|^{M_2}. \tag{2.1}$$

Hence, in the bounded case, $\sup G^*\| \|^{M_2} = \|G\|_{\mathcal{C}}^{(\mathfrak{D}_1\oplus\mathfrak{D}_2,\mathfrak{D}_1)|O}$ by Proposition 1.3.

Any $\mathfrak{M}$ VB map $(\Phi,1)\colon \pi_1 \to \pi_2$ restricts to a fiber preserving map on $\pi_1|O$. If O is bounded and X is bounded then this is a continuous linear map $L\colon \mathfrak{M}(L(\pi_1;\pi_2)) \to \mathfrak{M}((\pi_1|O)^*\pi_2)$. With respect to norms induced by $(L(\mathfrak{D}_1;\mathfrak{D}_2),\mathfrak{a})$ and $(\mathfrak{D}_1 \oplus \mathfrak{D}_2,\mathfrak{D}_1)|O$, $\|L\| \leq O^*(\sup \| \|^{M_1})$, by Lemma II.1.9 (b).

Now by a <u>standard triple</u> $(\mathfrak{M}_1,\mathfrak{M},\mathfrak{M}_2)$ we will mean a trio of standard function space type satisfying: (1) $\mathfrak{M}_1 \subset \mathfrak{M}$ and $\mathfrak{M}_1 \subset \mathfrak{M}_2 \subset \mathscr{L}$. (2) $\mathfrak{M}_1$ maps $\mathfrak{M}$ to $\mathfrak{M}$ in an $\mathfrak{M}_2$ way. (3) $\mathfrak{M}_2$ satisfies FS7 (Strong). Throughout the rest of this chapter $(\mathfrak{M}_1,\mathfrak{M},\mathfrak{M}_2)$ is assumed to be a standard triple of function space types, eg. $(\mathscr{L}^{r+s},\mathcal{C}^r,\mathscr{L}^s)$ or $(\mathscr{L}^{r+s+1},\mathscr{L}^r,\mathscr{L}^s)$ $r,s \geq 0$.

2 <u>DEFINITION</u>: By a <u>composition situation</u> we mean: (1) Regular $\mathfrak{M}_1$ bundles $\pi_i\colon \mathbb{E}_i \to X$ $(i = 1,2)$ and O open in E_1. (2) A bounded $\mathfrak{M}$ manifold X_0 and an $\mathfrak{M}$ map $f\colon X_0 \to X$. (3) An $\mathfrak{M}_2$ manifold Z and an $\mathfrak{M}_2$ map $H\colon Z \to \mathfrak{M}(f^*\pi_1)$ with $\Phi_f^1(H(z)(x)) \in O$ for all $(z,x) \in Z \times X_0$, i.e. the image of the section $H(z)$ lies in $f^*O = (\Phi_f^1)^{-1}O$.

By a <u>choice of atlases</u> for the composition situation we mean:

(4) $(\mathfrak{D}_1,\mathfrak{D}_2;\mathcal{A}) = \{V_\beta, g_\beta, \varphi_\beta^1, \varphi_\beta^2\}$ an admissible $\mathfrak{M}_1$ two-tuple on π_1,π_2. (5) $\mathcal{A}_0 = \{U_\alpha, h_\alpha\}$ an admissible $\mathfrak{M}$ atlas for X_0 with $f\colon \mathcal{A}_0 \to \mathcal{A}$ index preserving and $\mathcal{A}_0$ a subdivision of $f^{-1}\mathcal{A}$, i.e. $f^{-1}V_\beta = \cup\{U_\alpha\colon \beta = \beta(\alpha)\}$ for all β. (6) $\mathcal{A}_Z = \{W_\gamma, k_\gamma\}$ an admissible $\mathfrak{M}_2$ atlas on Z.

In general, given atlases as in (4) and (6) and $\{U_\alpha, h_\alpha\} = \mathcal{A}_0$ an arbitrary admissible $\mathfrak{M}$ atlas on X_0 we can define $\mathcal{A}_f = \mathcal{A}_0 \cap f^{-1}\mathcal{A} = \{U_\alpha \cap f^{-1}V_\beta, h_\alpha\}$ indexed by pairs (α,β). $\mathcal{A}_f$ satisfies (5).

Given an atlas choice for a composition situation we will denote by $\| \ \|^{M_i}$ the max Finsler on π_i and $f^*\pi_i$ induced by $(\mathfrak{D}_i,\mathcal{A})$ and $(f^*\mathfrak{D}_i,\mathcal{A}_0)$ respectively $(i = 1,2)$. Since the Finsler on $f^*\pi_i$ is just the pull back by Φ_f^i of the Finsler on π_i (equation III (5.2)) we will use the same symbol for both. On $\mathfrak{M}(f^*\pi_i)$ we have the norm induced by $(f^*\mathfrak{D}_i,\mathcal{A}_0)$ and we will denote by $\mathfrak{M}((f^*\mathfrak{D}_i,\mathcal{A}_0))$ the trivial atlas induced on $\mathfrak{M}(f^*\pi_i)$ by choosing this norm $(i = 1,2)$.

Given a composition situation, if $(G,1_X)\colon \pi_1|O \to \pi_2$ is a fiber preserving map, then we define for each section $s_1 = H(z)$ the section $\Omega_H^f(G)(z) = s_2$ such that the following commutes:

(2.2)

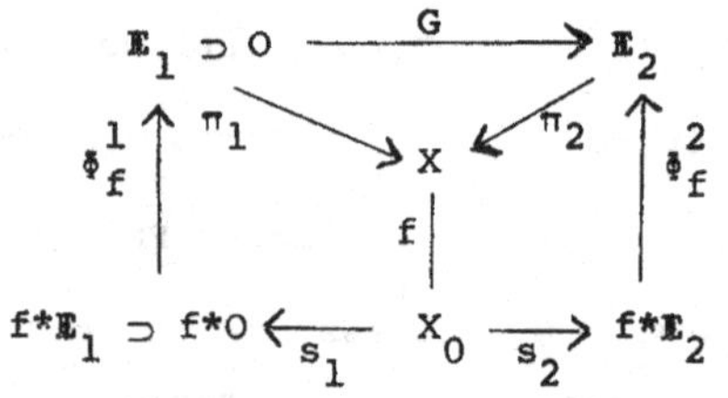

Thus, $\Omega_H^f(G)$ is defined by $\Phi_f^2 \cdot \Omega_H^f(G)(z) = G \cdot \Phi_f^1 \cdot H(z)$ or $\Omega_H^f(G)(z) = f^*(G \cdot \Phi_f^1 \cdot H(z))$. If G is an $\mathfrak{M}$ metric map then $\Omega_H^f(G)(z)$ is an $\mathfrak{M}$ metric section for all $z \in Z$ i.e. $\Omega_H^f(G)\colon Z \to \mathfrak{M}(f^*\pi_2)$.

3 <u>THEOREM</u>: Assume $(\mathfrak{M}_1,\mathfrak{M},\mathfrak{M}_2)$ is a standard triple. Let $\pi_i\colon \mathbb{E}_i \to X$ $(i = 1,2)$, $O \subset \mathbb{E}_1$, $f\colon X_0 \to X$ and $H\colon Z \to \mathfrak{M}(f^*\pi_1)$ be a composition situation. If $(G,1_X)\colon \pi_1|O \to \pi_2$ is an $\mathfrak{M}_1$ fiber preserving map then

$\Omega_H^f(G): Z \to \mathfrak{M}(f^*\pi_2)$ is an $\mathfrak{M}_2$ map.

Assume $(\mathfrak{D}_1,\mathfrak{D}_2;\mathfrak{A})$, $\mathfrak{A}_0$, $\mathfrak{A}_Z$ is an atlas choice for the composition situation. With respect to these atlas choices and the induced atlas and norms we have the following estimates:

(a) If ρ is an admissible bound on X and $K_G, n_G \geq 1$ are constants such that $\tilde{\rho}^{\mathfrak{M}_1}(G;\mathfrak{D}_1|O,\mathfrak{D}_2) \leq K_G \max(\pi_1{}^*\rho, \|\;\|^{M_1})^{n_G}$ then $\rho^{\mathfrak{M}_2}(\Omega_H^f(G);\mathfrak{A}_Z,\mathfrak{M}(f^*\mathfrak{D}_2,\mathfrak{A}_0)) \leq K_G(\sup f^*\rho)^{n_G} \times$
$O^*[\tilde{k}^{\mathfrak{M}}(f;\mathfrak{A}_0,\mathfrak{A}),K_G\rho^{\mathfrak{M}_2}(H;\mathfrak{A}_Z,\mathfrak{M}((f^*\mathfrak{D}_1,\mathfrak{A}_0)))^{n_G}]$.

(b) If O is vertically bounded, ρ is an admissible bound on X and K_G, n_G are constants such that $\tilde{\rho}^{\mathfrak{M}_1}(G;\mathfrak{D}_1|O,\mathfrak{D}_2) \leq K_G(\pi_1{}^*\rho)^{n_G}$ then $\rho^{\mathfrak{M}_2}(\Omega_H^f(G);\mathfrak{A}_Z,\mathfrak{M}(f^*\mathfrak{D}_2,\mathfrak{A}_0)) \leq K_G(\sup f^*\rho)^{n_G} O^*[\tilde{k}^{\mathfrak{M}}(f;\mathfrak{A}_0,\mathfrak{A}),$
$\rho^{\mathfrak{M}_2}(H;\mathfrak{A}_Z,\mathfrak{M}((f^*\mathfrak{D}_1,\mathfrak{A}_0)))]$.

(c) If Z and O are bounded then $\Omega_H^f: \mathfrak{M}_1((\pi_1|O)^*\pi_2) \to \mathfrak{M}_2(Z,\mathfrak{M}(f^*\pi_2))$ is a linear map with $\|\Omega_H^f\| \leq$
$O^*[\tilde{k}^{\mathfrak{M}}(f;\mathfrak{A}_0,\mathfrak{A}),k^{\mathfrak{M}_2}(H;\mathfrak{A}_Z,\mathfrak{M}((f^*\mathfrak{D}_1,\mathfrak{A}_0)))]$.

<u>PROOF</u>: We begin with the case Z and O bounded and prove the estimate in (c). This is a simple matter of localizing. With $\beta = \beta(\alpha)$ we define $H_\alpha^\gamma: k_\gamma(W_\gamma) \to \mathfrak{M}(h_\alpha(U_\alpha),E_\beta \times F_\beta^1)$ by $H_\alpha^\gamma(k_\gamma(z)) = f_\alpha \times P_\alpha(H(z))$, i.e. $H_\alpha^\gamma = c_{f_\alpha} \times P_\alpha \cdot H \cdot k_\gamma^{-1}$. Thus, $\|H_\alpha^\gamma\|_{\mathfrak{M}_2} \leq \max(\tilde{k}_f,k_H)$. For $z \in W_\gamma$, $H_\alpha^\gamma(k_\gamma(z))$ has image in $\varphi_\beta^1(O \cap \pi_1^{-1}U_\beta)$ $(\beta = \beta(\alpha))$ a bounded open subset of $E_\beta \times F_\beta^1$. The principal part of the fiber preserving map $(G,1_X)$ is $G_\beta: \varphi_\beta^1(O \cap \pi_1^{-1}U_\beta) \to F_\beta^2$ and $\|G_\beta\|_{\mathfrak{M}_1} \leq \tilde{k}_G = \|G\|_{\mathfrak{M}_1}$. Because $\mathfrak{M}_1$ maps $\mathfrak{M}$ to $\mathfrak{M}$ in an $\mathfrak{M}_2$ way, $\Omega_{H_\alpha^\gamma}(G_\beta): k_\gamma(W_\gamma) \to \mathfrak{M}(h_\alpha(U_\alpha),F_\beta^2)$ is an $\mathfrak{M}_2$ map with $\|\Omega_{H_\alpha^\gamma}(G_\beta)\|_{\mathfrak{M}_2} \leq \|G_\beta\| O^*(\|H_\alpha^\gamma\|_{\mathfrak{M}_2}) \leq \|G\| O^*(\tilde{k}_f,k_H)$. Finally, we have the commutative diagram:

$$\begin{array}{ccc} W_\gamma & \xrightarrow{\Omega_H^f(G)} & \mathfrak{M}(f^*\pi_2) \\ \Big\downarrow k_\gamma & & \Big\downarrow \hat{\Pi}\, P_\alpha \\ k_\gamma(W_\gamma) & \xrightarrow[\hat{\Pi}\Omega_{H_\alpha^\gamma}(G_\beta)]{} & \hat{\Pi}_\alpha \mathfrak{M}(h_\alpha(U_\alpha), F_\beta^2). \end{array}$$

By FS7 (Strong) $\hat{\Pi}\, \Omega_{H_\alpha^\gamma}(G_\beta)$ is an $\mathfrak{M}_2$ map and since $\hat{\Pi}\, P_\alpha$ is an isometry FS1 implies that $\Omega_H^f(G)\cdot k_\gamma^{-1}$ is an $\mathfrak{M}_2$ map with norm $\le \|G\| O^*(\tilde{k}_f, k_H)$. Thus, as an element of $\mathfrak{M}_2(Z,\mathfrak{M}(f^*\pi_2))$ with norm $\| \;\|^{\mathfrak{G}_Z}, \|\Omega_H^f(G)\| \le \|G\| O^*(\tilde{k}_f, k_H)$. This completes the proof of (c).

Now if we define $R(G): Z \to [0,\infty]$ by $R(G)(z) =$ $\inf\{\sup \tilde{\rho}^{\mathfrak{M}_1}(G;\mathfrak{D}_1|O,\mathfrak{D}_2)|O_1: O_1$ is bounded and open in O and Image $H(z_1) \subset f^*O_1$ for all z_1 in some neighborhood of z in $Z\}$. It easily follows for general O and Z that $\rho^{\mathfrak{M}_2}(\Omega_H^f(G);\mathfrak{G}_Z,\mathfrak{M}(f^*\mathfrak{D}_2,\mathfrak{G}_0)) \le$ $R(G)O^*[\tilde{k}_f, \rho(H;\mathfrak{G}_Z,\mathfrak{M}(f^*\mathfrak{D}_1,\mathfrak{G}_0))]$. For $z \in Z$ and $\epsilon > 0$ choose Z_1 a neighborhood z and O_1 open and bounded in O such that $\sup \tilde{\rho}^{\mathfrak{M}_1}(G;\mathfrak{D}_1|O,\mathfrak{D}_2)|O_1 \le (1+\epsilon)R(G)(z), \sup \rho_H|Z_1 \le (1+\epsilon)\rho_H(z)$ and Image $H(z_1)$ lies in f^*O_1 for all $z_1 \in Z_1$. Applying the estimate of (a) we get $\rho^{\mathfrak{M}_2}(\Omega_H^f(G),\mathfrak{G}_Z,\mathfrak{M}(f^*\mathfrak{D}_2,\mathfrak{G}_0))|Z_1 \le (1+\epsilon)R(G)(z)O^*(\tilde{k}_f,(1+\epsilon)\rho_H(z))$. Now let ϵ go to 0. We prove (a) and (b) by estimating $R(G)$. In particular, we will show $R(G) < \infty$ on Z.

$\max(\pi_1{}^*\rho, \|\;\|^{M_1})$ is an admissible bound on O and so there exist $K_G, n_G \ge 1$ with $K_G \max(\pi_1{}^*\rho,\|\;\|^{M_1})^{n_G} > \tilde{\rho}_G$. Note that for $z \in Z$, $\sup\|\;\|^{M_1}|\text{Image } \Phi_f^1\cdot H(z) = \sup\|\;\|^{M_1}|\text{Image } H(z) \le \|H(z)\|^{(f^*\mathfrak{D}_1,\mathfrak{G}_0)}$ by Proposition 1.3. Also, $\|H(z)\| \le \rho_H^{\mathcal{C}}(z) \le \rho_H(z)$. Thus, Image $H(z)$ $\subset (\Phi_f^1)^{-1}\{\|\;\|^{M_1} \le \rho_H(z)\}$. So by choosing a continuous partition of unity refining $\{V_\beta\}$, Image $H(z) \subset (\Phi_f^1)^{-1}\{\|\;\|^{\mathcal{P}_1} \le \rho_H^{\mathfrak{M}_2}(z)\}$. Now let d be an admissible metric on X and $1 > \epsilon > 0$. Define $Z_1 = \{\rho_H < (1+\epsilon)\rho_H(z)\}$ and $O_1 = \pi_1^{-1}\{\rho^d < 2\sup f^*\rho^d\} \cap \{\|\;\|^{\mathcal{P}_1} < 2\rho_H(z)\}$ $\cap \{\tilde{\rho}_G < K_G \max(\sup f^*\rho,(1+\epsilon)\rho_H(z))^{n_G}\}$. O_1 and Z_1 are open because

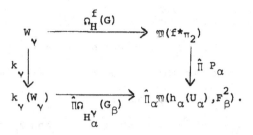

$\| \ \|^{\varphi_1}$ and ρ^d are continuous and $\tilde{\rho}_G$ and ρ_H are upper semicontinuous. The first two factors make O_1 bounded. Clearly, Image $H(z_1) \subset f^*O_1$ for $z_1 \in Z_1$. Thus, $R(G)(z) \leq K_G \max(\sup f^*\rho, (1+\epsilon)\rho_H(z))^{n_G}$. Letting $\epsilon \to 0$ we have (a). (b) is similar using the fact that $\pi_1{}^*\rho$ is an admissible bound on O if O is vertically bounded. Q.E.D.

The most important applications of this theorem occur with $Z \subset \mathfrak{M}(f^*O)$ (cf. Corollary 1.5 (a)) and H = the inclusion into $\mathfrak{M}(f^*\pi_1)$. In this case, Ω_H^f is denoted Ω_Z^f. If $(\mathfrak{M}_1, \mathfrak{M}, \mathfrak{M}_2)$ is a standard triple then so is $(\mathfrak{M}_1^1, \mathfrak{M}, \mathfrak{M}_2^1)$ and so if π_1, π_2, O, f is a composition situation for the latter triple and $(G, 1_X): \pi_1|O \to \pi_2$ is $\mathfrak{M}_1^1$ then $\Omega_Z^f(G): Z \to \mathfrak{M}(f^*\pi_2)$ is an $\mathfrak{M}_2^1$ map. In this case we can compute the derivative explicitly. Taking the derivative of G along the fibers of π_1 we get the <u>vertical tangent map</u> $(\partial_v G, 1_X): \pi_1|O \to L(\pi_1;\pi_2)$ which is an $\mathfrak{M}_2$ map. In fact, if $G_\beta: \varphi_\beta^1(O \cap \pi_1^{-1}V_\beta) \to F_\beta^2$ is a principal part of the fiber preserving map $(G,1)$ then the domain is open in $E_\beta \times F_\beta^1$ and it is easy to see that $(\partial_v G)_\beta = D_2(G_\beta): \varphi_\beta^1(O \cap \pi_1^{-1}V_\beta) \to L(F_\beta^1; F_\beta^2)$. By evaluation, we can define the related fiber preserving map $(vTG, 1_X): \pi_1 \oplus \pi_1|P_1^{-1}O \to \pi_2$ where $(P_1, 1_X): \pi_1 \oplus \pi_1 \to \pi_1$. By diagram II. (2.1) and the Jet Lemma II.1.7 applied to the family $\{P_{\beta(\alpha)}\}$, the following diagram commutes:

(2.3)
$$\begin{array}{ccc}
\mathfrak{M}(f^*\pi_1) \supset Z & \xrightarrow{\ \Omega_Z^f(\partial_v G)\ } & \mathfrak{M}(f^*L(\pi_1;\pi_2)) \\
\Big\downarrow {\scriptstyle D\,\Omega_Z^f(G)} & & \Vert \\
L(\mathfrak{M}(f^*\pi_1);\mathfrak{M}(f^*\pi_2)) & \xrightarrow[\ \lambda\ ]{} & \mathfrak{M}(L(f^*\pi_1;f^*\pi_2))
\end{array} .$$

We can also express this using vTG by the equation, for Z_0 any bounded open set in $\mathfrak{M}(f^*\pi_1)$:

(2.4)
$$\Omega_{Z \times Z_0}^f(vTG)(s, \dot{s}) = D\Omega_Z^f(G)(s)(\dot{s}).$$

Here $Z \times Z_0$ is open in $\mathfrak{M}(f^*\pi_1) \times \mathfrak{M}(f^*\pi_1) = \mathfrak{M}(f^*(\pi_1 \oplus \pi_1))$.

If X_0 is unbounded, we still have:

4 THEOREM: Assume $(\mathfrak{M}_1,\mathfrak{M},\mathfrak{M}_2)$ is a standard triple. Let $\pi_i\colon \mathbb{E}_i \to X$ be semicomplete $\mathfrak{M}_1$ bundles, X_0 an $\mathfrak{M}$ manifold, $f\colon X_0 \to X$ an $\mathfrak{M}$ map and $(G,1)\colon \pi_1|O \to \pi_2$ an $\mathfrak{M}_1$ fiber preserving map with O open in $\mathbb{E}_1$. If $Z = \{s \in \mathfrak{M}(f^*\pi_1)\colon \Phi_f \circ s(X_0) \subset\subset O\}$, then Z is open in $\mathfrak{M}(f^*\pi_1)$ and $\Omega_Z^f(G)\colon Z \to \mathfrak{M}(f^*\pi_2)$ is continuous.

PROOF: That Z is open is easy. Now choose a two-tuple $(\mathfrak{N}_1,\mathfrak{N}_2;\mathcal{G})$ for π_1,π_2 and $\mathcal{G}_0$ star bounded for X_0 such that $f\colon \mathcal{G}_0 \to \mathcal{G}$ is index preserving. Fix $s \in Z$. By restricting O to a small enough full neighborhood of $s(X_0)$ we can assume O is vertically bounded. Choose $\rho,\bar\rho$ admissible bounds on X_0, X such that $\pi_1^*\bar\rho \geq \rho(G;\mathfrak{N}_1|O,\mathfrak{N}_2)$, $\rho \geq \max(f^*\bar\rho,\tilde\rho(f;\mathcal{G}_0,\mathcal{G}))$. Now pick K_1,n_1 such that if $Z_\alpha(s) = \{g\colon \|s_\alpha - g\| < (K_1\rho_\alpha^{n_1})^{-1}\}$ and $Z(s) = \{s_1\colon s_{1\alpha} \in Z_\alpha(s) \text{ for all } \alpha\}$, then $Z(s) \subset Z$. Note that $Z_\alpha(s)$ and $Z(s)$ are open in $\mathfrak{M}(h_\alpha(U_\alpha),F^1_{\beta(\alpha)})$ and $\mathfrak{M}(f^*\pi_1)$ respectively. Finally, for $s_1 \in Z(s)$, $\Omega_Z^f(G)(s_1)_\alpha = \Omega_{Z_\alpha(s)}^{f_\alpha}(G_{\beta(\alpha)})(s_{1\alpha})$. Continuity on $Z(s)$ then follows from the fact that $\|\Omega_{Z_\alpha(s)}^{f_\alpha}(G_{\beta(\alpha)})\|_L \leq K_2\rho_\alpha^{n_2}$ for some constants K_2,n_2. Thus, if $\|s_{1\alpha} - s_{2\alpha}\| \leq (K\,K_2\rho_\alpha^{n+n_2})^{-1}$ then $\|\Omega_Z^f(s_1)_\alpha - \Omega_Z^f(s_2)_\alpha\| \leq (K\,\rho_\alpha^n)^{-1}$ for $s_1,s_2 \in Z(s)$. Q.E.D.

3. Manifolds of Mappings: For sets X_0 and X and map $\rho\colon X \to [1,\infty]$ let X^{X_0} be the set of maps from X_0 to X and $S_\rho\colon X^{X_0} \to [1,\infty]$ (also written $S[\rho](f)$) be defined by $S_\rho(f) = \sup f^*\rho$. Clearly, if X_0 and X are $\mathfrak{M}$ manifolds with X_0 bounded and ρ is an admissible bound on X then S_ρ is a bound (i.e. $S_\rho < \infty$) on the subset $\mathfrak{M}(X_0,X) \subset X^{X_0}$ consisting of $\mathfrak{M}$ maps. If $\mathcal{G}_0$ and $\mathcal{G}$ are admissible atlases on X_0 and X then $\sigma = \sigma_\rho(\mathcal{G}_0,\mathcal{G})$, defined by $\sigma_\rho(\mathcal{G}_0,\mathcal{G})(f) = \max(S_\rho(f),k(f;\mathcal{G}_0,\mathcal{G})) = \sup \max(f^*\rho,\rho(f;\mathcal{G}_0,\mathcal{G}))$, is a bound on $\mathfrak{M}(X_0,X)$.

1 LEMMA: (a) If $\rho_1 > \rho_2$ (or $\rho_1 \geq \rho_2$) on X then $S_{\rho_1} > S_{\rho_2}$ (or $S_{\rho_1} \geq S_{\rho_2}$) on X^{X_0}.

(b) If $F\colon X_1 \to X$ then, defining $F_*\colon X_1^{X_0} \to X^{X_0}$ by $F_*(f) = F\circ f$, $S_{F^*\rho} = S_\rho \circ F_* = (F_*)^* S_\rho$ on $X_1^{X_0}$. Let X_0, X, X_1 be $\mathfrak{M}$ manifolds with X_0 bounded, $\mathfrak{a}_0, \mathfrak{a}, \mathfrak{a}_1$ be admissible atlases on X_0, X, X_1 and ρ, ρ_1 be admissible bounds on X, X_1. If $F\colon X_1 \to X$ is an $\mathfrak{M}$ map then F_* maps $\mathfrak{M}(X_0,X_1)$ to $\mathfrak{M}(X_0,X)$ and $S_{\rho_1} > (F_*)^* S_\rho$, $\sigma_{\rho_1}(\mathfrak{a}_0,\mathfrak{a}_1) > (F_*)^*\sigma_\rho(\mathfrak{a}_0,\mathfrak{a})$.

(c) If $H\colon X_0 \to X_1$ then, defining $H^*\colon X^{X_1} \to X^{X_0}$ by $H^*(f) = f\circ H$, $S_\rho \geq (H^*)^* S_\rho$. Let X_0, X_1, X be $\mathfrak{M}$ metric manifolds with X_0, X_1 bounded, $\mathfrak{a}_0, \mathfrak{a}_1, \mathfrak{a}$ be admissible atlases on X_0, X_1, X and ρ be an admissible bound on X. If $H\colon X_0 \to X_1$ is an $\mathfrak{M}$ metric map then H^* maps $\mathfrak{M}(X_1,X)$ to $\mathfrak{M}(X_0,X)$ and $\sigma_\rho(\mathfrak{a}_1,\mathfrak{a}) > (H^*)^*\sigma_\rho(\mathfrak{a}_0,\mathfrak{a})$.

PROOF: The S_ρ results are obvious. The σ results follow from Proposition III.1.1 which implies:

$$\sigma_\rho(\mathfrak{a}_0,\mathfrak{a})(F_*f) \leq O^*(S_\rho(F_*f), S[\rho(F;\mathfrak{a}_1,\mathfrak{a})](f), \sigma_{\rho_1}(\mathfrak{a}_0,\mathfrak{a}_1)(f))$$

(3.1)

$$\sigma_\rho(\mathfrak{a}_0,\mathfrak{a})(H^*f) \leq O^*(k(H;\mathfrak{a}_0,\mathfrak{a}_1), \sigma_\rho(\mathfrak{a}_1,\mathfrak{a})(f)).$$

Q.E.D.

In particular, the bound equivalence classes of S_ρ and $\sigma_\rho(\mathfrak{a}_0,\mathfrak{a})$ on $\mathfrak{M}(X_0,X)$ are independent of the choices of admissible ρ, $\mathfrak{a}_0$ and $\mathfrak{a}$. Note that if V is a transversely bounded set in $X \times X$ then on

$\{(f,g) \in \mathfrak{M}(X_0,X) \times \mathfrak{M}(X_0,X)$ with $(f,g)(X_0) \subset V\}$

$\min(p_1{}^*S_\rho, p_2{}^*S_\rho) \sim \max(p_1{}^*S_\rho, p_2{}^*S_\rho)$.

Recall that $(\mathfrak{M}_1, \mathfrak{M}, \mathfrak{M}_2)$ is assumed to be a standard triple. If X_0 is a bounded $\mathfrak{M}$ manifold and X is semicomplete $\mathfrak{M}_1^1$ manifold admitting $\mathfrak{M}_1$ exponential maps $TX \supset D \overset{e}{\to} V \subset X \times X$ then $\mathfrak{M}(X_0,X)$ can be given the structure of a semicomplete $\mathfrak{M}_2$ metric manifold, as we will now demonstrate.

If $V \in \mathfrak{U}_X$ then let $V^{\mathfrak{M}} = \{(f,g) \in \mathfrak{M}(X_0,X) \times \mathfrak{M}(X_0,X)\colon (f\times g)(X_0) \subset\subset V\}$

and define for $f \in \mathfrak{M}(X_0,X)$:

$$V_f = V^{\mathfrak{M}}[f] = \{g \in \mathfrak{M}(X_0,X) : (f \times g)(X_0) \subset\subset V\}$$

(3.2) $$D_f = \{s \in \mathfrak{M}(f^*\tau_X) : \Phi_f \circ s(X_0) \subset\subset D\}$$

$$H_f \colon V_f \longrightarrow D_f \qquad \text{by} \qquad e \circ \Phi_f \circ H_f(g) = (f,g).$$

D_f is open by Lemma 1.6 and the remark thereafter. For $s \in \mathfrak{M}(f^*TX)$ we will write $s\colon X_0 \to TX$ for $\Phi_f \circ s$. $H_f(g) = s$ is then defined by the commutative diagram:

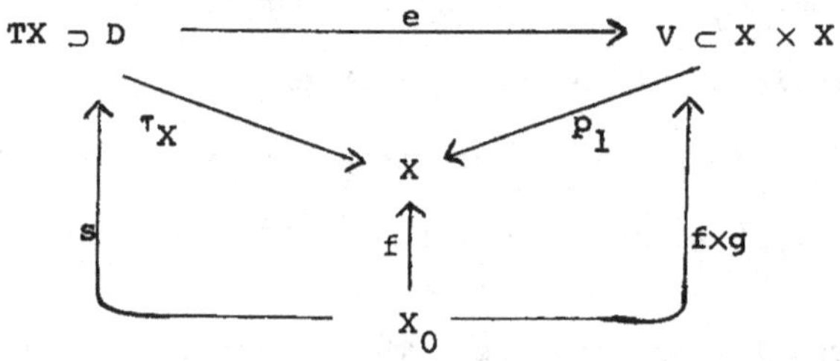

By Corollary III.7.8 (b), H_f is a bijection. $\{V_f,H_f\}$ is not quite an atlas on $\mathfrak{M}(X_0,X)$ because $\mathfrak{M}(f^*\tau_X)$ is not yet normed. This requires a choice $(\mathfrak{G},\rho,e)$ for X and $\mathfrak{G}_0$ for X_0 where: $\mathfrak{G},\mathfrak{G}_0$ are admissible atlases on X and X_0 ($\mathfrak{M}_1^1$ and $\mathfrak{M}$ respectively). ρ is an admissible bound. e is the exponential. Recall $\mathfrak{G}_f = \mathfrak{G}_0 \cap f^{-1}\mathfrak{G}$, the subdivision of $\mathfrak{G}_0$ induced by f (see p. 116), such that $f\colon \mathfrak{G}_f \to \mathfrak{G}$ is index preserving. $(f^*T\mathfrak{G},\mathfrak{G}_f)$ induces a norm on $\mathfrak{M}(f^*\tau_X)$.

2 <u>LEMMA</u>: Let X_i be a semicomplete $\mathfrak{M}_1^1$ manifold admitting $\mathfrak{M}_1$ exponentials and let $(\mathfrak{G}_i,\rho_i,e_i)$ be a choice of atlas, bound and exponential $(i = 1,2)$. Let X_0 be a bounded $\mathfrak{M}$ manifold with atlas $\mathfrak{G}_0$.

Let $f_i \in \mathfrak{M}(X_0,X_i)$ $(i = 1,2)$, $F \in \mathfrak{M}_1(U,X_2)$ with U open in X_1 and $m_1 \in \mathfrak{M}_2(G,\mathfrak{M}_1(f_1^*\tau_1))$ with G open and bounded in a B-space. Assume that $m_1(G) \subset D_{f_1}$ and that $H_{f_1}^{-1} \circ m_1(g)(X_0) \subset U$ for all $g \in G$.

$G_0 = m_1^{-1}(H_{f_1}(V_{f_1} \cap (F_*)^{-1}V_{f_2}))$ is open in G and

$m_2 = H_{f_2}\cdot F_*\cdot H_{f_1}^{-1}\cdot m_1 \in \mathfrak{M}_2(G_0;\mathfrak{M}(f_2*\tau_2))$. There exist constants depending on (G_i,ρ_i,e_i) $(i = 1,2)$ and F (as well as the function space types) such that $\|m_2\|_{\mathfrak{M}} \le K\max(\sigma_1(f_1),\sigma_2(f_2),\|m_1\|_{\mathfrak{M}})^n$. If V_i is transversely bounded $(i = 1,2)$, U is bounded and $K_i,n_i \ge 1$ are such that $K_i\rho_i^{n_i} \ge \rho_{G_i}^{\mathfrak{M}_1}$, $K_i(\tau_i*\rho_i)^{n_i} \ge \rho^{\mathfrak{M}_1}(e_i;TG_i|D_i,G_i\times G_i)$ and $K_i(p_1*\rho_i)^{n_i} \ge \rho^{\mathfrak{M}_1}(e_i^{-1};G_i\times G_i|V_i,TG_i)$ $(i = 1,2)$ then:

$$(3.3)\qquad \|m_2\|_{\mathfrak{M}_2} \le O^*(K_1\sigma_1(f_1)^{n_1},K_2\sigma_2(f_2)^{n_2},k(F;G_1|U,G_2),\|m_1\|_{\mathfrak{M}_2}).$$

<u>PROOF</u>: Let $\tau(F): X_1\times U\times X_2 \to X_1\times X_2\times X_2$ by $\tau(F)(y_1,y_2,z) = (y_1,z,F(y_2))$. If $s_1 = m_1(g)$ then $s_2 = m_2(g)$ is described by the following commutative diagram where $V_{12} = (V_1\times X_2)\cap \tau(F)^{-1}(X_1\times V_2)$ and $D_{12} = (e_1\times 1)^{-1}V_{12}$:

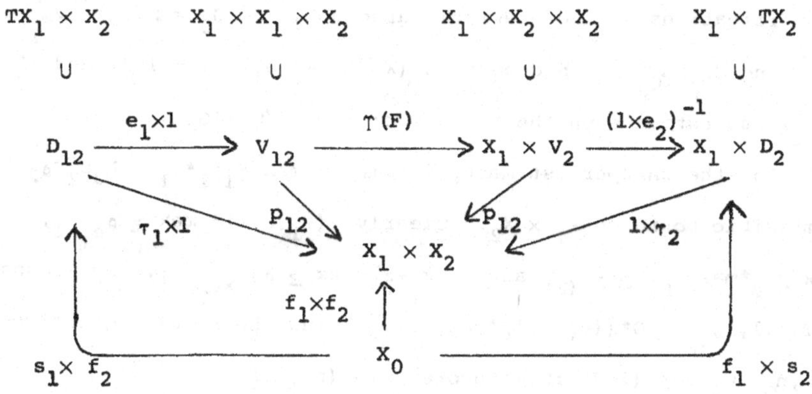

First, we prove that $G_0 \subset \tilde{G}_0 = \{g\colon (m_1(g)\times f_2)(X_0) \subset\subset D_{12}$ $(\mathrm{rel}(e_1\times 1)^{-1}(X_1\times U\times X_2))\}$. Note that $\tilde{G}_0$ is open in G by Lemma 1.6. Let $m_1(g) = s_1 \in D_{f_1}\cap H_{f_1}(F_*)^{-1}V_{f_2}$. Then $s_2 = m_2(g)$ is defined and lies in D_{f_2}. The result now follows from applications of Lemma III.6.3 and Corollary III.7.8 (b): $s_2(X_0) \subset\subset D_2$ (rel TX_2) $(s_2 \in D_{f_2})$.

$(f_1\times s_2)(X_0) \subset\subset X_1\times D_2$ (rel $X_1\times TX_2$) (III.6.3 (h)).

$(1\times e_2)^{-1}\cdot(f_1\times s_2)(X_0) \subset\subset X_1\times V_2$ (III.7.8 (b)).

$(H_{f_1}(s_1)\times f_2)(X_0) \subset \tau(F)^{-1}(1\times e_2)^{-1}\cdot(f_1\times s_2)(X_0) \subset\subset \tau(F)^{-1}(X_1\times V_2)$

$(\text{rel } X_1 \times U \times X_2)$ (III.6.3 (h)). On the other hand, $(s_1 \times f_2)(X_0) \subset\subset D_1 \times X_2$ $(\text{rel } TX_1 \times X_2)$ and $(H_{f_1}(s_1) \times f_2)(X_0) \subset\subset V_1 \times X_2$ $(\text{rel } X_1 \times X_1 \times X_2)$ in a similar way. $(H_{f_1}(s_1) \times f_2)(X_0) \subset\subset V_1 \times X_2 \cap X_1 \times U \times X_2$ $(\text{rel } X_1 \times U \times X_2)$ (III.6.3(h)). Intersecting and using III.6.3(d) we get $(H_{f_1}(s_1) \times f_2)(X_0) \subset\subset V_{12}$ $(\text{rel } X_1 \times U \times X_2)$. Finally, $(s_1 \times f_2)(X_0) \subset\subset D_{12}(\text{rel}(e_1 \times 1)^{-1}X_1 \times U \times X_2)$ (III.6.3 (h)).

Next, consider the identifications $(f_1 \times f_2)^*(\tau_1 \times 1) \cong (f_1 \times f_2)^* p_1^* \tau_1 \cong f_1^* \tau_1$ and $(f_1 \times f_2)^*(1 \times \tau_2) \cong f_2^* \tau_2$. Under the identification map the atlas $((f_1 \times f_2)^*[(T\mathcal{G}_1) \times \mathcal{G}_2], \mathcal{G}_{f_1 \times f_2})$ is a subdivision of $(f_1^* T\mathcal{G}_1, \mathcal{G}_{f_1})$ on $f_1^* \tau_1$ and so induces the same Finsler on $f_1^* \tau_1$ and norm on $\mathfrak{M}(f_1^* \tau_1)$. Similarly, for $f_2^* \tau_2$.

Now with $Q_F = (1 \times e_2)^{-1} \cdot \tau(F) \cdot (e_1 \times 1): D_{12} \to X_1 \times TX_2$. We apply Theorem 2.3 (a) and get the $\mathfrak{m}_2$ map $\Omega_{m_1}^{f_1 \times f_2}(Q_F): \tilde{G}_0 \to \mathfrak{M}(f_2^* \tau_2)$, using the identifications of the previous paragraph. So $G_0 = \Omega(Q_F)^{-1} D_{f_2}$ is open. Choosing K_{Q_F}, n_{Q_F} which depend on $(X_i, \mathcal{G}_i, \rho_i, e_i)$ $(i = 1,2)$ and F we get our first estimate on the $\mathfrak{M}_2$ norm of $m_2 = \Omega(Q_F)|G_0$.

For the sharper estimate, let $\max = \max(K_1 p_1^* \rho_1^{n_1}, K_2 p_2^* \rho_2^{n_2})$, an admissible bound on $X_1 \times X_2$. Clearly, $(\tau_1 \times 1)^* \max \geq \rho_{e_1 \times 1}$, $\max(p_{13}^* \max, k_F) \geq \rho_{\tau(F)}$ and $(1 \times \tau_2)^* \max \geq \rho_{(1 \times e_2)^{-1}}$. By Proposition III.1.1, $\rho_{Q_F} \leq O^*((\tau_1 \times 1)^* \max, k_F)$. Using the resulting choices for K_{Q_F}, n_{Q_F} we get (3.3) from Theorem 2.3 (b), as

$S_{\max}(f_1 \times f_2) = \max(K_1 S_{\rho_1}(f_1)^{n_1}, K_2 S_{\rho_2}(f_2)^{n_2})$. Q.E.D.

Applying Lemma 2 with F = identity, $(\mathcal{G}_i, \rho_i, e_i) = (\mathcal{G}, \rho, e)$ $(i = 1,2)$ and m_1 = inclusion of $D_f \cap B_N$ (B_N the N-ball in $\mathfrak{M}(f^* \tau_X)$), we get that $H_f(V_f \cap V_g)$ is open in $\mathfrak{M}(f^* \tau_X)$ and

$$(3.4) \qquad \| H_g H_f^{-1} | H_f(V_f \cap V_g) \cap B_N \|_{\mathfrak{M}_2} \leq K_1 \max(\sigma(f), \sigma(g), N)^{n_1} .$$

Here $K_1, n_1 \geq 1$ are constants depending on $(\mathcal{G}, \rho, e)$ and the function space types. With a similar choice of constants K_2, n_2, for $H_f(g) = s \in D_f$:

$$(3.5)\qquad \begin{aligned} \|s\| &\le K_2\, \sigma(f)^{n_2} O^*(\sigma(g)) \\ \sigma(g) &\le K_2 \max(\sigma(f),\|s\|)^{n_2} O^*(k_{\mathcal{G}_0}). \end{aligned}$$

These follow easily from Proposition III.1.1 and the inequalities: (1) $S_{\|\ \|^M}(s) \le \|s\|$, (2) $k(s;\mathcal{G}_f, T\mathcal{G}) \le O^*(k_{\mathcal{G}_0}, \sigma(f), \|s\|)$, (3) $\max(\tau^*\rho, \|\ \|^M) > \max(\rho_e, e^* \max(p_1^*\rho, p_2^*\rho))$ on D, (4) $\max(p_1^*\rho, p_2^*\rho) > \max(\rho_{e^{-1}}, e^{-1*}\max(\tau^*\rho, \|\ \|^M))$ on V.

3 <u>LEMMA:</u> There exist constants $\bar{K}, \bar{n} \ge 1$ depending on $(\mathcal{G},\rho,E)$ and the function space types such that $s \in \mathfrak{M}(f^*\tau_X)$ and $\|s\| < \Lambda(f) = (\bar{K}\sigma(f)^{\bar{n}})^{-1}$ implies $s \in D_f$ and $\sigma(f) \le \Lambda(g)^{-1} O^*(k_{\mathcal{G}_0})$ with $g = H_f^{-1}(s)$.

<u>PROOF</u>: Choose $V_0 \in \mathfrak{N}_X$ with $V_0^{-1} = V_0$ and $V_0 \cdot V_0 \subset\subset V$. Let $D_0 = e^{-1}V_0$. D_0 is a full neighborhood of the zero section and $D_0 \subset\subset D$ by Corollary III.7.8. Let K_3, n_3 be such that for $w \in TX$ $\|w\|^M < (K_3\rho(\tau w)^{n_3})^{-1}$ implies $w \in D_0$. So if $s \in \mathfrak{M}(f^*\tau)$ and $\|s\| < (K_3 S_\rho(f)^{n_3})^{-1}$ then $s(X_0) \subset D_0$ and so $s \in D_f$. In fact, if $s_1, s_2 \in \mathfrak{M}(f^*\tau)$ with $\|s_i\| < (K_3 S_\rho(f)^{n_3})^{-1}$ and $g_i = H_f^{-1}(s_i)$ $(i = 1,2)$ then $f \times g_i(X_0) \subset V_0$ and so $g_2 \times g_1(X_0) \subset V_0 \cdot V_0$ and so $g_1 \in V_{g_2}$. With $s_2 = s$, $g_2 = g$ and $N = 1$, we have that the ball $B(0,(K_3 S_\rho(f)^{n_3})^{-1}) \subset H_f(V_f \cap V_g) \cap B_1$ in $\mathfrak{M}(f^*\tau_X)$. Let $\bar{s} = H_g H_f^{-1}(0) = H_g(f)$. By (3.4), since $\mathfrak{M}_2 \subset \mathscr{L}$, $\|\bar{s}\| = \|\bar{s}-0\| = \|H_g H_f^{-1}(0) - H_g H_f^{-1}(s)\| \le K_1 \max(\sigma(f),\sigma(g))^{n_1}\|s\|$. So with $K_4 = \max(K_1,K_3)$, $n_4 = \max(n_1,n_3)$, $\|s\| < (K_4\sigma(f)^{n_4})^{-1}$ implies $\|\bar{s}\| \le K_1\sigma(g)^{n_1}$. Now by (3.5) applied to $\bar{s}$: $\sigma(f) = \sigma(H_g^{-1}(\bar{s}))$ $\le K_2 \max(\sigma(g),\|\bar{s}\|)^{n_2}\, O^*(k_{\mathcal{G}_0}) \le \bar{K}\,\sigma(g)^{\bar{n}} O^*(k_{\mathcal{G}_0})$ with $K = \bar{K}_2 K_4^{n_2}$ and $\bar{n} = n_2 n_4$. Q.E.D.

A choice of constants $\bar{K}, \bar{n}$ or equivalently a choice of $\Lambda(f) = (\bar{K}\sigma(f)^{\bar{n}})^{-1}$ is called a <u>Λ choice</u> for $(\mathcal{G},\rho,e)$ or for X. We then define

$$D_f^\Lambda = B(0,\Lambda(f)) \subset D_f$$

(3.6) $$V^\Lambda = \{(f,g) : H_f(g) \in D_f^\Lambda\} \subset V^{\mathfrak{M}}$$

$$\sigma^\Lambda(f) = \sup \sigma | (V^\Lambda)^{-1}[f] \quad \sigma_\Lambda(f) = \inf \sigma | V^\Lambda[f] \,(= V_f^\Lambda)$$

Lemma 3 says that $\sigma_\Lambda \leq \sigma^\Lambda \leq \Lambda^{-1} O^*(k_{\mathcal{G}_0}) = \bar{K}\sigma^{\bar{n}} O^*(k_{\mathcal{G}_0})$. (3.5) implies $\sigma \leq K_2 \sigma_\Lambda^{n_2} O^*(k_{\mathcal{G}_0})$.

4 <u>LEMMA</u>: Let X be an $\mathfrak{m}_1^1$ semicomplete manifold admitting $\mathfrak{m}_1$ exponentials with choice $(\mathcal{G},\rho,e)$. Let X_i be a bounded $\mathfrak{M}$ manifold with admissible atlas $\mathcal{G}_i$ $(i = 0,1)$.

Let $f_0 \in \mathfrak{m}(X_0,X)$, $h \in \mathfrak{m}(X_1,X_0)$ and $f_1 = h^*(f_0) = f_0 \circ h$. The map $H_{f_1} \circ h^* \circ H_{f_0}^{-1}: H_{f_0}(V_{f_0} \cap (h^*)^{-1} V_{f_1}) \cap B_N \to \mathfrak{m}(f_1^*\tau_X)$ is an $\mathfrak{m}_2$ metric map with norm $\leq O^*(k(h;\mathcal{G}_1,\mathcal{G}_0),N)$.

<u>PROOF</u>: Identifying $f_1^*\tau_X$ with $h^* f_0^* \tau_X$ we have the commutative diagram

$$\begin{array}{ccc}
V_{f_0} \cap (h^*)^{-1}(V_{f_1}) & \xrightarrow{\;H_{f_0}\;} & \mathfrak{M}(f_0^*\tau_X) \\
\Big\downarrow h^* & & \Big\downarrow h^* \\
V_{f_1} & \xrightarrow[\;H_{f_1}\;]{} & \mathfrak{M}(f_1^*\tau_X).
\end{array}$$

Thus, $H_{f_1} \circ h^* \circ H_{f_0}^{-1}$ is the restriction of the continuous linear map h^*. In particular, $H_{f_0}(V_{f_0} \cap (h^*)^{-1} V_{f_1}) = D_{f_0} \cap (h^*)^{-1} D_{f_1}$ is open. The norm on $\mathfrak{M}(f_0^*\tau_X)$ is given by $(f_0^*T\mathcal{G},\mathcal{G}_{f_0})$ and on $\mathfrak{M}(f_1^*\tau_X)$ by $(f_1^*T\mathcal{G},\mathcal{G}_{f_1})$. We can apply the pull back result to get $\|h^*\| \leq O^*(k(h;\mathcal{G}_1,\mathcal{G}_0))$ by noting that under the identification $h^* f_0^* \tau_X \approx f_1^*\tau_X$ the atlas $(h^* f_0^* T\mathcal{G}, (\mathcal{G}_{f_0})_h)$ is a subdivision of $(f_1^*T\mathcal{G},\mathcal{G}_{f_1})$. The result then follows from FS2 for $\mathfrak{M}_2$. Q.E.D.

5 <u>THEOREM</u>: Assume $(\mathfrak{M}_1,\mathfrak{M},\mathfrak{M}_2)$ is a standard triple. Let X_0 be a bounded $\mathfrak{M}$ manifold and X be a semicomplete $\mathfrak{M}_1^1$ manifold admitting $\mathfrak{M}_1$ exponentials. Choice of $(\mathcal{G},\rho,e,\Lambda)$ admissible atlas, bound, exponential

and Λ choice on X, and choice of $\mathcal{G}_0$ admissible on X_0 induce the atlas $\mathfrak{M}(\mathcal{G}_0,\mathcal{G}) = \{V_f^\Lambda, H_f\}$ on $\mathfrak{M}(X_0,X)$. $(\mathfrak{M}(\mathcal{G}_0,\mathcal{G}),\sigma)$ is a semicomplete $\mathfrak{M}_2$ adapted atlas on $\mathfrak{M}(X_0,X)$ and the resulting metric structure is independent of the choices.

If $F: X_1 \to X_2$ is an $\mathfrak{M}_1$ map of semicomplete $\mathfrak{M}_1^1$ manifolds admitting $\mathfrak{M}_1$ exponentials, then $F_*: \mathfrak{M}(X_0,X_1) \to \mathfrak{M}(X_0,X_2)$ is an $\mathfrak{M}_2$ map.

If $h: X_0 \to X_1$ is an $\mathfrak{M}$ map of bounded $\mathfrak{M}$ manifolds then $h^*: \mathfrak{M}(X_1,X) \to \mathfrak{M}(X_0,X)$ is an $\mathfrak{M}_2$ map.

<u>PROOF</u>: The atlas $\mathfrak{M}(\mathcal{G}_0,\mathcal{G})$ induces a Hausdorff topology on $\mathfrak{M}(X_0,X)$. For if $x \in X_0$ and U is open in X then $H_f^{-1}\{g \in V_f: g(x) \in U\}$ is easily seen to be open in $\mathfrak{M}(f^*\tau)$. Thus $ev_x: \mathfrak{M}(X_0,X) \to X$ is continuous for all $x \in X_0$. This implies $\mathfrak{M}(X_0,X)$ is Hausdorff because X is.

Since $\sigma_\Lambda \sim \sigma$ we can replace estimates involving $\max(\sigma(f_1),\sigma(f_2))$, for example, by $\max(\sigma_\Lambda(f_1),\sigma_\Lambda(f_2))$. Note that if $g \in V_{f_1}^\Lambda \cap V_{f_2}^\Lambda$ then $\sigma(g) \geq \max(\sigma_\Lambda(f_1),\sigma_\Lambda(f_2))$. That $(\mathfrak{M}(\mathcal{G}_0,\mathcal{G}),\sigma)$ is a semicomplete $\mathfrak{M}_2$ adapted atlas then follows from Lemma 2 and the Regularity Lemma III.7.3. That F_* is an $\mathfrak{M}_2$ metric map of adapted atlases follows from Lemmas 1 and 2. That h^* is an $\mathfrak{M}_2$ metric map of adapted atlases follows from Lemmas 1 and 4 and Corollary III.1.3. Q.E.D.

<u>ADDENDUM</u>: It is easy to sharpen the proof to show that if $A \subset \mathfrak{M}(X_0,X)$ and for $f \in A$, $\tilde{V}_f$ is open in V_f then $\{\tilde{V}_f, H_f\}$ is an admissible $\mathfrak{M}_2$ atlas on $\mathfrak{M}(X_0,X)$ if $\{\tilde{V}_f\}$ covers $\mathfrak{M}(X_0,X)$ and there exist constants K,n such that $g \in \tilde{V}_f$ implies $K\sigma(g)^n \geq \max(\sigma(f),\|H_f(g)\|)$.

If $(\mathfrak{M}_1,\tilde{\mathfrak{M}},\mathfrak{M}_2)$ is a standard triple with $\mathfrak{M} \subset \tilde{\mathfrak{M}}$ then the inclusion $i: \mathfrak{M}(X_0,X) \to \tilde{\mathfrak{M}}(X_0,X)$ is an $\mathfrak{M}_2$ metric map. In particular, for $\tilde{\mathfrak{M}} = \mathcal{C}$:

6 <u>PROPOSITION</u>: Assume $(\mathfrak{M}_1,\mathcal{C},\mathfrak{M}_2)$ is a standard triple. Let X be a semicomplete $\mathfrak{M}_1^1$ manifold admitting $\mathfrak{M}_1$ exponentials and X_0 be a bounded $\mathcal{C}$ manifold. Let d and β be admissible metric and bound on X. S_β

is an admissible bound on $\mathcal{C}(X_0,X)$ and $d^{\mathcal{C}}$ defined by $d^{\mathcal{C}}(f_1,f_2) = \sup\{d(f_1x,f_2x) : x \in X_0\}$ is an admissible metric on $\mathcal{C}(X_0,X)$. In particular, if X is semicompact then $\mathcal{C}(X_0,X)$ is semicompact.

<u>PROOF</u>: If $d_1 >_{\rho} d_2$ are pseudometrics on X with $\rho: X \to [1,\infty]$, then $d_1^{\mathcal{C}} >_{S_\rho} d_2^{\mathcal{C}}$ on X^{X_0}. Thus, the equivalence class of $d^{\mathcal{C}}$ is independent of the choice of admissible metric. S_ρ is an admissible bound on $\mathcal{C}$ because $S_\rho(f) > S_{\rho_{\mathcal{G}}}(f) \geq k^{\mathcal{C}}(f;\mathcal{G}_0,\mathcal{G})$.

Let $(\mathcal{G},\rho,e,\Lambda)$ and $\mathcal{G}_0$ be the choices defining $(\mathfrak{M}(\mathcal{G}_0,\mathcal{G}),\sigma)$ on $\mathfrak{M}(X_0,X)$, with X_0 a bounded $\mathfrak{M}$ metric manifold. Make sure that V is transversely bounded. Let $\mathcal{G}_e$ be the $\mathfrak{M}_1$ atlas of normal coordinates on X induced by e and the Finsler $\| \ \|^M$ (from $(T\mathcal{G},\mathcal{G})$) (cf. Proposition III.8.4). Let $d_{\mathfrak{M}}$ be the metric on $\mathfrak{M}(X_0,X)$ induced by the atlas $\mathfrak{M}(\mathcal{G}_0,\mathcal{G})$ and $d_{\mathcal{G}_e}$ the metric on X induced by $\mathcal{G}_e$.

$$(3.7) \qquad d_{\mathfrak{M}} \geq (d_{\mathcal{G}_e})^{\mathcal{C}} \quad \text{on} \quad \mathfrak{M}(X_0,X).$$

To prove (3.7), let $(g_1,f_1,\ldots,f_N,g_{N+1})$ be an $\mathfrak{M}(\mathcal{G}_0,\mathcal{G})$ chain i.e. $g_i,g_{i+1} \in V^{\Lambda}_{f_i}$ $i = 1,\ldots,N$, and let $x \in X_0$. Clearly, $\|H_{f_i}(g_i) - H_{f_i}(g_{i+1})\| \geq \|H_{f_i}(g_i)(x) - H_{f_i}(g_{i+1})(x)\|^M = \|e^{-1}(f_i(x),g_i(x)) - e^{-1}(f_i(x),g_{i+1}(x))\|^M$. Thus, $(g_1(x),f_1(x),\ldots,f_N(x),g_{N+1}(x))$ is an $\mathcal{G}_e$ chain of no greater length. Hence, for $g_1,g_2 \in \mathfrak{M}(X_0,X)$ and $x \in X_0$, $d_{\mathfrak{M}}(g_1,g_2) \geq d_{\mathcal{G}_e}(g_1x,g_2x)$.

Returning now to the case $\mathfrak{M} = \mathcal{C}$, it suffices to prove $(d_{\mathcal{G}_e})^{\mathcal{C}} >_{S_\rho} d_{\mathcal{C}}$ since the class of $d^{\mathcal{C}}$ is independent of the choice of admissible d. Thus, the proof is completed by proving:

<u>CLAIM</u>: There exist constants K,n such that $g_1,g_2,f \in \mathcal{C}(X_0,X)$ with $(d_{\mathcal{G}_e})^{\mathcal{C}}(g_i,f) < (KS_\rho(f)^n)^{-1}$ implies $g_i \in V^{\Lambda}_f$ $(i = 1,2)$ and $KS_\rho(f)^n(d_{\mathcal{G}_e})^{\mathcal{C}}(g_1,g_2) \geq \|H_f(g_1) - H_f(g_2)\| \geq d_{\mathcal{C}}(g_1,g_2)$.

<u>PROOF</u>: Convexity of D^{Λ}_f implies the right inequality. Since $\sigma_\rho \sim S_\rho$

there exist constants $K_1, n_1 \geq 1$ such that $\Lambda > (K_1(S_\rho)^{n_1})^{-1}$. So $s \in \mathcal{C}(f^*\tau_X)$ with $\|s\| \leq (K_1 S_\rho(f)^{n_1})^{-1}$ implies $s \in D_f^\Lambda$. By Proposition 1.3, $\|s\| = S[\| \ \|^M](s)$. Now choose $V_0 \in \mathfrak{U}_X$ with $V_0 \subset\subset V$. V_0 is still transverse bounded and $\mathcal{G}_e$ is s admissible. So by the Metric Estimate there exist constants $K_2, n_2 \geq 1$ such that $y_1, y_2, x \in X_0$ with $d_{\mathcal{G}_e}(y_i, x) \leq (K_2\rho(x)^{n_2})^{-1}$ implies $(x, y_i) \in V_0$ $(i = 1,2)$ and $K_2\rho(x)^{n_2} d_{\mathcal{G}_e}(y_1, y_2) \geq \|e^{-1}(x, y_1) - e^{-1}(x, y_2)\|^M$. Let $K = K_1 K_2$ and $n = n_1 + n_2$ and assume $(d_{\mathcal{G}_e})^{\mathcal{C}}(g_i, f) < (KS_\rho(f)^n)^{-1}$ $(i = 1,2)$. Since $g \times g_i(X_0) \subset V_0$, $g_i \in V_g^\Lambda$ and $(K_1 S_\rho(f)^{n_1})^{-1} = K_2 S_\rho(f)^{n_2}$. $(KS_\rho(f)^n)^{-1} \geq S[\| \ \|^M](H_f(g_i) - H_f(f)) = \|H_f(g_i)\|$. Thus, $H_f(g_i) \in D_f^\Lambda$. Similarly, $K_2 S_\rho(f)^{n_2} (d_{\mathcal{G}_e})^{\mathcal{C}}(g_1, g_2) \geq S[\| \ \|^M](H_f(g_1) - H_f(g_2)) = \|H_f(g_1) - H_f(g_2)\|$.
Q.E.D.

In turning to standard constructions, we pass over products quickly: If X_i is an $\mathfrak{M}_1^1$ semicomplete manifold admitting $\mathfrak{M}_1$ exponentials $(i = 1,2)$ and X_0 is a bounded $\mathfrak{M}$ manifold, then $p_{1_*} \times p_{2_*}: \mathfrak{M}(X_0, X_1 \times X_2) \to \mathfrak{M}(X_0, X_1) \times \mathfrak{M}(X_0, X_2)$ is an identification at the atlas level. In fact, with choices $(\mathcal{G}_i, \rho_i, e_i)$ on X_i $(i = 1,2)$, we choose $(\mathcal{G}_1 \times \mathcal{G}_2, \max(p_1^*\rho_1, p_2^*\rho_2), e_1 \times e_2)$ with the obvious twist identification in defining $e_1 \times e_2$ and get that $p_{1*} \times p_{2*}$ is an atlas identification between $\mathfrak{M}(\mathcal{G}_0, \mathcal{G}_1 \times \mathcal{G}_2)$ and $\mathfrak{M}(\mathcal{G}_0, \mathcal{G}_1) \times \mathfrak{M}(\mathcal{G}_0, \mathcal{G}_2)$. It is only necessary to make sure that the same Λ constants $\bar{K}, \bar{n}$ are chosen for X_1, X_2 and $X_1 \times X_2$. From the proof of Proposition 6 we see that V^Λ is a transversely bounded member of $\mathfrak{U}_{\mathfrak{M}(X_0,X)}$, i.e. a full neighborhood of the diagonal in $\mathfrak{M}(X_0, X) \times \mathfrak{M}(X_0, X)$.

7 <u>PROPOSITION</u>: Let X be a semicomplete $\mathfrak{M}_1^1$ manifold admitting $\mathfrak{M}_1$ exponentials with admissible metric and bound d, ρ. Let $\rho_1: X \to [1, \infty]$ with $\rho_1 > \rho$ and $\rho_1^d \sim \rho_1$, and $U = \{\rho_1 < \infty\}$. Define U_1 to be the semicomplete $\mathfrak{M}_1^1$ manifold consisting of U with the $\rho_1|U$ refinement of the restriction

metric structure. U_1 admits $\mathfrak{M}_1$ exponentials. Now if X_0 is a bounded $\mathfrak{M}$ metric manifold then the set $\mathfrak{M}(X_0,U_1)$ is the subset $\{S_{\rho_1} < \infty\}$ of $\mathfrak{M}(X_0,X)$. Let $d_{\mathfrak{M}},\sigma$ be admissible metric and bound on $\mathfrak{M}(X_0,X)$. $\max(S_{\rho_1},\sigma)^{d_{\mathfrak{M}}} \sim \max(S_{\rho_1},\sigma)$ and on the open set $\mathfrak{M}(X_0,U_1) = \{S_{\rho_1} < \infty\}$ $= \{\max(S_{\rho_1},\sigma) < \infty\}$ the $\max(S_{\rho_1},\sigma)$ induced refinement of the restriction $\mathfrak{M}_2$ structure is precisely the mapping space structure on $\mathfrak{M}(X_0,U_1)$.

PROOF: It is easy to see that $\rho_1^d \sim \rho_1$ implies $(S_{\rho_1})^{d^{\mathcal{C}}} \sim S_{\rho_1}$. Then regularity of $\mathfrak{M}(X_0,X)$ and (3.7) imply that $\max(S_{\rho_1},\sigma)^{d_{\mathfrak{M}}} =$ $\max((S_{\rho_1})^{d_{\mathfrak{M}}},\sigma^{d_{\mathfrak{M}}}) \sim \max(S_{\rho_1},\sigma)$. Hence, $\{S_{\rho_1} < \infty\} = \{\max(S_{\rho_1},\sigma) < \infty\}$ is open in $\mathfrak{M}(X_0,X)$ and $\max(S_{\rho_1},\sigma)$ induces a semicomplete refinement by Lemma III.7.9. Also, it is clear that if $f\colon X_0 \to X$ is an $\mathfrak{M}$ map then f factors through the $\mathfrak{M}_1^1$ inclusion $i\colon U_1 \to X$, i.e. $f\colon X_0 \to U_1$ is an $\mathfrak{M}$ map, iff $\sup \rho_1 \cdot f < \infty$. Thus as sets $\mathfrak{M}(X_0,U_1) = \{S_{\rho_1} < \infty\}$. Now if $(\mathcal{C},\rho,e,\Lambda)$ and $\mathcal{C}_0$ are choices for X_0 and X then $(\mathcal{C}|U,\rho_1|U,i{*}e,\Lambda)$ are choices for U_1, where $V_1 = (i \times i)^{-1}V = V \cap U \times U$, $D_1 = e^{-1}V_1$ and $i{*}e = e|D_1$. Since U_1 is semicomplete, Lemma III.6.3 (h) and Corollary III.7.8 (b) imply that $i{*}e$ is an $\mathfrak{M}_1$ metric exponential on U_1. Since $\rho_1 > \rho$, we can make the Λ choice, Λ_1 for U_1, such that $\Lambda_1 \leq \Lambda$ on $\mathfrak{M}(X_0,U_1)$. It then follows that the atlas $\mathfrak{M}(\mathcal{C}_0,\mathcal{C}|U)$ is a refinement of $\mathfrak{M}(\mathcal{C}_0,\mathcal{C})|\mathfrak{M}(X_0,U_1)$. Hence, $\mathfrak{M}(\mathcal{C}_0,\mathcal{C}|U)$ is admissible for the restriction $\mathfrak{M}_2$ metric structure on the open set $\mathfrak{M}(X_0,U_1)$ as well as for the mapping space structure. Since $\max(S_{\rho_1},\sigma)$ is an admissible bound for the mapping space structure that structure is the $\max(S_{\rho_1},\sigma)$ induced refinement of the restriction. Q.E.D.

Applying the above results to $\rho_U = \max(\rho,1/\delta^U(d))$ we get that $\mathfrak{M}(X_0,U) = \{f \in \mathfrak{M}(X_0,X_1)\colon f(X_0) \subset\subset U\}$ is an open set in $\mathfrak{M}(X_0,X_1)$ and on it $\max(S_{\rho_U},\sigma)$ induces a semicomplete refinement which agrees with the mapping space structure on $\mathfrak{M}(X_0,U)$ defined when U is given the canonical refinement structure. If $F: U \to X_2$ is an $\mathfrak{m}_1$ map when U is given the restriction structure, then $F_*: \mathfrak{m}(X_0,U) \to \mathfrak{m}(X_0,X_2)$ is an $\mathfrak{M}_2$ metric map when the domain has the restriction $\mathfrak{M}_2$ metric structure (and a fortiori when it has the $\max(S_{\rho_U},\sigma)$ refinement structure). This is proved by going back to Lemma 2. We can sharpen this result a bit. Let Z be an $\mathfrak{m}_2$ metric manifold and $m: Z \to \mathfrak{m}(X_0,X_1)$ an $\mathfrak{M}_2$ metric map with $m(z)(X_0) \subset U$ for all $z \in Z$. $m_F: Z \to \mathfrak{m}(X_0,X_2)$ defined by $m_F(z) = F \circ m(z)$ is an $\mathfrak{M}_2$ metric map. If U is bounded and $(\mathfrak{a}_i,\rho_i,e_i)$ are choices on X_i with V_i transversely bounded $(i = 1,2)$ then we give an explicit estimate. Choose $\mathfrak{a}_0,\mathfrak{a}_Z$ admissible atlases on X_0 and Z, constants K_i,n_i defined as before (3.3) and Λ choices $\Lambda_i = (\bar{K}_i\sigma_i^{\bar{n}_i})^{-1}$ $(i = 1,2)$. Let $\sigma_{12}(F) = \max(S_{\rho_2}(F),k(F,\mathfrak{a}_1|U,\mathfrak{a}_2))$.

$$\sigma_2 \circ m_F \le O^*(\sigma_1 \circ m,\sigma_{12}(F))$$

(3.8)

$$\rho(m_F;\mathfrak{a}_Z,\mathfrak{M}(\mathfrak{a}_0,\mathfrak{a}_2)) \le O^*[K(\sigma_1 \circ m)^n,K(k_{\mathfrak{a}_0})^n,\sigma_{12}(F),\rho(m;\mathfrak{a}_Z,\mathfrak{M}(\mathfrak{a}_0,\mathfrak{a}_1))]$$

where $n = \max(n_1,n_2)\max(\bar{n}_1,\bar{n}_2)$ and $K = \max(K_1,K_2,\bar{K}_1^{\bar{n}},\bar{K}_2^{\bar{n}})$. The first inequality comes from Proposition III.1.1. The second comes from (3.3) and Lemma 3.

The following is an easy exercise using (3.7) and we leave it to the reader:

8 <u>LEMMA</u>: Let U be open in X and $A \subset\subset U$. If $B \subset \mathfrak{m}(X_0,X)$ such that $f \in B$ implies $f(X_0) \subset A$ then $B \subset\subset \mathfrak{m}(X_0,U)$.

If X is a semicomplete $\mathcal{C}^\infty$ metric manifold admitting $\mathcal{C}^\infty$ exponentials, then for any standard function space type, $\mathfrak{M}$, $\mathfrak{m}(X_0,X)$ is a semicomplete $\mathcal{C}^\infty$ metric manifold. For any $s > 0$ there exists an r such that $(\mathcal{L}^r,\mathfrak{M},\mathcal{L}^s)$ is a standard triple. Hence, if $(\mathfrak{a},\rho,e)$ is a choice of $\mathcal{C}^\infty$ atlas, bound and exponential, the adapted atlas $(\mathfrak{m}(\mathfrak{a}_0,\mathfrak{a}),\sigma)$ is $\mathcal{L}^s$. Since this is true for all s the atlas is a $\mathcal{C}^\infty$ adapted atlas. The only point that has to be checked is that we can make a Λ choice which works for all (r,s)

pairs, but clearly if $\wedge$ is chosen for any such (r,s) pair it works for all the others, i.e. just which function spaces $\mathfrak{M}_1$ and $\mathfrak{M}_2$ are is irrelevant to the statement of Lemma 3. If F: $X_1 \to X_2$ is $\mathcal{C}^\infty$ then F_* is $\mathcal{C}^\infty$ and if h: $X_0 \to X_1$ is $\mathfrak{M}$ then h* is $\mathcal{C}^\infty$ because they are $\mathcal{L}^s$ for all s.

9 <u>THEOREM</u>: Let X_0 be an $\mathfrak{M}$ metric manifold and X be a semicomplete $\mathfrak{M}_1^1$ manifold admitting $\mathfrak{M}_1$ exponentials. Choice of exponential on X induces the "atlas" $\{V_f, H_f\}$ on $\mathfrak{M}(X_0,X)$ with $H_f(V_f) = D_f$ open in the topological group $\mathfrak{M}(f^*\tau_X)$. The topology induced on $\mathfrak{M}(X_0,X)$ is Hausdorff and is independent of the choice of exponential.

If F: $X_1 \to X_2$ is an $\mathfrak{M}_1$ metric map of semicomplete $\mathfrak{M}_1^1$ metric manifolds admitting $\mathfrak{M}_1$ exponentials, then F_* is continuous

If h: $X_0 \to X_1$ is an $\mathfrak{M}$ metric map of $\mathfrak{M}_3$ metric manifolds then h* is continuous.

<u>PROOF</u>: That F_* is continuous follows as in Lemma 2, using Theorem 2.4 in place of Theorem 2.3. Applied to F = identity, this implies the topology is independent of the choice of e. The Hausdorff property is proved as in Theorem 5. Continuity of h* follows as in Lemma 4. Q.E.D.

The generalization of Proposition 6 says that if d is an admissible metric on X and ρ is an admissible bound on X_0, then the topology on $\mathcal{C}(X_0,X)$ is given by the uniformity $\mathfrak{N}_{\mathcal{C}}$ generated by sets $\{(f,g): d(fx,gx) < (K_\rho(x)^n)^{-1}\}$. This uniformity has a countable base and so is metrizable [13; Thm. VI.13]. Thus, the topology of $\mathcal{C}(X_0,X)$ is metrizable. It follows that $\mathfrak{M}(X_0,X)$ is metrizable. For if i: $\mathfrak{M}(X_0,X) \to \mathcal{C}(X_0,X)$ is the inclusion, $V_f^{\mathfrak{M}} = i^{-1}(V_f^{\mathcal{C}})$ for $f \in \mathfrak{M}(X_0,X)$ and each $V_f^{\mathfrak{M}}$ is metrizable. Since the subset $i(\mathfrak{M}(X_0,X))$ of $\mathcal{C}(X_0,X)$ is metrizable and so is paracompact, the open cover $\{V_f^{\mathcal{C}} \cap i(\mathfrak{M}(X_0,X))\}$ has a locally finite refinement $\{U_\alpha\}$. Then $\{i^{-1}U_\alpha\}$ is a locally finite open cover of $\mathfrak{M}(X_0,X)$ consisting of metrizable sets. Hence, $\mathfrak{M}(X_0,X)$ is metrizable by the Smirnov Metrization Theorem [13; p. 130].

We close with an alternative description of the topology on $\mathfrak{M}(X_0,X)$. If $\{U_\alpha\}$ is a uniform open cover of a regular pseudometric space (X,d,ρ) and $W = W^{-1} \in \mathfrak{N}_X$ such that $\{W\circ W[y]: y \in X\}$ refines $\{U_\alpha\}$ then we define the W-trimming $\{\tilde{U}_\alpha\}$ of $\{U_\alpha\}$ by $\tilde{U}_\alpha = \cup\{W[y]: W\circ W[y] \subset U_\alpha\}$. Clearly, $\{W[y]: y \in X\}$ refines $\{\tilde{U}_\alpha\}$ and $W[\tilde{U}_\alpha] \subset U_\alpha$ for all α.

10 PROPOSITION: Let X be a semicomplete $\mathfrak{M}_1^1$ manifold admitting $\mathfrak{M}_1$ exponentials and X_0 be an $\mathfrak{M}$ manifold. Let $\mathcal{A} = \{U_\beta,h_\beta\}$ be an s admissible $\mathfrak{M}_1^1$ atlas on X, $W \in \mathfrak{N}_X$ inducing a W trimming $\tilde{\mathcal{A}} = \{\tilde{U}_\beta,h_\beta\}$ of $\mathcal{A}$. Let $\mathcal{A}_0 = \{U_\alpha^0,h_\alpha^0\}$ be a star bounded admissible $\mathfrak{M}$ atlas on X_0 and ρ be an admissible bound on X_0. Let $G \subset \mathfrak{M}(X_0,X)$ and $f \in \mathfrak{M}(X_0,X)$. G is a neighborhood of f in $\mathfrak{M}(X_0,X)$ iff there exist constants K,n such that $g \in W_f$ (i.e. $f \times g(X_0) \subset\subset W$) and $\|g_{\beta\alpha} - f_{\beta\alpha}\|_{\mathfrak{M}} \leq (K\rho_\alpha^n)^{-1}$ for all pairs (α,β) implies $g \in G$, where $g_{\beta\alpha} = h_\beta g(h_\alpha^0)^{-1}: h_\alpha^0(U_\alpha^0 \cap f^{-1}(\tilde{U}_\beta)) \to E_\beta$. Note that $g \in W_f$ implies $g(f^{-1}\tilde{U}_\beta) \subset U_\beta$. ($\rho_\alpha = \sup \rho|U_\alpha^0$.)

PROOF: Choose $e: D \to V$ an $\mathfrak{M}_1$ metric exponential with V transverse bounded and $V \subset\subset W$. Choose $\bar{\rho}$ an admissible bound on X such that $\tau^*\bar{\rho} \geq \rho_e$, $p_1^*\bar{\rho} \geq \rho_{e^{-1}}$, and $\bar{\rho} \geq (\rho_{\mathcal{A}})^d$, $d = d_{\mathcal{A}}$. By the Metric Estimate, we can choose constants $K_1,n_1 \geq 1$ such that $W[y] \subset U_\beta$ implies $h_\beta(V[y]) \supset B(h_\beta(y),(K_1\bar{\rho}(y)^{n_1})^{-1})$. With $\| \;\|^M$ defined by $(T\tilde{\mathcal{A}},\tilde{\mathcal{A}})$ on τ, we can choose K_1,n_1 such that $\|v\|^M \leq (K_1\bar{\rho}(\tau v)^{n_1})^{-1}$ implies $v \in D$. Since f is fixed and the result is clearly independent of the choice of admissible bound on X_0 choose $\rho > \max(\rho_f, f^*(\bar{\rho}^d))$.

Since $V \subset W$, $(p_1|V)^{-1}(\tilde{U}_\beta) = (\tilde{U}_\beta \times U_\beta) \cap V$ and so $(\tau|D)^{-1}(\tilde{U}_\beta) = e^{-1}(\tilde{U}_\beta \times U_\beta)$. Define $V_\beta = h_\beta \times h_\beta((p_1|V)^{-1}(\tilde{U}_\beta)) \subset h_\beta(\tilde{U}_\beta) \times E_\beta$ and $D_\beta = Th_\beta((\tau|D)^{-1}\tilde{U}_\beta) \subset h_\beta(\tilde{U}_\beta) \times E_\beta$ and $e_\beta = (h_\beta \times h_\beta)\circ e\circ (Th_\beta)^{-1}: D_\beta \to V_\beta$. Note that $p_1\circ e_\beta = p_1$. Now let $Z_{\beta\alpha} = B(0,(K_1\rho_\alpha^{n_1})^{-1})$ in $\mathfrak{M}(h_\alpha^0(U_\alpha^0 \cap f^{-1}\tilde{U}_\beta),E_\beta)$ and let $Z'_{\beta\alpha} = Z_{\beta\alpha} + f_{\beta\alpha}$. Define $D_{\beta\alpha}$ (or $V_{\beta\alpha}$) to be

D_β (resp. V_β) intersected with $h_\beta(\tilde{U}_\beta \cap \{\bar{\rho}^{-d} < \rho_\alpha\}) \times \{\| \ \| < \rho_\alpha + 1\}$. Note that for $g \in Z_{\beta\alpha}$ (or $Z'_{\beta\alpha}$), the image of $f_{\beta\alpha} \times g$ lies in $D_{\beta\alpha}$ (resp. $V_{\beta\alpha}$). $\|e_\beta | D_{\beta\alpha}\|_{\mathfrak{M}_1}$, $\|e_\beta^{-1} | V_{\beta\alpha}\|_{\mathfrak{M}_1}$ and $\|f_{\beta\alpha}\|_{\mathfrak{M}} < \rho_\alpha$ (from the choices of $\bar{\rho}$ and ρ). Hence, $\|\Omega_{Z_{\beta\alpha}}^{f_{\beta\alpha}}(e_\beta | D_{\beta\alpha})\|_L$ and $\|\Omega_{Z'_{\beta\alpha}}^{f_{\beta\alpha}}(e_\beta^{-1} | V_{\beta\alpha})\|_L \le K_2 \rho_\alpha^{n_2}$ for some constants $K_2, n_2 \ge 1$.

If $s \in \mathfrak{M}(f^*\tau)$ satisfies $\|s_{\beta\alpha}\| \le (KK_2\rho_\alpha^{n+n_2})^{-1}$ with $K \ge K_1$, $n \ge n_1$ then $s_{\beta\alpha} \in Z_{\beta\alpha}$ and $g_{\beta\alpha} = \Omega_{Z_{\beta\alpha}}^{f_{\beta\alpha}}(e_\beta | D_{\beta\alpha})(s_{\beta\alpha})$ for $g = H_f^{-1}(s)$. Hence, $\|g_{\beta\alpha} - f_{\beta\alpha}\| = \|H_f^{-1}(s)_{\beta\alpha} - H_f^{-1}(0)_{\beta\alpha}\| \le (K\rho_\alpha^n)^{-1}$. Similarly, if $g \in W_f$ satisfies $\|g_{\beta\alpha} - f_{\beta\alpha}\| < (KK_2\rho_\alpha^{n+n_2})^{-1}$ then $g_{\beta\alpha} \in Z'_{\beta\alpha}$ and $s_{\beta\alpha} = \Omega_{Z'_{\beta\alpha}}^{f_{\beta\alpha}}(e_\beta^{-1} | V_{\beta\alpha})(g_{\beta\alpha})$ for $s = H_f(g)$. Hence, $\|s_{\beta\alpha}\| = \|s_{\beta\alpha} - 0\| = \|H_f(g)_{\beta\alpha} - H_f(f)_{\beta\alpha}\| < (K\rho_\alpha^n)^{-1}$. Q.E.D.

<u>ADDENDUM</u>: In the case when X_0 is bounded and $(\tilde{\mathfrak{a}}, \bar{\rho}^{-d}, e)$ and Λ constants $\bar{K} \ge K_1$, $\bar{n} \ge n_1$ are chosen for X then on V_f^Λ the map $P_f: V_f^\Lambda \to \hat{\Pi}_{\beta\alpha}\mathfrak{M}(h_\alpha^0(U_\alpha^0 \cap f^{-1}\tilde{U}_\beta), E_\beta)$ is well defined by $P_f(g) = \{g_{\beta\alpha}\}$. Furthermore, $P_f H_f^{-1}$ is an $\mathfrak{M}_2$ map with $\|P_f H_f^{-1}\|_{\mathfrak{M}_2} \le O^*(\sigma(f))$. In the other direction, on $P_f(V_f^\Lambda) \cap B(\{f_{\beta\alpha}\}, (K_1\sigma(f)^{n_1})^{-1})$ the inverse map $H_f P_f^{-1}$ is Lipschitz with $L(H_f P_f^{-1}) \le O^*(\sigma(f))$.

4. <u>Linear Map Bundles</u>: For $\mathfrak{M}$ bundles $\pi_0: \mathbb{E}_0 \to X_0$, $\pi: \mathbb{E} \to X$ let $\mathfrak{M}\mathcal{L}(\mathbb{E}_0; \mathbb{E})$ denote the set of $\mathfrak{M}$ VB maps $(\Phi, f): \pi_0 \to \pi$. $\mathfrak{M}\mathcal{L}(\pi_0; \pi): \mathfrak{M}\mathcal{L}(\mathbb{E}_0; \mathbb{E}) \to \mathfrak{M}(X_0, X)$ is defined by $\mathfrak{M}\mathcal{L}(\pi_0; \pi)(\Phi, f) = f$.

We say that $\pi: \mathbb{E} \to X$ is a semicomplete $\mathfrak{M}_1$ bundle admitting fiber exponentials if π is a semicomplete $\mathfrak{M}_1$ bundle, X is a semicomplete $\mathfrak{M}_1^1$ manifold and π admits $\mathfrak{M}_1$ fiber exponentials. By Definition III.8.1 and the remarks thereafter, this is equivalent to a $\mathfrak{M}_1$ exponential $e: D \to V$ on X and an $\mathfrak{M}_1$ isomorphism $T: (\pi \times X)_V \to (X \times \pi)_V$ pulling back to the identity under $\Delta: X \to V$. A choice $(\mathfrak{D}, \mathfrak{a}, \rho, T, e)$ for π

consists of $(\mathfrak{D},\mathcal{A})$ $\mathfrak{M}_1$ admissible for π with $\mathcal{A}$ $\mathfrak{M}_1^1$ admissible for X, $e: D \to V$ an $\mathfrak{M}_1$ exponential for X and T isomorphism as above. If $\pi_0 : \mathbb{E}_0 \to X_0$ is a bounded $\mathfrak{M}$ metric vector bundle with admissible atlas $(\mathfrak{D}_0,\mathcal{A}_0)$ and $f \in \mathfrak{M}(X_0,X)$ then $(\mathfrak{D}_f,\mathcal{A}_f)$ is the subdivision of $(\mathfrak{D}_0,\mathcal{A}_0)$ induced by $f: \mathcal{A}_0 \to \mathcal{A}$. $(\mathfrak{D}_f, f^*\mathfrak{D};\mathcal{A}_f)$ is an admissible $\mathfrak{M}$ two-tuple for π_0, $f^*\pi$ and so $(L(\mathfrak{D}_f;f^*\mathfrak{D}),\mathcal{A}_f)$ is an admissible atlas for $L(\pi_0;f^*\pi)$. Define

$$T_f: \mathfrak{M}\mathcal{L}(\pi_0;\pi)^{-1}V_f \longrightarrow D_f \times \mathfrak{M}(L(\pi_0;f^*\pi))$$

$$T_f(\Phi,g) = (H_f(g),(f\times g)^*T^{-1}\circ[f\times(\Phi,g)]). \tag{4.1}$$

Thus, identifying a section S of $\mathfrak{M}(L(\pi_0;f^*\pi))$ with the associated VB map $(S,1): \pi_0 \to f^*\pi$ and then with the VB map $(\Phi_f\circ S,f): \pi_0 \to f^*\pi$, $T_f(\Phi,g) = (s,S)$ iff $s = H_f(g)$ and the following commutes:

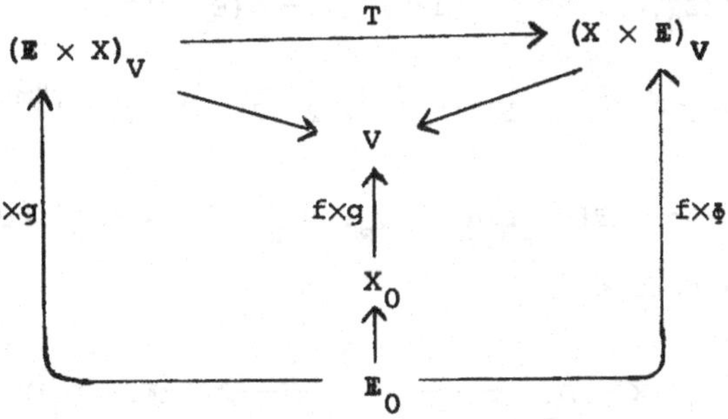

or if $(E_0,e): (\tau|D)^*\pi \to (p_1|V)^*\pi$ is the associated fiber exponential, $T_f(\Phi,g) = (s,S)$ iff the following commutes:

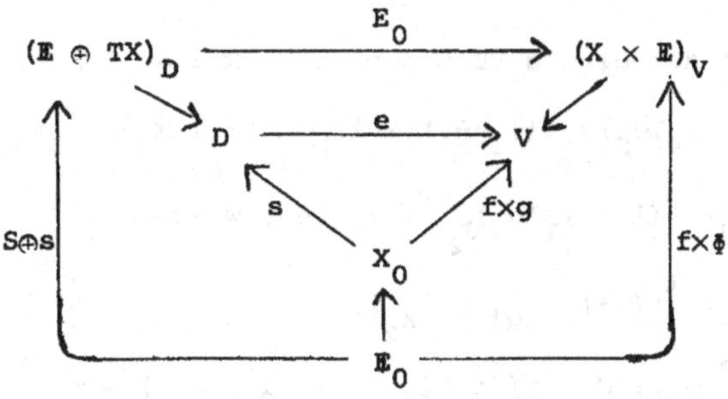

1 <u>LEMMA</u>: Let $\pi_i: \mathbb{E}_i \to X_i$ be an $\mathfrak{M}_1$ semicomplete bundle admitting fiber exponentials with choices $(\mathfrak{D}_i, \mathcal{G}_i, \rho_i, T_i, e_i)$ $(i = 1,2)$. Let $\pi_0: \mathbb{E}_0 \to X_0$ be a bounded $\mathfrak{M}$ bundle with admissible atlas $(\mathfrak{D}_0, \mathcal{G}_0)$.

Let $f_i \in \mathfrak{M}(X_0, X_i)$ $(i = 1,2)$, $(\Psi, F) \in \mathfrak{M}_1\mathcal{L}((\pi_1)_U;\pi_2)$ with U open in X_1 and $m \in \mathfrak{M}_2(G, \mathfrak{M}(f_1{}^*\tau_1))$ with G open and bounded in a B-space. Assume the image of m lies in $H_{f_1}(V_{f_1} \cap (F_*)^{-1}V_{f_2})$ and that $H_{f_1}^{-1}(m(g))(X_0) \subset U$ for all $g \in G$.

$M: G \to L(\mathfrak{M}(L(\pi_0;f_1{}^*\pi_1)); \mathfrak{M}(L(\pi_0;f_2{}^*\pi_2))$ defined by $(H_{f_2}F_*H_{f_1}^{-1}(m(g)), M(g)(S)) = T_{f_2}[(\Psi,F)\circ T_{f_1}^{-1}(m(g),S)]$ is an $\mathfrak{M}_2$ map. There exist constants K, n depending on $(\mathfrak{D}_i, \mathcal{G}_i, \rho_i, T_i, e_i)$ $(i = 1,2)$ and (Ψ, F) as well as the function space types such that $\|M\|_{\mathfrak{M}_2} \le K \max(\sigma_1(f_1), \sigma_2(f_2), \|m\|_{\mathfrak{M}_2})^n$. If V_i is transversely bounded $(i = 1,2)$, U is bounded and $K_i, n_i \ge 1$ are such that $K_i\rho_i^{n_i} \ge \max(\rho_{\mathcal{G}_i}^{\mathfrak{M}_1^1}, \rho_{(\mathfrak{D}_i,\mathcal{G}_i)}^{\mathfrak{M}_1})$, $K_i(\tau_i{}^*\rho_i)^{n_i} \ge \rho_{(E_0,e)}$ and $K_i(p_1{}^*\rho_i)^{n_i} \ge \rho_{(E_0,e)-1}$ $(i = 1,2)$ then:

(4.2) $$\|M\|_{\mathfrak{M}_2} \le O^*(K_1\sigma_1(f_1)^{n_1}, K_2\sigma_2(f_2)^{n_2}, k_{(\Psi,F)}, \|m\|_{\mathfrak{M}_2}).$$

<u>PROOF</u>: Recall the definition of $Q_F: D_{12} \to X_1 \times TX_2$ from Lemma 3.2. Analogously, we define:

$$\begin{array}{ccc} TX_1 \times X_2 \supset D_{12} & \xrightarrow{\ \bar{Q}_\Psi\ } & L(\mathbb{E}_1 \times X_2; X_1 \times \mathbb{E}_2) \\ & {\scriptstyle \tau_1 \times 1} \searrow \qquad \swarrow {\scriptstyle L(\pi_1 \times 1; 1 \times \pi_2)} & \\ & X_1 \times X_2 & \end{array}$$

by defining the associated VB map over $\tau_1 \times 1 = (1 \times \tau_2)\circ Q_F$. It is the composition of $(Q_\Psi, Q_F): ((\tau_1{}^*\pi_1) \times 1)_{D_{12}} \to (1 \times \tau_2{}^*\pi_2)_{D_2}$ and $(1 \times \Phi_{\tau_2}, 1 \times \tau_2): (1 \times \tau_2{}^*\pi_2)_{D_2} \to 1 \times \pi_2$, where $Q_\Psi = (1_{X_1} \times \mathbb{E}_{02})^{-1}\circ T(\Psi)\circ(\mathbb{E}_{01} \times 1_{X_2})$.

Now $\Omega_m^{f_1 \times f_2}(\bar{Q}_\Psi): G \to \mathfrak{M}((f_1 \times f_2)^*L(\pi_1 \times 1; 1 \times \pi_2)) \cong \mathfrak{M}(L(f_1{}^*\pi_1; f_2{}^*\pi_2))$.

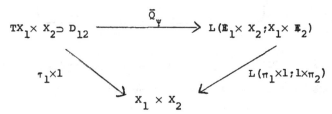

Let $L: L(f_1{}^*\pi_1; f_2{}^*\pi_2) \to L(L(\pi_0; f_1{}^*\pi_1); L(\pi_0; f_2{}^*\pi_2))$ be the composition map and recall the natural map λ (cf. p. 109). The result follows as in the second half of the proof of Lemma 3.2 because

$M = \lambda \circ L_* \circ \Omega_m^{f_1 \times f_2}(\bar{Q}_\Psi)$.   Q.E.D.

2 <u>LEMMA</u>: Let $\pi: \mathbb{E} \to X$ be an $\mathfrak{M}_1$ semicomplete bundle admitting fiber exponentials with choice $(\mathfrak{D}, \mathfrak{G}, \rho, T, e)$. Let $\pi_i: \mathbb{E}_i \to X_i$ be a bounded $\mathfrak{M}$ metric bundle with admissible atlas $(\mathfrak{D}_i, \mathfrak{G}_i)$ $(i = 0,1)$.

Let $f_0 \in \mathfrak{M}(X_0, X)$, $\Psi, h) \in \mathfrak{M}\ell(\pi_1; \pi_0)$ and $f_1 = h^*(f_0)$. Define $(\Psi^*, h^*): \mathfrak{M}\ell(\pi_0; \pi) \to \mathfrak{M}\ell(\pi_1; \pi)$ by $\Psi^*(\phi, g) = (\phi, g) \circ (\Psi, h)$. The local representative $(\Psi^*)_{f_1 f_0}: H_{f_0}[V_{f_0} \cap (h^*)^{-1} V_{f_1}] \cap B_N \to L(\mathfrak{M}(L(\pi_0; f_0{}^*\pi)); \mathfrak{M}(L(\pi_1; f_1{}^*\pi)))$ is an $\mathfrak{M}_2$ map with $\|(\Psi^*)_{f_1 f_0}\|_{\mathfrak{M}_2} \leq O^*(k_{(\Psi,h)})$.

<u>PROOF</u>: Similar to Lemma 3.4. The following commutes:

$$\begin{array}{ccc}
\mathfrak{M}\ell(\pi_0;\pi)^{-1}(V_{f_0} \cap (h^*)^{-1} V_{f_1}) & \xrightarrow{T_{f_0}} & \mathfrak{M}(f_0{}^*\tau_X) \times \mathfrak{M}(L(\pi_0; f^*\pi)) \\
 & & \downarrow h^* \times h^* \\
\Psi^* \downarrow & & \mathfrak{M}(f_1{}^*\tau_X) \times \mathfrak{M}(L(h^*\pi_0; f_1{}^*\pi)) \\
 & & \downarrow I \times L(\Psi,h)_* \\
\mathfrak{M}\ell(\pi_1;\pi)^{-1}(V_{f_1}) & \xrightarrow[T_{f_1}]{} & \mathfrak{M}(f_1{}^*\tau_X) \times \mathfrak{M}(L(\pi_1; f_1{}^*\pi))
\end{array}$$

where $(L(\Psi,h), 1): L(h^*\pi_0; f_1{}^*\pi) \to L(\pi_1; f_1{}^*\pi)$ is the VB map induced by $(h^*\Psi, 1): \pi_1 \to h^*\pi_0$. Thus, the local representative is constantly, $L(\Psi,h)_* \circ h^*$.   Q.E.D.

3 <u>THEOREM</u>: Assume $(\mathfrak{M}_1, \mathfrak{M}, \mathfrak{M}_2)$ is a standard triple. Let π_0 be a bounded $\mathfrak{M}$ bundle and π be a semicomplete $\mathfrak{M}_1$ bundle admitting fiber exponentials. Choice of $(\mathfrak{D}, \mathfrak{G}, \rho, T, e, \Lambda)$ (Λ is a $\wedge$ choice for $(\mathfrak{G}, \rho, e)$) and choice of $(\mathfrak{D}_0, \mathfrak{G}_0)$ admissible for π_0 induces the VB atlas $\{V_f^\wedge, H_f, T_f\}$ denoted $(\mathfrak{M}\ell(\mathfrak{D}_0; \mathfrak{D}), \mathfrak{M}(\mathfrak{G}_0, \mathfrak{G}))$ on $\mathfrak{M}\ell(\pi_0; \pi)$. $(\mathfrak{M}\ell(\mathfrak{D}_0; \mathfrak{D}), \mathfrak{M}(\mathfrak{G}_0, \mathfrak{G}), \sigma)$ is a semicomplete $\mathfrak{M}_2$ VB adapted atlas and the resulting metric structure is

independent of the choices.

If $(\Psi,F)\colon \pi_1 \to \pi_2$ is an $\mathfrak{M}_1$ VB map of semicomplete $\mathfrak{M}_1$ bundles admitting fiber exponentials, then $(\Psi_*,F_*)\colon \mathfrak{M}\mathcal{L}(\pi_0;\pi_1) \to \mathfrak{M}\mathcal{L}(\pi_0;\pi_2)$ is an $\mathfrak{M}_2$ metric VB map.

If $(\Psi,h)\colon \pi_0 \to \pi_1$ is an $\mathfrak{M}$ metric VB map of bounded $\mathfrak{M}$ metric bundles then $(\Psi^*,h^*)\colon \mathfrak{M}\mathcal{L}(\pi_1;\pi) \to \mathfrak{M}\mathcal{L}(\pi_0;\pi)$ is an $\mathfrak{M}_2$ metric VB map.

<u>PROOF</u>: Just like Theorem 3.5. Q.E.D.

With respect to the atlas $(\mathfrak{M}\mathcal{L}(\mathfrak{D}_0;\mathfrak{D}),\mathfrak{M}(\mathfrak{A}_0,\mathfrak{A}))$ we can define a Finsler $\| \;\|^{\mathfrak{M}}$ on $\mathfrak{M}\mathcal{L}(\pi_0;\pi)$ by $\|(\Phi,g)\|^{\mathfrak{M}} = \|p_2\circ T_g(\Phi,g)\| = \|(g^*\Phi,1)\|$ regarded as a section in $\mathfrak{M}(L(\pi_0;g^*\pi))$ with norm induced by $(L(\mathfrak{D}_g;g^*\mathfrak{D}),\mathfrak{A}_g)$. This Finsler is admissible because $\| \;\|^{m} \le \| \;\|^{\mathfrak{M}} \le \| \;\|^{M}$.

The most important special case of Theorem 3 is when $\pi_0 = \epsilon_R\colon X_0 \times R \to X_0$. We can identify: $\mathfrak{M}\mathcal{L}(\epsilon_R;\pi) \cong \pi_*\colon \mathfrak{M}(X_0,\mathbb{E}) \to \mathfrak{M}(X_0,X)$ by associating $(\Phi,f)\colon \epsilon_R \to \pi$ with $u = \Phi\circ s_1\colon X_0 \to \mathbb{E}$ where $s_1\colon X_0 \to X_0 \times R$ is the canonical section $s_1(x) = (x,1)$. Under the B-space identification $L(R;\mathbb{E}) \cong \mathbb{E}$, the local representatives of the VB map $(\Phi,f)\colon (\mathfrak{A}_0 \times R,\mathfrak{A}_0) \to (\mathfrak{D},\mathfrak{A})$ are the same as the local representatives of the map $u\colon \mathfrak{A}_0 \to \mathfrak{D}$. Thus, choices $(\mathfrak{D},\mathfrak{A},\rho,T,e)$ for π, and $\mathfrak{A}_0$ for X_0 induce for $f \in \mathfrak{M}(X_0,X)$ $T_f\colon (\pi_*)^{-1}V_f \to D_f \times \mathfrak{M}(f^*\pi)$. For $g = \pi\circ u$, $T_f(u) = (H_f(g),S)$ iff the following commutes:

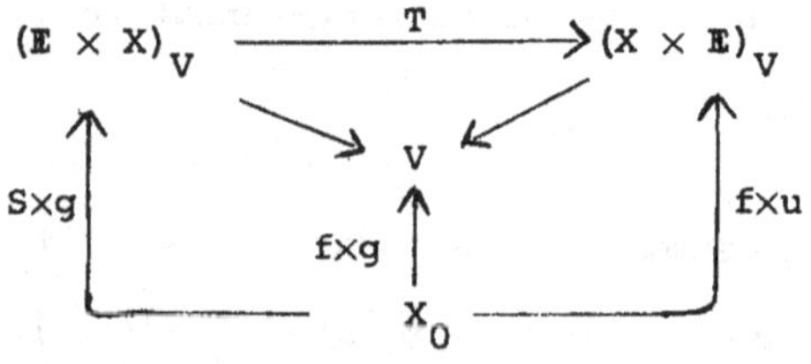

or equivalently, $T_f(u) = (s,S)$ iff the following commutes:

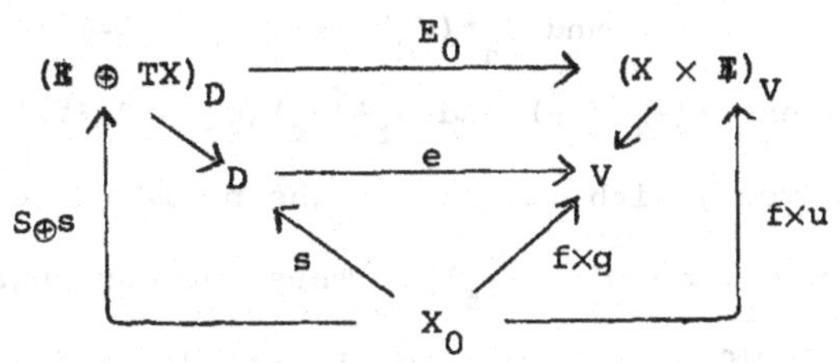

In particular, the space $\mathfrak{M}(X,\mathbb{E})$ receives a semicomplete $\mathfrak{M}_2$ metric manifold structure as the total space of π_*. Under reasonable conditions this total space structure agrees with the mapping space $\mathfrak{M}_2$ metric structure on $\mathfrak{M}(X_0,\mathbb{E})$. The bounds agree because:

$$\max(\|\ \|^{\mathfrak{M}}, \sigma_{\mathfrak{p}}(\mathfrak{G}_0,\mathfrak{G})\circ\pi_*) \sim \sigma_{\max(\|\ \|^M,\ \pi^*\mathfrak{p})}(\mathfrak{G}_0,\mathfrak{D}) \tag{4.2}$$

where $\|\ \|^M$ on π is defined by $(\mathfrak{D},\mathfrak{G})$. This follows from:
$S[\|\ \|^M](u) \le \|u\|^{\mathfrak{M}} \le \max(\|u\|^{\mathfrak{M}}, k(f;\mathfrak{G}_0,\mathfrak{G})) \le k(u;\mathfrak{G}_0,\mathfrak{D}) \le O^*(k_{\mathfrak{G}_0}, S_{\mathfrak{p}_{\mathfrak{G}}}(f),$
$k(f;\mathfrak{G}_0,\mathfrak{G}), \|u\|^{\mathfrak{M}})$.

4 <u>PROPOSITION</u>: If $\pi\colon \mathbb{E}\to X$ is a semicomplete $\mathfrak{M}_1^1$ bundle admitting $\mathfrak{M}_1$ bundle exponentials, then $\mathbb{E}$ is a semicomplete $\mathfrak{M}_1^1$ manifold admitting $\mathfrak{M}_1$ exponentials and the $\mathfrak{M}_2$ metric structure on $\mathfrak{M}(X_0,\mathbb{E})$ as the total space of π_* agrees with the mapping space structure.

<u>PROOF</u>: Choose $(E,e)\colon (T\pi)_D \overset{\cong}{\to} (\pi\times\pi)_V$ an $\mathfrak{M}_1$ exponential on π. Let (E_0,e) be the fiber exponential obtained by restricting to kernels (see p. 102) and T be the associated isomorphism. Let $(\mathfrak{D},\mathfrak{G},\mathfrak{p})$ be an admissible $\mathfrak{M}_1^1$ atlas for π and let $\bar{\mathfrak{p}} = \max(\pi^*\mathfrak{p}, \|\ \|^M)$ on $\mathbb{E}$. Choices $(\mathfrak{D},\mathfrak{G},\mathfrak{p},T,e)$ on π and $\mathfrak{G}_0$ on X_0 define the atlas $\{V_f,H_f,T_f\}$ on π_*. Choices $(\mathfrak{T},\bar{\mathfrak{p}},E)$ on $\mathbb{E}$ define the atlas $\{V_u,H_u\}$ on $\mathfrak{M}(X_0,\mathbb{E})$. Note that by Proposition III.7.12 $(u\times u_1)(X_0) \subset\subset (\mathbb{E}\times\mathbb{E})_V$ iff $(f\times g)(X_0) \subset\subset V$ where $f = \pi\circ u$ and $g = \pi\circ u_1$, i.e. for $f = \pi_* u$, $V_u = (\pi_*)^{-1}V_f$.

$T_f(\pi_*^{-1}V_f) = D_f \times \mathfrak{M}(f^*\pi)$ is open in $\mathfrak{M}(f^*\tau_X)\times\mathfrak{M}(f^*\pi) \cong \mathfrak{M}(f^*\tau_X \oplus f^*\pi) \cong \mathfrak{M}(f^*(\tau_X\oplus\pi))$. These isomorphisms are isometric with respect to the norms induced by $(f^*T\mathfrak{G},\mathfrak{G}_f)$, $(f^*\mathfrak{D},\mathfrak{G}_f)$ and $(f^*(T\mathfrak{G}\oplus\mathfrak{D}),\mathfrak{G}_f)$. The $\bar{J}$

identification between $\tau \oplus \pi$ and $0_\pi{}^*(\tau_{\mathbb{E}})$ (cf. pp. 12-13) pulls back to an identification between $f^*(\tau_X \oplus \pi)$ and $0_f{}^*(\tau_{\mathbb{E}})$ $(0_f = 0_\pi \circ f)$. On section spaces we get an isometry with respect to the norm induced by $(0_f{}^*T\mathfrak{D}, \mathcal{G}_{0_f}) = (0_f{}^*T\mathfrak{D}, \mathcal{G}_f)$ on $\mathfrak{M}(0_f{}^*(\tau_{\mathbb{E}}))$. Thus, we can regard $T_f(\pi_*^{-1}V_f)$ as an open subset of $\mathfrak{M}(0_f{}^*\tau_{\mathbb{E}})$. Because T was defined via (E,e) it is easy to check that with this identification $T_f = H_{0_f}$ on $\pi_*^{-1}V_f = V_{0_f}$.

If $f = \pi \circ u$ then $S_{\bar{\rho}}(u) \geq S_{\pi_*\rho}(u) = S_\rho(f) = S_{\bar{\rho}}(0_f)$ and $k(u;\mathcal{G}_0,\mathfrak{D}) \geq k(f;\mathcal{G}_0,\mathcal{G}) = k(0_f;\mathcal{G}_0,\mathfrak{D})$. Thus, $\bar{\sigma}(u) \geq \sigma(f) = \bar{\sigma}(0_f)$ where σ and $\bar{\sigma}$ are the mapping space bounds on $\mathfrak{M}(X_0,X)$ and $\mathfrak{M}(X_0;\mathbb{E})$ defined by our choices. By the identification of T_f with H_{0_f}, the atlas $\{\pi_*^{-1}(V_f^\wedge), H_{0_f}\}$ is admissible for the total space structure and we complete the proof by applying the Addendum to Theorem 3.5 to show that $\{\pi_*^{-1}(V_f^\wedge), H_{0_f}\}$ is admissible for the mapping space structure. By (4.2) we know that $(\{\pi_*^{-1}(V_f^\wedge), H_{0_f}\}, \bar{\sigma})$ is an adapted $\mathfrak{M}_2$ atlas. So there exist K_1, n_1 such that $K_1\bar{\sigma}(u)^{n_1} \geq \|H_{0_f}(u)\|$ for $u \in \pi_*^{-1}V_f^\wedge$. On the other hand there exist constants K_2, n_2 such that $g \in V_f^\wedge$ implies $K_2\sigma(g)^{n_2} \geq \sigma(f)$. But $\bar{\sigma}(u) \geq \sigma(g)$ if $\pi \circ u = g$. Q.E.D.

<u>THE STANDARD CONSTRUCTIONS:</u> (1) <u>Products:</u> As with spaces, if π_1 and π_2 are semicomplete $\mathfrak{M}_1$ bundles admitting fiber exponentials then $(P_{1*},P_{1*}) \times (P_{2*},P_{2*})_*: \mathfrak{M}\mathscr{L}(\pi_0;\pi_1 \times \pi_2) \to \mathfrak{M}\mathscr{L}(\pi_0;\pi_1) \times \mathfrak{M}\mathscr{L}(\pi_0;\pi_2)$ is an identification at the atlas level. Given choices $(\mathfrak{D}_i, \mathcal{G}_i, \rho_i, T_i, e_i)$ $(i = 1,2)$ use $((\mathfrak{D}_1,\mathcal{G}_1) \times (\mathfrak{D}_2,\mathcal{G}_2), \max(p_1{}^*\rho_1, p_2{}^*\rho_2), T_1 \times T_2, e_1 \times e_2)$ with the obvious twist identifications in defining $T_1 \times T_2$ as with $e_1 \times e_2$.
(2) <u>Subbundles:</u> Let $\pi: \mathbb{E} \to X$ be a semicomplete $\mathfrak{M}_1$ bundle admitting fiber exponentials and inducing a semicomplete $\mathfrak{M}_1$ subbundle structure on $\pi|\mathbb{E}_1$. If there exist T isomorphisms on π such that $T((\mathbb{E}_1 \times X)_V) = (X \times \mathbb{E}_1)_V$ then $\mathfrak{M}\mathscr{L}(\pi_0;\pi|\mathbb{E}_1)$ is a semicomplete $\mathfrak{M}_2$ subbundle of $\mathfrak{M}\mathscr{L}(\pi_0;\pi)$. Choose $(\mathfrak{D},\mathcal{G},\rho,T,e)$ for π with $(\mathfrak{D},\mathcal{G})$ inducing the subbundle atlas $(\mathfrak{D}|\mathbb{E}_1,\mathcal{G})$, and T preserving $\mathbb{E}_1$ as above. Then $(\mathfrak{M}\mathscr{L}(\mathfrak{D}_0;\mathfrak{D}_1), \mathfrak{M}(\mathcal{G}_0,\mathcal{G}))$

induces a subbundle atlas on $\mathfrak{M}\ell(\mathbb{E}_0;\mathbb{E}_1)$. (3) Direct Sums: If $\pi_i: \mathbb{E}_i \to X$ is a semicomplete $\mathfrak{M}_1$ bundle admitting fiber exponentials (i = 1,2), then $(P_{1*} \oplus P_{2*},1): \mathfrak{M}\ell(\pi_0;\pi_1 \oplus \pi_2) \cong \mathfrak{M}\ell(\pi_0;\pi_1) \oplus \mathfrak{M}\ell(\pi_0;\pi_2)$ is an identification at the atlas level. Choose $(\mathfrak{D}_1,\mathfrak{D}_2;\mathfrak{a})$ an admissible two-tuple on π_1,π_2 and e: $D \to V$ so that T_1,T_2 are defined over V. Then $(\mathfrak{D}_1 \oplus \mathfrak{D}_2,\mathfrak{a},\beta,e,T_1 \oplus T_2)$ yields the identification. (4) Trivial Bundle: If X is a semicomplete $\mathfrak{M}_1^1$ manifold admitting $\mathfrak{M}_1$ exponentials and F is a B-space, then $\epsilon_F: X \times F \to X$ admits fiber exponentials. In fact, there is a canonical isomorphism $T_F: \epsilon_F \times X \cong \overset{\times}{\epsilon}_F \cong X \times \epsilon_F$ where $\overset{\times}{\epsilon}_F$ is the trivial bundle on $X \times X$. If $(\mathfrak{a},\beta,e)$ is a choice for X then $(\mathfrak{a} \times F,\mathfrak{a},\beta,T_F,e)$ is a choice for ϵ_F yielding an identification at the atlas level between $\mathfrak{M}\ell(\pi_0;\epsilon_F)$ and $\epsilon_{\mathfrak{M}(L(\pi_0;\epsilon_F^0))}$ where $\epsilon_F^0: X_0 \times F \to X_0$. The identification comes from the identification $f^*\epsilon_F \cong \epsilon_F^0$ for all $f \in \mathfrak{M}(X_0,X)$. (5) Pull Backs: If $\pi: \mathbb{E} \to X$ is a semicomplete $\mathfrak{M}_1$ bundle admitting fiber exponentials, X_1 is a semicomplete $\mathfrak{M}_1^1$ manifold admitting exponentials and $F: X_1 \to X$ is an $\mathfrak{M}_1$ metric map, then $((F_*)^*(\phi_F)_*,1): \mathfrak{M}\ell(\pi_0;F^*\pi) \to (F_*)^*\mathfrak{M}\ell(\pi_0;\pi_1)$ is an identification at the atlas level. Given choices $(\mathfrak{D},\mathfrak{a},\beta,T,e,\Lambda)$ for π, and $\mathfrak{a}_1$ an $\mathfrak{M}_1^1$ atlas on X_1 so that $(F^*\mathfrak{D},\mathfrak{a}_F)$ is an atlas on $F^*\pi$. Choose $e_1: D_1 \to V_1$ with $V_1 \subset\subset (F \times F)^{-1}V$ and Λ choice Λ_1 so that $F_*^{-1}(V_{F_*f}^{\Lambda}) \supset V_f^{\Lambda_1}$. Then $F_*: \mathfrak{M}(\mathfrak{a}_0,\mathfrak{a}_1) \to \mathfrak{M}(\mathfrak{a}_0,\mathfrak{a})$ is index preserving. Defining F^*T in the obvious fashion we get choices for $F^*\pi$ yielding the atlas identification. In particular, in the situation of Proposition 3.7, if i: $U_1 \to X$ is the inclusion then $i^*\pi$ is the refinement of π_U induced by $\beta_1|U$. Hence, $\mathfrak{M}\ell(\pi_0;i^*\pi)$ is the refinement of $\mathfrak{M}\ell(\pi_0;\pi)_{\mathfrak{M}(X_0,U_1)}$ induced by $\max(S_{\beta_1},\sigma)$.

If $X_0 = \{p\}$, a single point, with atlas $\mathfrak{a}_0 = \{\{p\},0\}$, then $\pi_*: \mathfrak{M}(\{p\},\mathbb{E}) \to \mathfrak{M}(\{p\},X)$ can be identified with π. Choices $(\mathfrak{D},\mathfrak{a},\beta,T,e,\Lambda)$ for π then identifies $(\mathfrak{M}(\mathfrak{a}_0,\mathfrak{D}),\mathfrak{M}(\mathfrak{a}_0,\mathfrak{a}))$ with a refinement of the atlas of normal coordinates $(\mathfrak{D}_T,\mathfrak{a}_e)$ on π determined by the max Finslers on

τ_X and π (cf. Prop. III.8.4). So the identification is an $\mathfrak{M}_2$ VB isomorphism. Now for any bounded X_0 and $x \in X_0$ the maps $\{p\} \overset{i_x}{\to} X_0 \to \{p\}$ induce via this identification $\mathfrak{M}_2$ metric VB maps: $(c,c): \pi \to \pi_*$ and $(ev_x, ev_x): \pi_* \to \pi$. Similarly, the VB map $(i_v, i_x): \epsilon_R \to \pi_0$, where $\epsilon_R: \{p\} \times R \to \{p\}$ and $i_v(p,t) = tv$ ($v \in \mathbb{E}_0$ with $\pi_0(v) = x$) induces the $\mathfrak{M}_2$ metric VB map $(ev_v, ev_x): \mathfrak{M}\ell(\pi_0;\pi) \to \pi$.

Just as in the space case, if π is a semicomplete $\mathcal{C}^\infty$ bundle admitting $\mathcal{C}^\infty$ exponentials then $\mathfrak{M}\ell(\pi_0;\pi)$ and π_* are $\mathcal{C}^\infty$ vector bundles. $(\Psi,F)_*$ is $\mathcal{C}^\infty$ if (Ψ,F) is and $(\Psi,h)^*$ is $\mathcal{C}^\infty$ if (Ψ,h) is $\mathfrak{M}$.

5. Tangent Bundle Identifications: If X is a semicomplete $\mathfrak{M}_1^2$ manifold with $\mathfrak{M}_1^1$ exponential $e: D \to V$ then $(\tau e, e) = (Te \cdot \mathrm{T}, e): (T\tau_X)_D \to (\tau \times \tau)_V$ is an $\mathfrak{M}$ bundle exponential on the semicomplete $\mathfrak{M}_1^1$ bundle τ_X called the Jacobi exponential (cf. p. 102). By the commutative diagram below, the fiber exponential obtained by restricting to kernels is the vertical tangent map $(ve,e): (\tau_X|D)^*\tau_X \cong (p_1)_D \to (1 \times \tau_X)_V \cong (p_2|V)^*\tau_X$:

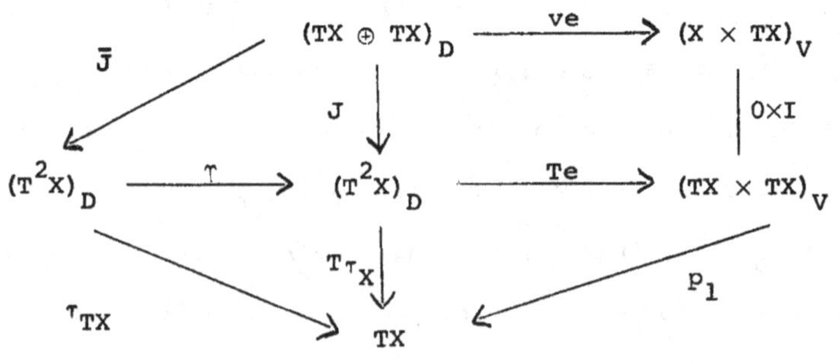

1 THEOREM: Let X be a semicomplete $\mathfrak{M}_1^2$ manifold admitting $\mathfrak{M}_1^1$ exponentials and X_0 be a bounded $\mathfrak{M}$ manifold. τ_X is a semicomplete $\mathfrak{M}_1^1$ bundle admitting $\mathfrak{M}_1$ bundle exponentials. $\mathfrak{M}(X_0,X)$ is a semicomplete $\mathfrak{M}_2^1$ manifold, $(\tau_X)_*$ is a semicomplete $\mathfrak{M}_2$ bundle and there is an isomorphism over the identity identifying $(\tau_X)_*: \mathfrak{M}(X_0,TX) \to \mathfrak{M}(X_0,X)$ with $\tau_{\mathfrak{M}}: T\mathfrak{M}(X_0,X) \to \mathfrak{M}(X_0,X)$. It is an identification at the atlas level: Choosing $(\mathfrak{a},\emptyset,e,\Lambda)$ for X,

(TG,G,ρ,ve,e,Λ) for τ_X, and G_0 for X_0, the identification relates $(\mathfrak{M}\ell(G_0 \times R,TG), \mathfrak{M}(G_0,G))$ with $(T\mathfrak{M}(G_0,G),\mathfrak{M}(G_0,G))$.

If $F: U \to X_2$ is an $\mathfrak{M}_1^1$ map with U open in X_1, X_1 and X_2 semicomplete $\mathfrak{M}_1^2$ manifolds admitting $\mathfrak{M}_1^1$ exponentials then F_* is an $\mathfrak{M}_2^1$ map and $(T(F_*),F_*): \tau_{\mathfrak{M}(X_0,U)} \to \tau_{\mathfrak{M}(X_0,X_2)}$ is identified with $((TF)_*,F_*): ((\tau_{X_1})_*)_{\mathfrak{M}(X_0,U)} \to (\tau_{X_2})_*$.

If $h: X_0 \to X_1$ is an $\mathfrak{M}$ map of bounded $\mathfrak{M}$ manifolds then h^* is an $\mathfrak{M}_2^1$ map and $(T(h^*),h^*): \tau_{\mathfrak{M}(X_1,X)} \to \tau_{\mathfrak{M}(X_0,X)}$ is identified with $(h^*,h^*): (\tau_X^1)_* \to (\tau_X^0)_*$ where $(\tau_X^i)_*: \mathfrak{M}(X_i,TX) \to \mathfrak{M}(X_i,X)$.

PROOF: $(\mathfrak{M}_1^1,\mathfrak{M},\mathfrak{M}_2^1)$ is a standard function space type by Theorem II.2.5. The proof is a matter of computing local representatives. When $F: U \to X_2$, let (G_i,ρ_i,e_i) be choices for X_i, $f_i \in \mathfrak{M}(X_0,X_i)$ $(i = 1,2)$ and consider $(F_*)_{f_2f_1}: H_{f_1}(V_{f_1} \cap F_*^{-1}(V_{f_2})) \to \mathfrak{M}(f_2{}^*\tau_2)$. (Recall that F_* is defined on the open set $\mathfrak{M}(X_0,U)$ in $\mathfrak{M}(X_0,X_1)$.) With the choices $(TG_i,G_i,\rho_i,ve_i,e_i)$ for τ_{X_i} the corresponding representative for (TF,F) is $((TF)_*)_{f_2f_1}: H_{f_1}(V_{f_1} \cap F_*^{-1}(V_{f_2})) \to L(\mathfrak{M}(f_1{}^*\tau_1);\mathfrak{M}(f_2{}^*\tau_2))$. We have to show that $((TF)_*)_{f_2f_1} = D[(F_*)_{f_2f_1}]$. Returning to the proofs of Lemmas 3.2 and 4.1, it is easy to check that the choice of fiber exponential implies equality of the VB maps $(Q_{TF},Q_F) = (vQ_F,Q_F): ((\tau_1{}^*\tau_1) \times 1)_{D_{12}} \to (1 \times \tau_2{}^*\tau_2)_{D_2}$. Hence, $\bar{Q}_{TF} = \partial_v Q_F: D_{12} \to L(TX_1 \times X_2;X_1 \times TX_2)$. So by diagram (2.3), $((TF)_*)_{f_2f_1} = \lambda\circ\Omega_G^{f_1\times f_2}(\bar{Q}_{TF}) = D\Omega_G^{f_1\times f_2}(Q_F) = D(F_*)_{f_2f_1}$ with G the domain of $(F_*)_{f_2f_1}$. Applying this result with $U = X_1 = X_2 = X$ and $F = 1$ we see that the transition maps of the atlases $(T\mathfrak{M}(G_0,G),\mathfrak{M}(G_0,G))$ and $(\mathfrak{M}\ell(G_0 \times R,TG),\mathfrak{M}(G_0,G))$ agree, inducing the identification at the atlas level. This computation also proves that $(TF)_* = T(F_*)$. The result for h follows from the proofs of Lemmas 3.4 and 4.2 which show that the representative $H_{f\circ h}\circ h^*\circ H_f^{-1}$ is the restriction of the linear map $h^*: \mathfrak{M}(f^*\tau) \to \mathfrak{M}((f\circ h)^*\tau)$ and the corresponding representative for (h^*,h^*) is constant at this same linear map. Q.E.D.

If $\pi\colon \mathbb{E} \to X$ is a semicomplete $\mathfrak{M}_1^2$ bundle with $\mathfrak{M}_1^1$ bundle exponential (E,e) on π, then it is easy to check that $(\tau E, \tau e)$ is an $\mathfrak{M}_1$ bundle exponential on the semicomplete $\mathfrak{M}_1$ bundle $T\pi\colon T\mathbb{E} \to TX$. If X_0 is a bounded $\mathfrak{M}$ metric manifold then Theorem 1 identifies the diagrams:

$$\begin{array}{ccc} \mathfrak{M}(X_0,T\mathbb{E}) & \xrightarrow{(T\pi)_*} & \mathfrak{M}(X_0,TX) \\ \downarrow{\scriptstyle (\tau_{\mathbb{E}})_*} & & \downarrow{\scriptstyle (\tau_X)_*} \\ \mathfrak{M}(X_0,\mathbb{E}) & \xrightarrow[\pi_*]{} & \mathfrak{M}(X_0,X) \end{array} \qquad \text{and} \qquad \begin{array}{ccc} T\mathfrak{M}(X_0,\mathbb{E}) & \xrightarrow{T(\pi_*)} & T\mathfrak{M}(X_0,X) \\ \downarrow{\scriptstyle \tau} & & \downarrow{\scriptstyle \tau} \\ \mathfrak{M}(X_0,\mathbb{E}) & \xrightarrow[\pi_*]{} & \mathfrak{M}(X_0,X). \end{array}$$

At the atlas level, a combination of the proofs of Theorem 1 and Proposition 4.4 identifies $(T\mathfrak{M}\ell(G_0 \times R,\mathfrak{H}),T\mathfrak{M}(G_0,G))$ with the refinement of $(\mathfrak{M}\ell(G_0 \times R,T\mathfrak{H}),\mathfrak{M}(G_0,TG))$ obtained from the refinement $\{(\tau_X)_*^{-1}V_f^{\wedge}\colon f \in \mathfrak{M}(X_0,X)\}$ of $\{V_u\colon u \in \mathfrak{M}(X_0,TX)\}$. In general, it is possible to identify $T\mathfrak{M}\ell(\pi_0;\pi)$ and with $\mathfrak{M}\ell(\pi_0;T\pi)$ but the proof is rather long. We will only need the section space results.

Theorem 1 applied to $p \overset{i_x}{\to} X_0 \to p$ yields the following commutative diagram:

(5.1)
$$\begin{array}{ccccc} & & (T\pi)_* & & \\ & \nearrow{\scriptstyle (c,c)} & & \searrow{\scriptstyle (ev_x,ev_x)} & \\ T\pi & & \| & & T\pi \\ & \searrow{\scriptstyle (Tc,c)} & & \nearrow{\scriptstyle (Tev_x,ev_x)} & \\ & & T(\pi_*) & & \end{array}$$

If X is a semicomplete $\mathfrak{M}_1^3$ manifold admitting $\mathfrak{M}_1^2$ exponentials then we can apply these results to the bundles τ_{TX} and $T\tau_X$. For $(T,1)\colon \tau_{TX} \to T\tau_X$ the twist isomorphism:

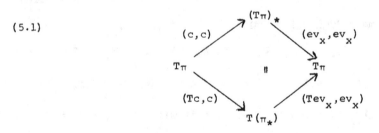

$$(5.2)\qquad \begin{array}{ccc} (\tau_{TX})_* & \xrightarrow{(T_*,1)} & (T\tau_X)_* \\ \| & & \| \\ \tau_{T\mathfrak{M}} & \xrightarrow[(T,1)]{} & T\tau_{\mathfrak{M}} \end{array}$$

commutes where $\mathfrak{M}$ stands for $\mathfrak{M}(X_0,X)$. The diagram commutes after applying ev_x and Tev_x by naturality of T. (5.2) then commutes because the family $\{ev_x: x \in X_0\}$ distinguishes points.

Now if X is an $\mathfrak{M}_1^2$ manifold admitting $\mathfrak{M}_1^1$ exponentials then $\mathfrak{M}(X_0,X)$ is an $\mathfrak{M}_2^1$ manifold admitting $\mathfrak{M}_2$ exponentials. For if $e: D \to V$ is an $\mathfrak{M}_1$ exponential on X then $e_*: \mathfrak{M}(X_0,D) \to \mathfrak{M}(X_0,V) = V^{\mathfrak{M}}$ is an exponential on $\mathfrak{M}(X_0,X)$ (see Lemma 3.8). Similarly, if π is an $\mathfrak{M}_1^2$ bundle admitting $\mathfrak{M}_1^1$ bundle exponentials, π_* (and $\mathfrak{M}\ell(\pi_0;\pi)$ in general) admit $\mathfrak{M}_2$ bundle exponentials.

2 <u>PROPOSITION</u>: Let X be a semicomplete $\mathfrak{M}_1^2$ manifold admitting $\mathfrak{M}_1^1$ exponentials and let X_0 be a bounded $\mathfrak{M}^1$ manifold. $\mathfrak{M}^1(X_0,X)$ is a semicomplete $\mathfrak{M}_2$ manifold and $\tau: \mathfrak{M}^1(X_0,X) \to \mathfrak{M}\ell(TX_0;TX)$ defined by $\tau(f) = (Tf,f)$ is an $\mathfrak{M}_2$ map.

<u>PROOF</u>: $(\mathfrak{M}_1^1,\mathfrak{M}^1,\mathfrak{M}_2)$ is a standard triple by Theorem II.2.4. Choose (G,ρ,e,Λ) for X, (TG,G,ρ,ve,e,Λ) for τ_X and (TG_0,G_0) for τ_{X_0}. Choosing the same Λ constants for $\mathfrak{M}$ and $\mathfrak{M}^1$, $\tau: \mathfrak{M}^1(G_0,G) \to \mathfrak{M}\ell(TG_0,TG)$ is index preserving and we need only look at the principal part $\tau_f = p_2 T_f \tau H_f^{-1}: D_f^\Lambda \to \mathfrak{M}(L(\tau_{X_0};f^*\tau_X))$ with $f \in \mathfrak{M}^1(X_0,X)$. If $s \in D_f^\Lambda$ and $g = H_f^{-1}(s)$ then $\tau_f(s)$ is characterized by the commutative diagram:

$$\begin{array}{ccccccc} (T^2X)_D & \xrightarrow{Te} & (TX \times TX)_V & \xrightarrow{\tau_X \times 1} & (X \times TX)_V & \xrightarrow{(ve)^{-1}} & (TX \oplus TX)_D \\ & {\scriptstyle Ts}\nwarrow & \uparrow{\scriptstyle Tf\times Tg} & & \uparrow{\scriptstyle f\times Tg} & \nearrow{\scriptstyle (s\circ\tau_{X_0})\oplus\tau_f(s)} & \\ & & & TX_0 & & & \end{array}$$

Thus, defining $(q_e, \tau_X): (\tau_{TX})_D \to \tau_X$ by $q_e = p_2(ve)^{-1}(\tau \times 1)Te$ then $(\tau_f(s), f)$ is the composition $(q_e, \tau_X)\circ(Ts, s)$. This is not quite a composition situation of the Theorem 2.3 type but the proof that $s \to \tau_f(s)$ is $\mathfrak{M}_2$ and the norm estimate is very similar, so we will just sketch the argument using the notation of that proof.

The principal parts of the section $\tau_X{*}q_e$ of $\mathfrak{M}_1(L(\tau_{TX};\tau_X{*}\tau_X)_D)$ are maps $q_\beta: D_\beta \to L(E_\beta \times E_\beta;E_\beta)$. If m_α is the inclusion of D_f^Λ into $\mathfrak{M}(f{*}\tau_X)$ followed by projection to $\mathfrak{M}(h_\alpha(U_\alpha), E_\beta)$ $(\beta = \beta(\alpha))$ then $\Omega_{m_\alpha}^{f_\alpha}(q_\beta): D_f^\Lambda \to \mathfrak{M}(h_\alpha(U_\alpha), L(E_\beta \times E_\beta;E_\beta))$ is an $\mathfrak{M}_2$ map. Let d_α denote the composition of $c_{f_\alpha} \times P_\alpha: D_f^\Lambda \to \mathfrak{M}(h_\alpha(U_\alpha), E_\beta \times E_\beta)$ and $D: \mathfrak{M}(h_\alpha(U_\alpha), E_\beta \times E_\beta) \to \mathfrak{M}(h_\alpha(U_\alpha), L(E_\alpha^0;E_\beta \times E_\beta))$. $P_\alpha \circ \tau_f = \text{comp}_*\circ(\Omega_{m_\alpha}^{f_\alpha}(g_\beta) \times d_\alpha)$. Q.E.D.

If X is $\mathfrak{M}_1^3$ with $\mathfrak{M}_1^2$ exponentials then using the identification $T\mathfrak{M}\ell(\tau_{X_0};\tau_X) = \mathfrak{M}\ell(\tau_{X_0};T\tau_X)$ it is possible to compute the tangent map of τ and show by applying (ev_v, ev_x) for $v \in T_xX_0$ that $(T\tau, \tau) = (\tau_*, 1)\circ(\tau, \tau)$.

6. Composition and Evaluation Results:

1 **LEMMA**: Assume that $(\mathfrak{M}_1, \mathfrak{M}, \mathfrak{M}_2)$ is a standard triple and $\mathfrak{M}_1$ satisfies FS7 (Strong). Let X_i be a semicomplete $\mathfrak{M}_1^1$ manifold admitting $\mathfrak{M}_1$ exponentials $(i = 1,2)$. With U open and bounded in X_1, let $\pi_1: \mathbb{E}_1 \to U$ be an $\mathfrak{M}_1$ bundle. Let $\pi_2: \mathbb{E}_2 \to X_2$ be a semicomplete $\mathfrak{M}_1$ bundle admitting fiber exponentials and $(\Psi, F): \pi_1 \to \pi_2$ be an $\mathfrak{M}_1$ VB map. Let X_0 be a bounded $\mathfrak{M}$ manifold. Let G be open and bounded in $\mathfrak{M}(X_0, U)$. Define:

$$\begin{array}{ccc}
\mathfrak{M}(X_0,U) \times \mathfrak{M}_1(\pi_1) & \xrightarrow{\ \text{Com}_{(\Psi,F)}\ } & \mathfrak{M}(X_0,\mathbb{E}_2) \\
\Big\downarrow{\scriptstyle \epsilon_{\mathfrak{M}_1}} & & \Big\downarrow{\scriptstyle \pi_{2*}} \\
\mathfrak{M}(X_0,U) & \xrightarrow[\ F_*\ ]{} & \mathfrak{M}(X_0,X_2)
\end{array}$$

$$\text{Sec}^G_{(\Psi,F)}: \mathfrak{M}_1(\pi_1) \longrightarrow \mathfrak{M}_2((F_*|G)^*(\pi_{2*}))$$

by $\mathrm{Com}_{(\psi,F)}(f,s) = \psi\circ s\circ f$ and $\mathrm{Sec}^G_{(\psi,F)}(s)(f) = (\psi\circ s\circ f, f)$. Then $(\mathrm{Com}_{(\psi,F)}, F_*)$ is an $\mathfrak{M}_2$ VB map and $\mathrm{Sec}^G_{(\psi,F)}$ is a bounded linear map.

PROOF: By Proposition 1.7 the composition of Ev: $U \times \mathfrak{M}_1(\pi_1) \to \mathbb{E}_1$ and ψ defines an $\mathfrak{M}_1$ VB map $(\psi\circ\mathrm{Ev}, F)\colon (\epsilon_{\mathfrak{M}_1})_U \to \pi_2$. By the trivial bundle identification, (4) of p. 141, this map induces the $\mathfrak{M}_2$ VB map:

$$\begin{array}{ccc}
\mathfrak{M}(X_0,U) \times \mathfrak{M}(X_0,\mathfrak{M}_1(\pi_1)) & \xrightarrow{\ (\mathrm{Ev}\circ\psi)_*\ } & \mathfrak{M}(X_0,\mathbb{E}_2) \\
\Big\downarrow \epsilon & & \Big\downarrow \pi_{2*} \\
\mathfrak{M}(X_0,U) & \xrightarrow[\ F_*\ ]{} & \mathfrak{M}(X_0,X_2)
\end{array}$$

Preceding by the map of trivial bundles induced by the linear map $c\colon \mathfrak{M}_1(\pi_1) \to \mathfrak{M}(X_0,\mathfrak{M}_1(\pi_1))$ we get $(\mathrm{Com}_{(\psi,F)}, F_*)$. Pulling back and restricting we get $((F_*)^*\mathrm{Com}_{(\psi,F)}, 1)\colon (\epsilon_{\mathfrak{M}_1})_G \to (F_*|G)^*(\pi_{2*})$. The associated map of sections $\mathfrak{M}_2(G,\mathfrak{M}_1(\pi_1)) \to \mathfrak{M}_2((F_*|G)^*(\pi_{2*}))$ when preceded by $c\colon \mathfrak{M}_1(\pi_1) \to \mathfrak{M}_2(G,\mathfrak{M}_1(\pi_1))$ is $\mathrm{Sec}^G_{(\psi,F)}$. Q.E.D.

Given choices $(\mathcal{G}_1,\rho_1,e_1,\Lambda_1)$ for X_1, $(\mathfrak{D}_2,\mathcal{G}_2,\rho_2,T_2,e_2,\Lambda_2)$ for π_2, atlas $(\mathfrak{D},\mathcal{G}_1|U)$ for π_1 and atlas $\mathcal{G}_0$ for X_0, such that V_1, V_2 are transversely bounded, choose constants $(K_i,n_i) \geq (\bar{K}_i,\bar{n}_i)$ $(i = 1,2)$ with K_1,n_1 satisfying the conditions of the hypothesis of Lemma 3.2 and $K_1\rho_1^{n_1} \geq \rho(\mathfrak{D},\mathcal{G}_1|U)$ on π_1, and with K_2,n_2 satisfying the analogous conditions of the hypothesis of Lemma 4.1. With respect to the induced atlases and norms:

$$(6.1)\qquad \rho(\mathrm{Com}_{(\psi,F)},F_*) \leq O^*(K\sigma_1{}^n, K(k_{\mathcal{G}_0})^n, \sigma_{12}(F), k_{(\psi,F)})$$

$$\|\mathrm{Sec}^G_{(\psi,F)}\| \leq O^*(K\sup(\sigma_1|G)^n, K(k_{\mathcal{G}_0})^n, S_{\rho_2}(F), k_{(\psi,F)}),$$

where $n = \max(n_1,n_2)^2$ and $K = \max(K_1,K_2)^n$. This follows from (3.8) and its analogue for bundle maps. Recall that with respect to the various induced norms the maps c are isometries.

If $X_1 = X_2$, $\pi_1 = \pi_{2U}$ and (ψ,F) is the inclusion we get the $\mathfrak{M}_2$ VB map

(Com,1): $\epsilon_{\mathfrak{M}_1} \to \pi_*$ by Com(f,s) = s∘f.

2 LEMMA: With the hypotheses of Lemma 1, let $\pi\colon \mathbb{E} \to X_1$ be a semicomplete $\mathfrak{M}_1$ bundle admitting fiber exponentials and let $\pi_0\colon \mathbb{E}_0 \to X_0$ be a bounded $\mathfrak{M}$ metric bundle. Define for $\epsilon_{\mathfrak{M}_1}\colon \mathfrak{M}(X_0,U) \times \mathfrak{M}_1(L(\pi_U;\pi_1)) \to \mathfrak{M}(X_0,U)$, $(L\ \mathrm{Com}_{(\Psi,F)},1)\colon \epsilon_{\mathfrak{M}_1} \to L[\mathfrak{M}\ell(\pi_0;\pi)_{\mathfrak{M}(X_0,U)};(F_*)^*\mathfrak{M}\ell(\pi_0;\pi_2)]$ and $L\ \mathrm{Sec}^G_{(\Psi,F)}\colon \mathfrak{M}_1(L(\pi_U;\pi_1)) \to \mathfrak{M}_2(L[\mathfrak{M}\ell(\pi_0;\pi)_G;(F_*|G)^*\mathfrak{M}\ell(\pi_0;\pi_2)])$ by $L\ \mathrm{Com}_{(\Psi,F)}(f,S)(\Phi,f) = (F^*[\Psi\circ S\circ\Phi],f)$ and $L\ \mathrm{Sec}^G_{(\Psi,F)}(S)(f)(\Phi,f) = (F^*[\Psi\circ S\circ\Phi],f)$. $L\ \mathrm{Com}_{(\Psi,F)}$ is an $\mathfrak{M}_2$ VB map and $L\ \mathrm{Sec}^G_{(\Psi,F)}$ is a bounded linear map.

PROOF: Proceed as in the proof of Lemma 1, beginning with the bilinear $\mathfrak{M}_1$ VB map: $\pi_U \oplus (\epsilon_{\mathfrak{M}_1}(L))_U \overset{I\oplus Ev}{\to} \pi_U \oplus L(\pi_U,\pi_1) \overset{ev}{\to} \pi_1 \overset{(\Psi,F)}{\to} \pi_2$. Q.E.D.

3 PROPOSITION: Assume $(\mathfrak{M}_1,\mathfrak{M},\mathfrak{M}_2)$ is a standard triple and $\mathfrak{M}_2 \subset \mathfrak{M}$.

(a) Let X be a semicomplete $\mathfrak{M}_1^1$ manifold admitting $\mathfrak{M}_1$ exponentials and X_0 be a bounded $\mathfrak{M}$ manifold. ev: $X_0 \times \mathfrak{M}(X_0,X) \to X$ defined by $ev(x,f) = f(x)$ is an $\mathfrak{M}$ map.

(b) Assume $\mathfrak{M}$ satisfies FS7 (Strong). Let π be a semicomplete $\mathfrak{M}_1$ bundle admitting fiber exponentials and π_0 be a bounded $\mathfrak{M}$ bundle. (Lev, ev): $\pi_0 \times \mathfrak{M}\ell(\pi_0;\pi) \to \pi$ defined by $\mathrm{Lev}(v,(\Phi,f)) = \Phi(v)$ is an $\mathfrak{M}$ metric bilinear VB map.

PROOF: (a): Choose $(\mathfrak{G},\rho,e,\Lambda)$ for X with V transverse bounded, $\mathfrak{G}_0$ for X_0, and constants K_1,n_1 satisfying the Lemma 3.2 conditions. Let $f \in \mathfrak{M}(X_0,X)$. By Proposition 1.7 and its proof, $X_0 \times \mathfrak{M}(f^*\tau_X) \overset{Ev}{\to} f^*TX \overset{\Phi_f}{\to} TX$ is an $\mathfrak{M}$ metric map and on the open subset $X_0 \times D_f^\Lambda$, $k_{\Phi_f\circ Ev} \leq O^*(k_{\mathfrak{G}_0},S_{\rho_{\mathfrak{G}}}(f)) \leq O^*(k_{\mathfrak{G}_0},K_1\sigma(f)^{n_1})$. $ev_f = ev\circ(1 \times H_f^{-1})$ is $\Phi_f\circ Ev\colon X_0 \times D_f^\Lambda \to D$ composed with $p_2\circ e\colon D \to X$. So $k_{ev_f} \leq O^*(k_{\mathfrak{G}_0},K_1\sigma(f)^{n_1})$. By the usual Λ choice argument:

(6.2) $$\rho \cdot ev \leq p_2{}^*S_\rho$$

$$\rho(ev, \mathfrak{a}_0 \times \mathfrak{M}(\mathfrak{a}_0,\mathfrak{a}),\mathfrak{a}) \leq O^*(K(k_{\mathfrak{a}_0})^n, Kp_2{}^*\sigma^n)$$

where $K = \max(K_1,\bar{K})^n$ and $n = n_1\bar{n}$.

(b): Similar to (a). Since $\mathfrak{M}$ satisfies FS7 (Strong), letting $\epsilon_{\mathfrak{M}(L)}: X_0 \times \mathfrak{M}(L(\pi_0;f^*\pi)) \to X_0$, $(ev,1): \epsilon_{\mathfrak{M}(L)} \to L(\pi_0;f^*\pi)$ is an $\mathfrak{M}$ VB map. So the composition

$$\pi_0 \oplus \epsilon_{\mathfrak{M}(L)} \xrightarrow{(I\oplus ev,1)} \pi_0 \oplus L(\pi_0;f^*\pi) \xrightarrow{(Ev,1)} f^*\pi \xrightarrow{(\Phi_f,f)} \pi$$

is an $\mathfrak{M}$ VB map. $Lev \cdot T_f^{-1}$ is this map preceded by projection off the $D_f^\wedge$ factor. Q.E.D.

We can replace the assumption that $\mathfrak{M}_2 \subset \mathfrak{M}$ by the assumption that $\mathfrak{M}_2 \subset \mathfrak{M}_3$ and $\mathfrak{M} \subset \mathfrak{M}_3$. Then ev in (a) is $\mathfrak{M}_3$. In (b) we can then assume that $\mathfrak{M}_3$ (instead of $\mathfrak{M}$) satisfies FS7 (Strong) or even $\mathfrak{M}$ interchanges into $\mathfrak{M}_3$ (cf. p. 25) and get that (Lev,ev) is an $\mathfrak{M}_3$ VB map.

4 <u>THEOREM</u>: Assume a and b are nonnegative integers such that $(\mathfrak{M}_1^a,\mathfrak{M}_1,\mathfrak{M}_1)$ and $(\mathfrak{M}_1^b,\mathfrak{M},\mathfrak{M}_1)$ are standard triples. Let $r \geq a$ be an integer.

(a) Let X_1 be a semicomplete $\mathfrak{M}_1^{b+1}$ manifold admitting $\mathfrak{M}_1^b$ exponentials, X_2 be a semicomplete $\mathfrak{M}_1^{b+r+2}$ manifold admitting $\mathfrak{M}_1^{b+r+1}$ exponentials, and X_0 be a bounded $\mathfrak{M}$ manifold. For U open and bounded in X_1 and G open and bounded in $\mathfrak{M}(X_0,U)$ the equation $\Omega_G(f) = f_*|G$ defines an $\mathfrak{M}_1^{r-a}$ function

$$\Omega_G: \mathfrak{M}_1^b(U,X_2) \longrightarrow \mathfrak{M}_1(G,\mathfrak{M}(X_0,X_2)).$$

(b) Let $\pi_1: \mathbb{E}_1 \to X_1$ be a semicomplete $\mathfrak{M}_1^b$ bundle admitting fiber exponentials, $\pi_2: \mathbb{E}_2 \to X_2$ be a semicomplete $\mathfrak{M}_1^{b+r+1}$ bundle admitting fiber exponentials and $\pi_0: \mathbb{E}_0 \to X_0$ be a bounded $\mathfrak{M}$ manifold. For U open and bounded in X_1 and G open and bounded in $\mathfrak{M}(X_0,U)$ the equation $L\Omega_G(\Phi,f) = (\Phi_*,f_*)|\mathfrak{ML}(\pi_0,\pi_1)_G$ defines an $\mathfrak{M}_1^{r-a}$ VB map

$$(L\Omega_G,\Omega_G) : \mathfrak{M}_1^b\mathcal{L}((\pi_1)_U;\pi_2) \longrightarrow \mathfrak{m}_1\mathcal{L}(\mathfrak{M}\mathcal{L}(\pi_0;\pi_1)_G;\mathfrak{M}\mathcal{L}(\pi_0;\pi_2)).$$

<u>PROOF</u>: (a): Choose $(\mathcal{G}_i,\rho_i,e_i,\Lambda_i)$ for X_i, with V_i transversely bounded, $(i = 1,2)$ and $\mathcal{G}_0$ for X_0. We get adapted atlases:
(1) $(\mathfrak{m}_1^b(\mathcal{G}_1|U,\mathcal{G}_2),\sigma_{\rho_2}(\mathcal{G}_1|U,\mathcal{G}_2))$, denoted $(\mathfrak{m}_{12}^U,\sigma_{12}^U)$, in the $\mathfrak{M}_1^{r-a}$ structure of $\mathfrak{M}_1^b(U,X_2)$. (2) $(\mathfrak{M}(\mathcal{G}_0,\mathcal{G}_1),\sigma_{\rho_1}(\mathcal{G}_0,\mathcal{G}_1))$, denoted $(\mathfrak{m}_{01},\sigma_{01})$, in the $\mathfrak{M}_1$ structure of $\mathfrak{M}(X_0,X_1)$. (3) $(\mathfrak{M}(\mathcal{G}_0,\mathcal{G}_2),\sigma_{\rho_2}(\mathcal{G}_0,\mathcal{G}_2))$, denoted $(\mathfrak{m}_{02},\sigma_{02})$, in the $\mathfrak{m}_1^{r+1}$ structure of $\mathfrak{M}(X_0,X_2)$ and exponential e_{2*} on $\mathfrak{M}(X_0,X_2)$.
(4) $(\mathfrak{M}_1(\mathfrak{m}_{01}|G,\mathfrak{m}_{02}),\sigma_{\sigma_{02}}(\mathfrak{m}_{01}|G,\mathfrak{m}_{02}))$, denoted $(\mathfrak{m}_{012}^G,\sigma_{012}^G)$, in the $\mathfrak{M}_1^{r-a}$ structure of $\mathfrak{M}_1(G,\mathfrak{M}(X_0,X_2))$.

For the rest of the proof we will denote by K,n constants depending on the choices $(\mathcal{G}_i,\rho_i,e_i,\Lambda_i)$ and $\mathcal{G}_0$ $(i = 1,2)$ as well as the function space types, but not on f,U,G etc. We will say A is dominated by B if $A \leq KB^n$ for such choice of constants.

First, note that by (3.8) and the remarks preceding it, Ω_G is a well defined set map and for some constants K_1,n_1:

$$\sigma_{012}^G\cdot\Omega_G \leq K_1 \max(\sup(\sigma_{01}|G),\ \sup(\rho_1|U),\sigma_{12}^U)^{n_1}. \tag{6.3}$$

Now if $f \in \mathfrak{M}_1^b(U,X_2)$ then the principal part $H_{f_*}\Omega_G H_f^{-1}$ maps $D_f(\subset \mathfrak{M}(f^*\tau_{X_2}))$ into $\mathfrak{M}_1((f_*)^*\tau_{\mathfrak{M}(X_0,X_2)})$. Applying the functor $\mathfrak{M}(X_0,\_)$ to the diagram on p. 122 describing $s = H_f(g)$ we get:

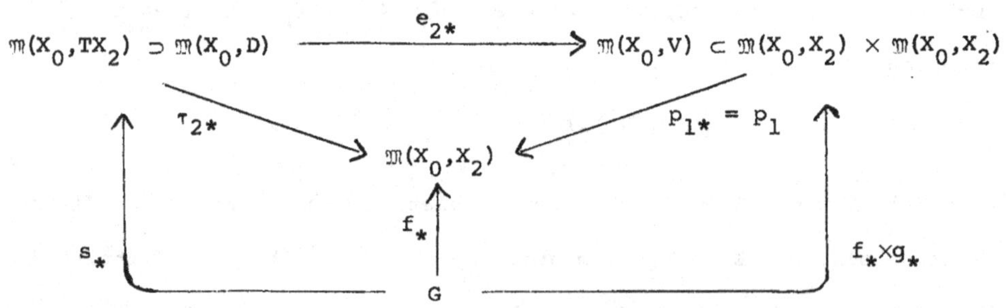

Identifying τ_{2*} with $\tau_{\mathfrak{M}(X_0,X_2)}$ we see that $H_{f_*}\cdot\Omega_G\cdot H_f^{-1}(s) = s_*$. Hence, $H_{f_*}\cdot\Omega_G\cdot H_f^{-1}$ is the restriction of $\mathrm{Sec}^G_{(\Phi_f,f)} : \mathfrak{m}_1^b(f^*\tau_2) \to \mathfrak{M}_1((f_*|G)^*(\tau_{2*}))$.

So $H_f(V_f^\wedge \cap \Omega_G^{-1}(V_{f_*}^{\wedge}))$ is an open subset of the unit ball containing 0 and on it $H_{f_*} \cdot \Omega_G \cdot H_f^{-1}$ is an $\mathfrak{M}_1^{r-a}$ map with norm dominated by $\max(\sup(\sigma_{01}|G), \sigma_{12}^U(f))$. Since $\wedge$ choices for $\mathfrak{M}_1^b(U,X_2)$ and $\mathfrak{M}_1(G,\mathfrak{M}(X_0,X_2))$ do not depend on U or G, σ_{12}^U dominates $\rho_{\mathfrak{M}_{12}^U}$ and σ_{012}^G dominates $\rho_{\mathfrak{M}_{012}^G}$. Thus, from (6.3), Corollary III.1.3 and the usual $\wedge$ choice argument:

$$(6.4) \qquad \rho(\Omega_G;\mathfrak{M}_{01}^U,\mathfrak{M}_{012}^G) \leq K_2 \max(\sup(\sigma_{01}|G), \sup(\rho_1|U), \sigma_{12}^U)^{n_2}$$

for some constants K_2, n_2.

(b): Analogous to (a). The key step is the computation that the principal part of $(L\Omega_G,\Omega_G)$ with respect to the index f is constantly the linear map $L\,\mathrm{Sec}^G_{(\Phi_f,f)}: \mathfrak{M}_1^b(L(\pi_1|U;f^*\pi_2)) \to \mathfrak{M}_1(L(\mathfrak{ML}(\pi_0;\pi_1)_G;(f_*|G)^*\mathfrak{ML}(\pi_0;\pi_2))$. Q.E.D.

The above arguments extend to the case when U is unbounded and show that $\Omega: \mathfrak{M}_1^b(U,X) \to \mathfrak{M}_1(\mathfrak{M}(X_0,U),\mathfrak{M}(X_0,X_2))$ is continuous. Also, the results hold if $r = \infty$, i.e. if X_2 is a semicomplete C^∞ manifold admitting C^∞ exponentials (or if π_2 is a semicomplete C^∞ bundle admitting C^∞ fiber exponentials) then Ω_G is a C^∞ map (resp. $(L\Omega_G,\Omega_G)$ is a C^∞ VB map). This follows by applying Theorem 4 to $\mathfrak{M}_1^r$ for $r = a, a+1, \ldots$.

5 <u>COROLLARY</u>: With the hypotheses of Theorem 4,

(a') Comp: $\mathfrak{M}(X_0,U) \times \mathfrak{M}_1^b(U,X_2) \to \mathfrak{M}(X_0,X_2)$ defined by $\mathrm{Comp}(f,g) = f\cdot g$ is an $\mathfrak{M}_1$ map.

(b') (L Comp,Comp): $\mathfrak{ML}(\pi_0;\pi_1)_{\mathfrak{M}(X_0,U)} \times \mathfrak{M}_1^b\mathcal{L}((\pi_1)_U;\pi_2) \to \mathfrak{ML}(\pi_0;\pi_2)$ defined by $L\,\mathrm{Comp}((\Phi,f),(\Psi,g)) = (\Phi\cdot\Psi, f\cdot g)$ is an $\mathfrak{M}_1$ VB map.

<u>PROOF</u>: (a'): Restricting to $G \times \mathfrak{M}_1^b(U,X)$, $\mathrm{Comp} = \mathrm{ev}\cdot(1_G \times \Omega_G)$ where ev: $G \times \mathfrak{M}_1(G,\mathfrak{M}(X_0,X_2)) \to \mathfrak{M}(X_0,X_2)$ is the map of Proposition 3 (a). The estimates (6.2), (6.3) and (6.4) show that Comp is $\mathfrak{M}_1$ on the entire product.

(b'): Just like (a'), using Proposition 3(b). Note that $\mathfrak{M}_1$ satisfies

FS7 (Strong).

The two main examples where Theorem 4 and Corollary 5 apply are:

(1) $\mathfrak{M}_1 = \mathscr{L}^s$, $\mathfrak{M} = \mathcal{C}^t$ (or $\mathfrak{M} = \mathfrak{N}^t$ and $t \geq 1$), $a = s + 1$, $b = t$, Ω_G: $\mathscr{L}^{s+t}(U,X_2) \to \mathscr{L}^s(G,\mathcal{C}^t(X_0,X_2))$ with G open in $\mathcal{C}^t(X_0,U)$, Comp: $\mathcal{C}^t(X_0,U) \times \mathscr{L}^{s+t}(U,X_2) \to \mathcal{C}^t(X_0,X_2)$. (2) $\mathfrak{M}_1 = \mathscr{L}^s$, $\mathfrak{M} = \mathscr{L}^t$, $a = s + 1$, $b = t + 1$, Ω_G: $\mathscr{L}^{s+t+1}(U,X_2) \to \mathscr{L}^s(G,\mathscr{L}^t(X_0,X_2))$ with G open in $\mathscr{L}^t(X_0,U)$, Comp: $\mathscr{L}^t(X_0,U) \times \mathscr{L}^{s+t+1}(U,X_2) \to \mathscr{L}^t(X_0,X_2)$.

Note that while the smoothness of Ω_G increases with that of the space X_2, Comp is no smoother than $\mathfrak{M}_1$.

6 PROPOSITION: Assume $(\mathfrak{M}_1,\mathfrak{M},\mathfrak{M}_2)$ is a standard triple. Let X be a semicomplete $\mathfrak{M}_1^1$ manifold admitting $\mathfrak{M}_1$ exponentials.

(a) Assume $\mathfrak{M}_2 \subset \mathfrak{M}$. If X_0 and X_1 are bounded $\mathfrak{M}$ manifolds and $F: X_0 \to \mathfrak{M}(X_1,X)$ is an $\mathfrak{M}$ map, then $C(F): X_0 \times X_1 \to X$ is an $\mathfrak{M}$ map well defined by $C(F)(x_0,x_1) = F(x_0)(x_1)$.

(b) If X_0 and X_1 are bounded $\mathfrak{M}_1$ manifolds and $F: X_0 \times X_1 \to X$ is an $\mathfrak{M}_1$ map then $\bar{C}(F): X_0 \to \mathfrak{M}(X_1,X)$ is an $\mathfrak{M}_2$ map well defined by $\bar{C}(F)(x_0)(x_1) = F(x_0,x_1)$.

PROOF: (a): $C(F) = ev\circ(p_2 \times F)$. Result by Proposition 3.

(b): With choices $(\mathcal{G},\rho,e,\Lambda)$ for X with V transversely bounded, and $\mathfrak{M}_1$ atlases $\mathcal{G}_0,\mathcal{G}_1$ for X_0,X_1 we can replace X_0 by $h_\alpha(U_\alpha)$ and F by $F\circ h_\alpha^{-1} \times 1$, provided that we make sure that the estimates obtained are uniform in α. So we can assume $X_0 = U$ is open and bounded in a B-space E and $\mathcal{G}_0 = \{U,1_U\}$. Let $f \in \mathfrak{M}(X_1,X)$ and let $U_f = \bar{C}(F)^{-1}(V_f) = \{u: \{(f(x),F(u,x)): x \in X_1\} \subset\subset V\}$, open in U. Pulling e^{-1} back by f we have

$$\begin{array}{ccccc}
 & & X_1 \times X & & \\
 & & \cup & & \\
X_1 \times E \supset X_1 \times U_f & \xrightarrow{p_1 \times F} & (f \times 1)^{-1}V & \xrightarrow{f^*e^{-1}} & f^*TX \\
 & \searrow^{\epsilon_E} & \downarrow p_1 & \swarrow_{f^*\tau_X} & \\
 & & X_1 & &
\end{array}$$

Let $H: U_f \to \mathfrak{M}(\epsilon_E)$ be the inclusion of constant sections.

$H_f \cdot \bar{C}(F)|U_f = \Omega_H^{1_{X_1}}((f^*e^{-1}) \cdot (p_1 \times F))$. Result by Theorem 2.3. Q.E.D.

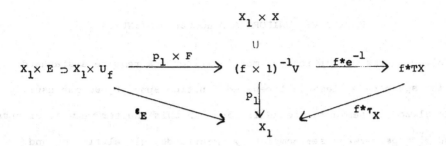

CHAPTER V: UNIFORM AND HÖLDER CONTINUITY

Although the definitions of metric structures require a standard function space type, less well behaved function space types can usually be globalized on semicomplete manifolds. In this chapter we will discuss $\mathcal{C}_u^r$ and $\mathcal{L}_a^r$ maps between semicomplete $\mathcal{L}^r$ manifolds, globalizing $\mathcal{C}_u^r$ and $\mathcal{L}ip_a^r$, respectively. For later applications to leaf immersions we will describe maps which are $\mathcal{C}_u^r$ or $\mathcal{L}_a^r$ with respect to pseudometrics coarser than the admissible metrics. This will require some structure relating the coarser pseudometrics to the original metric structure.

1. <u>Accessory Pseudometric Structures</u>: Let X be a regular metric manifold with admissible metric and bound d_X and ρ. A pseudometric d on X is called an <u>accessory pseudometric</u> (or an <u>apm</u>) if $d_X >_\rho d$ and $\rho^d \sim \rho$. These conditions are clearly independent of the choice of d_X and ρ. d is an apm on X iff (X,d,ρ) is a regular pseudometric space and the identity $(X,d_X,\rho) \to (X,d,\rho)$ is an $\mathcal{L}$ map of regular pseudometric spaces. Recall the uniformity $\mathfrak{U}_X$ on X induced by the metric structure. We will denote by $\mathfrak{U}_d$ the uniformity of (X,d,ρ). The uniformity (and topology) of d is coarser than that of d_X, i.e. of X. Thus, $\mathfrak{U}_d \subset \mathfrak{U}_X$ and d is continuous on $X \times X$. We will continue to think of X with its original topology and mention explicitly when we refer to the topology or uniformity of d, eg. (d) open set or $\mathcal{C}_u(d)$ map.

Any admissible metric on X is an apm. From this trivial example the other usual examples are generated by:

1 <u>LEMMA</u>: Let X_i be a regular metric manifold with admissible bound ρ_i $(i = 1,2)$. Let $f: X_1 \to X_2$ be an $\mathcal{L}$ map satisfying $\rho_1 \sim f^*\rho_2$ (such a

map is called <u>metricly proper</u>). If d is an apm on X_2 then f^*d is an apm on X_1.

<u>PROOF</u>: Let d_{X_i} be an admissible metric on X_i $(i = 1,2)$. The maps $f\colon (X_1,d_{X_1},\rho_1) \to (X_2,d_{X_2},\rho_2)$ and $1\colon (X_2,d_{X_2},\rho_2) \to (X_2,d,\rho_2)$ are $\mathcal{L}$ maps. So the pull back of the composition, $1\colon (X_1,d_{X_1},\rho_1) \to (X_1,f^*d,f^*\rho_2)$ is an $\mathcal{L}$ map of regular pseudometric spaces. Replace $f^*\rho_2$ by the equivalent bound ρ_1. Q.E.D.

Applying this lemma to the inclusion of an open subset U in X, we see that $d_X|U$ is an apm on the restriction metric structure of U. In this case the apm is admissible for some refinement metric structure (eg. the canonical refinement) and induces the original topology on U.

We need a generalization of the atlas concept. $\mathcal{a} = \{U_\alpha,h_\alpha\}$ is called a <u>multiatlas on X</u> if $\{U_\alpha\}$ is an open cover of X and h_α is a local homeomorphism of U_α onto an open subset of a B-space E_α. If U is open in $U_\alpha \cap U_\beta$ and $h_\alpha|U$ is injective then $h_{\beta\alpha U} = h_\beta \cdot (h_\alpha|U)^{-1}\colon h_\alpha(U) \to E_\beta$ is defined and continuous. $(\mathfrak{F},\mathcal{a}) = \{U_\alpha,h_\alpha,\varphi_\alpha\}$ is a <u>multiatlas on $\pi\colon E \to X$</u> if $\mathcal{a} = \{U_\alpha,h_\alpha\}$ is a multiatlas on X, $\varphi_\alpha\colon \pi^{-1}U_\alpha \to h_\alpha(U_\alpha) \times F_\alpha$ is a surjective local homeomorphism over h_α which is a linear isomorphism on fibers and the maps $\varphi_{\beta\alpha U}\colon h_\alpha(U) \to L(F_\alpha;F_\beta)$ are continuous. $\varphi_{\beta\alpha U}$ is defined when U is open in $U_\alpha \cap U_\beta$ and $h_\alpha|U$ is injective. Then $\varphi_\beta(\varphi_\alpha|\pi^{-1}U)^{-1}(u,v) = (h_{\beta\alpha U}(u),\varphi_{\beta\alpha U}(u)(v))$. For a standard function space type $\mathfrak{M}$ the definitions of bounded $\mathfrak{M}$ multiatlas and $\mathfrak{M}$ multiatlas are direct generalizations of the atlas terms. Thus, $k_{\mathcal{a}}$ is the infinum of constants $K \geq 1$ such that $K \geq \|h_{\beta\alpha U}\|_{\mathfrak{M}}$ for all β,α,U, or ∞ if no such constants exist. $\rho_{\mathcal{a}}(x) = \inf\{k_{\mathcal{a}|U}\colon U \text{ open, containing } x\}$. Similarly for bundles.

Given a multiatlas $\mathcal{a} = \{U_\alpha,h_\alpha\}$ we can always find atlas subdivisions, eg. define $\tilde{\mathcal{a}} = \{U_{\alpha x},h_\alpha|U_{\alpha x}\}$, induced by pairs with $x \in U_\alpha$, choosing $U_{\alpha x}$

open, containing x with $h_\alpha|U_{\alpha x}$ injective. Clearly, $\rho_{\tilde{\mathcal{G}}} = \rho_{\mathcal{G}}$. If X is a metric manifold then a multiatlas $\mathcal{G}$ is called admissible if it satisfies the equivalent conditions: (1) All atlas refinements of $\mathcal{G}$ are admissible. (2) There exists an admissible atlas subdivision of $\mathcal{G}$. (3) If $(\mathcal{G}_0,\rho)$ is an adapted atlas in the metric structure of X then $\mathcal{G} \cup \mathcal{G}_0$ is a multiatlas satisfying $\rho > \rho_{\mathcal{G}\cup\mathcal{G}_0}$. If $(\pi,\mathcal{G})$ is a multiatlas on π and $\tilde{\mathcal{G}}$ is an atlas refinement (or subdivision) of $\mathcal{G}$ then $\tilde{\mathcal{G}}$ induces $(\tilde{\pi},\tilde{\mathcal{G}})$ an atlas refinement (resp. subdivision) of $(\pi,\mathcal{G})$. $(\pi,\mathcal{G})$ is an admissible multiatlas for a metric bundle if it satisfies the three equivalent conditions analogous to the manifold conditions.

2 LEMMA: Let X be a uniformly complete, regular metric manifold and $\mathcal{G} = \{U_\alpha,h_\alpha\}$ be an admissible multiatlas. If $\{U_\alpha\}$ is a uniform cover of X, then X is semicomplete and $\mathcal{G}$ admits s admissible atlas subdivisions.

PROOF: Let $\tilde{\mathcal{G}} = \{V_\gamma,h_{\alpha(\gamma)}\}$ be an atlas subdivision of $\mathcal{G}$ and $d = d_{\tilde{\mathcal{G}}}$. Choose ρ an admissible bound with $\rho > \max(1/\kappa(d), 1/\delta_{\mathcal{G}}(d))$. For $x \in U_\alpha$ choose $K_{\alpha x} = \sup\{r \le 1\colon B^d[x;r] \subset U_\alpha$ and is complete$\}$. Note that for each x, there exists an α such that $1/K_{\alpha x} \le \rho(x)$. Now adapt the proof of inequality III.(7.2) to show that every ray from $h_\alpha x$ of length less than $K_{\alpha x}$ lifts back continuously under h_α to a path in U_α based at x. Since we are using $d_{\tilde{\mathcal{G}}}$ (instead of the--undefined--$d_{\mathcal{G}}$), the proof requires the following fact: If a segment $[h_\alpha x_1, h_\alpha x_2]$ in $h_\alpha(U_\alpha)$ lifts back to a continuous path between x_1 and x_2 then $\|h_\alpha x_1 - h_\alpha x_2\|_\alpha \ge d(x_1,x_2)$. But because $\tilde{\mathcal{G}}$ is a subdivision of $\mathcal{G}$, the continuous lifting can be used to define an $\tilde{\mathcal{G}}$ chain between x_1 and x_2 with indices $\gamma_1,\dots,\gamma_N$ satisfying $\alpha(\gamma_i) = \alpha$ and whose length is exactly $\|h_\alpha x_1 - h_\alpha x_2\|_\alpha$, proving the inequality. It then follows from Corollary II.3.6 that $h_\alpha\colon U_\alpha \to E_\alpha$ admits a unique local inverse $g_{\alpha x}$ defined on $B^{\|\ \|_\alpha}(h_\alpha x;\ K_{\alpha x})^{-1})$ with $g_{\alpha x}(h_\alpha x) = x$. Let $U_{\alpha x} \subset U_\alpha$ denote the image of this ball. $\tilde{\mathfrak{a}} = \{U_{\alpha x}, h_\alpha | U_{\alpha x}\}$ is an atlas subdivision of $\mathfrak{a}$. $\lambda_{\tilde{\mathfrak{a}}}(x) \geq \sup\{K_{\alpha x}: x \in U_\alpha\} \geq 1/\rho(x)$. Thus, X is semicomplete and $\tilde{\mathfrak{a}}$ is s admissible. Q.E.D.

We now define the structures which relate apm's to the metric theory. In preparation, note that if $\mathfrak{a} = \{U_\alpha, h_\alpha\}$ is a C^r multiatlas, then $j^r(h_{\beta\alpha}) \cdot h_\alpha : U_\alpha \cap U_\beta \to J^r(E_\alpha; E_\beta)$ is a continuous map well defined by $(j^r(h_{\beta\alpha}) \cdot h_\alpha)|U = j^r(h_{\beta\alpha U}) \cdot (h_\alpha|U)$ for U open in $U_\alpha \cap U_\beta$ with $h_\alpha|U$ injective. Similarly, if $(\mathfrak{B}, \mathfrak{a}) = \{U_\alpha, h_\alpha, \varphi_\alpha\}$ is a $\mathcal{C}^r$ bundle multiatlas then $j^r(\varphi_{\beta\alpha}) \cdot h_\alpha : U_\alpha \cap U_\beta \to J^r(F_\alpha; F_\beta)$ is a continuous map well defined by $(j^r(\varphi_{\beta\alpha}) \cdot h_\alpha)|U = j^r(\varphi_{\beta\alpha U}) \cdot (h_\alpha|U)$. Analogous expressions occur later with analogous definitions. Recall the definitions of Section III.6.

3 <u>DEFINITION</u>: Let X be a regular $\mathcal{L}$ manifold with admissible bound ρ and apm d. Let $\pi: E \to X$ be a $\mathcal{C}$ bundle on X. Let $r \geq 0$ be an integer, and $0 < a \leq 1$.

(a) A C^r multiatlas $\mathfrak{a} = \{U_\alpha, h_\alpha\}$ is a $\mathcal{C}_u^r(d)$ (or $\mathcal{L}^r(d)$) multiatlas on X if: (1) $\mathfrak{a}$ is an admissible $\mathcal{L}$ multiatlas. (2) $\{U_\alpha\}$ is a $\mathfrak{N}_d$ uniform, d open cover of X. (3) The family $\{j_1^r(h_{\beta\alpha}) \cdot h_\alpha : U_\alpha \cap U_\beta \to J_1^r(E_\alpha; E_\beta)\}$ is a $\mathcal{C}_u$ (resp. $\mathcal{L}$) family of maps where the domain is the regular family of pseudometric spaces obtained by restriction from (X, d, ρ) and the range is a regular family of B-spaces.

(b) A C^r multiatlas $(\mathfrak{B}, \mathfrak{a}) = \{U_\alpha, h_\alpha, \varphi_\alpha\}$ is a $\mathcal{C}_u^r(d)$ (or $\mathcal{L}_a^r(d)$) multiatlas on π if $\mathfrak{a}$ is a $\mathcal{C}_u^r(d)$ (resp. $\mathcal{L}^r(d)$) multiatlas on X and: (4) The family $\{j^r(\varphi_{\beta\alpha}) \cdot h_\alpha : U_\alpha \cap U_\beta \to J^r(E_\alpha; L(F_\alpha; F_\beta))\}$ is a $\mathcal{C}_u$ (resp. $\mathcal{L}_a$) family of maps.

Some points to note about the definition: In (a)(3) the range of the jet map is $J_1^r(E_\alpha; E_\beta) = L(E_\alpha; E_\beta) \times \ldots \times L_s^r(E_\alpha; E_\beta)$. So for $r = 0$ condition (3) is vacuous. The base space multiatlas $\mathfrak{a}$ of an $\mathcal{L}_a^r(d)$ multiatlas $(\mathfrak{B}, \mathfrak{a})$ is assumed to be $\mathcal{L}^r(d)$. Any $\mathcal{L}_a^r(d)$ multiatlas is a $\mathcal{C}_u^r(d)$ multiatlas because an $\mathscr{L}_a$ family is a $\mathcal{C}_u$ family. Any $\mathcal{C}_u^r(d)$ multiatlas is a $\mathcal{C}^r$ multiatlas with $\rho > \rho_{\mathfrak{A}}^{\mathcal{C}^r}$ because $d_X >_\rho d$. In fact, any $\mathcal{C}_u^r(d)$ (or $\mathscr{L}_a^r(d)$) multiatlas is a $\mathcal{C}_u^r(d_1)$ (or $\mathscr{L}_a^r(d_1)$) multiatlas for any apm d_1 on X with $d_1 >_\rho d$. In particular, this holds with $d_1 = d_X$, an admissible metric for X. Finally, while we can subdivide a $\mathcal{C}_u^r$ multiatlas to obtain an atlas satisfying the uniform continuity condition we usually lose uniformity of the cover. This is because although h_α is a local homeomorphism with respect to the X topology on U_α, it need not be locally injective with respect to the d topology. This is why multiatlases are needed.

4 DEFINITION: Let X_i be a regular $\mathscr{L}$ manifold with admissible metric and bound d_{X_i}, ρ_i and apm d_i. Let π_i: $\mathbb{E}_i \to X_i$ be a $\mathcal{C}$ bundle on X_i $(i = 1,2)$. Let (Φ,f): $\pi_1 \to \pi_2$ be a VB map. Let $r \geq 0$ be an integer and $0 < a \leq 1$.

(a) If $\mathfrak{A}_1 = \{U_\alpha, h_\alpha\}$ and $\mathfrak{A}_2 = \{V_\beta, g_\beta\}$ are $\mathcal{C}_u^r(d_1)$ and $\mathcal{C}_u^r(d_2)$ (resp. $\mathscr{L}^r(d_1)$ and $\mathscr{L}^r(d_2)$) multiatlases on X_1 and X_2 then f: $\mathfrak{A}_1 \to \mathfrak{A}_2$ is a $\mathcal{C}^r(d_1,d_2)$ (resp. $\mathscr{L}_a^r(d_1,d_2)$) map if: (1) f: $(X_1,d_{X_1},\rho_1) \to (X_2,d_{X_2},\rho_2)$ is a $\mathcal{C}_u$ (resp. $\mathscr{L}_a$) map. (2) f: $(X_1,d_1,\rho_1) \to (X_2,d_2,\rho_2)$ is $\mathscr{L}_a$) map. (3) The family $\{j_1^r(f_{\beta\alpha})\cdot h_\alpha$: $U_\alpha \cap f^{-1}V_\beta \to J_1^r(E_\alpha^1;E_\beta^2)\}$ is a $\mathcal{C}_u$ (resp. $\mathscr{L}_a$) family of maps.

(b) If $(\mathfrak{D}_1,\mathfrak{A}_1) = \{U_\alpha,h_\alpha,\varphi_\alpha\}$ and $(\mathfrak{D}_2,\mathfrak{A}_2) = \{V_\beta,g_\beta,\psi_\beta\}$ are $\mathcal{C}_u^r(d_1)$ and $\mathcal{C}_u^r(d_2)$ (resp. $\mathscr{L}_a^r(d_1)$ and $\mathscr{L}_a^r(d_2)$) multiatlases on π then (Φ,f): $(\mathfrak{D}_1,\mathfrak{A}_1) \to (\mathfrak{D}_2,\mathfrak{A}_2)$ is a $\mathcal{C}_u^r(d_1,d_2)$ (resp. an $\mathscr{L}_a^r(d_1,d_2)$) VB map if f: $\mathfrak{A}_1 \to \mathfrak{A}_2$ is a $\mathcal{C}_u^r(d_1,d_2)$ (resp. $\mathscr{L}_a^r(d_1,d_2)$) map and (4) the family $\{j^r(\Phi_{\beta\alpha})\cdot h_\alpha$: $U_\alpha \cap f^{-1}V_\beta \to J^r(E_\alpha;L(F_\alpha^1;F_\beta^1))\}$ is a $\mathcal{C}_u$ (resp. an $\mathscr{L}_a$) family of maps.

As in Definition 3, the domains in (3) and (4) are the regular pseudometric spaces obtained by restriction from (X_1,d_1,ρ_1). By Proposition III.5.3(b), (1) in the $\mathscr{L}^r$ case is equivalent to: f: $X_1 \to X_2$ is an $\mathscr{L}$ map of $\mathscr{L}$ metric manifolds.

5 PROPOSITION: (a) If $f\colon \mathfrak{G}_1 \to \mathfrak{G}_2$ is a $\mathcal{C}_u^r(d_1,d_2)$ (or $\mathcal{L}_a^r(d_1,d_2)$) map and $g\colon \mathfrak{G}_2 \to \mathfrak{G}_3$ is a $\mathcal{C}_u^r(d_2,d_3)$ (resp. $\mathcal{L}_b^r(d_2,d_3)$) map then $g\circ f\colon \mathfrak{G}_1 \to \mathfrak{G}_3$ is a $\mathcal{C}_u^r(d_1,d_3)$ (resp. $\mathcal{L}_{ab}^r(d_1,d_3)$) map.

(b) If $(\Phi,f)\colon (\mathfrak{D}_1,\mathfrak{G}_1) \to (\mathfrak{D}_2,\mathfrak{G}_2)$ is a $\mathcal{C}_u^r(d_1,d_2)$ (or $\mathcal{L}_a^r(d_1,d_2)$) VB map and $(\Psi,g)\colon (\mathfrak{D}_2,\mathfrak{G}_2) \to (\mathfrak{D}_3,\mathfrak{G}_3)$ is a $\mathcal{C}_u^r(d_2,d_3)$ (resp. $\mathcal{L}_b^r(d_2,d_3)$) VB map then $(\Psi\circ\Phi,g\circ f)\colon (\mathfrak{D}_1,\mathfrak{G}_1) \to (\mathfrak{D}_3,\mathfrak{G}_3)$ is a $\mathcal{C}_u^r(d_1,d_3)$ (resp. $\mathcal{L}_{ab}^r(d_1,d_3)$) VB map.

(c) If $(\Phi,f)\colon (\mathfrak{D}_1,\mathfrak{G}_1) \to (\mathfrak{D}_2,\mathfrak{G}_2)$ is an $\mathcal{L}_a^r(d_1,d_2)$ VB map with $f\colon \mathfrak{G}_1 \to \mathfrak{G}_2$ an $\mathcal{L}(d_1,d_2)$ map and $(\Psi,g)\colon (\mathfrak{D}_2,\mathfrak{G}_2) \to (\mathfrak{D}_3,\mathfrak{G}_3)$ is an $\mathcal{L}_a^r(d_2,d_3)$ VB map then $(\Psi\circ\Phi,g\circ f)\colon (\mathfrak{D}_1,\mathfrak{G}_1) \to (\mathfrak{D}_3,\mathfrak{G}_3)$ is an $\mathcal{L}_a^r(d_1,d_3)$ VB map.

PROOF: We have to verify (3) and (4) of Definition 4 for the compositions, i.e. we have to prove the families $\{j_1^r((g\circ f)_{\gamma\alpha})\circ h_\alpha\colon U_\alpha \cap f^{-1}g^{-1}W_\gamma \to J_1^r(E_\alpha^1;E_\gamma^3)\}$ and $\{j^r((\Psi\circ\Phi)_{\gamma\alpha})\circ h_\alpha\colon U_\alpha \cap f^{-1}g^{-1}W_\gamma \to J^r(E_\alpha^1;L(F_\alpha^1;F_\gamma^3))\}$ are $\mathcal{C}_u$, $\mathcal{L}_{ab}$ etc. families. Since f is $\mathcal{C}_u(d_1,d_2)$, $\{f^{-1}V_\beta\}$ is a $\mathfrak{U}_{d_1}$ uniform open cover of X_1 and so by Proposition III.6.7 it suffices to consider the family of maps on $U_\alpha \cap f^{-1}V_\beta \cap f^{-1}g^{-1}W_\gamma$. This step is the reason that $\mathfrak{U}_d$ uniform covers are essential. The result then follows from the chain rule equations. On $U_\alpha \cap f^{-1}V_\beta \cap f^{-1}g^{-1}W_\gamma$, $j_1^r((g\circ f)_{\gamma\alpha})\circ h_\alpha$ is the polynomial Comp composed with the product of $(j_1^r(g_{\gamma\beta})g_\beta)\circ f$ and $j_1^r(f_{\beta\alpha})\circ h_\alpha$. $j^r((\Psi\circ\Phi)_{\gamma\alpha})\circ h_\alpha$ is the bilinear map $\mathrm{comp}_\#$ applied to the product of $\mathrm{Comp}\circ[((j^r(\Psi_{\gamma\beta})g_\beta)\circ f) \times j_1^r(f_{\beta\alpha})\circ h_\alpha]$ and $j^r(\Phi_{\beta\alpha})\circ h_\alpha$. The result follows from the uniform $\mathcal{L}$ estimates on Comp and $\mathrm{comp}_\#$. Note that the $\mathcal{L}_{ab}^r$ results come from the composition with f. Thus, if f is an $\mathcal{L}$ map of regular pseudometric spaces we get the $\mathcal{L}_a^r$ result in (c).

Q.E.D.

ADDENDUM: A similar argument shows that if $f\colon \mathfrak{G}_1 \to \mathfrak{G}_2$ is index preserving we can replace Definition 4.(a)(3) by: (3') Tne family $\{j_1^r(f_\alpha)h_\alpha\colon U_\alpha \to J_1^r(E_\alpha^1;E_{\beta(\alpha)}^2)\}$ is a $\mathcal{C}_u$ (resp. $\mathcal{L}_a$) family of maps.

Similarly, for bundle maps.

As in Section III.5 we can now define equivalence: If X is a regular $\mathcal{L}$ manifold with admissible bound ρ and apm d, then $\mathcal{C}_u^r(d)$ (or $\mathcal{L}^r(d)$) multiatlases $\mathfrak{A}_1$ and $\mathfrak{A}_2$ on X are equivalent if they satisfy the equivalent conditions: (1) $\mathfrak{A}_1 \cup \mathfrak{A}_2$ is a $\mathcal{C}_u^r(d)$ (resp. $\mathcal{L}^r(d)$) multiatlas on X. (2) The identity maps $1: \mathfrak{A}_1 \to \mathfrak{A}_2$ and $1: \mathfrak{A}_2 \to \mathfrak{A}_1$ are $\mathcal{C}_u^r(d,d)$ (resp. $\mathcal{L}^r(d,d)$) maps. An equivalence class of multiatlases is a $\mathcal{C}_u^r(d)$ (resp. $\mathcal{L}^r(d)$) structure on X. X with such a structure is called a $\mathcal{C}_u^r(d)$ (resp. $\mathcal{L}^r(d)$) manifold and a multiatlas in the equivalence class is called an admissible multiatlas. A map $f: X_1 \to X_2$ between $\mathcal{C}_u^r(d_1)$, $\mathcal{C}_u^r(d_2)$ manifolds is called a $\mathcal{C}^r(d_1,d_2)$ map if it is a $\mathcal{C}_u^r(d_1,d_2)$ map for some, and hence any, choices of admissible multiatlases on X_1 and X_2. Similarly, for $\mathcal{L}_a^r(d_1,d_2)$ maps between $\mathcal{L}^r(d_1)$ and $\mathcal{L}^r(d_2)$ manifolds. The definitions of $\mathcal{C}_u^r(d)$ and $\mathcal{L}_a^r(d)$ structures and VB maps for bundles are completely analogous. That the atlas relations are equivalence relations and that the smoothness of a map is independent of multiatlas choices follow from Proposition 5 (a) and (c). Any $\mathcal{L}_a^r(d)$ structure is contained in a $\mathcal{C}_u^r(d)$ structure which is contained in a regular $\mathcal{C}^r$ metric structure. The $\mathcal{C}^r$ structure is contained in (or contains) the original $\mathcal{L}$ structure if $r > 0$ (resp. if $r = 0$). Since $\mathfrak{U}_d \subset \mathfrak{U}_X$ the $\mathcal{C}^r$ structure is semicomplete iff the original $\mathcal{L}$ structure is uniformly complete iff the original $\mathcal{L}$ structure is semicomplete. If d_1 is another apm on X and $d_1 >_\rho d$ (or $\mathfrak{U}_d \subset \mathfrak{U}_{d_1}$) then any $\mathcal{L}_a^r(d)$ structure is contained in a unique $\mathcal{L}_a^r(d_1)$ structure (resp. any $\mathcal{C}_u^r(d)$ structure is contained in a unique $\mathcal{C}_u^r(d_1)$ structure). Thus, if $d_1 \sim_\rho d$ then $\mathcal{L}_a^r(d)$ and $\mathcal{L}_a^r(d_1)$ structures agree and if $\mathfrak{U}_{d_1} = \mathfrak{U}_d$ then $\mathcal{C}_u^r(d)$ and $\mathcal{C}_u^r(d_1)$ structures agree. In particular, every $\mathcal{L}_a^r(d)$ structure is contained in a $\mathcal{L}_a^r(d_X)$ structure and every $\mathcal{C}_u^r(d)$ structure is contained in a $\mathcal{C}_u^r(d_X)$ structure and the d_X structures are independent of the choice of admissible metric d_X. We will call a

$\mathcal{C}_u^r(d_X)$ structure on X a $\mathcal{C}_u^r$ structure and a $\mathcal{C}_u^r(d_{X_1},d_{X_2})$ map a $\mathcal{C}_u^r$ map.

STANDARD CONSTRUCTIONS: (1) Refinements and Transfers of Atlas Construction: Any $\mathfrak{U}_d$ uniform, d open refinement of an admissible multiatlas is admissible and by Proposition 5 type arguments the transfer of atlas construction works for multiatlases to yield the analogue of Lemma III.5.1. So tangent bundles, products, trivial bundles, direct sum and linear map bundles work just as in the ordinary theory. (2) Open Sets: Let X be a $\mathcal{C}_u^r(d)$ (or $\mathcal{L}^r(d)$) manifold and U be open in X. The class $\{\mathfrak{a}|U\}$ with $\mathfrak{a}$ admissible for X generates a $\mathcal{C}_u^r(d|U)$ (resp. $\mathcal{L}^r(d|U)$) structure on U. Note that $d|U$ is an apm on U by Lemma 1. By Proposition III.6.6(e) and an easy transfer of atlas argument one can show that if $\pi\colon \mathbb{E} \to U$ is a $\mathcal{C}_u^r(d|U)$ (or $\mathcal{L}_a^r(d|U)$) bundle over U, then π has admissible multiatlases $(\mathfrak{M},\mathfrak{a}|U)$ with $\mathfrak{a}$ admissible for X.
(3) Sections: If $\pi\colon \mathbb{E} \to X$ is a $\mathcal{C}_u^r(d)$ (or $\mathcal{L}_a^r(d)$) bundle and s is a section of π then s is a $\mathcal{C}_u^r(d)$ (resp. $\mathcal{L}_a^r(d)$) section if for some (and hence any) admissible multiatlas $(\mathfrak{M},\mathfrak{a}) = \{U_\alpha,h_\alpha,\varphi_\alpha\}$ the family $\{j^r(s_\alpha)\cdot h_\alpha\colon U_\alpha \to J^r(E_\alpha;F_\alpha)\}$ is $\mathcal{C}_u$ (resp. $\mathcal{L}_a$). Independence of the atlas choice follows from Proposition 5 arguments as does: If $(\Phi,1)\colon \pi_1 \to \pi_2$ is a $\mathcal{C}_u^r(d,d)$ (resp. $\mathcal{L}_a^r(d,d)$) VB map and s is a $\mathcal{C}_u^r(d)$ (resp. $\mathcal{L}_a^r(d)$) section then $\Phi_*(s)$ is a $\mathcal{C}_u^r(d)$ (resp. $\mathcal{L}_a^r(d)$) section. (4) Pullbacks: If $\pi\colon \mathbb{E} \to X$ is an $\mathcal{L}_a^r(d)$ bundle and $f\colon X_0 \to X$ is an $\mathcal{L}_b^r(d_0,d)$ map then the class $\{(f^*\mathfrak{M},\mathfrak{a}_0)\}$ generates an $\mathcal{L}_{ab}^r(d_0)$ structure on $f^*\pi$ where $(\mathfrak{M},\mathfrak{a})$ is admissible on π, $\mathfrak{a}_0$ is admissible on X_0 and $f\colon \mathfrak{a}_0 \to \mathfrak{a}$ is index preserving. If s is an $\mathcal{L}_a^r(d)$ section of π, then f^*s is an $\mathcal{L}_{ab}^r(d_0)$ section of $f^*\pi$. If $(\Phi,1)\colon \pi \to \pi_1$ is an $\mathcal{L}_a^r(d,d)$ VB map then $(f^*\Phi,1)\colon f^*\pi \to f^*\pi_1$ is an $\mathcal{L}_{ab}^r(d_0,d_0)$ map. Similarly, for $\mathcal{C}_u^r(d)$ structures. (5) Jet Bundles: If $\pi\colon \mathbb{E} \to X$ is an $\mathcal{L}_a^{r+k}(d)$ bundle then the class $\{(J^k(\mathfrak{M}),\mathfrak{a})\}$ generates an $\mathcal{L}_a^r(d)$ structure on $J^k(\pi)\colon J^k(\mathbb{E}) \to X$. s is an $\mathcal{L}_a^{r+k}(d)$ section of π iff $j^k(s)$ is an $\mathcal{L}_a^r(d)$ section of $J^k(\pi)$. If $f\colon X_0 \to X$ is an $\mathcal{L}_b^{r+k}(d_0,d)$ map

then $(\Psi^k_f,1)\colon f^*J^k(\pi)\to J^k(f^*\pi)$ is an $\mathcal{L}^r_{ab}(d_0,d_0)$ VB map. Similarly, for $\mathcal{C}^r_u(d)$ structures.

2. Admissible APM's on Semicomplete Spaces: We will show that on a semicomplete $\mathcal{L}$ manifold an $\mathcal{L}^r(d)$ structure with d an admissible apm is essentially just a compatible $\mathcal{L}^r$ structure. We will also examine $\mathcal{C}^r_u$ structures on such spaces. By Lemma 1.2 we need only deal with atlases.

$\mathfrak{a} = \{U_\alpha,h_\alpha\}$ is a bounded $\mathcal{L}ip^r$ atlas on X if $\{h_{\beta\alpha}\}\in \hat{\Pi}\,\mathcal{L}ip^r(h_\alpha(U_\alpha\cap U_\beta),E_\beta)$ and in that case $k^{\mathcal{L}ip^r}_{\mathfrak{a}} = \max(\|\{h_{\beta\alpha}\}\|,1)$, $(\mathfrak{n},\mathfrak{a}) = \{U_\alpha,h_\alpha,\varphi_\alpha\}$ is a bounded $\mathcal{L}ip^r$ atlas on π if $\mathfrak{a}$ is a bounded $\mathcal{L}ip^r$ atlas on the base space and $\{\varphi_{\beta\alpha}\}\in\hat{\Pi}\,\mathcal{L}ip^r(h_\alpha(U_\alpha\cap U_\beta),L(F_\alpha;F_\beta))$. In that case $k^{\mathcal{L}ip^r}_{(\mathfrak{n},\mathfrak{a})} = \max(\|\{h_{\beta\alpha}\}\|,\|\{\varphi_{\beta\alpha}\}\|,1)$. $(\mathfrak{a},\rho)$ is an adapted $\mathcal{L}ip^r$ atlas on X if ρ is a bound on X with $\rho^{d_{\mathfrak{a}}}\sim\rho$ and there exist constants K,n such that U open and bounded implies $K\sup(\rho|U)^n \geq k^{\mathcal{L}ip^r}_{\mathfrak{a}|U}$. Similarly, for bundles. Clearly, a $\mathcal{L}ip^r$ adapted atlas is a regular $\mathcal{L}^r$ adapted atlas.

1 LEMMA: Let $(\mathfrak{n},\mathfrak{a}) = \{U_\alpha,h_\alpha,\varphi_\alpha\}$ be an atlas on $\pi\colon \mathbb{E}\to X$ such that $(\mathfrak{a},\rho)$ is a regular adapted $\mathcal{L}$ atlas on X. Let $m(x_1,x_2) = \max(\rho(x_1),\rho(x_2))$. $(\mathfrak{a},\rho)$ is a $\mathcal{L}ip^r$ adapted atlas iff it is a $\mathcal{C}^r$ adapted atlas and there exist constants K,n such that for $x_1,x_2\in U_\alpha\cap U_\beta$:

$$(3.1)\quad K\,m(x_1,x_2)^n\|h_\alpha x_1 - h_\alpha x_2\| \geq \min(\|j^r(h_{\beta\alpha})(h_\alpha x_1) - j^r(h_{\beta\alpha})(h_\alpha x_2)\|,1).$$

$(\mathfrak{n},\mathfrak{a},\rho)$ is a $\mathcal{L}ip^r$ adapted atlas iff it is a $\mathcal{C}^r$ adapted atlas and there exist constants K,n such that for $x_1,x_2\in U_\alpha\cap U_\beta$ (3.1) holds and:

$$(3.2)\quad K\,m(x_1,x_2)^n \geq \min(\|j^r(\varphi_{\beta\alpha})(h_\alpha x_1) - j^r(\varphi_{\beta\alpha})(h_\alpha x_2)\|,1).$$

If $\mathfrak{a}$ is either star bounded or convex then the same result holds with $m(x_1,x_2) = \min(\rho(x_1),\rho(x_2))$.

PROOF: If $(\mathcal{G},\rho)$ is $\mathcal{L}ip^r$ and $M > \max(\rho^d(x_1),\rho^d(x_2))$ then letting $U = \{\rho^d < M\}$, $KM^n \geq \|j^r(h_{\beta\alpha})|h_\alpha(U \cap U_\alpha \cap U_\beta)\|_{\mathcal{L}ip}$ which implies (3.1) with ρ replaced by ρ^d. Conversely, if (3.1) holds and $K_1, n_1 \geq 1$ are constants such that $K_1 \rho(x)^{n_1} \geq \|j^r(h_{\beta\alpha})(h_\alpha x)\|$ for $x \in U_\alpha$, then $2K_1K \max(\rho(x_1),\rho(x_2))^{n+n_1} \geq \|j^r(h_{\beta\alpha})(h_\alpha x_1) - j^r(h_{\beta\alpha})(h_\alpha x_2)\|/\|h_\alpha x_1 - h_\alpha x_2\|$. This implies $2K_1K \sup(\rho|U)^{n+n_1} \geq \|j^r(h_{\beta\alpha})|h_\alpha(U \cap U_\alpha \cap U_\beta)\|_{\mathcal{L}ip}$ and so $(\mathcal{G},\rho)$ is $\mathcal{L}ip^r$. If $\mathcal{G}$ is star bounded then max $\sim$ min on $\cup U_\alpha \times U_\alpha$. Finally, if (3.1) holds with m = max and $\mathcal{G}$ is convex then with $d = d_{\mathcal{G}}$, $K \max(\rho(x_1),\rho(x_2))^n \|h_\alpha x_1 - h_\alpha x_2\| \leq K \min(\rho^d(x_1),\rho^d(x_2))^n \|h_\alpha x_1 - h_\alpha x_2\|$ if the latter is less than 1, because $\|h_\alpha x_1 - h_\alpha x_2\| \geq d(x_1,x_2)$ by convexity and $d(x_1,x_2) < (\min(\rho^d(x_1),\rho^d(x_2)))^{-1}$ implies $\max(\rho(x_1,),\rho(x_2)) \leq \min(\rho^d(x_1),\rho^d(x_2))$. The proofs for $(\mathfrak{T},\mathcal{G},\rho)$ are similar. Q.E.D.

$\mathcal{G}$ is called a $\mathcal{L}ip^r$ atlas on a regular $\mathcal{L}$ manifold X if $(\mathcal{G},\rho)$ is an adapted $\mathcal{L}ip^r$ atlas in the $\mathcal{L}$ metric structure when ρ is an admissible bound on X. Similarly for bundles.

2 PROPOSITION: Let X be a semicomplete $\mathcal{L}^r$ manifold. If $\mathcal{G}$ is an s admissible atlas on X, then $\mathcal{G}$ admits convex, s admissible $\mathcal{L}ip^r$ refinements. Similarly, for semicomplete $\mathcal{L}^r$ bundles.

PROOF: Let $d = d_{\mathcal{G}}$. Begin by subdividing to obtain $\mathcal{G}_1 = \{U_x, h_x\}$ indexed by $x \in X$ satisfying: (1) $x \in U_x$ and $h_x(U_x) = B(h_x(x), r(x))$ with $1/r$ an admissible bound on X. (2) $\{U_x\}$ is star bounded. (3) Letting $d_x(x_1,x_2) = \|h_x(x_1) - (h_x(x_2)\|$ on U_x, $d_x \leq d \leq K(x)d_x$ with K an admissible bound on X. Let ρ be an admissible bound on X with $\rho > K/r$ and define $1/\varepsilon(x) = \sup\{\rho(y) : U_y \cap U_x \neq \emptyset\}$. By (2) $1/\varepsilon$ is an admissible bound on X. (4) $\varepsilon(x) < r(y)/K(y)$ if $U_y \cap U_x \neq \emptyset$.

Let $\tilde{U}_x = B^{d_x}(x, \varepsilon(x)^2/6)$. $\mathcal{G}_2 = \{\tilde{U}_x, h_x\}$ is a star bounded, convex, s admissible refinement of $\mathcal{G}$. To prove that $\mathcal{G}_2$ is $\mathcal{L}ip^r$, we will show that if $\tilde{U}_x \cap \tilde{U}_y \neq \emptyset$ there exists a convex open set C_{xy} in $h_x(U_x)$ such that

$\tilde{U}_y \subset h_x^{-1}(C_{xy}) \subset U_y$. Thus, $h_x(\tilde{U}_y) \subset C_{xy}$ and $h_y h_x^{-1} | h_x(\tilde{U}_x \cap \tilde{U}_y)$ is the restriction of $h_y h_x^{-1} | C_{xy}$ which is $\mathcal{L}ip^r$ because it is $\mathcal{L}^r$ and C_{xy} is convex. If $\tilde{U}_x \cap \tilde{U}_y \neq \emptyset$ then by (3), $\tilde{U}_y \subset B^d(y, \epsilon(y)^2/6) \subset B^d(x, (2\epsilon(y)^2 + \epsilon(x)^2)/6)$. By (4), $(2\epsilon(y)^2 + \epsilon(x)^2)/6 < r(x)/K(x)$ and so the right ball is in $B^d(x, r(x)/K(x)) \subset U_x$. Since $\epsilon(y), \epsilon(x) < 1/K(x)$, $\tilde{U}_y \subset B^{d_x}(x, (2\epsilon(y) + \epsilon(x))/6)$ which we define to be $h_x^{-1}(C_{xy})$. Thus, C_{xy} is an open ball centered at $h_x(x)$. Since $y \in \tilde{U}_y$, $h_x^{-1}(C_{xy}) \subset B^{d_x}(y, (4\epsilon(y) + 2\epsilon(x))/6) \subset B^d(y, (4\epsilon(y) + 2\epsilon(x))/6)$. Finally, $(4\epsilon(y) + 2\epsilon(x))/6 < r(y)/K(y)$ and so $h_x^{-1}(C_{xy}) \subset B^d(y, r(y)/K(y)) \subset U_y$. Q.E.D.

Next we relate $\mathcal{L}ip^r$ atlases to condition $\mathcal{L}$ (cf. p. 99).

3 <u>LEMMA</u>: Let X be a semicomplete $\mathcal{L}$ manifold with $\mathcal{A}$ an s admissible atlas. If $\mathcal{A}$ satisfies condition $\mathcal{L}$ then $\mathcal{A}$ is a $\mathcal{L}ip$ atlas. Conversely, if $\mathcal{A}$ is a convex, $\mathcal{L}ip$ atlas then $\mathcal{A}$ satisfies condition $\mathcal{L}$.

<u>PROOF</u>: Let $d = d_{\mathcal{A}}$. Assuming condition $\mathcal{L}$, i.e. the family $\{h_\alpha : (U_\alpha, d|U_\alpha, \rho|U_\alpha) \to (h_\alpha(U_\alpha), \| \|_\alpha, \rho \circ h_\alpha^{-1})\}$ and its inverse are $\mathcal{L}$ families of maps, then by restricting and composing we get that $\{h_{\beta\alpha} : (h_\alpha(U_\alpha \cap U_\beta), \| \|_\alpha, \rho \circ h_\alpha^{-1}) \to \{E_\beta\}$ is an $\mathcal{L}$ family. This implies (3.1) with $r = 0$. For the converse we need only show that $\{h_\alpha\}$ is an $\mathcal{L}$ family as the inequality $d(x_1, x_2) \leq \|h_\alpha x_1 - h_\alpha x_2\|$, $x_1, x_2 \in U_\alpha$, (this is where convexity of $\mathcal{A}$ is used) takes care of the inverse family. By (3.1), with $m = \min$, the family $\{h_{\beta\alpha}\}$ is an $\mathcal{L}$ family. Using the Metric Estimate and semicompleteness we can choose $V \in \mathfrak{U}_X$ such that for $x \in X$, $V[x] \subset U_{\alpha(x)}$ and the family $\{h_{\alpha(x)} : V[x] \to h_{\alpha(x)}(U_{\alpha(x)})\}$ is an $\mathcal{L}$ family. $h_\alpha | U_\alpha \cap V[x] = h_{\alpha\alpha(x)} \circ h_{\alpha(x)}$ and so the family $\{h_\alpha : U_\alpha \cap V[x] \to h_\alpha(U_\alpha)\}$ is an $\mathcal{L}$ family by composition. The result then follows from Proposition III.6.7. Q.E.D.

4 <u>THEOREM</u>: Let X be a semicomplete $\mathcal{L}$ manifold with admissible metric

and bound d_X and ρ. Let π be a semicomplete $\mathcal{C}$ bundle over X. Let $(\mathfrak{D},\mathcal{G})$ be admissible for π with $\mathcal{G}$ an s admissible $\mathcal{L}$ atlas satisfying condition $\mathcal{L}$. If $\mathcal{G}$ is either convex or star bounded then $\mathcal{G}$ (or $(\mathfrak{D},\mathcal{G})$) is an $\mathcal{L}^r(d_X)$ atlas iff it is a $\mathcal{L}ip^r$ atlas. Thus, an $\mathcal{L}^r(d_X)$ structure on X (resp. π) is essentially the same as an $\mathcal{L}^r$ metric structure. In particular, if d is an apm on X then any $\mathcal{L}^r(d)$ structure induces an $\mathcal{L}^r$ structure. Let s be a $\mathcal{C}$ section of π. s is an $\mathcal{L}^r(d_X)$ section iff it is an $\mathcal{L}^r$ section. Thus, if s is an $\mathcal{L}^r(d)$ section then it is an $\mathcal{L}^r$ section.

Let X_0 be another semicomplete $\mathcal{L}$ manifold with admissible metric d_{X_0} and apm d_0, and π_0 a $\mathcal{C}$ bundle over X_0. Let $(\phi,f): \pi \to \pi_0$ be a $\mathcal{C}$ VB map with $f: X \to X_0$ an $\mathcal{L}$ map. If f (or (ϕ,f)) is an $\mathcal{L}^r(d,d_0)$ map then it is an $\mathcal{L}^r$ metric map. Conversely, if f (or (ϕ,f)) is an $\mathcal{L}^r$ metric map then it is an $\mathcal{L}^r(d_X,d_{X_0})$ metric map and if X_0 (resp. π_0) has an $\mathcal{L}^r(d_0)$ structure, f (resp. (ϕ,f)) is an $\mathcal{L}^r(d_X,d_0)$ map iff $d_X >_{\rho} f^*d_0$.

<u>PROOF</u>: We will prove the space results as the bundle proofs are similar. Condition $\mathcal{L}$ allows us to replace the family $\{(U_\alpha,d_X|U_\alpha,\rho)\}$ by the $\mathcal{L}$ equivalent family $\{(h_\alpha(U_\alpha),\| \ \|_\alpha,\rho\cdot h_\alpha^{-1})\}$. It follows from Definition 1.3 and Lemma 3 that $\mathcal{G}$ is an $\mathcal{L}^r(d_X)$ atlas iff $(\mathcal{G},\rho)$ is an adapted $\mathcal{C}^r$ atlas and (3.1) holds with m = min. Apply Lemma 1. The convex, $\mathcal{L}ip^r$ atlases in a semicomplete $\mathcal{L}^r$ structure generate an $\mathcal{L}^r(d_X)$ structure. The convex atlases, satisfying condition $\mathcal{L}$, in an $\mathcal{L}^r(d_X)$ structure generate the $\mathcal{L}^r$ structure. Let $f: \mathcal{G} \to \mathcal{G}_0$ be an $\mathcal{L}$ map of convex, $\mathcal{L}ip^r$ atlases. If f is an $\mathcal{L}^r(d_X,d_{X_0})$ map of atlases, then condition $\mathcal{L}$, Definition 1.4 and Proposition III.7.16 easily imply that f is an $\mathcal{L}^r$ map of atlases. If f is an $\mathcal{L}^r$ map of atlases, let $\mathcal{G}_1$ be a convex refinement of $\mathcal{G} \cap f^{-1}\mathcal{G}_0$. Thus, $\mathcal{G}_1$ is a convex, $\mathcal{L}ip^r$ atlas and $f: \mathcal{G}_1 \to \mathcal{G}_0$ is index preserving. By convexity the $\mathcal{L}^r$ estimates on the principal parts become $\mathcal{L}ip^r$ estimates and so condition $\mathcal{L}$ and the Addendum to Proposition 1.5 imply that

$f\colon \mathcal{G}_1 \to \mathcal{G}_0$ is $\mathcal{L}^r(d_X, d_{X_0})$. If $f\colon \mathcal{G} \to \mathcal{G}_0$ is an $\mathcal{L}^r(d, d_0)$ map of admissible multiatlases, then by Definition 1.4 f is $\mathcal{L}^r(d_X, d_{X_0})$. Hence, $f\colon X \to X_0$ is an $\mathcal{L}^r$ map. Also, if $\mathcal{G}_0$ is an $\mathcal{L}^r(d_0)$ multiatlas and $f\colon \mathcal{G} \to \mathcal{G}_0$ is an $\mathcal{L}^r(d_X, d_{X_0})$ map then by Definition 1.4 f is $\mathcal{L}^r(d_X, d_0)$ iff $d_X >_\rho f^*d_0$.
Q.E.D.

The identification between $\mathcal{L}^r(d_X)$ and $\mathcal{L}^r$ structures for semicomplete spaces and bundles is natural with respect to pullbacks, products, direct sums, etc.

5 __COROLLARY:__ Let X be a semicomplete $\mathcal{L}^r$ manifold with admissible metric d_X. Let U be open in X and π be a $\mathcal{C}$ bundle over U (restriction $\mathcal{L}^r$ structure on U). The restriction $\mathcal{L}^r(d_X|U)$ structure is generated by $\{\mathcal{G}|U\}$ with $\mathcal{G}$ s admissible $\mathcal{L}ip^r$ atlases on X which are either convex or star bounded and satisfy condition $\mathcal{L}$ (on X). This class also generates the restriction $\mathcal{L}^r$ metric structure on U. Any $\mathcal{L}^r(d_X|U)$ structure on π with base the restriction $\mathcal{L}^r(d_X|U)$ structure is generated by $\mathcal{L}^r(d_X|U)$ atlases $\{(\mathfrak{D}, \mathcal{G}|U)\}$ with $\mathcal{G}|U$ in the above class. Such atlases are $\mathcal{L}^r$ and generate an $\mathcal{L}^r$ structure on π. Any $\mathcal{L}^r(d_X|U)$ section of π is $\mathcal{L}^r$.

Let $\pi_0\colon E_0 \to X_0$ be a semicomplete $\mathcal{L}^r(d_0)$ bundle. If $(\Phi, f)\colon \pi \to \pi_0$ is an $\mathcal{L}^r(d_X|U, d_0)$ VB map then (Φ, f) is an $\mathcal{L}^r$ VB map. Similarly, for $f\colon U \to X_0$.

__PROOF:__ In restricting the atlas to U we retain the uniform $\mathcal{L}$ equivalence between $(U_\alpha \cap U, d_X|U_\alpha \cap U, \rho)$ and $(h_\alpha(U_\alpha \cap U), \|\ \|_\alpha, \rho \cdot h_\alpha^{-1})$. Thus, the condition $\mathcal{L}$ arguments of the previous theorem go through here yielding $\mathcal{L}^r$ structures from $\mathcal{L}^r(d_X|U)$ structures. However, we lose the converses. The new type of result here is that any $\mathcal{L}^r(d_X|U)$ structure on π is generated as stated. Let $(\mathfrak{D}_1, \mathcal{G}_1)$ be a $\mathcal{L}^r(d_X|U)$ multiatlas on π with $\mathcal{G}_1$ admissible for the restriction $\mathcal{L}^r(d_X|U)$ structure. By Proposition

III.6.6(e) the $\mathcal{G}_1$ cover is the intersection with U of some $\mathfrak{U}_X$ uniform open cover of X. So we can choose $\mathcal{G}$ a convex s admissible $\mathcal{L}ip^r$ atlas on X with $\mathcal{G}|U$ refining the open cover of $\mathcal{G}_1$. Then refine $(\mathfrak{F}_1,\mathcal{G}_1)$ in a corresponding manner and apply the transfer of atlas construction to get $(\mathfrak{F},\mathcal{G}|U)$ in the $\mathcal{L}^r(d_X|U)$ structure. Q.E.D.

6 PROPOSITION: Let X be a semicomplete $\mathcal{L}^r$ manifold with admissible metric and bound d_X and ρ. Let U be open in X and π be a $\mathcal{C}$ bundle over U (restriction metric structure on U). Assume $(\mathfrak{F},\mathcal{G}|U)$ is an admissible $\mathcal{C}$ atlas on π with $\mathcal{G}$ a star bounded, s admissible $\mathcal{L}ip^r$ atlas satisfying condition $\mathcal{L}$ on X. $(\mathfrak{F},\mathcal{G}|U)$ is an $\mathcal{L}^r_a(d_X)$ atlas for π iff $\rho > \rho^{\mathcal{L}ip^r_a}(\mathfrak{F},\mathcal{G}|U)$ on U where $\rho^{\mathcal{L}ip^r_a}(\mathfrak{F},\mathcal{G}|U)(x) = \sup\{\|\varphi_{\beta\alpha}\|_{\mathcal{L}ip^r_a}: x \in U_\alpha \cap U_\beta\}$. Any $\mathcal{L}^r_a(d_X)$ bundle structure on π over the restriction $\mathcal{L}^r(d_X)$ structure on U is generated by such atlases. If s is a section of such a bundle, s is an $\mathcal{L}^r_a(d_X)$ section iff $\rho > \tilde{\rho}^{\mathcal{L}ip^r_a}(s,(\mathfrak{F},\mathcal{G}|U))$ where $\tilde{\rho}^{\mathcal{L}ip^r_a}(s)(x) = \sup\{\|s_\alpha\|_{\mathcal{L}ip^r_a}: x \in U_\alpha\}$.

Let X_0 be a semicomplete $\mathcal{L}^r$ metric manifold with admissible metric d_{X_0} and π_0 be an $\mathcal{L}^r_a(d_{X_0})$ bundle over X (with the associated $\mathcal{L}^r(d_{X_0})$ structure) with admissible atlas $(\mathfrak{F}_0,\mathcal{G}_0)$. Let $(\phi,f)\colon \pi \to \pi_0$ be a $\mathcal{C}$ VB map with $f\colon U \to X_0$ an $\mathcal{L}$ map. f is an $\mathcal{L}^r_a(d_X|U,d_{X_0})$ map iff $\rho > \rho^{\mathcal{L}ip^r_a}(f;\mathcal{G}|U,\mathcal{G}_0)$ with $\rho^{\mathcal{L}ip^r_a}(f)(x) = \sup\{\|f_{\beta\alpha}\|_{\mathcal{L}ip^r_a}: x \in U_\alpha \cap f^{-1}V_\beta\}$. (ϕ,f) is an $\mathcal{L}^r_a(d_X|U,d_{X_0})$ map iff $\rho > \rho^{\mathcal{L}ip^r_a}(\phi,f;(\mathfrak{F},\mathcal{G}|U),(\mathfrak{F}_0,\mathcal{G}_0))$ with $\rho^{\mathcal{L}ip^r_a}(\phi,f)(x) = \max(\rho^{\mathcal{L}ip^r_a}(f)(x),\sup\{\|\varphi_{\beta\alpha}\|_{\mathcal{L}ip^r_a}: x \in U_\alpha \cap f^{-1}V_\beta\})$.

PROOF: The arguments are similar to those sketched in the proof of Corollary 5. Star boundedness and condition $\mathcal{L}$ allow us to exchange the defining conditions of an $\mathcal{L}^r_a(d)$ atlas or map to conditions using the function space type $\mathcal{L}ip^r_a$. For f: the "j^r_1 condition" and "f: $(U,d_X|U,\rho) \to (X_0,d_{X_0},\rho_0)$ is $\mathcal{L}_a$" hypothesis are combined into one j^r condition by

Proposition III.7.16. That the atlases $(\mathfrak{D},\mathcal{G}|U)$ generate is the transfer of atlas construction together with Proposition III.6.6(e) again. Q.E.D.

Now recall that a $\mathcal{C}_u^r$ structure on X means a $\mathcal{C}_u^r(d_X)$ structure on X where d_X is an admissible metric on a regular $\mathcal{L}$ manifold.

7 PROPOSITION: Let $r \geq 1$ and $\pi\colon \mathbb{E} \to X$ be a semicomplete $\mathcal{C}^r$ bundle with admissible bound ρ. Let $(\mathfrak{D},\mathcal{G})$ be an s admissible $\mathcal{C}^r$ atlas for π. If $(\mathfrak{D},\mathcal{G})$ is $\mathcal{L}ip^{r-1}$ and convex, then $\mathcal{G}$ is a $\mathcal{C}_u^r$ atlas on X iff the family $\{D^r h_{\beta\alpha}\colon (h_\alpha(U_\alpha \cap U_\beta), \| \ \|_\alpha, \rho\cdot h_\alpha^{-1}) \to L_s^r(E_\alpha;E_\beta)\}$ is a $\mathcal{C}_u$ family of maps. $(\mathfrak{D},\mathcal{G})$ is a $\mathcal{C}_u^r$ atlas on π iff, in addition, the family $\{D^r \varphi_{\beta\alpha}\colon (h_\alpha(U_\alpha \cap U_\beta), \| \ \|_\alpha, \rho\cdot h_\alpha^{-1}) \to L_s^r(E_\alpha;L(F_\alpha;F_\beta))\}$ is $\mathcal{C}_u$. Any $\mathcal{C}_u^r$ structure on X or π compatible with the original $\mathcal{C}^r$ structure is generated by such atlases. The atlases $\mathfrak{D}$ on the total space are then $\mathcal{L}ip^{r-1}$, convex, $\mathcal{C}_u^r$ atlases generating a $\mathcal{C}_u^r$ structure on the total space compatible with the $\mathcal{C}^r$ total space structure. A section $\mathcal{C}^r$ s of π is $\mathcal{C}_u^r$ iff the family $\{D^r s_\alpha\colon (h_\alpha(U_\alpha), \| \ \|_\alpha, \rho\cdot h_\alpha^{-1}) \to L_s^r(E_\alpha;F_\alpha)\}$ is $\mathcal{C}_u$. This is true iff $s\colon X \to \mathbb{E}$ is a $\mathcal{C}_u^r$ map.

Let $\pi_0\colon \mathbb{E}_0 \to X_0$ be a semicomplete $\mathcal{C}^r$ manifold with $(\mathfrak{D}_0,\mathcal{G}_0)$ admissible for a compatible $\mathcal{C}_u^r$ structure and $(\Phi,f)\colon \pi \to \pi_0$ be a $\mathcal{C}^r$ VB map. f is $\mathcal{C}_u^r$ iff the family $\{D^r f_{\beta\alpha}\colon (h_\alpha(U_\alpha \cap f^{-1}V_\beta), \| \ \|_\alpha, \rho\cdot h_\alpha^{-1}) \to L_s^r(E_\alpha^0;E_\beta)\}$ is a $\mathcal{C}_u$ family. (Φ,f) is $\mathcal{C}_u^r$ iff, in addition, the family $\{D^r \Phi_{\beta\alpha}\colon (h_\alpha(U_\alpha \cap f^{-1}V_\beta), \| \ \|_\alpha, \rho\cdot h_\alpha^{-1}) \to L_s^r(E_\alpha;L(F_\alpha;F_\beta^0))\}$ is a $\mathcal{C}_u$ family. If (Φ,f) is $\mathcal{C}_u^r$ then $\Phi\colon \mathbb{E} \to \mathbb{E}_0$ is $\mathcal{C}_u^r$.

PROOF: Since $\mathcal{C}^r \subset \mathcal{L}^{r-1}$ we can use Theorem 4 and are reduced to checking uniform continuity of the highest derivatives. The rest is the usual condition $\mathcal{L}$ exchange and transfer of atlas argument to get generators. If $f\colon \mathcal{G} \to \mathcal{G}_0$ is index preserving we need only consider the family of principal parts in testing whether f is $\mathcal{C}_u^r$. This implies the section result. The only new type of result is the claims about the total space.

If $(\mathfrak{P},\mathcal{G})$ is a $\mathscr{L}ip^r$ or $\mathcal{C}_u^r$ atlas then $\mathfrak{P}$ is a $\mathscr{L}ip^r$ or $\mathcal{C}_u^r$ atlas. Similarly, for maps. This is because if $\{j^r(\varphi_{\beta\alpha}): (h_\alpha(U_\alpha \cap U_\beta), \| \|_\alpha, \rho\cdot h_\alpha^{-1}) \to L_s^r(E_\alpha; L(F_\alpha;F_\beta))\}$ is an $\mathscr{L}$ or $\mathcal{C}_u$ family then so is $\{j^r(\varphi_\beta\varphi_\alpha^{-1}): (h_\alpha(U_\alpha \cap U_\beta) \times F_\alpha, \| \|_\alpha, \max(p_1{*}(\rho\cdot h_\alpha^{-1}), p_2{*}\| \|_\alpha)) \to L_s^r(E_\alpha \times F_\alpha, F_\beta)\}$ by direct computation using partial derivatives. Similarly, for maps. Q.E.D.

8 PROPOSITION: Let $r \geq 0$ and X be a semicomplete $\mathscr{L}^r$ manifold with admissible bound and metric ρ and d_X. Let $U \subset X$ be open. The $\mathscr{L}^r(d_X|U)$ structure on U induces a $\mathcal{C}_u^r(d_X|U)$ structure on U which is the restriction of the similar $\mathcal{C}_u^r$ structure on X. Let π be a $\mathcal{C}$ bundle over U and $(\mathfrak{P},\mathcal{G}|U)$ an admissible $\mathcal{C}$ atlas with $\mathcal{G}$ an s admissible convex, $\mathscr{L}ip^r$ atlas on X. $(\mathfrak{P},\mathcal{G}|U)$ is a $\mathcal{C}_u(d_X)$ atlas iff the family $\{j^r(\varphi_{\beta\alpha}): (h_\alpha(U_\alpha \cap U_\beta \cap U), \| \|_\alpha, \rho\cdot h_\alpha^{-1}) \to J^r(E_\alpha;F_\alpha)\}$ is $\mathcal{C}_u$. Let π_0 be a $\mathcal{C}_u^r$ vector bundle and $(\phi,f): \pi \to \pi_0$ be a $\mathcal{C}$ VB map. Assume $f: (U,d_X|U,\rho) \to (X_0,d_{X_0},\rho_0)$ is $\mathcal{C}_u$. Let $(\mathfrak{P}_0,\mathcal{G}_0)$ be an admissible $\mathcal{C}_u^r$ atlas on π_0. f is a $\mathcal{C}_u^r$ map iff the family $\{j^r(f_{\beta\alpha}): (h_\alpha(U_\alpha \cap f^{-1}V_\beta \cap U), \| \|_\alpha, \rho\cdot h_\alpha^{-1}) \to J^r(E_\alpha,E_\beta^0)\}$ is $\mathcal{C}_u$. (ϕ,f) is a $\mathcal{C}_u^r$ map iff f is and the family $\{j^r(\varphi_{\beta\alpha}): (h_\alpha(U_\alpha \cap f^{-1}V_\beta \cap U, \| \|_\alpha, \rho\cdot h_\alpha^{-1}) \to J^r(E_\alpha;L(F_\alpha;F_\beta^0))\}$ is $\mathcal{C}_u$.

PROOF: Similar to the preceding proofs. Note that if f is $\mathcal{C}_u$ from $d_X|U$ then the open cover $f^{-1}\mathcal{G}_0$ is $\mathfrak{A}_{d_X|U}$ uniform. Also, if $U = X$ this result overlaps with the previous one. In that case, if $(\mathfrak{P},\mathcal{G})$, s, (ϕ,f) are $\mathcal{C}^r$ then the conditions involving j^r can be replaced by similar conditions on just the highest derivative D^r. Q.E.D.

9 LEMMA: Let $\pi: \mathbb{E} \to X$ be a semicomplete $\mathscr{L}^r$ manifold. Let X_1 be a $\mathcal{C}_u^r(d_1)$ (or $\mathscr{L}^r(d_1)$) manifold and $f: X_1 \to X$ be a $\mathcal{C}_u^r(d_1,d_X)$ (resp. $\mathscr{L}_a^r(d_1,d_X)$) map so that $f^*\pi$ is a $\mathcal{C}_u^r(d_1)$ (resp. $\mathscr{L}_a^r(d_1)$) bundle. Let s be a section of $f^*\pi$. $\phi_f\cdot s: X_1 \to \mathbb{E}$ is a $\mathcal{C}_u^r(d_1,d_{\mathbb{E}})$ (resp. $\mathscr{L}_a^r(d_1,d_{\mathbb{E}})$) map iff f^*s is a $\mathcal{C}_u^r(d_1)$ (resp. $\mathscr{L}_a^r(d_1)$) section of $f^*\pi$.

PROOF: Let $(\mathfrak{D},\mathfrak{a})$ be a convex, $\mathcal{L}ip^r$ s admissible atlas on π. $\mathfrak{D}$ is an atlas of the same type on $\mathbb{E}$. Assume $\mathfrak{a}_1$ is an admissible multiatlas on X_1 such that $f\colon \mathfrak{a}_1 \to \mathfrak{a}$ is index preserving. $(f^*\mathfrak{D}_1, \mathfrak{a}_1)$ is an admissible multiatlas on $f^*\pi$. The result follows because the principal parts of the map $\Phi_f \cdot s$ are the products of the principal parts of f and those of the section s. The discrepancy between the j^r condition on the section and the conditions of Definition 1.4 are filled in using Proposition III. 7.16, and condition $\mathcal{L}$ on the range. Q.E.D.

3. Section Spaces:

1 PROPOSITION: Let X be a regular $\mathcal{L}$ manifold with apm d and let π be a $\mathcal{C}_u^r(d)$ bundle over X. The vector space $\mathcal{C}_u^r(d)(\pi)$ of $\mathcal{C}_u^r(d)$ sections of π is a closed subspace of the topological group $\mathcal{C}^r(\pi)$.

PROOF: Let $(\mathfrak{D},\mathfrak{a}) = \{U_\alpha, h_\alpha, \varphi_\alpha\}$ be a star bounded, admissible $\mathcal{C}^r(d)$ multi-atlas on π. With ρ an admissible bound let $\rho_\alpha = \sup \rho|U_\alpha$. If $\{s^n\}$ is a sequence in $\mathcal{C}_u^r(d)(\pi)$ converging to $s \in \mathcal{C}^r(\pi)$, then we can adapt the usual proof that uniform convergence preserves uniform continuity as follows: Given $K,n \geq 1$ choose N such that $\|j^r(s_\alpha^N)\cdot h_\alpha - j^r(s_\alpha)\cdot h_\alpha\|_0 < (3\,K\rho_\alpha^{n})^{-1}$. Choose K_1, n_1 such that $x_1, x_2 \in U_\alpha$ and $d(x_1,x_2) < (K_1\rho_\alpha^{n_1})^{-1}$ implies $\|j^r(s_\alpha^N)(h_\alpha x_1) - j^r(s_\alpha^N)(h_\alpha x_2)\| < (3\,K\rho_\alpha^{n})^{-1}$. It follows that $\|j^r(s_\alpha)(h_\alpha x_1) - j^r(s_\alpha)(h_\alpha x_2)\| < (K\,\rho_\alpha^{n})^{-1}$. Hence the family $\{j^r(s_\alpha)\cdot h_\alpha\}$ is $\mathcal{C}_u$. Q.E.D.

Let $(\mathfrak{M}_1, \mathcal{C}^r, \mathfrak{M}_2)$ be a standard triple with $\mathfrak{M}_1 \subset \mathcal{L}^r$ and X_1 be a semi-complete $\mathfrak{M}_1^1$ manifold admitting $\mathfrak{M}_1$ exponentials. If X is a bounded $\mathcal{C}_u^r(d)$ manifold then it follows that $\mathcal{C}_u^r(d)(X,X_1)$, the set of $\mathcal{C}_u^r(d,d_{X_1})$ maps is a closed submanifold of the $\mathfrak{M}_2$ manifold $\mathcal{C}^r(X,X_1)$. In fact, if $f \in \mathcal{C}_u^r(d)(X,X_1)$ and $H_f(g) = s$, $s \in D_f \subset \mathcal{C}^r(f^*\tau_{X_1})$ it follows from Lemma 2.9 that $s \in \mathcal{C}^r(d)(f^*\tau_{X_1})$ iff $\Phi_f \cdot s$ is a $\mathcal{C}^r(d, d_{TX_1})$ map. Since

$\Phi_f \circ s(X) \subset\subset D$ and $e: D \to V$ is an $\mathcal{L}^r$ isomorphism it easily follows that $\Phi_f \circ s$ is a $\mathcal{C}^r(d, d_{TX_1})$ map iff g is a $\mathcal{C}^r(d, d_{X_1})$ map. Thus, $H_f(V_f \cap \mathcal{C}^r_u(d)(X, X_1)) = D_f \cap \mathcal{C}^r_u(d)(f^*\tau_{X_1})$.

Uniform continuity allows us to avoid FS7 (Strong) assumptions. By a <u>composition situation</u> for $(\mathcal{C}^{p+q}_u, \mathcal{C}^p, \mathcal{C}^q_u)$ we mean if $p + q \geq 1$ (or $p = q = 0$): (1) Semicomplete $\mathcal{C}^{p+q}_u$ (resp. $\mathcal{L}$) bundles $\pi_i: \mathbb{E}_i \to X$ $(i = 1,2)$ and O open in $\mathbb{E}_1$. (2) A bounded $\mathcal{C}^p$ manifold X_0 and a $\mathcal{C}^p$ map $f: X_0 \to X$. (3) A semicomplete $\mathcal{L}^q$ manifold Z, Z_0 open in Z and a $\mathcal{C}^q_u(d_Z|Z_0, \| \ \|)$ map $H: Z_0 \to \mathcal{C}^p(f^*\pi_1)$ such that $\Phi^1_f(H(z)(x)) \in O$ for all $(z,x) \in Z_0 \times X_0$.

2 <u>THEOREM</u>: Given a composition situation for $(\mathcal{C}^{p+q}_u, \mathcal{C}^p, \mathcal{C}^q_u)$, let $G: O \to \mathbb{E}_2$ be a $\mathcal{C}^{p+q}_u(d_{\mathbb{E}_1}|O, d_{\mathbb{E}_2})$ map with $\pi_2 \circ G = \pi_1|O$. $\Omega^f_H(G): Z_0 \to \mathcal{C}^p(f^*\pi_2)$ is a $\mathcal{C}^q_u(d_Z|Z_0, \| \ \|)$ map.

<u>PROOF</u>: Choose $(\mathfrak{M}_1, \mathfrak{M}_2; \mathfrak{A})$ an s admissible convex, $\mathcal{L}ip^{p+q-1}, \mathcal{C}^{p+q}_u$ (resp. convex, $\mathcal{L}ip$) two-tuple for $\pi_1, \pi_2, \mathfrak{A}_0$ an admissible $\mathcal{C}^p$ atlas on X_0 with $f: \mathfrak{A}_0 \to \mathfrak{A}$ index preserving, and $\mathfrak{A}_Z$ an admissible convex, $\mathcal{L}ip^q$ s admissible atlas on Z. Then follow the proof of Theorem IV.2.3. At the point when FS7 (Strong) for $\mathfrak{M}_2$ is used there, apply Proposition II.2.8. Q.E.D.

<u>ADDENDUM</u>: If X_0 is a bounded $\mathcal{C}^p_u(d)$ manifold, f is a $\mathcal{C}^p_u(d_0, d_X)$ map and $H(Z_0) \subset \mathcal{C}^p_u(d)(f^*\pi_1)$ then by Lemma 2.9, $\Omega^f_H(G)(Z_0) \subset \mathcal{C}^p_u(d)(f^*\pi_2)$ and so $\Omega^f_H(G)$ is a $\mathcal{C}^q_u(d_Z|Z_0, \| \ \|)$ map into this closed subspace.

Let $\pi: \mathbb{E} \to X$ be a bounded $\mathcal{L}^r_a(d)$ bundle with admissible multiatlas $(\mathfrak{M}, \mathfrak{A}) = \{U_\alpha, h_\alpha, \varphi_\alpha\}$. For $s \in \mathcal{L}^r_a(d)(\pi)$ the norm

$$\max(\|s\|^{(\mathfrak{M},\mathfrak{A})}_{\mathcal{C}^r}, \sup\{\|j^r(s_\alpha)(h_\alpha x_1) - j^r(s_\alpha)(h_\alpha x_2)\|/d(x_1,x_2)^a : x_1 \neq x_2 \in U_\alpha\})$$

makes $\mathcal{L}^r_a(d)(\pi)$ into a Banach space. As in Section IV.1, the Banachable space structure is independent the choice of multiatlases. If $f: X_1 \to X$ is an $\mathcal{L}^r_b(d_1, d)$ map then $f^*: \mathcal{L}^r_a(d)(\pi) \to \mathcal{L}^r_{ab}(d_1)(f^*\pi)$ is a bounded linear map. If $(\Phi, 1_X): \pi \to \pi_0$ is an $\mathcal{L}^r_a(d,d)$ VB map then

$\Phi_*: \mathscr{L}_a^r(d)(\pi) \to \mathscr{L}_a^r(d)(\pi_0)$ is a bounded linear map. Finally, let X be a semicomplete $\mathscr{L}^r$ manifold, U be a bounded open set in X and $\pi: \mathbb{E} \to U$ be an $\mathscr{L}_a^r(d_X|U)$ bundle. If $(\mathfrak{D},\mathfrak{G}|U)$ is an admissible $\mathscr{L}_a^r$ atlas with $\mathfrak{G}$ a convex, $\mathscr{L}ip^r$ s admissible atlas on X, then $\|s\| = \sup \tilde{\rho}^{\mathscr{L}ip_a^r}(s;(\mathfrak{D},\mathfrak{G}|U))$ defines an equivalent norm on $\mathscr{L}_a^r(d)(\pi)$ by Proposition 2.6.

4. $\mathfrak{M}$ Mappings: If X_{dis} is a set with the discrete topology and ρ is any bound on X_{dis} then with ρ as admissible bound X_{dis} is a semicomplete $\mathscr{C}^\infty$ manifold with an, essentially, unique atlas $\{\{x\},0_x\}$ indexed by X_{dis}, where 0_x maps $\{x\}$ to the zero B-space. Any atlas $\mathfrak{G}$ on X_{dis} is a subdivision of this one of the form $\{\{x_\alpha\},0_{x_\alpha}\}$ where $\alpha \to x_\alpha$ is a map of the index set onto X. In any case, $\rho_{\mathfrak{G}} = 1$, $d_{\mathfrak{G}}(x_1,x_2) = 0$ (or ∞) if $x_1 = x_2$ (resp. $x_1 \neq x_2$) and $\lambda_{\mathfrak{G}} = 1$. A bundle $\pi: \mathbb{E} \to X_{dis}$ with atlas $(\mathfrak{D},\mathfrak{G}) = \{\{x_\alpha\},0_{x_\alpha},\varphi_\alpha\}$ consists of a Banachable space over each point and for each x, $\{\varphi_\alpha: x_\alpha = x\}$ is essentially a collection of norms $\|\ \|_\alpha$ on E_x. $(\mathfrak{D},\mathfrak{G},\rho)$ is an adapted $\mathscr{C}$ atlas iff it is an adapted $\mathscr{C}^\infty$ atlas iff $\sup\{\text{dist}(\|\ \|_{\alpha_1},\|\ \|_{\alpha_2}): x_{\alpha_1} = x_{\alpha_2} = x\} \leq n \log \rho(x) + \log K$ for some constants K,n (cf. Sec. III.3). In which case $K\rho^n \geq \rho_{(\mathfrak{D},\mathfrak{G})}$ and $\|\ \|^M \leq K(\pi^*\rho)^n\|\ \|^m$. If X_1 is any $\mathfrak{M}$ manifold with admissible atlas $\mathfrak{G}_1$ and $f: X_{dis} \to X_1$ is any set map then $\rho^{\mathfrak{M}}(f;\mathfrak{G},\mathfrak{G}_1) = f^*\rho^{\mathscr{C}}_{\mathfrak{G}_1}$. Hence, f is an $\mathfrak{M}$ map iff $\rho > f^*\rho_1$.

Now let X be an $\mathfrak{M}$ manifold with admissible atlas $\mathfrak{G} = \{U_\alpha,h_\alpha\}$ and admissible metric and bound d_X and ρ. Let X_{dis} be the same set with the discrete topology and dis: $X_{dis} \to X$ be the identity map. Let $\mathfrak{G}_{dis}$ be indexed by pairs αx with $x \in U_\alpha$ and let $x_{\alpha x} = x$. We will write αx for $\{x_{\alpha x}\}$. X_{dis} is a semicomplete $\mathscr{C}^\infty$ manifold with apm d_X inducing an $\mathscr{L}^r(d_X)$ structure on X_{dis} for all r. The associated multiatlas is $\{U_\alpha,0_\alpha\}$. If $\pi: \mathbb{E} \to X$ is an $\mathfrak{M}$ bundle with atlas $(\mathfrak{D},\mathfrak{G})$ then $\text{dis}^*\pi: E \to X_{dis}$ is a semicomplete $\mathscr{C}^\infty$ bundle with admissible atlas

$(\text{dis}^*\mathfrak{H}, \mathcal{G}_{\text{dis}})$. $\rho(\text{dis}^*\mathfrak{D}, \mathcal{G}_{\text{dis}}) \leq \rho^{\mathcal{C}}_{(\mathfrak{H},\mathcal{G})}$. It is an $\mathcal{L}^r(d_X)$ bundle with admissible atlas $\{U_\alpha, \mathcal{O}_\alpha, \mathcal{O} \times p_2 \circ \varphi_\alpha\}$ for all r. The max Finslers $\| \|^M$ induced on $\mathbf{E} = \text{dis}^*\mathbf{E}$ by $(\mathfrak{H}, \mathcal{G})$ and $(\text{dis}^*\mathfrak{D}, \mathcal{G}_{\text{dis}})$ are the same. Now if π is bounded define $\mathfrak{M}(\pi) = \mathcal{C}(\text{dis}^*\pi)$. Clearly, $\mathcal{C}(\pi) = \mathcal{C}_u(d_X)(\text{dis}^*\pi)$ is a closed subspace of $\mathfrak{M}(\pi)$ with the norm $\| \|^{(\mathfrak{H},\mathcal{G})}$ just the restriction of the norm $\| \|^{(\text{dis}^*\mathfrak{H}, \mathcal{G}_{\text{dis}})}$. If $(\mathfrak{M}_1, \mathcal{C}, \mathfrak{M}_2)$ is a standard triple and X_1 is a semi-complete $\mathfrak{M}_1^1$ manifold admitting $\mathfrak{M}_1$ exponentials then define the $\mathfrak{M}_2$ manifolds $\mathfrak{M}(X, X_1) = \mathcal{C}(X_{\text{dis}}, X_1)$. Then $\mathcal{C}(X, X_1) = \mathcal{C}_u(d_X)(X_{\text{dis}}, X_1)$ is a closed submanifold of $\mathfrak{M}(X, X_1)$.

CHAPTER VI: IMMERSIONS AND SUBMERSIONS

1. Linear Injections and Surjections: By an injection $j\colon E_1 \to E_2$ we mean an injective continuous linear map of B-spaces with closed image. By the closed graph theorem a linear map is an injection iff the restriction $\bar{j}\colon E_1 \to j(E_1)$ is a linear isomorphism iff $m(j) > 0$ where $m(j) = \sup\{r\colon \|jv\| \geq r\|v\| \text{ for all } v \in E_1\}$. Denote by $L\bar{i}(E_1;E_2)$ the set of injections in $L(E_1;E_2)$ and define $\bar{\theta}(j) = \max(\|j\|, m(j)^{-1}, 1)$ on $L\bar{i}$. In applications, we will be most interested in split injections, namely continuous linear maps $j\colon E_1 \to E_2$ admitting left inverses, i.e. admitting $P \in L(E_2;E_1)$ such that $Pj = I_1$. Denote by $Li(E_1;E_2) \subset L\bar{i}(E_1;E_2)$ the set of split injections in $L(E_1;E_2)$ and define $\theta(j) = \max(\|j\|, 1, \inf\{\|P\|\colon P \in L(E_2;E_1) \text{ with } Pj = I_1\})$ on Li. If E_1 is a closed subspace of E_2 and j is the inclusion then a left inverse for j is a projection P with image E_1. In this case, $\theta(j)$ is denoted $\theta(E_1)$ and equals $\max(1, \inf\{\|P\|\colon \text{projections on } E_1\})$.

1 PROPOSITION: Let $j\colon E_1 \to E_2$ be an injection.

(a) $\|\bar{j}^{-1}\| = m(j)^{-1}$. Hence, $\bar{\theta}(j) = \theta(\bar{j}) = \max(1, \|j\|, \|\bar{j}^{-1}\|)$. j is a split linear injection iff $j(E_1)$ is a split subspace of E_2. In that case: $\bar{\theta}(j) \leq \theta(j)$, $\theta(j(E_1)) \leq \theta(j)^2$ and $\theta(j) \leq O^*[\bar{\theta}(j), \theta(j(E_1))]$.

(b) j is a split injection if there exists a linear map $j_1\colon E_3 \to E_2$ with $j + j_1\colon E_1 \times E_3 \to E_2$ an isomorphism. In this case, $\theta(j) \leq \theta(j + j_1)$. Conversely, if j is a split injection then j_1 can be chosen the inclusion of a closed subspace $E_3 \subset E_2$ with $j + j_1$ an isomorphism and $\theta(j + j_1) \leq O^*(\theta(j))$.

(c) If $j_1\colon E_2 \to E_3$ is an injection then $j_1 \circ j$ is an injection and $\theta(j_1 \circ j) \leq \bar{\theta}(j_1)\bar{\theta}(j)$. If j_1 and j are split then $j_1 \circ j$ is and

$\theta(j_1 \cdot j) \le \theta(j_1)\theta(j)$.

PROOF: (a): That $m(j) = \|\bar{j}^{-1}\|^{-1}$ is easy. If P is a left inverse then $\|P\| \ge \|\bar{j}^{-1}\|$. Also, jP is a projection onto $j(E_1)$. Conversely, if P_1 is a projection onto $j(E_1)$ then $(\bar{j}^{-1})P_1$ is a left inverse for j. (b) and (c) are easy exercises. Q.E.D.

By a surjection $P: E_1 \to E_2$ we mean a surjective, continuous linear map of B-spaces. By the open mapping theorem P is a surjection iff $m(P) > 0$ where $m(P) = \sup\{r: B^{\|\ \|_2}(0,r) \subset P(B^{\|\ \|_1}(0,1))\}$. Denote by $L\bar{s}(E_1;E_2)$ the set of surjections in $L(E_1;E_2)$ and define $\bar{\theta}(P) = \max(\|P\|, m(P)^{-1}, 1)$ on $L\bar{s}$. A split surjection is a continuous linear map P admitting right inverses. Denote by $Ls(E_1;E_2) \subset L\bar{s}(E_1;E_2)$ the set of split surjections in $L(E_1;E_2)$ and define $\theta(P) = \max(\|P\|, 1, \inf\{\|j\|: j \in L(E_2;E_1)$ with $Pj = I_2\})$ on Ls.

2 PROPOSITION: Let $P: E_1 \to E_2$ be a surjection with kernel $K(P)$.

(a) P is a split surjection iff $K(P)$ is a split subspace of E_1. In that case: $\bar{\theta}(P) \le \theta(P)$, $\theta(K(P)) \le 2\theta(P)^2$ and $\theta(P) \le O^*(\bar{\theta}(P), \theta(K(P)))$.

(b) P is a split surjection if there exists a surjection $P_1: E_1 \to E_3$ such that $P \times P_1: E_1 \to E_2 \times E_3$ is an isomorphism. In that case $\theta(P) \le \theta(P \times P_1)$. Conversely, if P is a split surjection then P_1 can be chosen a projection onto $K(P)$ with $P \times P_1$ an isomorphism and $\theta(P \times P_1) \le O^*(\theta(P))$.

(c) If $P_1 : E_2 \to E_3$ is a surjection then $P_1 \circ P$ is and $\bar{\theta}(P_1 \circ P) \le \bar{\theta}(P_1)\bar{\theta}(P)$. If P and P_1 are split then $P_1 \circ P$ and $\theta(P_1 \circ P) \le \theta(P_1)\theta(P)$.

(d) Let E_0 be a closed subspace of E and $q: E \to E/E_0$ the quotient map. E_0 is a split subspace of E iff q is a split surjection. In that case, $\theta(q) \le 1 + \theta(E_0)$ and $\theta(E_0) \le 1 + \theta(q)$.

PROOF: (a): If j is a right inverse then $j(B^{\|\ \|_2}(0,\|j\|^{-1})) \subset B^{\|\ \|_1}(0,1)$. So $B^{\|\ \|_2}(0,\|j\|^{-1}) \subset P(B^{\|\ \|_1}(0,1))$. Also, $I - j \cdot P$ is a projection onto $K(P)$. Conversely, if P_1 is a projection onto $K(P)$ then $P|K(P_1): K(P_1) \to E_2$ is an isomorphism with inverse j, a right inverse for P. $\|j\| \leq 2\|P_1\|/m(P)$ because if $\|v\|_2 < 1$ then $Pu = v$ for some u with $\|u\| < m(P)^{-1}$. Hence, $j(v) = u - P_1 u$ and $\|j(v)\| < 2\|P_1\|/m(P)$. (b) and (c) are easy. (d): Let i be the inclusion of E_0. If P is a projection onto E_0 then $I - P$ factors through q to define a map $j: E/E_0 \to E$ with the same norm as $I - P$ and $qj = I$. If j is a right inverse for q then $I - jq = P$ maps E into E_0 and $Pi = I$. Q.E.D.

3 PROPOSITION: $L\bar{i}(E_1;E_2) \cap L\bar{s}(E_1;E_2) = Li(E_1;E_2) \cap Ls(E_1;E_2) = Lis(E_1;E_2)$ and for $T \in Lis\ (E_1;E_2)$, $\bar{\theta}(T) = \theta(T) = \max(\|T\|, \|T^{-1}\|, 1)$.

PROOF: The intersection result is clear. If $T \in Lis$, then $m(T) = \|T^{-1}\|^{-1}$ by either the $L\bar{i}$ or $L\bar{s}$ definition of m. Q.E.D.

Thus, $\bar{\theta}$ (or θ) is well defined on $L\bar{i} \cup L\bar{s}$ (resp. $Li \cup Ls$). We extend the definitions by $\bar{\theta} = \infty$ on $L - (L\bar{i} \cup L\bar{s})$ (resp. $\theta = \infty$ on $L - (Li \cup Ls)$). Note that the 1's occuring in the definitions of θ and $\bar{\theta}$ are superfluous except when E_1 or $E_2 = 0$. We will usually just leave them out.

Any covariant B-space functor takes split injections to split injections and split surjections to split surjections. A contravariant functor interchanges split injections and split surjections. For functors in infinitely many variables side conditions may be necessary, eg. if $\{T_\alpha: E_\alpha \to F_\alpha\}$ is a family of split injections (or surjections) then $\hat{\Pi}_\alpha T_\alpha$ is a well defined split injection (resp. surjection) iff $\sup \theta(T_\alpha) < \infty$ in which case $\theta(\hat{\Pi}_\alpha T_\alpha) = \sup \theta(T_\alpha)$.

4 LEMMA: On a B-space E, $s,d: E \times E \to E$ defined by $s(v_1,v_2) = v_1 + v_2$,

$d(v_1,v_2) = v_1 - v_2$ are split surjections with $\theta(s) = \theta(d) = 2$. $\Delta, \bar{\Delta}: E \to E \times E$ defined by $\Delta(v) = (v,v)$, $\bar{\Delta}(v) = (v,-v)$ are split injections with $\theta(\Delta) = \theta(\bar{\Delta}) = 1$. $K = (\sqrt{2})^{-1}(s \times d): E \times E \to E \times E$ is an idempotent isomorphism with $\theta(K) = \|K\| = \sqrt{2}$.

PROOF: Obvious. Note that K identifies the direct sum diagrams:

$$E \underset{p_1}{\overset{i_1}{\rightleftarrows}} E \times E \underset{i_2}{\overset{p_2}{\rightleftarrows}} E \quad \text{and} \quad E \underset{as}{\overset{a\Delta}{\rightleftarrows}} E \times E \underset{a\bar{\Delta}}{\overset{ad}{\rightleftarrows}} E$$

with $a = (\sqrt{2})^{-1}$. Q.E.D.

ADDENDUM: Let $q: E \times E \to E \times E/\Delta(E)$ be the quotient map. d factors through q to an isomorphism $D: E \times E/\Delta(E) \to E$ with inverse $q \circ i_1$. Thus, $\|D\| = \theta(D) = 2$. The following clearly commutes:

$$\begin{array}{ccc} E \times E & \xrightarrow{\ q\ } & E \times E/\Delta(E) \\ {\scriptstyle I\times(-I)}\downarrow & & \downarrow{\scriptstyle D} \\ E \times E & \xrightarrow[\ s\ ]{} & E \end{array} .$$

Any injection j is the composition of an isomorphism and an inclusion, i.e. an isometric injection. This breaks the problem of estimating $\theta(j)$ into two parts as in Proposition 1 (a). First, the splitting problem of a closed subspace E_0 of E. For example, if E is a Hilbert space the orthogonal projection to E_0 has norm 1 and so $\theta(E_0) = 1$. In general, we want to know does E_0 split in E and if so what is the "most orthogonal" projection to E_0? Second, the isomorphism problem. $\bar{\theta}(j) = \theta(\bar{j})$ measures how the norm on E_1 matches the norm pulled back from E_2 by j. In general, if E_1 and E_2 are isomorphic B-spaces, what is the best fit between the norms, what is $\inf\{\theta(T): T \in \mathrm{Lis}(E_1;E_2)\}$? For example, isomorphic Hilbert spaces are isometric. For if $\langle\ ,\ \rangle_T$ is the pull back by T of the inner product

$\langle\ ,\ \rangle_2$ on E_2 then $\langle\ ,\ \rangle_T$ is equivalent to $\langle\ ,\ \rangle_1$ and so there exists a positive defininte map $A: E_1 \to E_1$ with $\langle Tv_1, Tv_2\rangle_2 = \langle Av_1, v_2\rangle_1$. Let P be the positive definite square root of A, then $\langle Tv_1, Tv_2\rangle_2 = \langle Pv_1, Pv_2\rangle_1$, i.e. TP^{-1} is an isometric isomorphism. We consider these problems for injections with finite dimensional domains.

5 __PROPOSITION__: Let E_0 be a k-dimensional subspace of E. Let $\epsilon > 0$. With the sup norm on R^k there is a split injection $j: R^k \to E$ with image E_0 and $\theta(j) \leq 2^{k-1} + \epsilon$.

__PROOF__: We must choose a basis of E_0. Recall Riesz' Theorem [27;p.84]: If E_1 is a closed subspace of E then we can choose $v \in E$ with $\|v\| = 1$ and $\|v - E_1\| > (1 + \epsilon_1)^{-1}$. This means the projection of the subspace $[v,E_1]$ spanned by v and E_1 onto $[v]$ has norm $\leq 1 + \epsilon_1$. The complimentary projection to E_1 then has norm $\leq 2 + \epsilon_1$. Choose v_1 a unit vector in E_0 and apply Riesz' theorem inductively to $V_i = [v_1,\ldots,v_i]$ in E_0 $(i < k)$ to choose v_{i+1}. Then we have a diagram of projections:

$$\begin{array}{ccccccccc} E_0 = V_k & \longrightarrow & V_{k-1} & \longrightarrow \cdots \longrightarrow & V_2 & \longrightarrow & V_1 & = [v_1] \\ \downarrow & & \downarrow & & \downarrow & & & \\ [v_k] & & [v_{k-1}] & & [v_2] & & & \end{array}$$

where each horizontal map has norm $\leq 2 + \epsilon_1$ and each vertical map has norm $\leq 1 + \epsilon_1$. So the composed projections $P_i: E_0 \to [v_i]$ are bounded by $2^{k-1} + \epsilon$ where $\epsilon \to 0$ with ϵ_1. By the Hahn-Banach Theorem there is a linear functional $T_i: E \to R$ with $\|T_i\| \leq 2^{k-1} + \epsilon$ and $T_i(v)v_i = P_i(v)$ for $v \in E_0$. $T = T_1 \times\ldots\times T_k: E \to R^k$ has right inverse $j: R^k \to E$ defined by $j(x_1,\ldots,x_k) = x_1v_1 +\ldots+ x_kv_k$. $\|j\| \leq k \leq 2^{k-1}$. Q.E.D.

6 __COROLLARY__: (a) Let E_1 and E_2 be k-dimensional B-spaces. $\inf\{\theta(T): T \in Lis(E_1;E_2)\} \leq 2^{2k-2}$.

(b) Let E_0 be a k-dimenisonal subspace of a B-space E.

$\theta(E_0) \leq 2^{2k-2}$.

(c) Let $j: E_0 \to E$ be an injection with $\dim E_0 = k$. j is split and $\theta(j) \leq O^*(\bar{\theta}(j), 2^k)$.

(d) Let E_0 be a closed subspace of codimension k. $\theta(E_0) \leq 2^{2k-1}$.

(e) Let $P: E \to E_0$ be a surjection with $\dim E_0 = k$. P is split and $\theta(P) \leq O^*(\bar{\theta}(P), 2^k)$.

(f) If E is finite dimensional, let det: $L(E;E) \to R$ be the determinant function. On $Lis(E;E)$, $\theta \sim \max(\| \|, 1/|\det|)$.

<u>PROOF</u>: (a) - (c) follow from Propositions 1 and 5. (d): Let $q: E \to E/E_0$ be the quotient map. By the proof of Proposition 5 there is an isomorphism $j: R^k \to E/E_0$ with $\theta(j)$ close to 2^{k-1} and j maps the standard basis to unit vectors. Lifting these vectors to vectors in E of norm $< 1 + \varepsilon_1$ we get a lifting $j_1: R^k \to E$ of norm $\leq k(1 + \varepsilon_1)$. $I - j_1 j^{-1} q$ is a projection to E_0 with norm close to $k2^{k-1} + 1 \leq 2^{2k-1}$. (e) follows from (d) and Proposition 2. (f): Picking a basis in E associates $L(E;E)$ with the matrix space R^{k^2}. If $T \to (a_{ij})$ let $\|T\|_1 = \max|a_{ij}|$. Clearly, $\| \| \sim \| \|_1 > |\det|$ and so $\|T^{-1}\| \sim \|T^{-1}\|_1 > 1/|\det(T)|$ on Lis. Conversely, Cramer's rule implies $\max(\|T\|_1, 1/|\det(T)|) > \|T^{-1}\|_1$. Q.E.D.

The dependence on $k = \dim E_0$ cannot be eliminated. The definition of $\theta(E_0)$ is due to Murray [17] who showed that ℓ_p contains nonsplit, closed subspaces ($p \neq 2$). That $\theta(E_0)$ can blow up with $\dim E_0$ follows from a result of Davis, Dean and Singer [3; Thms. 1,2]: Theorem: If E is a reflexive B-space then every closed subspace of E splits iff $\sup\{\theta(E_0)\} < \infty$ where E_0 varies over the finite dimensional subspaces of E.

I is a regular value of comp: $L(E;E_0) \times L(E_0;E) \to L(E_0;E_0)$. For if $P \circ j = I$, then P is a split surjection, j is a split injection and $D\,comp(P,j) = j^\# + P_\#$ is a split surjection. Note that $\max(\|P\|, \|j\|) = \max(\theta(P), \theta(j))$. Thus, $comp^{-1}(I)$ is a submanifold of the product and

p_1 (resp. p_2) maps $\text{comp}^{-1}(I)$ onto $Ls(E;E_0)$ (resp. $Li(E_0;E)$).

Let $P.j = I$ and $K(P)$ be the kernel of P. Identify $L(E_0;K(P))$ with $\{T \in L(E_0;E): PT = 0\}$ and $L(E/j(E_0);E_0)$ with $\{\bar{T} \in L(E;E_0): \bar{T}j = 0\}$ by the obvious isometric isomorphisms. Define affine maps $L(E_0;K(P)) \to L(E_0;E)$ and $\to L(E;E)$ by $T \to j_T = j + T$ and $\to Q_T = I + TP$. Since $PT = 0$, $Pj_T = I$ and $j_{(\ )}$ maps into $Li(E_0;E)$. In fact, it maps onto $(P_\#)^{-1}(I)$. $Q_{T_1+T_2} = Q_{T_1} \circ Q_{T_2}$ and so Q is a homomorphism of the additive group $L(E_0;K(P))$ into the multiplicative group $Lis(E)$. In particular, $Q_{-T} = (Q_T)^{-1}$. Also, $j_T = Q_T \circ j$.

(1.1) $\quad \theta(Q_T) \leq 1 + \|TP\| \leq 1 + \|T\|\|P\|, \quad \theta(j_T) \leq \theta(Q_T)\theta(j).$

Similarly, define affine maps $L(E/j(E_0);E_0) \to L(E;E_0)$ and $\to L(E;E)$ by $\bar{T} \to P_{\bar{T}} = P + \bar{T}$ and $\to \bar{Q}_T = I + j\bar{T}$. $P_{\bar{T}} \circ j = I$ and so $P_{(\ )}$ maps onto $(j^\#)^{-1}(I)$ in $Ls(E;E_0)$. $\bar{Q}$ is a group homomorphism with $P_{\bar{T}} = P \circ \bar{Q}_{\bar{T}}$.

(1.2) $\quad \theta(\bar{Q}_{\bar{T}}) \leq 1 + \|j\bar{T}\| \leq 1 + \|j\|\|\bar{T}\|, \quad \theta(P_{\bar{T}}) \leq \theta(\bar{Q}_{\bar{T}})\theta(P).$

7 PROPOSITION: Assume $P \circ j = I$ and let $1 > \epsilon > 0$.

(a) On the ball $B_\epsilon = B(j, \epsilon/\max(\|P\|,1))$ in $L(E_0;E)$ there is an analytic function $J \times T: B_\epsilon \to Lis(E_0) \times L(E_0;K(P))$ characterized by $(J,T) = (J(j_1),T(j_1))$ iff $j_1 = j_T J = Q_T jJ$. $\theta(J) < (1 - \epsilon)^{-1}$ and for any standard function space type $\mathfrak{M}$, $J \times T \in \mathfrak{M}(B_\epsilon, L(E_0;E_0) \times L(E_0;K(P)))$ with $\|J \times T\|_{\mathfrak{M}} \leq O^*(\|j\|,\|P\|, (1 - \epsilon)^{-1})$. $J^{-1}P$ is a left inverse for j_1 and so $\theta(j_1) \leq (1 - \epsilon)^{-1}\max(\|P\|,1)$.

(b) On the ball $\bar{B}_\epsilon = B(P;\epsilon/\max(\|j\|,1))$ in $L(E;E_0)$ there is an analytic function $\bar{J} \times \bar{T}: \bar{B}_\epsilon \to Lis(E_0) \times L(E/j(E_0);E_0)$ characterized by $(\bar{J},\bar{T}) = (\bar{J}(P_1),\bar{T}(P_1))$ iff $P_1 = \bar{J}P_{\bar{T}} = \bar{J}P\bar{Q}_{\bar{T}}$. $\theta(\bar{J}) \leq (1 - \epsilon)^{-1}$ and $\bar{J} \times \bar{T} \in \mathfrak{M}(B_\epsilon, L(E_0;E_0) \times L(E/j(E_0);E_0))$ with $\|\bar{J} \times \bar{T}\|_{\mathfrak{M}} \leq O^*(\|j\|,\|P\|,(1 - \epsilon)^{-1})$. $j\bar{J}^{-1}$ is a right inverse for P_1 and so $\theta(P_1) \leq (1 - \epsilon)^{-1}\max(\|j\|,1)$.

PROOF: (a): $j_1 = j_T J$ iff $J = Pj_1$ and $T = j_1 J^{-1} - j$. If $j_1 \in B_\epsilon$ then $\|I - Pj_1\| < \epsilon$ and so $J \in Lis(E_0)$ with $\theta(J) < (1 - \epsilon)^{-1}$. The rest is obvious. (b) is similar. Q.E.D.

8 COROLLARY: Let $S: E_1 \to E_2$ be a split injection (or split surjection) and $1 > \epsilon > 0$. If $\|S - S_1\| < \epsilon\theta(S)^{-1}$ then S_1 is a split injection (resp. split surjection) with $\theta(S_1) < (1 - \epsilon)^{-1}\theta(S)$. Thus, θ is upper semicontinuous on $L(E_1;E_2)$ and induces a semicomplete C^∞ refinement on the open set $\{\theta < \infty\} = Li \cup Ls$. The group $Lis(E_2) \times Lis(E_1)$ acts on $Li \cup Ls(E_1;E_2)$ by $(T_2,T_1,S) \to T_2 S T_1^{-1}$. The orbits of the action are open and closed in $Li \cup Ls$. In particular, as unions of orbits, Li and Ls are open and closedin $Li \cup Ls$. $Lis = Li \cap Ls$ is either empty or a single orbit.

PROOF: If $S = j$ and $S = j_1$ choose P with $Pj = I$ and $\|j - j_1\| < \epsilon\theta(P)^{-1}$. Apply Proposition 8 (a) and let $\max(\|P\|,\|j\|)$ approach $\theta(S)$ to get the estimate on $\theta(j_1)$. By Proposition 8 (a) the ball lies in a single orbit. Similarly, for $S = P \in Ls$. Thus, the orbits are open and θ induces a semicomplete refinement by Lemma III.7.9. Q.E.D.

9 PROPOSITION: Let $S: E_1 \to E_2$ be an injection (or a surjection) and $0 < \epsilon < 1$. If $\|S - S_1\| < \epsilon\bar{\theta}(S)^{-1}$ then S_1 is an injection (resp. surjection) and $\bar{\theta}(S_1) \leq (1 - \epsilon)^{-1}\theta(S)$. Thus, $\bar{\theta}$ is upper semicontinuous on $L(E_1;E_2)$ and induces a semicomplete C^∞ refinement on the open set $\{\bar{\theta} < \infty\} = L\bar{i} \cup L\bar{s}$. $L\bar{i}$ and $L\bar{s}$ are each open and closed in the union and so $Lis = L\bar{i} \cap L\bar{s}$ is open and closed in $L\bar{i} \cup L\bar{s}$.

PROOF: If j is an injection and $\|j - j_1\| < m(j)$, then $\|j_1(v)\| \geq m(j)\|v\| - \|j - j_1\|\|v\|$ and so $m(j_1) \geq m(j) - \|j - j_1\|$. If T is a surjection and $\|T - T_1\| < m(T)$, then let $t = m(T)^{-1}$, $\delta = \|T - T_1\|$ and $\epsilon = m(T) - \|T - T_1\|$. Let $v \in E_2$ with $\|v\| < \epsilon$. There exists

$u_1 \in E_1$ with $\|u_1\| < t\epsilon$ and $Tu_1 = v$. Hence, $\|v - T_1u_1\| \leq \delta t\epsilon$. Inductively, choose $u_n \in E_1$ with $\|u_n\| < \delta^{n-1}t^n\epsilon$ and $T_1u_n = v - T_1(u_1 +\ldots+ u_{n-1})$. Then $\|v - T_1(u_1 +\ldots+ u_n)\| < \delta^n t^n \epsilon$. $u = \Sigma u_i$ exists and $\|u\| < t\,\epsilon\,\Sigma(t\delta)^n = t\,\epsilon\,(1 - t\delta)^{-1} = 1$ and $T_1u = v$. Hence, $m(T_1) \geq \epsilon$. Thus, $L\bar{i}$ and $L\bar{s}$ are open and $\bar{\theta} = \max(\| \;\|, m^{-1})$ induces a semicomplete refinement on each of them and on the union. Each is then open and closed in the union by Corollary III.7.8 (a) applied to $X = Li \cup Ls$ with admissible bound $\bar{\theta}$. Q.E.D.

2. <u>VB Injections and Surjections</u>: Let $\pi: \mathbb{E} \to X$ be a metric bundle with admissible atlas $(\mathfrak{N},\mathfrak{G}) = \{U_\alpha, h_\alpha, \varphi_\alpha\}$ and let O be open in $\mathbb{E}$. $(\mathfrak{N},\mathfrak{G})$ <u>induces an open subbundle atlas</u> $(\mathfrak{N}|O,\mathfrak{G})$ on $\pi|O$ iff for all α, there is an open set O_α in F_α such that $\varphi_\alpha(O \cap \pi^{-1}U_\alpha) = h_\alpha(U_\alpha) \times O_\alpha$. $\pi|O$ is an <u>open subbundle</u> of π if such admissible $(\mathfrak{N},\mathfrak{G})$ exist. Admissible atlases inducing the open subbundle are called <u>admissible atlases for $\pi|O$</u>. The open subbundle is called semicomplete if π is semicomplete and there are s-admissible atlases for $\pi|O$, i.e. admissible atlases for $\pi|O$ which are s-admissible for π.

1 <u>PROPOSITION</u>: Let $\pi: \mathbb{E} \to X$ be a semicomplete metric bundle with admissible Finsler and bound $\| \;\|$ and ρ. Let $d_{\mathbb{E}}$ be an admissible metric on $\mathbb{E}$. If $\pi|O$ is a semicomplete open subbundle, then $\max(\pi^*\rho, \| \;\|, 1/\delta^O(d)) \sim \max(\pi^*\rho, \| \;\|, 1/\delta^O(\| \;\|))$.

<u>PROOF</u>: Let $(\mathfrak{N},\mathfrak{G})$ be an s-admissible atlas for $\pi|O$. Let $\| \;\|^m$ be the associated min Finsler. $\lambda_{\mathfrak{N}|O} \geq \min(\pi^*\lambda_{\mathfrak{G}}, \delta^O(\| \;\|^m))$. For if $v \in O$ and $\epsilon < \lambda_{\mathfrak{G}}(\pi v)$, $\delta^O(\| \;\|^m)(v)$ then there exists an α with $B(h_\alpha \pi v, \epsilon) \subset h_\alpha(U_\alpha)$. Also, $B^{\| \;\|^m}(v,\epsilon) \subset O$ (ball in the fiber $\mathbb{E}_{\pi v}$). Hence, $B(p_2\varphi_\alpha v, \epsilon) \subset O_\alpha$. Thus, $\varphi_\alpha(O \cap \pi^{-1}U_\alpha) = h_\alpha(U_\alpha) \times O_\alpha \supset B(h_\alpha\pi v,\epsilon) \times B(p_2\varphi_\alpha v,\epsilon) = B(\varphi_\alpha v,\epsilon)$. Hence, $\max(\pi^*\rho, \| \;\|^m, 1/\delta^O(\| \;\|^m)) > \max(\pi^*\rho, \| \;\|^m, 1/\lambda_{\mathfrak{N}|O})$. The result now

follows from independence of Finsler choice, Proposition III.7.7 and Lemma III.7.14. Q.E.D.

It follows that if $\pi|O$ is a semicomplete open subbundle then the vertical fullness question has a positive answer. (cf. p. 98), i.e. $A \subset\subset O$ iff $\max(\pi^*\beta, \|\ \|)|A \sim \max(\pi^*\beta, \|\ \|, 1/\delta^O(\|\ \|))$. Recall the discussion around Lemmas III.7.13 and III.7.14.

2 <u>LEMMA</u>: Let $\pi: \mathbb{E} \to X$ be a metric bundle with admissible atlas $(\mathfrak{N},\mathfrak{G}) = \{U_\alpha, h_\alpha, \varphi_\alpha\}$ and admissible bound β. Assume that on each B-space F_α there is a map $N_\alpha: F_\alpha \to [1,\infty]$. Assume that: (1) $N_\alpha >_1 N_\alpha^{\|\ \|_\alpha}$ uniformly in α. (2) $N_\alpha >_{\beta(x)} \varphi_{\beta\alpha}(h_\alpha x)^* N_\beta$ uniformly in x,α,β with $x \in U_\alpha \cap U_\beta$. Define $N^M, N^m: \mathbb{E} \to [1,\infty]$ by $N^M(v) = \sup$ (resp. $N^m(v) = \inf$) $\{N_\alpha(p_2\varphi_\alpha(v)): \pi v \in U_\alpha\}$. $N^M \sim_\beta N^m$ and N^M, N^m are vertically regular. $O = \{N^M < \infty\} = \{N^m < \infty\}$ is open and $(\mathfrak{N},\mathfrak{G})$ induces an open subbundle atlas on $\pi|O$. If π is semicomplete and $(\mathfrak{N},\mathfrak{G})$ is s admissible then $\pi|O$ is semicomplete and $\max(\pi^*\beta, \|\ \|^M, N^M) \sim \max(\pi^*\beta, \|\ \|^m, N^m)$ induce a semicomplete refinement on O.

<u>PROOF</u>: (1) and (2) mean there exist constants $K, n \geq 1$ such that $K\beta(x)^n N_\alpha \geq N_\beta \circ (\varphi_{\beta\alpha}(h_\alpha x))$ whenever $x \in U_\alpha \cap U_\beta$, and $\|u_1 - u_2\|_\alpha < (KN_\alpha(u_1))^{-1}$ implies $KN_\alpha(u_1) \geq N_\alpha(u_2)$ for $u_1, u_2 \in F_\alpha$. It follows that $K(\pi^*\beta)^n N^m \geq N^M$, i.e. $N^m \sim_\beta N^M$. If $v_1, v_2 \in \mathbb{E}_x$ and $\|v_1 - v_2\|^M < (KN^M(v_1))^{-1}$ then $\|v_1 - v_2\|^\alpha < [KN_\alpha(p_2\varphi_\alpha(v_1))]^{-1}$ and so $KN^M(v_1) \geq N_\alpha(p_2\varphi_\alpha(v_2)) \geq N^m(v_2)$. Since $N^m \sim_\beta N^M$, they are both vertically regular. $\varphi_\alpha(\{N^M < \infty\}) = \varphi_\alpha(\{N^m < \infty\}) = h_\alpha(U_\alpha) \times \{N_\alpha < \infty\}$ is open by (1). Hence, $(\mathfrak{N},\mathfrak{G})$ induces an open subbundle atlas on $\pi|O$. In the semicomplete case the Regularity Lemma III.7.3 shows that $(\mathfrak{N}|O, \max(\pi^*\beta, \|\ \|^M, N^M))$ is a semicomplete adapted atlas. Q.E.D.

3 <u>PROPOSITION</u>: Let $\pi_i: \mathbb{E}_i \to X$ be a metric bundle $(i = 1,2)$ and let

$(\mathfrak{T}_1, \mathfrak{T}_2; \mathfrak{A}) = \{U_\alpha, h_\alpha, \varphi^1_\alpha, \varphi^2_\alpha\}$ be an admissible two-tuple for π_1, π_2. $(L(\mathfrak{T}_1; \mathfrak{T}_2), \mathfrak{A}) = \{U_\alpha, h_\alpha, L(\varphi^1, \varphi^2)_\alpha\}$ is an admissible atlas on the linear map bundle $L(\pi_1; \pi_2): L(\mathbb{E}_1; \mathbb{E}_2) \to X$. Identifying $L(\mathbb{E}_1; \mathbb{E}_2)_x = L(\mathbb{E}_{1x}; \mathbb{E}_{2x})$ for $x \in X$ we define: $L\bar{i}(\mathbb{E}_1; \mathbb{E}_2) = \{\varphi$ is an injection$\}$, $Li(\mathbb{E}_1; \mathbb{E}_2) = \{\varphi$ is a split injection$\}$, $L\bar{s}(\mathbb{E}_1; \mathbb{E}_2) = \{\varphi$ is a surjection$\}$, $Ls(\mathbb{E}_1; \mathbb{E}_2) = \{\varphi$ is a split surjection$\}$ and $Lis(\mathbb{E}_1; \mathbb{E}_2) = \{\varphi$ is an isomorphism$\}$. $(L(\mathfrak{T}_1; \mathfrak{T}_2), \mathfrak{A})$ induces an open subbundle atlas on each of these sets. Defining θ^M, θ^m, $\bar{\theta}^M$ and $\bar{\theta}^m$ as in Lemma 2 using θ_α and $\bar{\theta}_\alpha$ on $L(F^1_\alpha; F^2_\alpha)$, $\theta^M \sim_\rho \theta^m$ and $\bar{\theta}^M \sim_\rho \bar{\theta}^m$. All four are vertically regular. $\theta^M = \bar{\theta}^M$ and $\theta^m = \bar{\theta}^m$ on Lis. If π_1, π_2 are semicomplete then, with $\pi = L(\pi_1; \pi_2)$, $\max(\pi^*\rho, \theta^M) \sim \max(\pi^*\rho, \theta^m)$ induce semicomplete refinements on Li, Ls and Lis and $\max(\pi^*\rho, \bar{\theta}^M) \sim \max(\pi^*\rho, \bar{\theta}^m)$ induce semicomplete refinements on $L\bar{i}$ and $L\bar{s}$.

<u>PROOF</u>: Consider Li. Define $N_\alpha = \theta_\alpha = \theta$ on $Li(F^1_\alpha; F^2_\alpha)$ and $= \infty$ on L - Li. Condition (1) of Lemma 2 holds by Corollary 1.8. Let $j_\alpha \in Li(F^1_\alpha; F^2_\alpha)$ and $j_\beta = L(\varphi^1; \varphi^2)_{\beta\alpha}(h_\alpha x)(j_\alpha) = \varphi^2_{\beta\alpha}(h_\alpha x) \cdot j_\alpha \cdot (\varphi^1_{\beta\alpha}(h_\alpha x))^{-1}$. $N_\beta[j_\beta] \leq \rho_{(\mathfrak{T}_1, \mathfrak{A})}(x) \rho_{(\mathfrak{T}_2, \mathfrak{A})}(x) N_\alpha(j_\alpha)$ by Proposition 1.1 (c), since $\rho^{\mathcal{C}}_{(\mathfrak{T}_1, \mathfrak{A})}(x) \geq \theta(\varphi^1_{\beta\alpha}(h_\alpha x))$. Condition (2) of Lemma 2 follows. Clearly, $N^M = \theta^M$ and $N^m = \theta^m$ on Li. Note that $\theta^M \geq \| \ \|^M$. The remaining arguments are similar applications of Lemma 2. Q.E.D.

<u>ADDENDUM</u>: If $\| \ \|_1$ and $\| \ \|_2$ are admissible Finslers on π_1 and π_2 then with these norms $\mathbb{E}_{1x}$ and $\mathbb{E}_{2x}$ are B-spaces. So θ_{12} and $\bar{\theta}_{12}$ can be defined on $L(\mathbb{E}_1; \mathbb{E}_2)_x = L(\mathbb{E}_{1x}; \mathbb{E}_{2x})$. It is easy to check that the $\sim_\rho$ equivalence classes of θ_{12} and $\bar{\theta}_{12}$ are independent of the Finsler choices and with $\| \ \|_i = \| \ \|^M$ induced by $(\mathfrak{T}_i, \mathfrak{A})$, $\theta_{12} \sim_\rho \theta^M$ and $\bar{\theta}_{12} \sim_\rho \bar{\theta}^M$. Thus, the $\sim_\rho$ classes of θ^M and $\bar{\theta}^M$ are independent of the atlas choices. In general, if atlases or Finslers are not specified we will denote by θ (or $\bar{\theta}$) any element of the $\sim_\rho$ equivalence class of θ^M (resp. $\bar{\theta}^M$).

Let $(\Phi,1)\colon \pi_1 \to \pi_2$ be an $\mathfrak{M}$ VB map. $(\Phi,1)$ is called an <u>$\mathfrak{M}$ VB injection</u> or <u>metric VB injection</u> (resp. <u>$\mathfrak{M}$ VB surjection</u> or <u>metric VB surjection</u>) if there exists an admissible two-tuple $(\mathfrak{D}_1,\mathfrak{D}_2;\mathfrak{a}) = \{U_\alpha,h_\alpha,\varphi_\alpha^1,\varphi_\alpha^2\}$ with F_α^1 a closed subspace of F_α^2 (resp. F_α^2 a quotient space of F_α^1) and the principal part $\Phi_\alpha\colon h_\alpha(U_\alpha) \to L(F_\alpha^1;F_\alpha^2)$ constant at the inclusion (resp. at the quotient map). $(\Phi,1)$ is called <u>locally split</u> if $F_\alpha^2 = F_\alpha^1 \times \tilde{F}_\alpha$ and Φ_α is constant at the inclusion of the first factor (resp. $F_\alpha^1 = \tilde{F}_\alpha \times F_\alpha^2$ and Φ_α is constant at the projection to the second factor). $(\Phi,1)$ is called <u>semicomplete</u> if π_1 and π_2 are semicomplete and the two-tuple can be chosen s admissible. In any of these cases a two-tuple displaying the appropriate properties is called an <u>admissible two-tuple</u> for $(\Phi,1)$. $(\Phi,1)$ is an $\mathfrak{M}$ VB injection iff it is an $\mathfrak{M}$ VB isomorphism onto an $\mathfrak{M}$ subbundle of π_2. In fact, if $(\mathfrak{D}_1,\mathfrak{D}_2;\mathfrak{a})$ is admissible for $(\Phi,1)$ then $(\mathfrak{D}_2,\mathfrak{a})$ induces a subbundle atlas on the image. If $(\Phi,1)$ is an $\mathfrak{M}$ VB surjection then the kernel $K(\Phi)$ is an $\mathfrak{M}$ subbundle of π_1 with $(\mathfrak{D}_1,\mathfrak{a})$ inducing a subbundle atlas. If $f\colon X_0 \to X$ is an $\mathfrak{M}$ map then $(f^*\Phi,1)\colon f^*\pi_1 \to f^*\pi_2$ is an $\mathfrak{M}$ VB injection (resp. $\mathfrak{M}$ VB surjection) if $(\Phi,1)$ is. If $(\mathfrak{D}_1,\mathfrak{D}_2;\mathfrak{a})$ is admissible for $(\Phi,1)$ and $f\colon \mathfrak{a}_0 \to \mathfrak{a}$ is index preserving, then $(f^*\mathfrak{D}_1,f^*\mathfrak{D}_2;\mathfrak{a}_0)$ is admissible for $(f^*\Phi,1)$. Clearly, $(f^*\Phi,1)$ is locally split iff $(\Phi,1)$ is. If $(\Phi,1)$ and X_0 are semicomplete and $\mathfrak{M} \subset \mathscr{L}$ then $(f^*\Phi,1)$ is semicomplete. $(\Phi,1)$ is an $\mathfrak{M}$ VB isomorphism iff it is an $\mathfrak{M}$ VB injection or surjection which is an isomorphism on fibers. An $\mathfrak{M}$ VB isomorphism is trivally locally split.

It is clear that if $(\Phi,1)$ is an $\mathfrak{M}$ VB injection/surjection/locally split injection/locally split surjection then the $\mathfrak{M}$ section Φ of $L(\pi_1;\pi_2)$ maps into $L\bar{i}/L\bar{s}/Li/Ls$ respectively with $\rho > \Phi^*\bar{\theta}$ in the first two cases and $\rho > \Phi^*\theta$ in the latter two. In fact, if $\bar{\theta}^m,\theta^m$ are defined by an admissible two-tuple for $(\Phi,1)$ then $\Phi^*\bar{\theta}^m = 1$ and in the locally split cases $\Phi^*\theta^m = 1$. The importance of assuming locally split comes from the following converse.

4 <u>THEOREM</u>: Let $(\Phi,1)\colon \pi_1 \to \pi_2$ be an $\mathfrak{M}$ VB map. If the section Φ of $L(\pi_1;\pi_2)$ maps into $Li(\mathbb{E}_1;\mathbb{E}_2)$ (resp. $Ls(\mathbb{E}_1;\mathbb{E}_2)$) and $\rho > \Phi^*\theta$, then $(\Phi,1)$ is an $\mathfrak{M}$ locally split VB injection (resp. $\mathfrak{M}$ locally split VB surjection). If π_1 and π_2 are semicomplete and $\mathfrak{M} \subset \mathcal{L}$, then $(\Phi,1)$ is semicomplete.

<u>PROOF</u>: We will do the semicomplete injection case. The rest are similar. Let $(\bar{\mathfrak{D}}_1,\bar{\mathfrak{D}}_2;\bar{\mathfrak{A}}) = \{U_\alpha,h_\alpha,\bar{\varphi}^1_\alpha,\bar{\varphi}^2_\alpha\}$ be an s admissible two-tuple. Let $d = d_{\bar{\mathfrak{A}}}$ and define θ^M and $\tilde{\rho}_\Phi$ using the two-tuple. Fix $x \in X$ and let $\epsilon^{-1} = 3\max(1/\lambda_{\bar{\mathfrak{A}}},\Phi^*\theta^M,\tilde{\rho}_\Phi,\rho_{(\bar{\mathfrak{D}}_1,\bar{\mathfrak{A}})},\rho_{(\bar{\mathfrak{D}}_2,\bar{\mathfrak{A}})})^d(x)$. Choose $\alpha = \alpha(x)$ with $B = B(h_\alpha x,\epsilon^2) \subset h_\alpha(U_\alpha)$. $\|\Phi_\alpha|B\|_{\mathfrak{M}} < \epsilon^{-1}$ and $\epsilon < [2\theta(\Phi_\alpha(h_\alpha x))]^{-1}$. Choose P_x a left inverse for $\Phi_\alpha(h_\alpha x)$ with $\epsilon < [2\max(\|P_x\|,1)]^{-1}$. For $u \in B$, $\|\Phi_{\beta\alpha}(u) - \Phi_{\beta\alpha}(h_\alpha x)\| < \epsilon$. Note that $(1-\epsilon)^{-1} < 2$. Hence, by Proposition 1.7 we can define functions $J \in \mathfrak{M}(B,L(F^1_\alpha;F^1_\alpha))$, $Q \in \mathfrak{M}(B,L(F^2_\alpha;F^2_\alpha))$ by $J(u) = J(\Phi_\alpha(u))$ and $Q(u) = Q_{T(\Phi_\alpha(u))}$ so that $\Phi_\alpha(u) = Q(u)\cdot\Phi_\alpha(h_\alpha x)\circ J(u)$. $\theta(J(u)) \le 2$ and $\theta(Q(u)) \le O^*(\epsilon^{-1})$. By FS9 and the estimates in Proposition 1.7 and (1.1), $\|J\|_{\mathfrak{M}},\|inv\circ J\|_{\mathfrak{M}}, \|Q\|_{\mathfrak{M}}, \|inv\circ Q\|_{\mathfrak{M}} \le O^*(\epsilon^{-1})$. Let $U_x = h^{-1}_\alpha(B)$. Let $\mathfrak{A}$ be the subdivision $\{U_x,h_{\alpha(x)}\}$ of $\bar{\mathfrak{A}}$ and let $(\bar{\bar{\mathfrak{D}}}_1,\bar{\bar{\mathfrak{D}}}_2;\mathfrak{A})$ be the corresponding subdivision of $(\bar{\mathfrak{D}}_1,\bar{\mathfrak{D}}_2;\bar{\mathfrak{A}})$ now define $(\mathfrak{D}_1,\mathfrak{D}_2;\mathfrak{A}) = \{U_x,h_{\alpha(x)},\varphi^1_x,\varphi^2_x\}$ with $F^1_x = \Phi_\alpha(h_\alpha x)(F^1_\alpha)$ and $F^2_\alpha = F^1_x \times K(P_x)$, so that the U_x principal part of $(I^1,1)\colon (\bar{\bar{\mathfrak{D}}}_1,\mathfrak{A}) \to (\mathfrak{D}_1,\mathfrak{A})$ is given by $I^1_x(u) = \Phi_\alpha(h_\alpha x)\cdot J(u)$ and of $(I^2,1)\colon (\bar{\bar{\mathfrak{D}}}_2,\mathfrak{A}) \to (\mathfrak{D}_2,\mathfrak{A})$ is given by $I^2_x(u) = [(\Phi_\alpha(h_\alpha x)\cdot P_x) \times (I - P_x)]\circ Q(u)^{-1}$. Since $\epsilon(x)^{-1} \le 2\epsilon(y)^{-1}$ for $y \in U_x \subset B^d(x,3^{-1}\epsilon(x))$, $(\mathfrak{D}_1,\mathfrak{D}_2;\mathfrak{A})$ is an s admissible atlas for $(\Phi,1)$ with:

$$\text{(2.1)}\qquad \max(1/\lambda_{\mathfrak{A}},\rho_{(\mathfrak{D}_1,\mathfrak{A})\cup(\bar{\mathfrak{D}}_1,\bar{\mathfrak{A}})},\rho_{(\mathfrak{D}_2,\mathfrak{A})\cup(\bar{\mathfrak{D}}_2,\bar{\mathfrak{A}})})$$
$$\le O^*[1/\lambda_{\bar{\mathfrak{A}}},(\Phi^*\theta^M)^d,(\tilde{\rho}_\Phi)^d,(\rho_{(\bar{\mathfrak{D}}_1,\bar{\mathfrak{A}})})^d,(\rho_{(\bar{\mathfrak{D}}_2,\bar{\mathfrak{A}})})^d].$$

Q.E.D.

Let X be a semicomplete $\mathcal{L}$ manifold, π_1 and π_2 be $\mathcal{C}_u$ bundles over X and Φ be a $\mathcal{C}_u$ section, i.e. $(\Phi,1)$ is a $\mathcal{C}_u$ VB map. If $(\Phi,1)$ is a locally split VB injection (resp. surjection) then $(\Phi,1)$ is semicomplete. The proof is essentially the same. One begins with an admissible $\mathcal{C}_u$ convex , two-tuple $(\bar{\mathfrak{D}}_1,\bar{\mathfrak{D}}_2;\bar{\mathfrak{a}})$ with $\bar{\mathfrak{a}}$ a $\mathcal{L}$ip atlas and replaces the use of $\tilde{\mathfrak{p}}^{\mathcal{L}}_{\Phi}$ by uniform continuity of $\{\Phi_\alpha\}$. One loses the estimate of $1/\lambda_{\mathfrak{a}}$ in (2.1). In general, Theorem 4 implies that if $(\Phi,1)$ is an $\mathfrak{M}$ VB map and a locally split $\mathcal{C}$ VB injection (resp. surjection) then $(\Phi,1)$ is a locally split $\mathfrak{M}$ VB injection (resp. surjection), i.e. if $i\colon \mathfrak{M}(L(\pi_1;\pi_2)) \to \mathcal{C}(L(\pi_1;\pi_2))$ is the inclusion:

$$(2.2)\qquad \mathfrak{M}(Li(\mathbb{E}_1;\mathbb{E}_2)) = i^{-1}\mathcal{C}(Li(\mathbb{E}_1;\mathbb{E}_2)) \quad\text{and}\quad \mathfrak{M}(Ls(\mathbb{E}_1;\mathbb{E}_2)) = i^{-1}\mathcal{C}(Ls(\mathbb{E}_1;\mathbb{E}_2)).$$

5 <u>COROLLARY</u>: Let $\pi_i\colon \mathbb{E}_i \to X$ be $\mathfrak{M}$ bundles $(i = 1,2)$. The subset $\mathfrak{M}(Li(\mathbb{E}_1;\mathbb{E}_2))$ (resp. $\mathfrak{M}(Ls(\mathbb{E}_1;\mathbb{E}_2))$) of $\mathfrak{M}(L(\pi_1;\pi_2))$ consisting of locally split $\mathfrak{M}$ VB injections (resp. locally split $\mathfrak{M}$ VB surjections) is open in $\mathfrak{M}(L(\pi_1;\pi_2))$. If π_1 and π_2 are bounded, $\|\ \|$ is a norm on $\mathfrak{M}(L(\pi_1;\pi_2))$ and θ is defined as usual on $Li \cup Ls$, then $\max(\|\ \|,S_\theta)$ induces a semicomplete refinement on $\mathfrak{M}(Li(\mathbb{E}_1;\mathbb{E}_2))$ (resp. $\mathfrak{M}(Ls(\mathbb{E}_1;\mathbb{E}_2))$).

<u>PROOF</u>: Letting $N_i = \theta$ on Li and $= \infty$ on $L - Li$, the result follows from Proposition 3 and Proposition IV.1.4 in the bounded case. Similarly, for Ls. The unbounded case follows from the analogue of Proposition IV.1.4. Q.E.D.

In the bounded case, if $(\mathfrak{D}_1,\mathfrak{D}_2;\mathfrak{a})$ is an admissible two-tuple for π_1, π_2 and $\theta = \theta^M$ and $\|\ \|$ are defined by $(L(\mathfrak{D}_1,\mathfrak{D}_2),\mathfrak{a})$, then it is not hard to show that

$$(2.3)\qquad \max(\|\ \|,S_\theta)^{d_{\|\ \|}} \le O^*(k_{(\mathfrak{D}_1,\mathfrak{a})},k_{(\mathfrak{D}_2,\mathfrak{a})},\|\ \|,S_\theta),$$

and the same estimate holds with θ replaced by N_i or N_s.

6 <u>COROLLARY</u>: Let π_1, π_2 be semicomplete $\mathfrak{m}$ bundles over X. Let $(\phi,1): \pi_1 \to \pi_2$ be an $\mathfrak{m}$ VB map. If $A \subset X$ and the section ϕ of $L(\pi_1;\pi_2)$ maps A into $Li(\mathbb{E}_1;\mathbb{E}_2)$ (resp. into $Ls(\mathbb{E}_1;\mathbb{E}_2)$) with $\rho > \phi^*\theta$ on A then there exists U open in X with $A \subset\subset U$ such that $(\phi,1)_U: (\pi_1)_U \to (\pi_2)_U$ is a locally split $\mathfrak{m}$ VB injection (resp. locally split $\mathfrak{m}$ VB surjection). $(\phi,1)_U$ is semicomplete with respect to the canonical refinement structure.

<u>PROOF</u>: Let $\| \ \|$ be an admissible Finsler on $L(\pi_1;\pi_2)$ and let $N_i = \theta$ on Li and $= \infty$ on L - Li. By hypothesis $\max(\pi^*\rho, N_i) \sim \max(\pi^*\rho, \| \ \|)$ on $\phi(A)$ in $Li(\mathbb{E}_1;\mathbb{E}_2)$. By Lemma III.7.10 there exists V open in $Li(\mathbb{E}_1;\mathbb{E}_2)$ with $\phi(A) \subset\subset V$ and $\max(\pi^*\rho, N_i) \sim \max(\pi^*\rho, \| \ \|)$ on V. Let $U = \phi^{-1}(V)$. On U, $\max(\rho, \phi^*N_i) \sim \max(\rho, \phi^*\| \ \|) \sim \rho$ since ϕ is a metric section. Since ϕ is $\mathcal{L}$, $A \subset\subset U$. $(\phi,1)_U$ is a locally split $\mathfrak{M}$ VB injection by Theorem 4. Q.E.D.

Alternatively, we can repeat the first part of the proof of Theorem 4 using points $x \in A$ and defining U to be the union of $h_\alpha^{-1}(B)$ over all $x \in A$. With ρ assumed $\geq \rho_{(\mathfrak{D}_i, \mathfrak{G})}$ $(i = 1,2)$ and $\geq 1/\lambda_{\bar{\mathfrak{G}}}$ we choose $\epsilon^{-1} = 3\max(\phi^*\theta^M, (\tilde{\rho}_\phi)^d, \rho^d)(x)$. If A is bounded then we get that U is bounded with

$$(2.4) \quad \max(\sup(1/\lambda_{\bar{\mathfrak{G}}|U})|A, \sup \phi^*\theta^M|U, \sup \rho^d|U)$$
$$\leq O^*(\sup \phi^*\theta^M|A, \sup(\tilde{\rho}_\phi)^d|A, \sup \rho^d|A).$$

Again, if $(\phi,1)$ is a $\mathcal{C}_u$ VB map then Corollary 6 still holds. In (2.3) we lose the estimate on $\sup(1/\lambda_{\bar{\mathfrak{G}}|U})|A$.

An $\mathfrak{M}$ VB map $(\phi,1)$ is called a <u>split $\mathfrak{M}$ VB injection</u> (resp. <u>split $\mathfrak{M}$ VB surjection</u>) if it has a left inverse (resp. right inverse). Let $(\psi,1): \pi_2 \to \pi_1$ with $\psi\phi = I$. If we define $\| \ \|_{12}$, θ_{12} on $L(\pi_1;\pi_2)$ and $\| \ \|_{21}$, θ_{21} on $L(\pi_2;\pi_1)$ by Finslers $\| \ \|_1$ and $\| \ \|_2$ on π_1 and π_2, then

$\theta > \max(\Phi^*\| \|_{12},\Psi^*\| \|_{21}) = \max(\Phi^*\theta_{12},\Psi^*\theta_{21})$. Hence, by Theorem 4, $(\Phi,1)$ (resp. $(\Psi,1)$) is a locally split $\mathfrak{M}$ VB injection (resp. surjection). The split injections are open in $\mathfrak{M}(\mathrm{Li}(\mathbb{E}_1;\mathbb{E}_2))$ because if $(\Psi\circ\Phi,1)$ is an $\mathfrak{M}$ VB isomorphism, then $(\Phi,1)$ is a split $\mathfrak{M}$ injection. Thus, if $\Psi\circ\Phi = I$, $(\Psi_*)^{-1}\mathfrak{M}(\mathrm{Lis})(\mathbb{E}_1;\mathbb{E}_1))$ is an open neighborhood of Φ in $\mathfrak{M}(L(\pi_1;\pi_2))$ and consists of split injections. Similarly, for split surjections.

Let $\pi_0\colon \mathbb{E}_0 \to X_0$ and $\pi\colon \mathbb{E} \to X$ be $\mathfrak{M}$ bundles and $(\Phi,f)\colon \pi_0 \to \pi$ be an $\mathfrak{M}$ VB map. Let $\| \|_0$ and $\| \|$ be admissible Finslers on π_0 and π. $\Phi_f^*\| \|$ is an admissible Finsler on $f^*\pi$. If $(f^*\Phi,1)\colon \pi_0 \to f^*\pi$ is a locally split $\mathfrak{M}$ injection (or surjection) then $\Phi_x\colon \mathbb{E}_{0x} \to \mathbb{E}_{fx}$ is a split injection (resp. surjection) for all x and using $\| \|_0$ and $\| \|$ we can define $\theta(\Phi_x)$. On the other hand, $\| \|_0$ and $\Phi_f^*\| \|$ define θ on the total space of $L(\pi_0;f^*\pi)$. Clearly, the latter θ evaluated at $(f^*\Phi)(x)$ (the value of the section at x) is the same as $\theta(\Phi_x)$. If π_0 is bounded, then we can define $S_\theta(\Phi,f)$ to be the sup as x varies on X_0 of either of these functions.

7 PROPOSITION: Assume $(\mathfrak{M}_1,\mathfrak{M},\mathfrak{M}_2)$ is a standard triple. Let π be a semicomplete $\mathfrak{M}_1$ bundle admitting fiber exponentials and π_0 be a bounded $\mathfrak{M}$ bundle. Define: $\mathfrak{M}\mathcal{L}i(\mathbb{E}_0;\mathbb{E}) = \{(\Phi,f)\colon (f^*\Phi,1)$ is a locally split $\mathfrak{M}$ VB injection$\}$, $\mathfrak{M}\mathcal{L}s(\mathbb{E}_0;\mathbb{E}) = \{(\Phi,f)\colon (f^*\Phi,1)$ is a locally split $\mathfrak{M}$ VB surjection$\}$, $\mathfrak{M}\mathcal{L}ix(\mathbb{E}_0;\mathbb{E}) = \{(\Phi,f)\colon (f^*\Phi,1)$ is a split $\mathfrak{M}$ injection$\}$, $\mathfrak{M}\mathcal{L}sx(\mathbb{E}_0;\mathbb{E}) = \{(\Phi,f)\colon (f^*\Phi,1)$ is a split $\mathfrak{M}$ surjection$\}$ and $\mathfrak{M}\mathcal{L}is(\mathbb{E}_0;\mathbb{E}) = \{(\Phi,f)\colon (f^*\Phi,1)$ is an $\mathfrak{M}$ VB isomorphism$\}$. The $\mathfrak{M}_2$ bundle $\mathfrak{M}\mathcal{L}(\pi_0;\pi)$ restricted to each of these is a semicomplete open subbundle. Define S_θ on $\mathfrak{M}\mathcal{L}(\mathbb{E}_0;\mathbb{E})$ using admissible Finslers on π_0 and π with $S_\theta = \infty$ on $\mathfrak{M}\mathcal{L} - (\mathfrak{M}\mathcal{L}i \cup \mathfrak{M}\mathcal{L}s)$. S_θ is vertically regular. If $\| \|$ is an admissible Finsler on $\mathfrak{M}\mathcal{L}(\pi_0;\pi)$ and σ is an admissible bound on $\mathfrak{M}(X_0,X)$, then $\max(\mathfrak{M}\mathcal{L}(\pi_0;\pi)^*\sigma,\| \|,S_\theta)$ induces a semicomplete refinement on $\mathfrak{M}\mathcal{L}i,\mathfrak{M}\mathcal{L}s$

and $\mathfrak{M}\ell$is.

<u>PROOF</u>: With choices $(\mathfrak{D},\mathfrak{G},T,e,\Lambda)$ for π and $(\mathfrak{D}_0,\mathfrak{G}_0)$ for π_0, we pull back the diagram of page 135 and see that $T_f(\phi,g) = (s,S)$ iff $H_f(g) = s$ and the following diagram commutes:

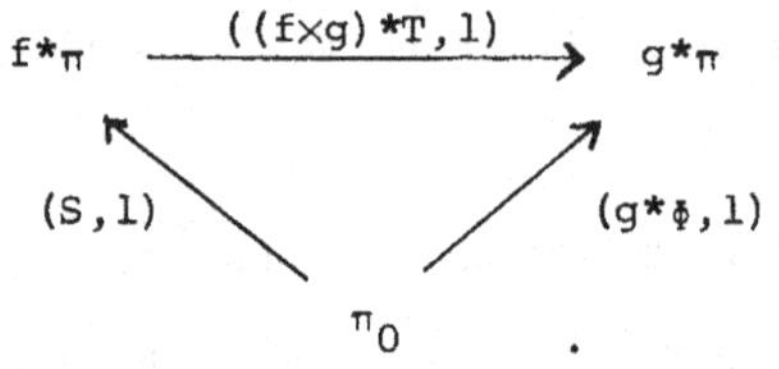

Since $((f \times g)^*T,1)$ is an $\mathfrak{M}$ VB isomorphism, $(S,1)$ is a locally split injection, locally split surjection, etc. iff $(g^*\phi,1)$ is. Thus, $T_f(\mathfrak{M}\ell i(\mathbf{E}_0;\mathbf{E}) \cap \mathfrak{M}\ell(\pi_0;\pi)^{-1}V_f) = D_f \times \mathfrak{M}(Li(\pi_0;f^*\pi))$ and similarly for the rest. This proves that the semicomplete $\mathfrak{M}_2$ atlas $(\mathfrak{M}\ell(\mathfrak{D}_0;\mathfrak{D}),\mathfrak{M}(\mathfrak{G}_0,\mathfrak{G}))$ induces an open subbundle atlas on each of the five sets. The S_θ results for $\mathfrak{M}\ell i$ and $\mathfrak{M}\ell s$ follow from Lemma 2. We sketch the verification of hypotheses (1) and (2) of that lemma. Since the $\sim_{S_\rho}$ class of S_θ is independent of the choice of Finsler we can choose them to be the max Finslers induced by $(\mathfrak{D}_0,\mathfrak{G}_0)$ and $(\mathfrak{D},\mathfrak{G})$. Note that the Finsler on the pull back is then the max Finsler induced by $(f^*\mathfrak{D},\mathfrak{G}_f)$. Hypothesis (1) then follows from (2.3) with $k_{(f^*\mathfrak{D},\mathfrak{G}_f)}$ estimated by $\sigma(f)$ and so for $g \in U_f^\Lambda$ by $\sigma(g)$, where $(S_\theta)_f$ is defined by the Finslers on $\mathfrak{M}(Li(E_0;f^*E))$. Hypothesis (2) is checked by using the above diagram which allows us to compare $(S_\theta)_f(S)$ with $S_\theta(g^*\phi)$ by using $S_\theta((f \times g)^*T)$. The latter factor is dominated by $\max(\sigma(f),\sigma(g))$ and so for $g \in U_f^\Lambda$ by $\sigma(g)$. This also shows that S_θ on $\mathfrak{M}\ell i$ is equivalent to $(S_\theta)^M$ and $(S_\theta)^m$ which arise in Lemma 2. Q.E.D.

3. <u>Immersions and Embeddings</u>: Let X_i be metric manifolds with admissible bounds ρ_i $(i = 1,2)$. Let $f\colon X_1 \to X_2$ be a metric map. f is called

metrically proper if $\rho_1 \sim f^*\rho_2$. The name comes from the analogy between compact subsets and bounded subsets. If f is metricly proper and A is bounded in X_2 then $f^{-1}A$ is bounded in X_1. Clearly, the composition of metricly proper maps is metricly proper. If $f: X_1 \to X_2$ and $g: X_3 \to X_4$ then $f \times g: X_1 \times X_3 \to X_2 \times X_4$ is metricly proper iff f and g are. If U is open in $X \times X$ then $p_1|U$ and $p_2|U$ are both metricly proper iff U is transversely bounded. Note that if $V \subset X_2 \times X_2$ is transversely bounded, and $f,g: X_1 \to X_2$ with $f \times g(X_1) \subset V$ and f metricly proper then g is metricly proper. For then $\max(f^*\rho_2, g^*\rho_2) \sim \min(f^*\rho_2, g^*\rho_2)$. Thus, if $(\mathfrak{M}_1, \mathfrak{M}, \mathcal{L})$ is a standard triple and X_2 is a semicomplete $\mathfrak{M}_1^1$ manifold admitting $\mathfrak{M}_1$ exponentials then the metricly proper maps are open and closed in the topological space $\mathfrak{M}(X_1, X_2)$. This is trivial in the bounded case since if X_1 is bounded any metric map is metricly proper. In general, if $e: D \to V$ is chosen with V transversely bounded and V_f meets the set of metricly proper maps then it is contained in that set.

An $\mathfrak{M}$ map $f: X_1 \to X_2$ is an $\underline{\mathfrak{M} \text{ immersion}}$ if there exist atlases $\mathcal{G}_1 = \{U_\alpha, h_\alpha\}$ and $\mathcal{G}_2 = \{V_\beta, g_\beta\}$ admissible for X_1 and X_2 with $f: \mathcal{G}_1 \to \mathcal{G}_2$ index preserving and such that $E^2_{\beta(\alpha)} = E^1_\alpha \times \tilde{E}_{\beta(\alpha)}$ and $f_\alpha: h_\alpha(U_\alpha) \to E^2_{\beta(\alpha)}$ is the restriction of the inclusion of the first factor. Such an atlas pair $(\mathcal{G}_1, \mathcal{G}_2)$ is called admissible for f. If $(\mathcal{G}_1, \mathcal{G}_2)$ is admissible for f then:

$$(3.1) \qquad f^*\rho_{\mathcal{G}_2} \geq \max(\rho_{\mathcal{G}_1}, \rho(f;\mathcal{G}_1,\mathcal{G}_2)) \quad \text{and} \quad d_{\mathcal{G}_1} \geq f^*d_{\mathcal{G}_2}.$$

The first because the transition maps of $\mathcal{G}_1$ and f are restrictions of those of $\mathcal{G}_2$. The second because any $\mathcal{G}_1$ chain maps to an $\mathcal{G}_2$ chain of the same length. In particular, if $\rho: X_2 \to [1,\infty]$ and $\rho^{d_{\mathcal{G}_2}} \sim \rho$ then $f^*\rho \sim f^*(\rho^{d_{\mathcal{G}_2}}) \geq (f^*\rho)^{f^*d_2} \geq (f^*\rho)^{d_{\mathcal{G}_1}}$, i.e. $f^*\rho \sim (f^*\rho)^{d_{\mathcal{G}_1}}$. It follows that if f is an $\mathfrak{M}$ immersion then the class $\{(\mathcal{G}_1, f^*\rho_2)\}$ with $(\mathcal{G}_1, \mathcal{G}_2)$ an admissible pair for f and ρ_2 an admissible bound for X_2 generates an $\mathfrak{M}$ metric structure on X_1, regular if X_2 is regular. With respect to this metric structure f is a metricly proper $\mathfrak{M}$ immersion. Furthermore, the original metric structure is a refinement of this f induced one. Thus, f is an $\mathfrak{M}$ immersion iff X_1 is a refinement of a metric manifold on which f is a metricly proper immersion. The metric structure of the domain of a metricly proper immersion is thus uniquely characterized by the metric structure of the range of the map f.

1 <u>THEOREM</u>: Let $f: X_1 \to X_2$ be an $\mathfrak{M}^1$ metricly proper map. The following are equivalent: (1) f is an $\mathfrak{M}^1$ immersion. (2) f is a $\mathcal{C}^1$ immersion. (3) $(f^*Tf,1): \tau_{X_1} \to f^*\tau_{X_2}$ is a locally split $\mathfrak{M}$ VB injection.

<u>PROOF</u>: If $(\mathcal{G}_1,\mathcal{G}_2)$ is admissible for f then $(T\mathcal{G}_1, f^*T\mathcal{G}_2; \mathcal{G}_1)$ is an admissible two-tuple for the $\mathfrak{M}$ VB injection $(f^*Tf,1)$. Thus, (1) implies (3) and (2) implies $(f^*Tf,1)$ is a $\mathcal{C}$ VB injection which implies (3) by equation (2.2). Clearly, (1) implies (2). Now assume (3). Let ρ be an admissible bound on X_2. We can choose star bounded, admissible atlases $\bar{\mathcal{G}}_1 = \{U_x, h_x\}$ and $\bar{\mathcal{G}}_2 = \{V_y, g_y\}$ on X_1 and X_2, indexed by the points of the underlying spaces such that: (1) $x \in U_x$, $h_x(x) = 0$ and $B(0,\lambda_1(x)) \subset h_x(U_x)$ for all $x \in X_1$. (2) $y \in V_y$, $g_y(y) = 0$ and $B(0,\lambda_2(y)) \subset g_y(V_y)$. (3) $f(U_x) \subset V_{f(x)}$. Defining θ^M using $(T\bar{\mathcal{G}}_1, f^*T\bar{\mathcal{G}}_2; \bar{\mathcal{G}}_1)$ we can choose a left inverse P_x for $Df_x(0)$ with $\|P_x\| \leq 2\theta^M(f)(x)$. Composing g_{fx} with the linear isomorphism $P_x \times [I - (Df_x(0))\circ P_x]$ we replace $\bar{\mathcal{G}}_2$ by an equivalent atlas because f is metricly proper and $\bar{\mathcal{G}}_2$ is star bounded. Thus, we can assume: (4) $E^2_{fx} = E^1_x \times \tilde{E}_x$ and $Df_x(0)$ is the inclusion of the first factor. Note that if X_1 and X_2 are semicomplete we can assume $\lambda = \min(\lambda_1, f^*\lambda_2)$ satisfies $f^*\rho > 1/\lambda$. Now let $B_x = B(0,\lambda(x))$ in $E^1_x \times \tilde{E}_x$ and define $F_x \in \mathfrak{M}^1(B_x, \lambda(x))$ by $f_x \circ p_1 + i_2 \circ p_2$. $DF_x(0) = I$ and $\|F_x\|_{\mathfrak{M}^1} \leq \|f_x\|_{\mathfrak{M}^1} + 1$. Note that $\rho(fx) > \|f_x\|_{\mathfrak{M}^1}$ because f is metricly proper and $\bar{\mathcal{G}}_1$ is star bounded. There exists $\lambda(x) > \epsilon(x) > 0$ such that for u

in $B(0,\epsilon(x))$, $\|DF_x(u) - I\| < 1/2$ and so $\theta(DF_x(u)) \leq 2$. In fact, if $\mathfrak{M} \subset \mathscr{L}$ we can choose $\epsilon(x) = \lambda(x)/\|F_x\|_{\mathfrak{M}^1}$. It follows from the Inverse Function Theorem and the Size Estimate, Propositions II.3.4 and II.3.7, that $F_x|B(0,\epsilon(x)/4)$ is an $\mathfrak{M}^1$ isomorphism onto an open subset of $B(0,\epsilon(x)/2)$ with $\|F_x^{-1}\|_{\mathfrak{M}^1} \leq O^*(\|f_x\|_{\mathfrak{M}^1})$. Let $\mathcal{G}_1 = \{h_x^{-1}[B(0,\epsilon(x)/4)],h_x\}$ and let $\mathcal{G}_2$ be $\bar{\mathcal{G}}_2$ with the charts $\{g_{fx}^{-1} F_x(B(0,\epsilon(x)/4)),F_x^{-1}\cdot g_{fx}\}$ adjoined. $\mathcal{G}_1$ is a refinement of $\bar{\mathcal{G}}_1$. $\mathcal{G}_2$ is admissible because $\bar{\mathcal{G}}_2$ is star bounded and f is metricly proper. The latter condition is crucial here because of the estimate on $\|F_x\|_{\mathfrak{M}^1}$. Clearly, $(\mathcal{G}_1,\mathcal{G}_2)$ is admissible for f. Q.E.D.

In general, if $f\colon \mathcal{G}_1 \to \mathcal{G}_2$ is index preserving, we define $\lambda_{(\mathcal{G}_1,\mathcal{G}_2)}(x) = \sup\{r \leq 1\colon B(h_\alpha x,r) \subset h_\alpha(U_\alpha)$ and $B(g_\beta fx,r) \subset g_\beta(V_\beta)$ for some index α with $\beta = \beta(\alpha)\}$. Clearly, $\lambda_{(\mathcal{G}_1,\mathcal{G}_2)} \leq \min(\lambda_{\mathcal{G}_1},f^*\lambda_{\mathcal{G}_2})$. If we begin with arbitrary atlases $\mathcal{G}$ and $\mathcal{G}_2$ and let $\mathcal{G}_f = \mathcal{G} \cap f^{-1}\mathcal{G}_2$ then because the index set of $\mathcal{G}_f$ is the product, with $\mathcal{G}_1 = \mathcal{G}_f$, $\lambda_{(\mathcal{G}_1,\mathcal{G}_2)}$ equals min. Let X_1 and X_2 be semicomplete $\mathfrak{M}$ manifolds and $f\colon X_1 \to X_2$ be an $\mathfrak{M}$ immersion. If there exist pairs $(\mathcal{G}_1,\mathcal{G}_2)$ admissible for f with $\rho_1 > 1/\lambda_{(\mathcal{G}_1,\mathcal{G}_2)}$ then f is called <u>semicomplete</u> and the pairs are called <u>s admissible</u> for f.

2 <u>LEMMA</u>: Let $f\colon X_1 \to X_2$ be a metricly proper $\mathfrak{M}^1$ immersion with X_2 semicomplete. Let ρ_1 induce a semicomplete refinement on X_1. If $\mathfrak{M} \subset \mathscr{L}$ or f is a $\mathcal{C}_u^1$ map then f is a semicomplete $\mathfrak{M}^1$ immersion when the domain is given the ρ_1 induced refinement. In particular, if X_1 is semicomplete to begin with then f is a semicomplete $\mathfrak{M}^1$ immersion with the original metric structure.

<u>PROOF</u>: This follows from the proof of Theorem 1, since we could choose λ_1 and λ_2 with $\rho_1 > \max(1/\lambda_1,1/f^*\lambda_2)$. Q.E.D.

3 <u>COROLLARY</u>: Let $(\mathfrak{M}_1,\mathfrak{M},\mathfrak{M}_2)$ be a standard triple. Let X_2 be a semicomplete $\mathfrak{M}_1^2$ manifold admitting $\mathfrak{M}_1^1$ exponentials and let X_1 be an $\mathfrak{M}^1$ manifold.

$\mathfrak{M}^1 i(X_1,X_2)$, the set of metricly proper $\mathfrak{M}^1$ immersions, is an open set in the topological space $\mathfrak{M}^1(X_1,X_2)$. If X_1 is bounded and S_θ is defined on $\mathfrak{M}\mathscr{L}(\tau_{X_1};\tau_{X_2})$ as in Proposition 2.7, then $\max(\sigma, S_\theta(Tf,f))$ induces a semi-complete refinement on $\mathfrak{M}^1 i(X_1,X_2)$.

<u>PROOF</u>: In the bounded case, $\mathfrak{M}^1 i(X_1,X_2) = \tau^{-1}(\mathfrak{M}\mathscr{L}i(\tau_{X_1};\tau_{X_2}))$ under the map τ of Proposition IV.5.2 and the result follows from Proposition 2.7 and Lemma III.7.11. In the unbounded case, it is easy to argue directly using Proposition IV.3.10. Q.E.D

4 <u>COROLLARY</u>: Assume $\mathfrak{M} \subset \mathscr{L}$. Let X_1, X_2 be semicomplete $\mathfrak{M}^1$ manifolds, O be open in X_1 and $f: O \to X_2$ be a metricly proper $\mathfrak{M}_1^1$ map (restriction metric structure in O). Let $A \subset\subset O$ be such that the section f^*Tf maps A into $Li((TX_1)_O; f^*TX_2)$ with $\rho_1 > \theta(f) = (f^*Tf)^*\theta$ on A. There exists U open in X_1 with $A \subset\subset U$ (rel X_1) and $f|U: U \to X_2$ is a metricly proper $\mathfrak{M}^1$ immersion with the restriction metric structure on U. $f|U$ is a semicomplete (but not metricly proper) $\mathfrak{M}^1$ immersion with the canonical refinement structure on U.

<u>PROOF</u>: By Corollary 2.6 applied to $(f^*Tf,1): (\tau_{X_1})_O \to f^*\tau_{X_2}$ with O given the canonical refined structure, we can choose U so $A \subset\subset U$ (rel O) such that $(f^*Tf,1)$ restricted to U is an $\mathfrak{M}^1$ VB locally split injection with metric structure obtained by restriction from the canonical refinement on O. By shrinking U, if necessary, we can assume $U \subset\subset O$ (rel X), i.e. $\rho_{1O} \sim \rho_1$ on U and so $A \subset\subset U$ (rel X_1) and the restriction metric structure from the canonical refinement on O agrees with the restriction structure from X_1. With respect to the latter $f|U$ is metricly proper. Theorem 1 applies and so $f|U$ is an $\mathfrak{M}^1$ immersion. With the canonical refinement on U $f|U$ is semicomplete by Lemma 2. Q.E.D.

5 PROPOSITION: Assume $\mathfrak{M} \subset \mathcal{L}$. Let $f: X_1 \to X_2$ be a semicomplete $\mathfrak{M}$ immersion. Let d_i, ρ_i be admissible metric and bound for X_i $(i = 1,2)$. There exists $V \in \mathcal{N}_{X_1}$, i.e. a full neighborhood of the diagonal such that $d_1|V \sim_{\rho_1} (f^*d_2)|V$.

PROOF: We are extending the definition of $\sim_{\rho_1}$ in the obvious way. Recall that $d_1 \succ_{\rho_1} f^*d_2$ on all of $X_1 \times X_1$. Choose $(\mathcal{G}_1,\mathcal{G}_2)$ s admissible for f. By refining if necessary we can assume that $\mathcal{G}_1$ and $\mathcal{G}_2$ satisfy condition $\mathcal{L}$. For $x_1,x_2 \in U_\alpha$ and $\beta = \beta(\alpha)$ this allows us to compare $d_1(x_1,x_2)$ with $\|h_\alpha x_1 - h_\alpha x_2\|_\alpha$ and $d_2(fx_1,fx_2)$ with $\|g_\beta fx_1 - g_\beta fx_2\|_\beta =$ $\|f_\alpha(h_\alpha x_1) - f_\alpha(h_\alpha x_2)\|_\beta$. But because f_α is an inclusion, $\|h_\alpha x_1 - h_\alpha x_2\|_\alpha = \|f_\alpha(h_\alpha x_1) - f_\alpha(h_\alpha x_2)\|_\beta$. Thus, we can compare d_1 and f^*d_2 on $V = \cup\{U_\alpha \times U_\alpha\}$. The reader can supply the details. Q.E.D.

Now we consider injective immersions and embeddings. Recall the definition of $S(f)$ (cf. p. 37).

6 PROPOSITION: Assume $\mathfrak{M} \subset \mathcal{L}$. Let $f: X_1 \to X_2$ be a semicomplete $\mathfrak{M}$ immersion. $\Delta_{X_1} \subset\subset X_1 \times X_1 - S(f)$. If A is a bounded subset of X_1 then there exists U with $A \subset\subset U$ and $f|U$ injective iff $A \times A \subset\subset X_1 \times X_1 - S(f)$.

PROOF: The first result follows from Proposition 5. The second from Corollary III.6.4. Q.E.D.

A $\mathcal{L}$ map $f: (X_1,d_1,\rho_1) \to (X_2,d_2,\rho_2)$ is an _embedding_ of regular pseudometric spaces if $d_1 \sim_{\rho_1} f^*d_2$. An $\mathfrak{M}$ immersion of regular $\mathfrak{M}$ manifolds $f: X_1 \to X_2$ is called an _$\mathfrak{M}$ embedding_ if it is an embedding of the underlying regular pseudometric spaces, i.e. if $d_1 \sim_{\rho_i} f^*d_2$ for d_i, ρ_i admissible metric and bound on X_i $(i = 1,2)$.

7 LEMMA: Let $f: (X_1,d_1,\rho_1) \to (X_2,d_2,\rho_2)$ be an embedding with d_1 and d_2

metrics, i.e. X_1 and X_2 are Hausdorff.

(a) f is a homeomorphism onto its image and if $V \in \mathfrak{U}_{X_1}$ then there exist constants K,n such that $d_2(fx_1,fx_2) < (K \min(\rho_1 x_1, \rho_2 x_2)^n)^{-1}$ implies $(x_1,x_2) \in V$.

(b) If (X_1,d_1,ρ_1) is uniformly complete and A is a closed, bounded subset of X_1, then f(A) is a closed, bounded subset of X_2.

(c) If (X_1,d_1,ρ_1) is uniformly complete and f is metricly proper, i.e. $\rho_1 \sim f^*\rho_2$ then f is a closed map and $f(X_1)$ is a closed subset of X_2.

PROOF: (a) is clear. (b): A is d_1 complete and hence f^*d_2 complete, i.e. f(A) is d_2 domplete and hence closed in X_2. (c): By (a) f is a closed map iff $f(X_1)$ is closed in X_2. If y is a limit point of $f(X_1)$ and B is a closed, bounded neighborhood of y then $A = f^{-1}(B)$ is closed and bounded in X_1 and y is a limit point of f(A). By (b), y is in f(A). Q.E.D.

8 LEMMA: Let $f: (X_1,d_1,\rho_1) \to (X_2,d_2,\rho_2)$ be an $\mathcal{L}$ map. If there exists $V \in \mathfrak{U}_{X_1}$ and constants K,n such that $f^*d_2|V >_\rho d_1|V$ and $d_2(fx_1,fx_2) < (K \min(\rho x_1,\rho x_2)^n)^{-1}$ implies $(x_1,x_2) \in V$ then f is an embedding.

PROOF: We can assume K and n are such that $K \min(\rho x_1,\rho x_2)^n d_2(fx_1,fx_2) \geq \min(d_1(x_1,x_2),1)$ for $(x_1,x_2) \in V$. By the hypothesized condition on K,n this also holds for $(x_1,x_2) \notin V$. Q.E.D.

If $f: X_1 \to X_2$ is a metricly proper $\mathfrak{M}$ embedding and $\mathcal{A}_i,\rho_i$ are admissible atlas and bound for X_i $(i = 1,2)$, then with $d_i = d_{\mathcal{A}_i}$ there exist $K,n \geq 1$ such that $K \min(\rho_2(fx_1),\rho_2(fx_2))^n d_2(fx_1,fx_2) \geq \min(d_1(x_1,x_2),1)$. Hence, if x_1,x_2 lie in different components of X_1, i.e. $d_1(x_1,x_2) = \infty$ then $d_2(fx_1,fx_2) \geq (K \min(\rho_2(fx_1),\rho_2(fx_2))^n)^{-1}$. Let $V_1 = \{(y_1,y_2): d_2(y_1,y_2) < (K \min(\rho_2(y_1),\rho_2(y_2))^n)^{-1}\} \in \mathfrak{U}_{X_2}$ and let $V = V^{-1} \in \mathfrak{U}_{X_2}$ such that $V \circ V \subset V_1$. If A_1 and A_2 are distinct components

X_1 then $V[f(A_1)] \cap V[f(A_2)] = \emptyset$.

To understand embeddings we define a function related to $\lambda_{(\mathcal{G}_1,\mathcal{G}_2)}$. Let $f\colon \mathcal{G}_1 \to \mathcal{G}_2$ be index preserving and let $A \subset X_1$. Define $\lambda^A_{(\mathcal{G}_1,\mathcal{G}_2)}(x) = \sup\{r \le 1\colon B(h_\alpha x,r) \subset h_\alpha(U_\alpha)$ and $B(g_\beta fx,r) \subset g_\beta(V_\beta - f(A - U_\alpha))$ for some index α of $\mathcal{G}_1$ with $\beta = \beta(\alpha)\}$. Clearly, $\lambda^A_{(\mathcal{G}_1,\mathcal{G}_2)} \le \lambda_{(\mathcal{G}_1,\mathcal{G}_2)}$ and $\lambda^A_{(\mathcal{G}_1,\mathcal{G}_2)}(x) > 0$ iff for some U_α containing x, fx is not a limit point of $f(A - U_\alpha)$. Thus, if $f\colon A \to f(A)$ is a homeomorphism then $\lambda^A_{(\mathcal{G}_1,\mathcal{G}_2)} > 0$ on A. Conversely, if f is an $\mathfrak{M}$ immersion, $(\mathcal{G}_1,\mathcal{G}_2)$ is an atlas pair for f and $\lambda^A_{(\mathcal{G}_1,\mathcal{G}_2)} > 0$ on A then $f\colon A \to f(A)$ is a homeomorphism. For if not then there exists B closed in A and $x \notin B$ with fx a limit point of fB. Let $x \in U_\alpha$. Since $f\colon U_\alpha \to f(U_\alpha)$ is a homeomorphism fx is a limit point of $f(B - U_\alpha)$ and hence of $f(A - U_\alpha)$.

9 <u>THEOREM</u>: Assume $f\colon X_1 \to X_2$ is a semicomplete $\mathfrak{M}$ immersion ($\mathfrak{M} \subset \mathcal{L}$) and $A \subset X_1$. Let d_i, ρ_i be admissible metric and bound on X_i $(i = 1,2)$. The following are equivalent: (1) $d_1 \sim_{\rho_1} f^*d_2$ on A. (2) For $V \in \mathfrak{U}_{X_1}$ there exist constants K,n such that $x_1,x_2 \in A$ and $d_2(fx_1,fx_2) < (K \min(\rho_1 x_1,\rho_1 x_2)^n)^{-1}$ imply $(x_1,x_2) \in V$. (3) There exists a full open neighborhood U of A with $d_1 \sim_{\rho_1} f^*d_2$ on U. (4) There exists a full open neighborhood U of A with $f|U\colon U \to X_2$ an $\mathfrak{M}$ embedding (canonical refinement structure on U). (5) For every s admissible atlas pair $(\mathcal{G}_1,\mathcal{G}_2)$ for f, $\rho_1 > 1/\lambda^A_{(\mathcal{G}_1,\mathcal{G}_2)}$ on A. (6) There exists an admissible atlas pair $(\mathcal{G}_1,\mathcal{G}_2)$ for f with $\rho_1 > 1/\lambda^A_{(\mathcal{G}_1,\mathcal{G}_2)}$. (7) There exist atlases $\mathcal{G}_1$ and $\mathcal{G}_2$ for X_1 and X_2 with $f\colon \mathcal{G}_1 \to \mathcal{G}_2$ index preserving and constants K,n such that $\rho_1 > 1/\lambda^A_{(\mathcal{G}_1,\mathcal{G}_2)}$ on A and $K \min(\rho x_1,\rho x_2)^n \|g_\beta fx_1 - g_\beta fx_2\| \ge \min(\|h_\alpha x_1 - h_\alpha x_2\|,1)$ for all α and $x_1,x_2 \in U_\alpha$ $(\beta = \beta(\alpha))$.

<u>PROOF</u>: (1) $\to$ (2) by Lemma 7(a) applied to $f|A$. (3) $\to$ (4) is clear and to see that (4) $\to$ (3) choose U_1 with $A \subset\subset U_1 \subset\subset U$ (rel X_1). Since d_1 is an admissible metric on U (canonical refinement) and $\rho_U \sim \rho_1$ on U_1,

(4) for U implies (3) for U_1. (3) → (1) and (5) → (6) → (7) are clear. It suffices to prove (2) → (5) and (7) → (3). We will prove (2) → (5) and (6) → (3), leaving to the reader the adjustments to get (7) → (3).

(2) → (5): Assume $d_i = d_{G_i}$ and $\rho_1 > \max((\rho_{G_1})^{d_1}, 1/\lambda_{(G_1,G_2)})$. For $x \in A$ choose $\alpha = \alpha(x)$ with $B(h_\alpha x, \rho_1(x)^{-1}) \subset h_\alpha(U_\alpha)$ and $B(g_\beta fx, \rho_1(x)^{-1}) \subset g_\beta(V_\beta)$ $(\beta = \beta(\alpha))$. By the Metric Estimate $h_\alpha^{-1}(B(h_\alpha x, \rho_1(x)^{-1}) \supset B^{d_1}(x, (4\rho_1(x)^2)^{-1})$. Using (2), choose $K, n \geq 4$ such that $d_2(fx_1, fx_2) < (K \min(\rho_1 x_1, \rho_1 x_2)^n)^{-1}$ and $x_1, x_2 \in A$ imply $d_1(x_1,x_2) < (4\rho_1(x_1)^2)^{-1}$ and so $x_2 \in U_\alpha$ $(\alpha = \alpha(x_1))$. Hence, $\lambda^A_{(G_1,G_2)}(x) \geq (K\rho_1(x)^n)^{-1}$

(6) → (3): Assume that $d_i = d_{G_i}$, $\rho_2 > (\rho_{G_2})^{d_2}$, $\rho_1 > \max(f^*\rho_2, (\rho_{G_1})^{d_1})$, $\rho_1 > 1/\lambda^A_{(G_1,G_2)}$ (on A) and $1/\rho_1$ is d_1 Lipschitz with Lipschtz constant ≤ 1. By (3.1) $d_1 \geq f^*d_2$. For $x \in A$ choose $\alpha = \alpha(x)$ and $\beta = \beta(\alpha)$: $B(h_\alpha x, \rho_1(x)^{-1}) \subset h_\alpha(U_\alpha)$ and $B(g_\beta fx, \rho_1(x)^{-1}) \subset g_\beta(V_\beta - f(A - U_\alpha))$. By the Metric Estimate: $B_1 = B^{d_1}(x, (4\rho_1(x)^2)^{-1}) \subset h_\alpha^{-1}(B(h_\alpha x, (2\rho_1(x))^{-1}))$ and $B_2 = B^{d_2}(fx, (4\rho_1(x)^2)^{-1}) \subset g_\beta^{-1}(B(g_\beta fx, (2\rho_1(x))^{-1}))$. For $x_1, x_2 \in B_1$ $d_1(x_1,x_2) \geq \rho(x)^{-1}\|h_\alpha x_1 - h_\alpha x_2\|$. For $y_1, y_2 \in B_2$, $d_2(y_1,y_2) \geq \rho(x)^{-1}\|g_\beta y_1 - g_\beta y_2\|$. Let $U = \cup(B^{d_1}(x, (24\rho_1(x)^3)^{-1})$. Clearly, $A \subset\subset U$. Now let $x_i' \in B^{d_1}(x_i, (24\rho_1(x_i)^3)^{-1})$ with $x_i \in A$ $(i = 1,2)$ and let $K = \rho_1(x_1) \leq \rho_1(x_2)$. Thus, $d_1(x_i, x_i') < (24\ K^3)^{-1}$. Assume that $d_2(fx_1', fx_2') < (24\ K^3)^{-1}$. Because $d_1 \geq f^*d_2$, $d_2(fx_i, fx_i') < (24\ K^3)^{-1}$ and so $d_2(fx_1, fx_2) < (8\ K^3)^{-1}$. Hence, with $\alpha = \alpha(x_1)$, $f(x_2) \in V_\beta - f(A - U_\alpha)$, i.e. $x_2 \in U_\alpha$. Furthermore, $(8\ K^2)^{-1} \geq \|g_\beta fx_1 - g_\beta fx_2\|$. Now since $x_1, x_2 \in U_\alpha$ and f_α is an inclusion $\|g_\beta fx_1 - g_\beta fx_2\| = \|h_\alpha x_1 - h_\alpha x_2\| \geq d_1(x_1,x_2)$. Thus, x_1' and $x_2' \in B^{d_1}(x_1, (4\ K^2)^{-1}) \subset U_\alpha$. As $d_1 \geq f^*d_2$, fx_1' and $fx_2' \in B^{d_2}(fx_1, (4\ K^2)^{-1})$. Now use the Metric estimates and the fact that f_α is an inclusion: $d_2(fx_1', fx_2') \geq K^{-1}\|g_\beta fx_1 - g_\beta fx_2\| = K^{-1}\|h_\alpha x_1 - h_\alpha x_2\| \geq K^{-1} d_1(x_1', x_2')$. To sum up, symmetry implies: $24 \min(\rho_1(x_1), \rho_1(x_2))^3 d_2(fx_1', fx_2') \geq \min(d_1(x_1', x_2'), 1)$. Since ρ_1 is d_1 Lipschitz, $2\rho(x_i') \geq \rho(x_i)$ and so $192 \min(\rho_1(x_1'), \rho(x_2'))^3 d_2(fx_1', fx_2') \geq$

$\min(d_1(x_1',x_2'),1)$ for $x_1',x_2' \in U$. Q.E.D.

ADDENDUM: If X_1 is an open subset of a semicomplete manifold X with the canonical refinement structure and $A \subset\subset X_1$ (rel X), then U in (3) and (4) can be chosen with $A \subset\subset U$ (rel X). It suffices to choose $U \subset\subset X_1$ (rel X).

10 COROLLARY: Assume $\mathfrak{M} \subset \mathcal{L}$. Let $f: X_1 \to X_2$ be a semicomplete $\mathfrak{M}$ immersion. The following are equivalent: (1) f is an $\mathfrak{M}$ embedding. (2) There exist admissible atlas pairs $(\mathfrak{a}_1,\mathfrak{a}_2)$ for f with $\rho_1 > 1/\lambda^{X_1}_{(\mathfrak{a}_1,\mathfrak{a}_2)}$. (3) For all s admissible atlas pairs $(\mathfrak{a}_1,\mathfrak{a}_2)$ for f, $\rho_1 > 1/\lambda^{X_1}_{(\mathfrak{a}_1,\mathfrak{a}_2)}$. (4) There exist s admissible atlas pairs $(\mathfrak{a}_1,\mathfrak{a}_2)$ for f with $U_\alpha = f^{-1}(V_{\beta(\alpha)})$ for all indices α in $\mathfrak{a}_1$. If f is metricly proper we can add: (5) The uniformity $\mathfrak{U}_{X_1}$ is generated by $(f \times f)^{-1}\mathfrak{U}_{X_2}$.

PROOF: The equivalence of (1) - (3) is immediate from Theorem 9. If f is metricly proper (5) is a restatement of Theorem 9 (2) with $A = X_1$. If $(\mathfrak{a}_1,\mathfrak{a}_2)$ is an atlas pair as described in (4) then $\lambda^{X_1}_{(\mathfrak{a}_1,\mathfrak{a}_2)} = \lambda_{(\mathfrak{a}_1,\mathfrak{a}_2)}$. Hence, (4) implies (3). Finally, if $(\bar{\mathfrak{a}}_1,\bar{\mathfrak{a}}_2)$ is an s admissible atlas pair for f then choose for each x an α with $B(h_\alpha x,(K\rho_1(x)^n)^{-1}) \subset h_\alpha(U_\alpha)$ and $B(g_\beta fx,(K\rho_1(x)^n)^{-1} \subset g_\beta(V_\beta - f(X_1 - U_\alpha))$. Let $\mathfrak{a}_1$ be the refinement of $\bar{\mathfrak{a}}_1 = \{h_\alpha^{-1}B(h_\alpha x,(K\rho_1(x)^n)^{-1}),h_\alpha\}$ indexed by $x \in X_1$ and to $\bar{\mathfrak{a}}_2$ adjoin the chart $\{g_\beta^{-1}(B(g_\beta fx,(K\rho_1(x)^n)^{-1})),g_\beta\}$ to obtain $\mathfrak{a}_2$. Since $f^{-1}g_\beta^{-1}(B(g_\beta fx,(K\rho_1(x)^n)^{-1})) \subset U_\alpha$ and f_α is an inclusion, it equals $h_\alpha^{-1}f_\alpha^{-1}(B(g_\beta fx,(K\rho_1(x)^n)^{-1})) = h_\alpha^{-1}(B(h_\alpha x,(K\rho_1(x)^n)^{-1}))$. Q.E.D.

It follows that if $(\mathfrak{a}_1,\mathfrak{a}_2)$ is an admissible pair for an $\mathfrak{M}$ immersion f with $f: X_1 \to f(X_1)$ a homeomorphism, then $\max(\rho_1,1/\lambda^{X_1}_{(\mathfrak{a}_1,\mathfrak{a}_2)})$ induces a semicomplete refinement on X_1 with respect to which f is a semicomplete $\mathfrak{M}^1$ embedding.

If in Corollary 10 f is metricly proper then we can by Lemma 7(c) replace the original charts in $\mathcal{G}_2$ by $\{V_\beta - f(X_1), h_\beta\}$ and so get an s admissible atlas pair $(\mathcal{G}_1,\mathcal{G}_2)$ satisfying the conditions of (4) with $\mathcal{G}_2$ s admissible for X_2 and $V_\beta \cap f(X_1) = \emptyset$ if $\beta \neq \beta(\alpha)$ for any α.

11 **PROPOSITION**: Assume $(\mathfrak{M}_1,\mathfrak{M},\mathfrak{M}_2)$ is a standard triple with $\mathfrak{M} \subset \mathcal{L}$. Let X_2 be a semicomplete $\mathfrak{M}_1^2$ manifold admitting $\mathfrak{M}_1^1$ exponentials and let X_1 be a semicomplete $\mathfrak{M}_1^1$ manifold. $\mathfrak{M}^1 e(X_1,X_2)$, the set of metricly proper $\mathfrak{M}^1$ embeddings, is an open set in the topological space $\mathfrak{M}^1(X_1,X_2)$.

PROOF: We give only a sketch. For $f \in \mathfrak{M}^1 e(X_1,X_2)$ choose an s admissible pair $(\mathcal{G}_1,\mathcal{G}_2)$ for f and construct V as in Proposition 5. Choose $V_1 \in \mathcal{U}_{X_1}$ with $V_1 \subset\subset V$. If g is $\mathcal{C}^1$ close to f then $d_1|V_1 \sim_{\rho_1} (g^*d_2)|V_1$ for $d_i = d_{\mathcal{G}_i}$ $(i = 1,2)$ by an adjustment of the proof of Proposition 5. With $m(x_1,x_2) = \min(\rho_1(x_1),\rho_1(x_2))$ we can choose $K,n \geq 1$ such that
$\min(d_1,1)/\min(f^*d_2,1)$ is bounded by Km^n on $X_1 \times X_1 - \Delta_{X_1}$,
$\min(d_1,1)/\min(g^*d_2,1)$ is bounded by Km^n on $V_1 - \Delta_{X_1}$ for g is a $\mathcal{C}^1$ neighborhood of f, and $\{d_1 < (Km^n)^{-1}\} \subset V_1$. If g is $\mathcal{C}$ close to f, in fact, if $d_2(fx,gx) < (4\, K^2\rho_1(x)^{2n})^{-1}$ then on $X_1 \times X_1 - V_1$
$d_2(gx_1,gx_2) \geq d_2(fx_1,fx_2) - (4\, K^2\rho_1(x_1)^{2n})^{-1} - (4\, K^2\rho_1(x_2)^{2n})^{-1}$
$\geq (K^2m(x_1,x_2)^{2n})^{-1} - 2(4\, K^2m(x_1,x_2)^{2n})^{-1} = (2\, K^2m(x_1,x_2)^{2n})^{-1}$, i.e.
$\min(d_1,1)/\min(g^*d_2,1)$ is bounded by $2\, K^2m^{2n}$ on $X_1 \times X_1 - V_1$ and hence on $X_1 \times X_1 - \Delta_{X_1}$. Q.E.D.

If X_1 is bounded and hence is semicompact we can define $(d_1:d_2)\colon \mathfrak{M}^1 e(X_1,X_2) \to (0,\infty)$ by $(d_1:d_2)(f)$ equal the sup over $X_1 \times X_1 - \Delta_{X_1}$ of $\min(d_1,1)/\min(f^*d_2,1)$. By sharpening the proof of Proposition 5 one can show that if d_i is an admissible metric on X_i $(i = 1,2)$ then $\max(\sigma,S_\theta,(d_1:d_2))$ induces a semicomplete refinement on $\mathfrak{M}^1 e(X_1,X_2)$ (compare Corollary 3).

If $f_1\colon X_1 \to X_2$ and $f_2\colon X_3 \to X_4$ are $\mathfrak{M}^1$ immersions with admissible atlas pairs $(\mathcal{G}_1,\mathcal{G}_2)$ and $(\mathcal{G}_3,\mathcal{G}_4)$ then $(\mathcal{G}_1 \times \mathcal{G}_2,\mathcal{G}_2 \times \mathcal{G}_4)$ is an admissible atlas pair for the $\mathfrak{M}^1$ immersion $f_1 \times f_2: X_1 \times X_3 \to X_2 \times X_4$. Thus, if f_1, f_2 are semicomplete or embeddings then $f_1 \times f_2$ is (by Corollary 10 in the latter case). If $f_1: X_1 \to X_2$ and $f_2: X_2 \to X_3$ are $\mathfrak{M}^1$ immersions and f_2 is metricly proper then $f_2 \cdot f_1$ is an $\mathfrak{M}^1$ immersion. Reduce to the case when f_1 is also metricly proper and apply Theorem 1. If $f: X_1 \to X_2$ is a semicomplete $\mathfrak{M}^1$ immersion ($\mathfrak{M} \subset \mathcal{L}$) with s admissible pairs $(\mathfrak{a}_1, \mathfrak{a}_2)$ then $Tf: TX_1 \to TX_2$ is a semicomplete $\mathfrak{M}$ immersion with admissible pair $(T\mathfrak{a}_1, T\mathfrak{a}_2)$. If f is metricly proper or an embedding then Tf is. Let $\pi_i: E_i \to X_i$ be semicomplete $\mathfrak{M}^1$ bundles $(i = 1,2)$ and $(\phi, f): \pi_1 \to \pi_2$ be an $\mathfrak{M}^1$ VB map. If f is a metricly proper $\mathfrak{M}^1$ immersion (or embedding) and $(f^*\phi, 1): \pi_1 \to f^*\pi_2$ is a locally split $\mathfrak{M}^1$ VB injection then $\phi: E_1 \to E_2$ is a metricly proper $\mathfrak{M}^1$ immersion (resp. embedding). In fact, if $(\mathfrak{D}_i, \mathfrak{a}_i)$ is s admissible for π_i $(i = 1,2)$ with $f: \mathfrak{a}_1 \to \mathfrak{a}_2$ index preserving we can apply the proof of Theorem 1 to the pair $(\mathfrak{a}_1, \mathfrak{a}_2)$ and apply the transfer of atlas construction to assume that $(\mathfrak{a}_1, \mathfrak{a}_2)$ is an admissible pair for f and then apply the proof of Theorem 2.4 to assume that $(\mathfrak{D}_1, f^*\mathfrak{D}_2; \mathfrak{a}_1)$ is an admissible two-tuple for $(f^*\phi, 1)$. It then follows that $(\mathfrak{D}_1, \mathfrak{D}_2)$ is an admissible pair for ϕ. If $f: X_1 \to X_2$ is a metricly proper $\mathfrak{M}^1$ map and there exists $g: X_2 \to X_1$ an $\mathfrak{M}^1$ map with $g \cdot f = 1$ then f is an $\mathfrak{M}^1$ embedding. In fact, $(f^*Tf, 1)$ is a split $\mathfrak{M}$ VB injection and $f^*d_2 >_{f^*\beta} f^*g^*d_1 = d_1$. In particular, if $f: X_1 \to X_2$ is an $\mathfrak{M}^1$ map then the graph $\Gamma_f: X_1 \to X_1 \times X_2$ is a metricly proper $\mathfrak{M}^1$ embedding. If $\pi: E \to X$ is a semicomplete $\mathfrak{M}^1$ bundle and $s: X \to E$ is an $\mathfrak{M}^1$ section then s is a metricly proper $\mathfrak{M}^1$ embedding.

Let X_0 be a closed subset of an $\mathfrak{M}$ manifold X. An admissible atlas $\mathfrak{a} = \{U_\alpha, h_\alpha\}$ on X <u>induces a submanifold atlas</u> $\mathfrak{a}|X_0 = \{U_\alpha \cap X_0, \bar{h}_\alpha\}$ if for all α, either $U_\alpha \cap X_0 = \emptyset$ or $h_\alpha(U_\alpha \cap X_0) = h_\alpha(U_\alpha) \cap E_\alpha^0$ for some closed subspace E_α^0 of E_α. Define $\bar{h}_\alpha: U_\alpha \cap X_0 \to E_\alpha^0$ to be the restriction. Such an $\mathfrak{a}$ is called <u>admissible</u> for X_0. In general, if the latter condition is satisfied for $\alpha \in I_0$ a subset of the index set I and $\{U_\alpha: \alpha \in I_0\}$ covers X_0, then we can refine $\mathfrak{a}$ by replacing U_α by $U_\alpha - X_0$ for $\alpha \in I - I_0$. $\mathfrak{a}|X_0$ is called <u>locally split</u> if $E_\alpha = E_\alpha^0 \times E_\alpha^1$ for all α with $U_\alpha \cap X_0 \neq \emptyset$. The class $\{(\mathfrak{a}|X_0, \rho|X_0)\}$ induces a metric structure on X_0. Clearly, $d_{\mathfrak{a}|X_0} \geq d_{\mathfrak{a}}|X_0$. Hence, X_0 is regular if X is. X_0 is a <u>semicomplete</u> submanifold if X is semicomplete and there exist $\mathfrak{a}$ s admissible for X and admissible for X_0. $\rho|X_0 > 1/\lambda_{\mathfrak{a}|X_0}$. So X_0 is semicomplete with $\mathfrak{a}|X_0$. So X_0 is semicomplete with $\mathfrak{a}|X_0$ s admissible.

12 <u>PROPOSITION</u>: Let X be a semicomplete $\mathfrak{M}^1$ manifold.

(a) If X_0 is the image of a metricly proper, semicomplete $\mathfrak{M}^1$ embedding then X_0 is a semicomplete locally split submanifold of X. Conversely, if X_0 is a semicomplete locally split submanifold of X then the inclusion of X_0 is a metricly proper, semicomplete $\mathfrak{M}^1$ embedding. In particular, if d is an admissible metric for X then $d|X_0$ is an admissible metric for X_0.

(b) Assume $\mathfrak{M} \subset \mathcal{L}$. If X_0 is a semicomplete locally split submanifold of X and X_1 is a semicomplete locally split submanifold of X_0, then X_1 is a semicomplete locally split submanifold of X.

<u>PROOF</u>: (a) follows from Corollary 10 and the remarks thereafter. In fact, if $f: X_1 \to X_2$ is an $\mathfrak{M}^1$ immersion and $(\mathfrak{a}_1, \mathfrak{a}_2)$ is an atlas pair with $U_\alpha = f^{-1}(V_{\beta(\alpha)})$ and $V_\beta \cap f(X_1) = \emptyset$ for $\beta \neq \beta(\alpha)$ for any a, then $\mathfrak{a}_1$ induces a locally split submanifold atlas on X_0. Conversely, if $\mathfrak{a}$ induces a locally split submanifold atlas on X_0 then $(\mathfrak{a}|X_0, \mathfrak{a})$ is such a pair for the inclusion. (b) follows from (a) by composition of embeddings. Q.E.D.

13 <u>FACTORING LEMMA</u>: Let $f: X_1 \to X_2$ be an $\mathfrak{M}$ immersion. Let X be an $\mathfrak{M}$ manifold and $g_i: X \to X_i$ $(i = 1,2)$ with $fg_1 = g_2$. g_1 is an $\mathfrak{M}$ map iff it is continuous, $\rho > g_1{}^*\rho_1$ and g_2 is an $\mathfrak{M}$ map. In particular, if

f is metricly proper and g_1 is continuous, g_1 is $\mathfrak{M}$ iff g_2 is.

<u>PROOF</u>: Let $(\mathcal{A}_1,\mathcal{A}_2)$ be an admissible atlas pair for f and $\mathcal{A} = \{W_\gamma, k_\gamma\}$ be an admissible atlas for X. $(g_1)_{\alpha\gamma}: k_\gamma(W_\gamma \cap g_1^{-1}U_\alpha) \to E^1_\alpha$ is the restriction of $(g_2)_{\beta\gamma}: k_\gamma(W_\gamma \cap g_2^{-1}(V_\beta) \to E^2_\beta = E^1_\alpha \times \bar{E}_\alpha$ $(\beta = \beta(\alpha))$ followed by the projection to E^1_α. Hence, $\rho(g_1;\mathcal{A},\mathcal{A}_1) \leq \rho(g_2;\mathcal{A},\mathcal{A}_2)$. Q.E.D

If $f: X_1 \to f(X_1)$ is a homeomorphism (<u>a fortiori</u> if f is an embedding) then continuity of g_1 follows from that of g_2. For more general injective immersions this may not be true. It is useful to have topological conditions which apply more generally.

14 <u>LEMMA</u>: Let $f: X_1 \to X_2$ be a continuous map such that for all $x \in X_1$ there exist neighborhoods U of x and V of fx satisfying: (1) U is path connected. (2) $f(U)$ is the fx path component of $V \cap f(X_0)$. (3) $f: U \to f(U)$ is a homeomorphism. Let X be a locally path connected space and $g: X \to X_1$. If $f \circ g$ is continuous, then g is continuous.

<u>PROOF</u>: Let $y \in X$, $x = gy$. Choose U and V as in the hypothesis. Let W be a path connected neighborhood of y in $(f \circ g)^{-1}V$. $fg(W) \subset V \cap f(X_0)$ and is connected. Hence, $fg(W) \subset f(U)$ and $g|W = (f_U)^{-1} \circ (f \circ g)|W$ where f_U is the homeomorphism of (3). Q.E.D.

4. <u>Submersions</u>: An $\mathfrak{M}$ map $f: X_1 \to X_2$ is called an $\mathfrak{M}$ <u>submersion</u> if there exist admissible atlases $\mathcal{A}_1 = \{U_\alpha, h_\alpha\}$ and $\mathcal{A}_2 = \{V_\beta, g_\beta\}$ with $f: \mathcal{A}_1 \to \mathcal{A}_2$ index preserving and such that $E^1_\alpha = \tilde{E}_\alpha \times E^2_\beta$ $(\beta = \beta(\alpha))$ and $f_\alpha: h_\alpha(U_\alpha) \to E^2_\beta$ is the restriction of the projection to the second factor. Such an atlas pair is called <u>admissible</u> for f. f is called semicomplete if $\mathfrak{M} \subset \mathcal{L}$, X_1 and X_2 are semicomplete and there exist admissible atlas pairs $(\mathcal{A}_1,\mathcal{A}_2)$ for f with $\mathcal{A}_1,\mathcal{A}_2$ s admissible. Such pairs are then called <u>s admissible</u> for f. Note that if $\mathcal{A}_1$ is

s admissible then we can add on any s admissible atlas on X_2 to make $\mathcal{G}_2$ s admissible. If $(\mathcal{G}_1,\mathcal{G}_2)$ is admissible for f then

(4.1) $$\rho_{\mathcal{G}_1} \geq \max(f^*\rho_{\mathcal{G}_2}, \rho(f;\mathcal{G}_1,\mathcal{G}_2)) \qquad d_{\mathcal{G}_1} \geq f^*d_{\mathcal{G}_2} .$$

The first is clear. f maps any $\mathcal{G}_1$ chain to an $\mathcal{G}_2$ chain of no greater length implying the second. If $\pi\colon E \to X$ is an $\mathfrak{M}$ bundle then π is an $\mathfrak{M}$ submersion. If $(\mathfrak{D},\mathcal{G})$ is an admissible atlas for the bundle π then $(\mathfrak{D},\mathcal{G})$ is an admissible atlas pair for the submersion π.

1 LEMMA: Let $f\colon X_1 \to X_2$ be an $\mathfrak{M}$ submersion. f is an open map and $f(X_1)$ is open in X_2. If f is semicomplete and metricly proper then $f(X_1)$ is open and closed in X_2.

PROOF: Since f is locally a projection, it is an open map and so $f(X_1)$ is open. If $(\mathcal{G}_1,\mathcal{G}_2)$ is admissible for f then clearly

(4.2) $$f^*\lambda_{\mathcal{G}_2|f(X_1)} \geq \lambda_{\mathcal{G}_1} = \lambda_{(\mathcal{G}_1,\mathcal{G}_2)} .$$

If $\mathcal{G}_1$ is s admissible and f is metricly proper then for ρ_2 an admissible bound on X_2, $f^*\rho_2 > 1/\lambda_{\mathcal{G}_1} \geq 1/f^*\lambda_{\mathcal{G}_2|f(X_1)}$. Hence, $(\mathcal{G}_2|f(X_1),\rho_2)$ is a semicomplete adapted atlas. $f(X_1)$ is open and closed by Proposition III.7.8(a). Q.E.D.

2 THEOREM: Let $f\colon X_1 \to X_2$ be an $\mathfrak{M}^1$ map. The following are equivalent:
(1) f is an $\mathfrak{M}^1$ submersion. (2) f is a $\mathcal{C}^1$ submersion.
(3) $(f^*Tf,1)\colon \tau_{X_1} \to f^*\tau_{X_2}$ is a locally split $\mathfrak{M}$ surjection. If X_1 and X_2 are semicomplete and either $\mathfrak{M} \subset \mathcal{L}$ or f is a $\mathcal{C}^1_u$ map, then f is semicomplete.

PROOF: Completely analogous to the proofs of Theorem 3.1 and Lemma 3.2. Since charts are adjusted on the domain in this case we don't need f to be metricly proper. We leave the details to the reader. Q.E.D.

3 **LEMMA:** Let X be a regular $\mathcal{L}$ manifold. $\mathfrak{U}_X$ is generated by $\{V \in \mathfrak{U}_X: V$ open and $V[x]$ is path connected for all $x \in X\}$.

PROOF: Let $\mathcal{G}$ be an admissible atlas and ρ be an admissible bound. Let $V = \{(x_1,x_2); d_{\mathcal{G}}(x_1,x_2) < (K\rho(x_1)^n)^{-1}\}$. $V[x] = B(x,(K\rho(x)^n)^{-1})$ is connected: If $d_{\mathcal{G}}(x_1,x_2) < r$, choose an $\mathcal{G}$ chain between x_1 and x_2 of length $< r$. The associated p.l. path lies in $B(x_1,r)$. Q.E.D.

4 **COROLLARY:** Let $(\mathfrak{M}_1,\mathfrak{M},\mathfrak{M}_2)$ be a standard triple. Let X_2 be a semicomplete $\mathfrak{M}_1^2$ manifold admitting $\mathfrak{M}_1^1$ exponentials and let X_1 be an $\mathfrak{M}^1$ manifold. $\mathfrak{M}^1 s(X_1,X_2)$, the set of $\mathfrak{M}^1$ submersions, is open in the topological space $\mathfrak{M}^1(X_1,X_2)$. The set of metricly proper $\mathfrak{M}^1$ submersions is open and closed in $\mathfrak{M}^1 s(X_1,X_2)$. If X_1 is semicomplete and $\mathfrak{M} \subset \mathcal{L}$ then the set of surjective, metricly proper $\mathfrak{M}^1$ submersions is open and closed in $\mathfrak{M}^1 s(X_1,X_2)$. If X_1 is bounded and S_θ is defined on $\mathfrak{M}\mathcal{L}(\tau_{X_1};\tau_{X_2})$ as in Proposition 2.7, then $\max(\sigma,S_\theta)$ induces a semicomplete refinement on $\mathfrak{M}^1 s(X_1,X_2)$.

PROOF: Since metricly proper maps are open and closed in $\mathfrak{M}^1(X_1,X_2)$ this follows just as Corollary 3.3 did. For the surjective case note that if $f \times g: X_1 \to V$ with V as in Lemma 3 and f,g are metricly proper, semicomplete submersions with f surjective then by Lemma 1, g is surjective. Q.E.D.

5 **COROLLARY:** Assume $\mathfrak{M} \subset \mathcal{L}$. Let X_1,X_2 be semicomplete $\mathfrak{M}^1$ manifolds, O be open in X_1 and $f: O \to X_2$ be an $\mathfrak{M}_1^1$ map (restriction metric structure on O). Let $A \subset\subset O$ be such that the section f^*Tf maps A into $Ls((TX_1)_O;f^*TX_2)$ with $\rho_1 > \theta(f) = (f^*Tf)^*\theta$ on A. There exists U open in X_1 with $A \subset\subset U$ (rel X_1) and $f|U: U \to X_2$ is an $\mathfrak{M}^1$ submersion with the restriction metric structure on U. With the canonical refinement structure on U $f|U$ is semicomplete.

PROOF: Analogous to Corollary 3.4.

6 LEMMA: Let $f\colon X_1 \to X_2$ be a semicomplete submersion with admissible atlas pair $(\mathcal{G}_1,\mathcal{G}_2)$. Let d_i,ρ_i be admissible metric and bound on X_i $(i = 1,2)$. Let U be open in X_2.

$$\max(\rho_1, 1/\delta^{f^{-1}U}(d_1)) \sim \max(\rho_1, 1/\lambda_{\mathcal{G}_1|f^{-1}U})$$

$$\sim \max(\rho_1, 1/f^*\lambda_{\mathcal{G}_2|U}) \sim \max(\rho_1, 1/f^*\delta^U(d_2)).$$

If f is metricly proper we can add on the equivalences:

$$\sim f^*\max(\rho_2, 1/\lambda_{\mathcal{G}_2|U}) \sim f^*\max(\rho_2, 1/f^*\delta^U(d_2)).$$

PROOF: Since $\rho_1 > f^*\rho_2$ the equivalence between the λ terms and the corresponding δ terms is given by Proposition III.7.7. That $\max(\rho_1, 1/f^*\delta^U(d_2)) > \max(\rho_1, 1/\delta^{f^{-1}U}(d_1))$ follows from III.(6.3). In fact, if $d_i = d_{\mathcal{G}_i}$, $\delta^{f^{-1}U}(d_1) \geq f^*\delta^U(d_2)$ by (4.1). On the other hand (4.2) generalizes to $f^*\lambda_{\mathcal{G}_2|U} \geq \lambda_{\mathcal{G}_1|f^{-1}U}$. Q.E.D.

7 COROLLARY: Let $f\colon X_1 \to X_2$ be a semicomplete $\mathfrak{M}$ submersion and let U be open in X_2. $f_U\colon f^{-1}U \to U$ is a semicomplete $\mathfrak{M}$ submersion with respect to the canonical refinement structures on U and $f^{-1}U$. If f is metricly proper then f_U is.

PROOF: By Lemma 6 $\rho_{f^{-1}U} > f^*\rho_U$. If $(\mathcal{G}_1,\mathcal{G}_2)$ is an s admissible pair for f then $(\mathcal{G}_1|f^{-1}U, \mathcal{G}_2|U)$ is an s admissible pair for f_U (cf. Prop. III.7.7). Q.E.D.

We now generalize Proposition III.7.12. Let $f\colon X_1 \to X_2$ be an $\mathfrak{M}$ submersion. $A \subset X_1$ is called vertically bounded if $\rho_1 \sim f^*\rho_2$ on A. Thus, f is metricly proper iff X_1 is vertically bounded.

8 PROPOSITION: Let $f\colon X_1 \to X_2$ be a semicomplete $\mathfrak{M}$ submersion, $A,O \subset X_1$ and $U \subset X_2$. If $f(A) \subset\subset U$ then $A \subset\subset f^{-1}U$. If A is vertically bounded and $A \subset\subset f^{-1}U$ then $f(A) \subset\subset U$. If A is vertically bounded and

$A \subset\subset O$ then $f(A) \subset\subset f(O)$.

PROOF: The first follows from Lemma III 6.3(g), the second from Lemma 7 (cf. Prop. III.7.12) and the third from the second since $A \subset\subset O \subset f^{-1}f(O)$ then implies $f(A) \subset\subset f(O)$. Q.E.D.

The product, composition and tangent maps of $\mathfrak{M}^1$ submersions are submersions. The details are similar to those for immersions on pages 200 and 201. If $(\Phi, f): \pi_1 \to \pi_2$ is an $\mathfrak{M}^1$ VB map with f a submersion and $(f^*\Phi, 1)$ a locally split VB surjection then $\Phi: E_1 \to E_2$ is an $\mathfrak{M}^1$ submersion.

9 FACTORING LEMMA: Let $f: X_1 \to X_2$ be a metricly proper surjective $\mathfrak{M}$ submersion. Let X be an $\mathfrak{M}$ manifold and $g_i: X_i \to X$ $(i = 1,2)$ with $g_2 \cdot f = g_1$. g_1 is an $\mathfrak{M}$ map iff g_2 is.

PROOF: If g_2 is $\mathfrak{M}$ then $g_1 = g_2 \cdot f$ is $\mathfrak{M}$. If g_1 is $\mathfrak{M}$, we can choose star bounded atlases $\mathcal{G}_1 = \{U_\alpha, h_\alpha\}$ and $\mathcal{G}_2 = \{V_\alpha, g_\alpha\}$ for X_1 and X_2 with $h_\alpha(U_\alpha) = g_\alpha(V_\alpha) \times O_\alpha$ for some open set O_α and $f_\alpha = p_1$. If $v_\alpha \in O_\alpha$ for any α then $g_{2\alpha} = g_{1\alpha} \cdot (1 \times c_{v_\alpha})$. Since f is metricly proper this easily implies g_2 is $\mathfrak{M}$. Q.E.D.

5. Local Isomorphisms: An $\mathfrak{M}$ map $f: X_1 \to X_2$ is called an $\mathfrak{M}$ local isomorphism if there exist admissible atlases $\mathcal{G}_1 = \{U_\alpha, h_\alpha\}$ and $\mathcal{G}_2 = \{V_\beta, g_\beta\}$ with $f: \mathcal{G}_1 \to \mathcal{G}_2$ index preserving and such that $E^1_\alpha = E^2_{\beta(\alpha)}$ and f_α is the restriction of the identity map. Such an atlas pair is called admissible for f. Semicompleteness of local isomorphisms and s admissibility of atlas pairs are defined as for general submersions.

1 PROPOSITION: Let $f: X_1 \to X_2$ be an $\mathfrak{M}^1$ map. The following are equivalent: (1) f is an $\mathfrak{M}^1$ local isomorphism. (2) f is a $\mathcal{C}^1$ local isomorphism. (3) f is an $\mathfrak{M}^1$ submersion and $T_x f$ is injective for all $x \in X_1$. (4) f is an $\mathfrak{M}^1$ immersion and $T_x f$ is surjective for all x. (5) $(f^*Tf, 1): \tau_{X_1} \to$

$f^*\tau_{X_2}$ is an $\mathfrak{M}$ VB isomorphism. If X_1 and X_2 are semicomplete and either $\mathfrak{M} \subset \mathscr{L}$ or f is a $\mathcal{C}^1_u$ map then f is semicomplete.

PROOF: Equivalence of (1), (2), (3), (5) by Theorem 4.2. (1) → (4) and (4) → (5) are clear. Q.E.D.

In particular, all of the immersion and submersion results carry over to local isomorphisms. For example, a local isomorphism is an open map and so $f(X_1)$ is open in X_2. If f is a local isomorphism, U is open in X_1 and $f|U$ is injective then $f: U \to f(U)$ is a homeomorphism. In fact,

(5.1) $$\lambda_{\mathcal{G}_1}|U \le \lambda^U_{(\mathcal{G}_1,\mathcal{G}_2)} \le \lambda_{\mathcal{G}_1}$$

if $(\mathcal{G}_1,\mathcal{G}_2)$ is admissible for f and $f|U$ is injective.

2 PROPOSITION: (a) An $\mathfrak{M}^1$ map $f: X_1 \to X_2$ is an $\mathfrak{M}^1$ isomorphism, i.e. has an $\mathfrak{M}^1$ inverse iff it is a bijective, metricly proper $\mathfrak{M}^1$ local isomorphism.

(b) If $f: X_1 \to X_2$ is an injective, metricly proper, semicomplete $\mathfrak{M}^1$ local isomorphism then f is an $\mathfrak{M}^1$ isomorphism onto an open and closed subset of X_2.

(c) If $f: X_1 \to X_2$ is a metricly proper semicomplete $\mathfrak{M}$ local isomorphism then f is a covering space map over $f(X_1)$ which is open and closed in X_2.

(d) If $f: X_1 \to X_2$ is a local homeomorphism onto an open subset of an $\mathfrak{M}$ manifold X_2 then there is a unique $\mathfrak{M}$ metric structure on X_1 such that f is a metricly proper $\mathfrak{M}$ local isomorphism. X_1 is regular if X_2 is. If f is a covering space map and X_2 is semicomplete then X_1 and f are semicomplete.

PROOF: (a): Proposition III.5.5. (b): Lemma 4.1 and (a). (c): Let $(\mathcal{G}_1,\mathcal{G}_2)$ be an s admissible atlas pair for f. For convenience assume f is surjective. Since f is metricly proper, Lemma 4.6 and (4.2) imply that we can choose ρ admissible on X_2 with $f^*\rho > \max(\rho_{G_1}, f^*\rho_{G_2}, 1/\lambda_{(G_1,G_2)})$. If $y \in X_2$ choose β such that $B(g_\beta y, \rho(y)^{-1}) \subset g_\beta(V_\beta)$. Let $d = d_{G_2}$. Let $B = g_\beta^{-1}(B(g_\beta y, (4\rho^d(y)^2)^{-1}))$. Let $x \in f^{-1}B$. $\rho^d(y) \geq \rho(fx) > 1/\lambda_{(G_1,G_2)}(x)$ so we can choose α with $B(h_\alpha x, \rho^d(y)^{-1}) = B(g_{\beta(\alpha)} fx, \rho^d(y)^{-1}) \subset h_\alpha(U_\alpha)$. Since $d(fx,y) < (2\rho^d(y))^{-1}$, $2\rho^d(y) \geq \rho^d(fx)$ $> \rho_{G_2} | g_{\beta(\alpha)}^{-1}(B(g_{\beta(\alpha)} fx, (2\rho^d(y))^{-1}))$. So by the Size Estimate $B \subset g_{\beta(\alpha)}^{-1}(B(g_{\beta(\alpha)} fx, (2\rho^d(y))^{-1})) \subset f(U_\alpha)$. $f|U_\alpha$ is a homeomorphism. So by Lemma II.3.5, $g_\beta \circ f\colon f^{-1}B \to g_\beta B$ is a homeomorphism on each component of the domain. Thus, f is a covering space map. (d): Let $G_2 = \{V_\beta, g_\beta\}$ be admissible for X_2 and ρ be an admissible bound for X_2. Let $\bar{G}_1 = \{f^{-1}(V_\beta), g_\beta \circ f\}$. $(\bar{G}_1, f^*\rho)$ is an $\mathfrak{M}^1$ multiatlas generating the metric structure on X_1 and if G_1 is any atlas subdivision of $\bar{G}_1$ then (G_1, G_2) is an admissible atlas pair for f. Regularity and uniqueness of the X_1 structure follow from the arguments of page 191. If f is a covering space map and G_2 is a convex, s admissible atlas then for each β let $\{U_{(\beta,i)}\}$ range over the components of $f^{-1}(V_\beta)$. $G_1 = \{U_{(\beta,i)}, g_\beta \circ f\}$ is an atlas subdivision of $\bar{G}_1$ with $\lambda_{G_1} = f^*\lambda_{G_2}$. Q.E.D.

3 **COROLLARY:** Assume $\mathfrak{M} \subset \mathcal{L}$. Let X_1, X_2 be semicomplete $\mathfrak{M}^1$ manifolds and O be open in X_1. Let $i\colon X_0 \to X_1$ be a metricly proper $\mathfrak{M}^1$ embedding with $i(X_0) \subset\subset O$. Let $f\colon O \to X_2$ be an $\mathfrak{M}^1$ map (restriction metric strcuture on O). If $(i^*f^*Tf, 1)\colon i^*\tau_{X_1} \to i^*f^*\tau_{X_2}$ is an $\mathfrak{M}$ VB isomorphism then there exists U open in O with $i(X_0) \subset\subset U$ (rel X) such that $f|U$ is an $\mathfrak{M}^1$ local isomorphism. If $f \circ i$ is an $\mathfrak{M}^1$ embedding then U can be chosen so that $f|U$ is injective. In particular, if f is metricly proper and $f \circ i$ is an $\mathfrak{M}^1$ embedding then U can be chosen so that $f|U$ is an $\mathfrak{M}^1$ isomorphism onto an open subset of X_2.

PROOF: By Corollary 4.5 U can be chosen so that $f|U$ is a submersion. Since $Lis(TO; f^*TX_2)$ is open and closed in $Ls(TO; f^*TX_2)$ and f^*Tf maps $i(X_0)$

into the former we can shrink U, if necessary, to get $f|U$ a local isomorphism. If $f \circ i$ is an embedding then $d_1 \sim_{\rho_1} f^*d_2$ on $i(X_0)$ and so by Theorem 3.9 and its Addendum we can shrink U, if necessary, to get $f|U$ injective. If f is metricly proper then $f|U$ is an $\mathfrak{M}^1$ isomorphism by Proposition 2 (a). Q.E.D.

By Proposition 3.12, this applies with i the inclusion of a semicomplete, locally split submanifold of X_1.

An $\mathfrak{M}^1$ immersion $f: X_1 \to X_2$ is called a <u>split immersion</u> if $(f^*Tf,1): \tau_{X_1} \to f^*\tau_{X_2}$ is a split $\mathfrak{M}$ VB injection, i.e. if it has a left inverse. The kernel of such a left inverse is an $\mathfrak{M}$ subbundle π of $f^*\tau_{X_2}$ and $(f^*Tf + i,1): \tau_{X_1} \oplus \pi \to f^*\tau_{X_2}$ is an $\mathfrak{M}$ VB isomorphism. If $\mathfrak{M} \subset \mathcal{L}$ and X_1 and X_2 are semicomplete then π is semicomplete. An $\mathfrak{M}^1$ immersion $f: X_1 \to X_2$ admits a <u>semicomplete $\mathfrak{M}^1$ splitting</u> π if X_1 and X_2 are semicomplete $\mathfrak{M}^2$ manifolds, and so τ_{X_1} and $f^*\tau_{X_2}$ are semicomplete $\mathfrak{M}^1$ bundles and π is a semicomplete $\mathfrak{M}^1$ subbundle of $f^*\tau_{X_2}$ such that $(f^*Tf + i,1): \tau_{X_1} \oplus \pi \to f^*\tau_{X_2}$ is an $\mathfrak{M}$ VB isomorphism. Thus, any split $\mathfrak{M}^2$ immersion of semicomplete $\mathfrak{M}^2$ manifolds admits semicomplete $\mathfrak{M}^1$ splittings.

In order to prove the tubular neighborhood theorem we will need to normalize our exponential maps. If $e: D \to V$ is an $\mathfrak{M}^1$ exponential for a semicomplete $\mathfrak{M}^2$ manifold X, then the vertical tangent (vTe,e) is defined by taking the derivative of $p_2 \circ e$ along the fibers of $\tau_X|D$. It is an $\mathfrak{M}$ VB isomorphism $(vTe,e): (\tau_X|D)^*\tau_X \to (p_2|V)^*\tau_X$. The pull back along $0_X: X \to D$ and $\Delta_X: X \to V$ yields an $\mathfrak{M}$ VB isomorphism $(\gamma_e,1): \tau_X \to \tau_X$. e is called a <u>normalized exponential</u> if $(\gamma_e,1)$ is the identity. In general, $e \circ (\gamma_e)^{-1}: \gamma_e(D) \to V$ is a normalized $\mathfrak{M}$ exponential on X. Thus, we can always normalize if we assume additional smoothness. The exponentials that arise from sprays are normalized.

4 <u>THEOREM</u>: Assume $\mathfrak{M} \subset \mathcal{L}$. Let X_1, X_2 be semicomplete $\mathfrak{M}^2$ manifolds with X_2

admitting normalized $\mathfrak{M}^1$ exponentials. Assume $f\colon X_1 \to X_2$ is an $\mathfrak{M}^1$ immersion admitting a semicomplete $\mathfrak{M}^1$ splitting $\pi\colon E \to X_1$. There exists O open in E with $0_\pi(X_1) \subset\subset O$ and an $\mathfrak{M}^1$ local isomorphism $F\colon O \to X_2$ such that $F\cdot 0_\pi = f$ (0_π is the zero section of π, restriction metric structure on O). If f is an $\mathfrak{M}^1$ embedding then O can be chosen so that $F|O$ is injective. If f is metricly proper then O can be chosen so that $F|O$ is metricly proper. If f is an $\mathfrak{M}^1$ metricly proper embedding then O can be chosen so that F is an $\mathfrak{M}^1$ isomorphism of O onto an open subset of X_2.

<u>PROOF</u>: Let $e\colon D \to V$ be an $\mathfrak{M}^1$ exponential for X_2 with V transversely bounded and so with D vertically bounded. Let $(i,1)\colon \pi \to f^*\tau_{X_2}$ be the inclusion. Let $D_1 = i^{-1}\Phi_f^{-1}D$ and $F = p_2\cdot e\cdot \Phi_f\cdot i\colon D_1 \to X_2$. Clearly, $0_\pi(X_1) \subset\subset D_1$ and F is an $\mathfrak{M}^1$ map with $F\cdot 0_\pi = f$. To apply Corollary 3 and complete the proof it suffices to show that $(0_\pi^*F^*TF,1)\colon 0_\pi^*\tau_E \to f^*\tau_{X_2}$ is an $\mathfrak{M}$ VB isomorphism. We have the commutative diagram:

$$\begin{array}{ccccccc}
(TE)_{D_1} & \xrightarrow{T(\Phi_f\cdot i)} & (T^2X_2)_D & \xrightarrow{Te} & (TX_2\times TX_2)_V & \xrightarrow{Tp_2} & TX_2 \\
\uparrow{\scriptstyle \Phi_{0_\pi}} & & \uparrow{\scriptstyle \Phi_{0_{X_2}}} & & \uparrow{\scriptstyle \Phi_{\Delta_{X_2}}} & & \\
TX_1\oplus E & \xrightarrow[Tf\oplus\Phi_f\cdot i]{} & TX_2\oplus TX_2 & \xrightarrow[\begin{pmatrix} I & I \\ 0 & \Psi_e\end{pmatrix}]{} & TX_2\oplus TX_2 & \xrightarrow{p_2} & TX_2
\end{array}$$

i.e. the tangent map of F on the zero section of π is given by Tf on the tangents along the zero section and $\Psi_e\cdot\Phi_f\cdot i$ on the tangents along the fiber of $\pi|O$. Since e is normalized $\Psi_e = I$ and $(0_\pi^*TF,f)\colon 0_\pi^*\tau_E \to \tau_{X_2}$ pulls back to the VB isomorphism $(f^*Tf + i,1)\colon \tau_{X_2} \oplus \pi \to f^*\tau_{X_2}$ under the identification $\tau_{X_2} \oplus \pi \cong 0_\pi^*\tau_E$. Q.E.D.

6. <u>Transversality</u>: Let $f\colon X_1 \to X_2$ be an $\mathfrak{M}^1$ map and W be a locally split

$\mathfrak{M}^1$ submanifold of X_2. f is <u>transverse to W</u> if there exist admissible atlases $\mathcal{A}_1 = \{U_\alpha, h_\alpha\}$, $\mathcal{A}_2 = \{V_\beta, g_\beta\}$ with $f\colon \mathcal{A}_1 \to \mathcal{A}_2$ index preserving and such that for all α, $(\beta = \beta(\alpha))$ either $V_\beta \cap W = \emptyset$ or $E^1_\alpha = \tilde{E}^1_\alpha \times F_\alpha$ and $E^2_\beta = \tilde{E}^2_\beta \times F_\alpha$ with $g_\beta(V_\beta \cap W) = g_\beta(V_\beta) \cap (\tilde{E}^2_\beta \times 0)$ and the following diagram commutes:

$$\begin{array}{ccc} \tilde{E}^1_\alpha \times F_\alpha \supset h_\alpha(U_\alpha) & \xrightarrow{\ f_\alpha\ } & \tilde{E}^2_\beta \times F_\alpha \\ & {\scriptstyle p_2}\searrow \quad \swarrow{\scriptstyle p_2} & \\ & F_\alpha & \end{array} \tag{6.1}$$

Such an atlas pair $(\mathcal{A}_1, \mathcal{A}_2)$ is said to <u>display the transversality</u>. f is <u>properly transverse to W</u> if X_1 and X_2 are semicomplete, W is a semicomplete locally split $\mathfrak{M}^1$ submanifold and $(\mathcal{A}_1, \mathcal{A}_2)$ exist with $\mathcal{A}_2$ s admissible, inducing a semicomplete submanifold atlas on W and $\rho_1 > 1/\lambda_{(\mathcal{A}_1,\mathcal{A}_2)}$. The last condition is the crucial one. If it holds, we can always throw away the charts of $\mathcal{A}_2$ not related to any α and add in an s admissible atlas on X_2 inducing a semicomplete submanifold atlas on W to get the first two conditions. The demand on $\lambda_{(\mathcal{A}_1,\mathcal{A}_2)}$ is quite strong, as we shall now see.

1 <u>LEMMA</u>: Let $(\mathcal{A}_1, \mathcal{A}_2)$ display the transversality of $f\colon X_1 \to X_2$ to W. If X_1 is regular then

$$\max(\rho_1, 1/\lambda_{\mathcal{A}_1|(X_1 - f^{-1}W)}, 1/\lambda_{(\mathcal{A}_1,\mathcal{A}_2)}) > 1/f^*\lambda_{\mathcal{A}_2|(X_2-W)}. \tag{6.2}$$

If $(\mathcal{A}_1, \mathcal{A}_2)$ displays the proper transversality of $f\colon X_1 \to X_2$ to W then with d_i an admissible metric on X_i:

$$\begin{aligned} \max(\rho_1, 1/\lambda_{\mathcal{A}_1|(X_1 - f^{-1}W)}) &\sim \max(\rho_1, 1/\delta^{d_1}(X_1 - f^{-1}W)) \\ &\sim \max(\rho_1, 1/f^*\lambda_{\mathcal{A}_2|(X_2-W)}) \sim \max(\rho_1, 1/f^*\delta^{d_2}(X_2 - W)). \end{aligned} \tag{6.3}$$

PROOF: In (6.3) the equivalence between the λ and θ torus follows from Proposition III.7.7. That $\max(\rho_1, 1/f^*\theta^{d_2}(X_2-W)) >$ $\max(\rho_1, 1/\theta^{d_1}(X_1-f^{-1}W))$ follows from III.(6.3) because $X_1 - f^{-1}W =$ $f^{-1}(X_2 - W)$. Finally, if $\rho_1 > 1/\lambda_{(G_1,G_2)}$, (6.2) implies the reverse inequality. To prove (6.2) choose $\rho_1 \geq \rho_{G_1}$, let $d_1 = d_{G_1}$ and fix $x \in X_1$. Let $K = \max(\rho_1^{d_1}, 2/\lambda_{G_1|(X_1-f^{-1}W)}, 2/\lambda_{(G_1,G_2)})(x)$ and choose α_1 such that $B(h_{\alpha_1}x, K^{-1}) \subset h_{\alpha_1}(U_{\alpha_1} - f^{-1}W)$ and α such that $B(h_\alpha x, K^{-1}) \subset h_\alpha(U_\alpha)$ and $B(g_\beta fx, K^{-1}) \subset g_\beta(V_\beta)$ with $\beta = \beta(\alpha)$. Since $\rho_{G_1}^{\mathcal{L}} \leq K$ on each ball, it follows that $h_\alpha^{-1}(B(h_\alpha x, K^{-2})) \subset h_{\alpha_1}^{-1}(B(h_{\alpha_1}x, K^{-1})) \subset X_1 - f^{-1}W$. I claim that $B(h_\beta fx, K^{-2}) \subset g_\beta(V_\beta - W)$ implying (6.2). If $V_\beta \cap W = \emptyset$ this is clear. If not then by (6.1), $p_2 B(h_\alpha x, K^{-2}) = p_2 B(g_\beta fx, K^{-2}) = B(p_2 h_\alpha x, K^{-2}) \subset F_\alpha$. The domain ball misses $h_\alpha(f^{-1}W) = h_\alpha(U_\alpha) \cap \tilde{E}_\alpha^1 \times 0$. Hence, $0 \notin B(\alpha_2 h_\alpha x, K^{-2})$ and $B(g_\beta fx, K^{-2}) \cap \tilde{E}_\alpha^2 \times 0 = \emptyset$. Q.E.D.

2 LEMMA: Let $f: X_1 \to X_2$ be an $\mathfrak{M}^1$ map and W be an $\mathfrak{M}^1$ submanifold of X_2. Assume there exist admissible atlases G_1, G_2 with $f: G_1 \to G_2$ index preserving and for ρ_1 an admissible bound on X_1 there exist constants K_1, n_1 such that with $\beta = \beta(\alpha)$ either $V_\beta \cap W = \emptyset$ or $E_\beta^2 = \tilde{E}_\beta^2 \times F_\beta$ with $g_\beta(V_\beta \cap W) = g_\beta(V_\beta) \cap \tilde{E}_\beta^2 \times 0$ and $p_2 Df_\alpha(h_\alpha x): E_\alpha^1 \to F_\beta$ is a split surjection with $K_1\rho_1(x)^{n_1} \geq \theta(p_2 Df_\alpha(h_\alpha x))$. Then f is transverse to W. If $\mathfrak{M} \subset \mathcal{L}$, X_1 and X_2 are semicomplete, W is a semicomplete submanifold of X_2 and $\rho_1 > 1/\lambda_{(G_1,G_2)}$ for the above atlas pair, then f is properly transverse to W.

PROOF: For convenience assume $\rho_1(x) > \max(\theta(p_2 Df_\alpha(h_\alpha x)), \rho_{G_1}(x),$ $\rho(f:G_1,G)(x))$. Let $d_1 = d_{G_1}$. Fix x and let $\epsilon^{-1} =$ $\max(\rho_1^{d_1}(x), 2/\lambda_{(G_1,G_2)}(x))$. Choose α so that $B = B(h_\alpha x, \epsilon) \subset h_\alpha(U_\alpha)$ and $(\beta = \beta(\alpha))$ $B(g_\beta fx, \epsilon) \subset g_\beta(V_\beta)$. Let $K = \rho_1^{d_1}(x)$. If $V_\beta \cap W = \emptyset$ then let $U_x = U_\alpha$, $h_x = h_\alpha$ and $\beta(x) = \beta(\alpha)$. If not, choose j_x a right inverse for $p_2 Df_\alpha(h_\alpha x)$ with $\|j_x\| < K$. Let $P_x = I - j_x \cdot Df_\alpha(h_\alpha x)$ be the projection with image $\tilde{E}^1_x \subset E^1_\alpha$. $\|P_x\| < 2K^2$, $\|f_\alpha|B\|_{\mathfrak{M}^1} < K$, $\|1_B\| < K$ and so defining $F_x = P_x \times p_2 \cdot f_\alpha$ we have $\|F_x\|_{\mathfrak{M}^1} \leq O^*(K)$ and $DF_x(h_\alpha x)$ is an isomorphism with $\theta(DF_x(h_\alpha x)) \leq O^*(K)$. By the Inverse Function Theorem: for some $0 < r_1 < \epsilon$, $0 < r_2$, F_x is an $\mathfrak{M}^1$ isomorphism from $B(h_\alpha x, r_1)$ onto an open set containing $B(F_x(h_\alpha x), r_2)$ with $\|F_x^{-1}\|_{\mathfrak{M}^1} \leq O^*(K)$. If $\mathfrak{M} \subset \mathcal{L}$ then by the Size Estimate we can choose r_1 and r_2 so that $r_1^{-1}, r_2^{-1} \leq O^*(\epsilon^{-1})$. Let $U_x = h_\alpha^{-1} F_x^{-1} B(F_x(h_\alpha x), r_2)$, $h_x = F_x \cdot h_\alpha$ and $\beta(x) = \beta(\alpha)$. $\bar{\mathfrak{a}}_1 = \{U_x, h_x\}$ and $\mathfrak{a}_2$ display the transversality. If $\rho_1 > 1/\lambda_{(\mathfrak{a}_1, \mathfrak{a}_2)}$ and $\mathfrak{M} \subset \mathcal{L}$ then $\rho_1 > 1/\lambda_{(\bar{\mathfrak{a}}_1, \mathfrak{a}_2)}$ and we can adjust $\mathfrak{a}_2$ as mentioned above to make sure that $(\bar{\mathfrak{a}}_1, \mathfrak{a}_2)$ properly displays. Q.E.D.

Clearly, if f is (properly) transverse to W then $f^{-1}W$ is a locally split (semicomplete) $\mathfrak{M}^1$ submanifold of X_1. In fact, if $(\mathfrak{a}_1, \mathfrak{a}_2)$ (properly) displays the transversality then $\mathfrak{a}_1$ induces a locally split (semicomplete) $\mathfrak{M}^1$ submanifold atlas on $f^{-1}W$. Also:

$$(6.4) \qquad T(f^{-1}W) = (Tf)^{-1}(TW) \subset TX_1.$$

Furthermore, if $(q,1)\colon \tau_{X_2}|W \to \tau_{X_2}/\tau_W$ is the quotient bundle map then $(f^*(q \cdot Tf), 1)\colon \tau_{X_1}|f^{-1}W \to f^*(\tau_{X_2}/\tau_W)$ is a locally split $\mathfrak{M}$ VB surjection with kernel $\tau_{f^{-1}W}$.

3 <u>THEOREM</u>: Assume $(\mathfrak{M}_1, \mathfrak{M}, \mathcal{L})$ is a standard triple with $\mathfrak{M} \subset \mathcal{L}$. Let X_2 be a semicomplete $\mathfrak{M}_1^2$ manifold admitting $\mathfrak{M}_1^1$ exponentials and X_1 be a semicomplete $\mathfrak{M}^1$ manifold. Let W be a semicomplete, locally split $\mathfrak{M}^1$ submanifold of X_2. If $f\colon X_1 \to X_2$ is properly transverse to W and $f^{-1}W \subset\subset U$ in X_1 then $\{g \in \mathfrak{M}^1(X_1, X_2)\colon g$ is properly transverse to W and $g^{-1}W \subset\subset U\}$ is open in the topological space $\mathfrak{M}^1(X_1, X_2)$. In fact, it is open in the topology induced from $\mathcal{C}^1(X_1, X_2)$.

<u>PROOF</u>: First, we dispose of the side condition involving U. Let U_1 be

open with $f^{-1}W \subset\subset U_1 \subset\subset U$. Let d_i be an admissible metric on X_i $(i = 1,2)$. Since $f^{-1}W \subset\subset U_1$, $X_1 - U_1 \subset\subset X_1 - f^{-1}W$. Hence, $\rho_1 \sim \max(\rho_1, \delta^{d_1}(X_1 - f^{-1}W)^{-1})$ on $X_1 - U_1$ and so by (6.3) there exist constants K,n such that $K\rho_1^n \geq 1/f^*\delta^{d_2}(X_2 - W)$ on $X_1 - U_1$. If $d_2(fx,gx) < (K\rho_1(x)^n)^{-1}$ for all $x \in X_1$ then $gx \in X_2 - W$ for all $x \in X_1 - U_1$, i.e. $g^{-1}W \subset U_1 \subset\subset U$. Thus, $\{g \in \mathfrak{M}^1(X_1,X_2): g^{-1}W \subset\subset U\}$ is a $\mathcal{C}$ neighborhood of f. It suffices to show that $\{g \in \mathfrak{M}^1(X_1,X_2): g$ is properly transverse to $W\}$ is a $\mathcal{C}^1$ neighborhood of f. We will apply Proposition IV.3.10. Let $(\mathcal{G}_1,\mathcal{G}_2)$ properly display the transversality with $\mathcal{G}_1$ star bounded (subdivide $\mathcal{G}_1$ if necessary) and let $V = V^{-1} \in \mathfrak{U}_{X_2}$ induce the V-trimming $\{\tilde{V}_\beta\}$ of $\mathcal{G}_2$. Recall $\mathcal{G}_f = \{U_\alpha \cap f^{-1}\tilde{V}_\beta, h_\alpha\}$ indexed by pairs (α,β). Let $d_i = d_{\mathcal{G}_i}$ and choose $\rho_1 > \max(\rho_f^{d_1}, \rho_{\mathcal{G}_1}^{d_1}, f^*\rho_{\mathcal{G}_2}^{d_2}, 1/\lambda_{(\mathcal{G}_1,\mathcal{G}_2)}, 1/f^*\lambda_{\tilde{\mathcal{G}}_2})$. Fix $x \in X_1$ and let $K = \rho_1(x)$. Choose α,β such that with $\beta_1 = \beta(\alpha)$, $B(h_\alpha x, K^{-1}) \subset h_\alpha(U_\alpha)$, $B(g_\beta fx, K^{-1}) \subset g_\beta(\tilde{V}_\beta)$ and $B(g_{\beta_1} fx, K^{-1}) \subset g_{\beta_\alpha}(V_{\beta_1})$. Let $\rho_\alpha = \sup \rho_1|U_\alpha$ and let $g \in V_f^{\mathfrak{M}}$ with $\|g_{\beta\alpha} - f_{\beta\alpha}\|_{\mathcal{C}^1} < (2\rho_\alpha^2)^{-1}$ for all α,β. The $\mathcal{G}_2$ transition map $g_{\beta_1\beta}$ has $\mathcal{L}$ norm less than K on the balls and so maps $B(g_\beta fx, K^{-2})$ into $B(g_{\beta_1} fx, K^{-1})$. Hence, $B(g_\beta gx, (2K^2)^{-1}) \subset B(g_\beta fx, K^{-2})$ pulls back under g_β^{-1} into V_{β_1}. Since $g_{\beta\alpha}$ has $\mathcal{L}$ norm $< 2K$ it maps $B(h_\alpha x, (4K^3)^{-1})$ into $B(g_\beta gx, (2K^2)^{-1})$. Let $U_x = (B(h_\alpha x, (4K^3)^{-1})$, $h_x = h_\alpha$ and $\beta(x) = \beta_1$. $U_x \subset U_\alpha \cap f^{-1}\tilde{V}_\beta \cap f^{-1}V_{\beta_1}$ and on $h_\alpha(U_x)$ $\|g_{\beta_1\alpha} - f_\alpha\|_{\mathfrak{M}^1} < (2\rho_\alpha^2)^{-1}$ by the Gluing Property. Either $V_{\beta_1} \cap W = \emptyset$ or $p_2 \cdot f_\alpha = p_2$. In the latter case $\|p_2 \cdot Dg_{\beta_1\alpha} - p_2\| < 1/2$ and so $p_2 \cdot Dg_{\beta_1\alpha}$ is a split surjection with $\rho(p_2 \cdot Dg_{\beta_1\alpha}) < 2$. $\mathcal{G}_g = \{U_x, h_x\}$ and $\mathcal{G}_2$ thus satisfy the hypotheses of Lemma 2. Q.E.D.

We now relate the atlas version of transversality to the classical definition.

4 <u>THEOREM</u>: Let $f: X_1 \to X_2$ be an $\mathfrak{M}^1$ map. Let W be an $\mathfrak{M}^1$ locally split submanifold of X_2 and $A \subset f^{-1}W$. Let $\| \|_i$, ρ_i be admissible Finsler and

bound for τ_{X_i}. Assume that for $x \in A$ the composition:

$$T_x X_1 \xrightarrow{T_x f} T_{fx} X_2 \xrightarrow{q_{fx}} T_{fx}X_2/T_{fx}W$$

is a split surjection (q_{fx} is the quotient map) and $\rho_1(x) \succ \theta(q_{fx} \cdot T_x f)$ as x varies in A (θ defined via $\| \ \|_1, \| \ \|_2$).

(a) If $A = f^{-1}W$ then f is transverse to W.

(b) If $\mathfrak{M} \subset \mathcal{L}$, X_1 and X_2 are semicomplete and W is a semicomplete submanifold then there exist U open in X_1 with $A \subset\subset U$ and such that $f|U: U \to X_2$ is properly transverse to W with the canonical refinement structure on U.

(c) If $\mathfrak{M} \subset \mathcal{L}$, $A = f^{-1}W$, X_i is semicomplete with admissible metric d_i ($i = 1,2$) and W is a semicomplete submanifold of X_2, then f is properly transverse to W iff $\max(\rho_1, 1/\theta^{d_1}(X_1 - f^{-1}W)) \sim \max(\rho_1, 1/f^*\theta^{d_2}(X_2 - W))$.

<u>PROOF</u>: We begin with (b). Let $G_2 = \{V_\beta, g_\beta\}$ be an s admissible atlas on X_2 inducing a semicomplete submanifold atlas on W, i.e. either $V_\beta \cap W = \emptyset$ or $E^2_\beta = \tilde{E}^2_\beta \times F_\beta$ with $g_\beta(V_\beta \cap W) = g_\beta(V_\beta) \cap (\tilde{E}^2_\beta \times 0)$. Let $G_1 = \{U_\alpha, h_\alpha\}$ be a star-bounded, s admissible atlas on X_1. Let $d_i = d_{G_i}$ ($i = 1,2$). Choose $\rho_1 \geq 2 \max(\rho_f^{d_1}, 1/\lambda_{G_1}, 1/\lambda_{G_2})$ and such that if $x \in A \cap U_\alpha \cap f^{-1}V_\beta$ $\rho_1(x) > \theta(p_2 \cdot Df_{\beta\alpha}(h_\alpha x))$. This is possible because $\max(\rho_1(x), \theta q_{fx} \cdot T_x f)) \succ \sup\{\max(\rho_1(x), \theta(p_2 \cdot Df_{\beta\alpha}(h_\alpha x))): x \in U_\alpha \cap f^{-1}V_\beta\}$ as x varies in A. Fix $x \in X_1$ and let $K = \rho_1(x)$. We can choose α and β such that $B(h_\alpha x, K^{-2}) \subset h_\alpha(U_\alpha)$, $B(g_\beta fx, K^{-1}) \subset g_\beta(V_\beta)$ and so $f_{\beta\alpha}(B(h_\alpha x, K^{-2})) \subset B(g_\beta fx, K^{-1})$. Now assume $x \in A$. Let j_x be a right inverse for $p_2 \cdot Df_{\beta\alpha}(h_\alpha x)$ with $\|j_x\| < K$. For $u \in B(h_\alpha x, K^{-2})$ $\|p_2 \cdot Df_{\beta\alpha}(u) \cdot j_x - I\| < K^{-1} \leq 1/2$. Hence, $p_2 \cdot Df_{\beta\alpha}(u)$ is a split surjection with $\theta(p_2 \cdot Df_{\beta\alpha}(u)) < 2K$. Let $U = \cup\{h_\alpha^{-1}(B(h_\alpha x, (2K^2)^{-1})): x \in A\}$, $U_x = h_\alpha^{-1}(B(h_\alpha x, K^{-2}))$, $h_x = h_\alpha$ and $\beta(x) = \beta$ for $x \in A$. If $\bar{G}_1 = \{U_x, h_x\}|U$ and $y \in h_\alpha^{-1}(B(h_\alpha x, (2K^2)^{-1})$, it is easy to check that

$B(h_\alpha y, \min((2K^2)^{-1}, \delta^{d_1}(U)(y))) \subset B(h_\alpha x, K^{-2}) \cap h_\alpha(U)$ and $B(g_\beta fy, (2K)^{-1}) \subset g_\beta(V_\beta)$. Hence, $\bar{\mathfrak{A}}_1$ is s admissible on U and $\rho_U > 1/\lambda_{(\bar{\mathfrak{A}}_1, \mathfrak{A}_2)}$. By Lemma 2, $f|U$ is properly transverse to W, proving (b). If $A = f^{-1}W$, let $U_x = h_\alpha^{-1}(B(h_\alpha x, K^{-2})) - f^{-1}W$, $h_x = h_\alpha$ for $x \in X_1 - f^{-1}W$. Replacing V_β by $V_\beta - W$ for $\beta = \beta(x)$ we get the atlas $\bar{\mathfrak{A}}_2$. $\bar{\bar{\mathfrak{A}}}_1 = \{U_x, h_x\}$ indexed by $x \in X_1$ and $\bar{\mathfrak{A}}_2$ imply the transversality of f with W by Lemma 2 and some easy adjustments of the above construction proves (a). It is easy to prove that on $X - f^{-1}W$, $\lambda_{(\bar{\bar{\mathfrak{A}}}_1, \bar{\mathfrak{A}}_2)} \geq \min(1/\rho_1, \delta^{d_1}(X_1 - f^{-1}W), f^*\delta^{d_2}(X_2 - W))$. Let $f^{-1}W \subset\subset U_1 \subset\subset U$ and so $X_1 - U_1 \subset\subset X_1 - f^{-1}W$. On U_1, $\rho_1 \sim \rho_U > 1/\lambda_{(\bar{\bar{\mathfrak{A}}}_1, \bar{\mathfrak{A}}_2)}$ by the proof of (b). On $X_1 - U_1$, $\max(\rho_1, 1/f^*\delta^{d_2}(X_2 - W)) \sim \max(\rho_1, 1/\delta^{d_1}(X_1 - f^{-1}W), 1/f^*\delta^{d_2}(X_2 - W)) \geq 1/\lambda_{(\bar{\bar{\mathfrak{A}}}_1, \bar{\mathfrak{A}}_2)}$. So with the hypotheses of (c) $\rho_1 > 1/\lambda_{(\bar{\bar{\mathfrak{A}}}_1, \bar{\mathfrak{A}}_2)}$ on $X_1 - U_1$, too. Thus, the inequality holds on X_1 and f is properly transverse to W by Lemma 2 again. Q.E.D.

5 <u>COROLLARY</u>: Assume $\mathfrak{M} \subset \mathcal{L}$. Let $f: X_1 \to X_2$ be an $\mathfrak{M}^1$ map transverse to W with X_1, X_2 semicomplete and W a semicomplete submanifold of X_2. There exists U open in X_1 with $f^{-1}W \subset\subset U$ such that $f|U: U \to X_2$ is properly transverse to W with the canonical refinement structure on U. If f is metricly proper, then f is properly transverse to W iff $\{f^{-1}V: W \subset\subset V\}$ is a base for the full neighborhoods of $f^{-1}W$.

<u>PROOF</u>: The $f|U$ result from Theorem 4(b) and the obvious converse of Theorem 4(a). If f is metricly proper and properly transverse then $\max(\rho_1, 1/\delta^{d_1}(X_1 - f^{-1}W)) \sim f^*\max(\rho_2, 1/\delta^{d_2}(X_2 - W))$. This easily implies that if $f^{-1}W \subset\subset U_1$ and so $X_1 - U_1 \subset\subset X_1 - f^{-1}W$ then $f(X_1 - U_1) \subset\subset X_2 - W$ and so $W \subset\subset X_2 - f(X_1 - U_1)$ (by Lemma III.6.3(b)). But $U_1 \supset f^{-1}(X_2 - f(X_1 - U_1))$. Conversely, if $f|U$ is properly transverse,

$W \subset\subset V$ and $f^{-1}V \subset\subset U$ then on $f^{-1}V$, $\max(\rho_1, 1/\theta^{d_1}(X_1 - f^{-1}W)) \sim \max(\rho_U, 1/\theta^{d_1}(U - f^{-1}W)) \sim \max(\rho_U, 1/f^*\theta^{d_2}(X_2 - W)) \sim \max(\rho_1, 1/f^*\theta^{d_2}(X_2 - W))$ while on $X_1 - f^{-1}V$, $\rho_1 > 1/\theta^{d_1}(X_1 - f^{-1}W)$ and $f^*\rho_2 > 1/f^*\theta^{d_2}(X_2 - W)$. Result by Theorem 4(c). Q.E.D.

6 <u>COROLLARY</u>: Let X be an $\mathfrak{M}^1$ manifold, W_j be an $\mathfrak{M}^1$ submanifold of X with inclusion map i_j $(j = 1,2)$. We say W_1 intersects W_2 transversely if the following equivalent conditions hold: (1) i_1 is transverse to W_2. (2) i_2 is transverse to W_1. (3) Defining θ via an admissible Finsler on τ_X, $T_xi_1 + T_xi_2\colon T_xW_1 \times T_xW_2 \to T_xX$ is a split surjection and $\rho(x) > \theta(T_xi_1 + T_xi_2)$ as x varies over $W_1 \cap W_2$. If $\mathfrak{M} \subset \mathcal{L}$, X and the submanifolds are semicomplete then i_1 is properly transverse to W_2 iff i_2 is properly transverse to W_1 iff $\{V_1 \cap V_2\colon W_1 \subset\subset V_1$ and $W_2 \subset\subset V_2\}$ is a base for the full neighborhoods of $W_1 \cap W_2$. In this case, we say W_1 intersects W_2 properly transversely.

<u>PROOF</u>: Consider the commutative diagram:

$$\begin{array}{ccc}
T_xW_1 \times T_xW_2 & \xrightarrow{\;T_xi_1 + T_xi_2\;} & T_xX \\
\downarrow{\scriptstyle p_1} & & \downarrow{\scriptstyle q_2} \\
T_xW_1 & \xrightarrow[\;q_2\cdot T_xi_1\;]{} & T_xX/T_xW_2
\end{array}$$

p_1 and q_2 are split surjections with isomorphic kernels. So if j is a right inverse for $T_xi_1 + T_xi_2$ then $p_1\cdot j$ factors through q_2 to get a right inverse for $q_2\cdot T_xi_1$. Conversely, if j_1 is a right inverse for $q_2\cdot T_xi_1$ then $j_1\cdot q\colon T_xX \to T_xW_1 \subset T_xW_1 \times T_xW_2$ fails to be a right inverse for $T_xi_1 + T_xi_2$ by an error map of T_xX into T_xW_2. Subtracting the error term we get a right inverse. Thus, by Theorem 4(a), (1) is equivalent to (3). Since (3) is symmetric, (2) is equivalent to (3).

For the proper condition note that i_1 is metricly proper and so

Corollary 5 applies. By Proposition 3.11(a) $\{V \cap W_1 : W_1 \cap W_2 \subset\subset V\}$ is a base for the full neighborhoods of $W_1 \cap W_2$ in W_1. Hence, the given condition implies that i_1 is properly transverse to W_2. Conversely, if i_1 is properly transverse to W_2, let $W_1 \cap W_2 \subset\subset U$ in X. By Corollary 5 we can choose $V \in \mathfrak{N}_X$ such that $V = V^{-1}$ and $(V \cdot V[W_2]) \cap W_1 \subset\subset U$. Hence, by shrinking V we can assume $V \cdot [W_1 \cap V \cdot V[W_2]] \subset U$. Now if $x \in V[w_1] \cap V[w_2]$, $w_i \in W_i$ $(i = 1,2)$ then $w_1 \in W_1 \cap V \cdot V[W_2]$ and so $x \in U$, i.e. $V[W_1] \cap V[W_2] \subset U$. This proves the converse. Finally, note that this condition is symmetric. Q.E.D.

Let $f_i : X_i \to X_0$ $(i = 1,2)$ be $\mathfrak{M}^1$ maps. f_1 is (<u>properly</u>) <u>transverse</u> to f_2 if $f_1 \times f_2 : X_1 \times X_2 \to X_0 \times X_0$ is (properly) transverse to the diagonal Δ_{X_0}. By Theorem 4 and the addendum to Lemma 1.4, f_1 is transverse to f_2 iff for all $(x_1, x_2) \in X_{12} = f_1 \times f_2^{-1}(\Delta_{X_0})$, $T_{x_1}f_1 - T_{x_2}f_2 : T_{x_1}X_1 \times T_{x_2}X_2 \to T_xX_0$ $(x = f_ix_i \ i = 1,2)$ is a split surjection and with respect to admissible Finslers $\| \ \|_i$ on T_{X_i} $(i = 0,1,2)$ $\max(\rho_1(x_1), \rho_2(x_2)) > \theta(T_{x_1}f_1 - T_{x_2}f_2)$. Clearly, transversality is symmetric in f_1 and f_2. Foregoing the symmetry we define:

$$\text{(6.5)} \qquad \begin{array}{ccc} X_{12} & \xrightarrow{\ F_{f_1}\ } & X_2 \\ {\scriptstyle f_1 * f_2}\downarrow & & \downarrow{\scriptstyle f_2} \\ X_1 & \xrightarrow[\ f_1\ ]{} & X_0 \end{array}$$

by $X_{12} = (f_1 \times f_2)^{-1}(\Delta) \subset X_1 \times X_2$. $f_1 * f_2$, F_{f_1} are the restrictions of the projections. The inclusion of X_{12} is an $\mathfrak{M}^1$ locally split, metricly proper immersion which is a homeomorphism onto its image. In the proper case, it is an $\mathfrak{M}^1$ metricly proper, semicomplete embedding. It follows from the Factoring Lemma 3.12 that the above is a pull back diagram, i.e. if $g_i : X \to X_i$ $(i = 1,2)$ are $\mathfrak{M}^1$ maps with $f_1 \cdot g_1 = f_2 \cdot g_2$ then there is a unique $\mathfrak{M}^1$ map $g_{12} : X \to X_{12}$ projecting to g_1 and g_2.

7 **PROPOSITION**: (a) Let $f_1: X_1 \to X_2$ be (properly) transverse to W and $f_0: X_0 \to X_1$. $f_1 \circ f_0$ is (properly) transverse to W iff f_0 is (properly) transverse to $f_1^{-1}(W)$.

(b) If f_1 is transverse to f_2 and f_2 is metricly proper then $f_1 * f_2$ is metricly proper.

(c) If f_1 is transverse to f_2 and f_2 is an immersion then $f_1 * f_2$ is an immersion.

(d) If f_2 is a (semicomplete) submersion (the domain of f_1 is semicomplete and $\mathfrak{M} \subset \mathcal{L}$) then f_1 is (properly) transverse to f_2 and $f_1 * f_2$ is a (semicomplete) submersion.

(e) If f_2 is the inclusion of a (semicomplete) submanifold W then f_1 is (properly) transverse to f_2 iff f_1 is properly transverse to W. In that case, $f_1 * f_2$ is an isomorphism onto $f_1^{-1}(W)$.

(f) Let $f: X_1 \to X_2$ be a (semicomplete) submersion and W_i be a locally split (semicomplete) submanifold of X_i $(i = 1,2)$. $f|W_1: W_1 \to X_2$ is (properly) transverse to W_2 iff W_1 intersects $f^{-1}(W_2)$ (properly) transversely.

PROOF: We will sketch the arguments, leaving the θ estimates to the reader. (a): Let $x_0 \in (f_1 \circ f_0)^{-1}W$, $x_1 = f_0 x_0$ and $x_2 = f_1 x_1$. Consider the diagram

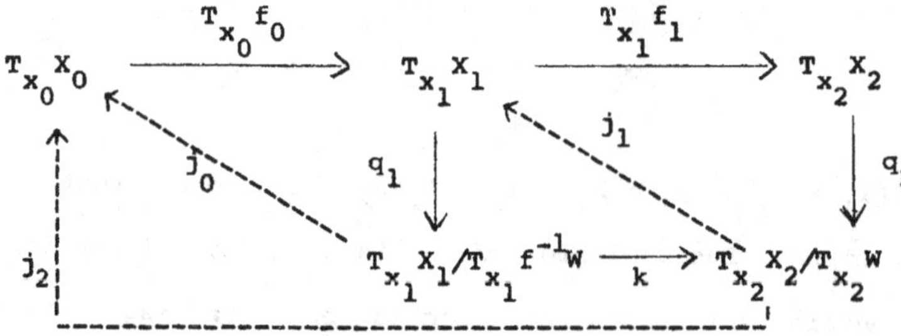

Since f_1 is transverse to W, k exists and is an isomorphism with $\theta(k) \leq \theta(q_2 \circ T_{x_1} f_1)$ by (6.4) and $q_2 \circ T_{x_1} f_1$ has a right inverse j_1. If f_0 is transverse to $f_1^{-1}W$ and j_0 is a right inverse for $q_1 \circ T_{x_0} f_0$ then

$j_2 = j_0 \cdot q_1 \cdot j_1 = j_0 \cdot k^{-1}$ is a right inverse for $q_2 \cdot T_{x_0} f_1 \cdot f_0$. If $f_1 \cdot f_0$ is transverse to W and j_2 is a right inverse for $q_2 \cdot T_{x_0}(f_1 \cdot f_0)$ then $j_2 \cdot k$ is a right inverse for $q_1 \cdot T_{x_0} f_0$. In the proper case, $\max(\rho_1, 1/\delta^{d_1}(X_1 - f_1^{-1}W)) \sim \max(\rho_1, 1/f_1{}^*\delta^{d_2}(X_2 - W))$ and so pulling back and using $\rho_0 > f_0{}^*\rho_1$, $\max(\rho_0, 1/f_0{}^*\delta^{d_1}(X_1 - f_1^{-1}W)) \sim \max(\rho_0, 1/(f_1 \cdot f_0)^*\delta^{d_2}(X_2 - W))$. Thus, one is equivalent to $\max(\rho_0, 1/\delta^{d_0}(X_0 - (f_1 \cdot f_0)^{-1}W))$ iff the other is.

(b): On X_{12}, $p^*\rho_1 > p_1{}^*f_1{}^*\rho_0 = p_2{}^*f_2{}^*\rho_0 \sim p_2{}^*\rho_2$. (c) and (d): Assume f_1 is transverse to f_2. For $(x_1, x_2) \in X_{12}$ let P be a projection of $T_{x_1}X_1 \times T_{x_2}X_2$ onto $T_{(x_1,x_2)}X_{12} = K(T_{x_1}f_1 - T_{x_2}f_2)$. If $x = f_i x_i$ $(i = 1,2)$ and $S\colon T_x X_0 \to T_{x_2}X_2$ is a left inverse for $T_{x_2}f_2$ then $v_1 \to P(v_1, ST_{x_1}f(v_1))$ is a left inverse for $T_{(x_1,x_2)}f_1{}^*f_2$ = the restriction of p_1 to $T_{(x_1,x_2)}X_{12}$. For if $(v_1, v_2) \in T_{(x_1,x_2)}X_{12}$, $P(v_1, ST_{x_1}f_1(v_1)) = P(v_1, ST_{x_2}f_2(v_2)) = (v_1, v_2)$. If $j\colon T_x X_0 \to T_{x_2}X_2$ is a right inverse for $T_{x_2}f_2$ then $v_1 \to (v_1, jT_{x_1}f_1(v_1))$ is a right inverse for $T_{(x_1,x_2)}f_1{}^*f_2$. For $T_{x_1}f_1(v_1) = T_{x_2}f_2(jT_{x_1}f_1(v_1))$. Thus, if f_1 is transverse to f_2 and f_2 is an immersion or submersion then $f_1{}^*f_2$ is an immersion or submersion. Now if $f\colon X_1 \to X_2$ is a submersion and W is a submanifold of X_2 then $q \cdot T_x f\colon T_x f_1 \to T_{x_2} f/T_{x_2}W$ is a split surjection with $\theta(q \cdot T_x f) \le \theta(q)\theta(T_x f)$. Hence, f is transverse to W. If $\mathfrak{M} \subset \mathcal{L}$ and everyone is semicomplete then f is properly transverse to W by Theorem 4 (c) and Lemma 4.6. To prove the transversality of maps result, note that $f_1 \times f_2\colon X_1 \times X_2 \to X_0 \times X_0$ is the composition of $1 \times f_2\colon X_1 \times X_2 \to X_1 \times X_0$ and $f_1 \times 1\colon X_1 \times X_0 \to X_0 \times X_0$. $(f_1 \times 1)^{-1}\Delta_{X_0}$ = graph f_1. Since, $1 \times f_2$ is a (semicomplete) submersion it is (properly) transverse to graph f_1. By (a), it suffices to show that $f_1 \times 1$ is (properly) transverse to Δ_{X_0}, i.e. f_1 is (properly) transverse to 1_{X_0}. $I - T_{x_1}f_1\colon T_{f_1x_1}X_0 \times T_{x_1}X_1 \to T_{f_1x_1}X_0$ is the composition of the isomorphism $(I - T_{x_1} f_1) \times p_2$ with projection p_1. Transversality follows. If $fx_1 \notin B^{d_0}(X_0,r)$, i.e. $\{x_1\} \times B^{d_0}(x_0,r) \cap \text{graph } f = \emptyset$ then $[B^{d_0}(fx_1,r/2) \times B^{d_0}(x_0,r/2)] \cap \Delta_{X_0} = \emptyset$. Thus, in the semicomplete case, proper transversality follows. (e): Consider the commutative diagrams (Γ_{f_1} is the graph map of f_1):

$$\begin{array}{ccccc} & & X_1 \times X_0 & & \\ & \overset{\Gamma_{f_1}}{\nearrow} & & \overset{p_2}{\searrow} & \\ X_1 & & \xrightarrow{\quad f_1 \quad} & & X_0 \end{array} \qquad \begin{array}{ccccc} & & X_1 \times X_0 & & \\ & \overset{1\times f_2}{\nearrow} & & \overset{f_1\times 1}{\searrow} & \\ X_1 \times W & & \xrightarrow{\quad f_1 \times f_2 \quad} & & X_0 \times X_0 . \end{array}$$

By (d), p_2 is (properly) transverse to W and $f_1 \times 1$ is (properly) transverse to Δ. Thus, by (a) f_1 is (properly) transverse to W iff Γ_{f_1} is (properly) transverse to $X_1 \times W$ and $f_1 \times f_2$ is (properly) transverse to Δ iff $1 \times f_2$ is (properly) transverse to graph (f_1). Since Γ_{f_1} differs by an isomorphism from the inclusion of graph (f_1), the result follows from Corollary 6. (f): From (a) and (d). Q.E.D.

It follows from (d) that if $f_1: X \to X_1$ and $f_2: \bar{X}_1 \to X_2$ are $\mathfrak{M}^1$ submersions and $(G,g): f_1 \to f_2$ is an $\mathfrak{M}^1$ map (i.e. $G: X \to \bar{X}$ and $g: X_1 \to X_2$ are $\mathfrak{M}^1$ maps with $f_2 \circ G = g \circ f_1$) then (G,g) factors to an $\mathfrak{M}^1$ map $(g^*G,1): f_1 \to g^*f_2$ of $\mathfrak{M}^1$ submersions. Note that g^*G is metricly proper iff $f_1 \times G$ is.

Define for $f: X_1 \to X_2$ an $\mathfrak{M}^1$ submersion the $\mathfrak{M}$ subbundle τ_f of τ_{X_1} to be the vertical tangent bundle, i.e. τ_f is the kernel of locally split VB surjection $(f^*Tf,1): \tau_{X_1} \to f^*\tau_{X_2}$.

8 <u>LEMMA:</u> Let $f_2: X_2 \to X_0$ be an $\mathfrak{M}^1$ submersion and $f_1 : X_1 \to X_0$ be an $\mathfrak{M}^1$ map. $((F_{f_1})^*TF_{f_1},1)$ is an $\mathfrak{M}$ VB isomorphism of $\tau_{f_1^*f_2}$ with $(F_{f_1})^*\tau_f$ (cf. diagram (6.5)).

<u>PROOF:</u> Let $x_i \in X_i$ $(i = 0,1,2)$ with $x_0 = f_i x_i$ $(i = 1,2)$. In the pull

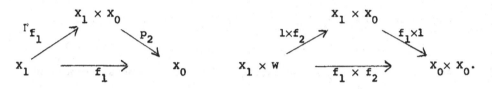

back diagram:

$$\begin{array}{ccc} T_{(x_1,x_2)}X_{12} & \xrightarrow{TF_{f_1}} & T_{x_2}X_2 \\ {\scriptstyle Tf_1*f_2}\downarrow & & \downarrow{\scriptstyle Tf_2} \\ T_{x_1}X_1 & \xrightarrow[Tf_1]{} & T_{x_0}X_0 \end{array}$$

$K(Tf_1*f_2) = 0 \times K(Tf_2)$ and so TF_{f_1} is an isometric isomorphism on the kernels once admissible Finslers are chosen. Q.E.D.

If $f_1x_1 = x_0$ then $F_{f_1}: (f_1*f_2)^{-1}(x_1) \to f_2^{-1}(x_0)$ is a metricly proper, bijective $\mathfrak{M}^1$ map. So by the above lemma it is an $\mathfrak{M}^1$ isomorphism.

9 <u>PROPOSITION</u>: Let $f_1: X \to X_1$ and $f_2: \bar{X} \to X_2$ be $\mathfrak{M}^1$ submersions. Let $(G,g): f_1 \to f_2$ be an $\mathfrak{M}^1$ map.

(a) If g is a metricly proper, $\mathfrak{M}^1$ immersion and G is metricly proper, then G is a $\mathfrak{M}^1$ immersion iff $(G^*TG,1): \tau_{f_1} \to G^*\tau_{f_2}$ is a locally split $\mathfrak{M}$ VB injection.

(b) If g is a $\mathfrak{M}^1$ submersion, then G is an $\mathfrak{M}^1$ submersion iff $(G^*TG,1)$ is a locally split $\mathfrak{M}$ VB surjection.

(c) If g is an $\mathfrak{M}^1$ local isomorphism, then G is an $\mathfrak{M}^1$ local isomorphism iff $(G^*TG,1)$ is an $\mathfrak{M}$ VB isomorphism.

<u>PROOF</u>: By Lemma 8 and Proposition 7 (c),(d) we reduce to the case where g is the identity by factoring $G = F_g \cdot (g^*G)$. In that case, Finsler choices make T_xG a map of split B-space surjections:

$$\begin{array}{ccc} T_xX & \xrightarrow{T_xf_1} & T_{f_1x}X_1 \\ {\scriptstyle T_xG}\downarrow & & \updownarrow{\scriptstyle =} \\ T_{Gx}\bar{X} & \xrightarrow[T_{Gx}f_2]{} & T_{f_2x}X_2 \end{array}$$

It is easy to check that T_xG is a split injection iff its restriction $(T_xG)_K$ to the kernels is a split injection with $\theta(T_xG) \le O^*(\theta((T_xG)_K), \theta(T_xf_1), \theta(T_{Gx}f_2))$ and $\theta((T_xG)_K) \le O^*(\theta(T_xG), \theta(T_xf_1), \theta(T_{Gx}f_2))$. Similarly, for surjections. Q.E.D.

We now prove the stability of the inverse manifold under perturbation.

10 <u>THEOREM</u>: Assume $(\mathfrak{M}_1, \mathfrak{M}, \mathcal{L})$ is a standard triple with $\mathfrak{M} \subset \mathcal{L}$. Let $f: X_1 \to X$ be a semicomplete $\mathfrak{M}^1$ submersion. Let W be a semicomplete, locally split $\mathfrak{M}^1$ submanifold of a semicomplete $\mathfrak{M}_1^2$ manifold X_2, admitting $\mathfrak{M}_1^1$ exponentials. $\{G \in \mathfrak{M}^1(X_1, X_2)$: G is properly transverse to W and $f: G^{-1}W \to X$ is an $\mathfrak{M}^1$ isomorphism$\}$ is open in $\mathfrak{M}^1(X_1, X_2)$. In fact, it is open in the topology induced from $\mathcal{C}^1(X_1, X_2)$.

<u>PROOF</u>: Let G be properly transverse to W with $f|W_1 (= G^{-1}W)$ an $\mathfrak{M}^1$ isomorphism onto X. We can regard f as a retraction of X_1 onto W_1 by identifying X with W_1 via $f|W_1$ and identifying f with $(f|W_1)^{-1} \circ f$.

The proof is an application of Proposition IV.3.10 once suitable atlases have been chosen. Ideally, we want atlases $\mathfrak{a}, \mathfrak{a}_1$ and $\mathfrak{a}_2$ on X, X_1 and X_2 with $\tilde{\mathfrak{a}}_2$ a trimming of $\mathfrak{a}_2$ such that $(\mathfrak{a}_1, \tilde{\mathfrak{a}}_2)$ properly displays the transversality and $(\mathfrak{a}_1, \mathfrak{a})$ an s admissible atlas pair for the submersion f. Really we only need the interlocking of charts about W_1 and that is what we construct.

Begin by choosing an atlas pair $(\mathfrak{a}_1, \mathfrak{a}_2)$ properly displaying the transversality of G with W with $\mathfrak{a}_1$ star bounded and an element $V_2 = V_2^{-1} \in \mathfrak{U}_{X_2}$ inducing the trimming $\tilde{\mathfrak{a}}_2$ of $\mathfrak{a}_2$ and such that $V_2 \circ V_2[f(U_\alpha)] \subset V_\beta$ $(\beta = \beta(\alpha))$ and so $f(U_\alpha) \subset \tilde{V}_\beta$. Define $\tilde{U}_\alpha = U_\alpha \cap f^{-1}(U_\alpha \cap W_1)$ when the latter set is nonempty. Since $\Delta_{W_1} \subset\subset \cup\{U_\alpha \times U_\alpha\}$, Lemma III.6.3(g) applied to $1 \times f: X_1 \to X_1 \times X_1$ shows that $W_1 \subset\subset \cup\{\tilde{U}_\alpha\}$ and hence that $\{\tilde{U}_\alpha\} \cup \{X - W_1\}$ is a uniform open cover of X_1. The principal part $f_\alpha: h_\alpha(\tilde{U}_\alpha) \to h_\alpha(U_\alpha \cap W_1) = h_\alpha(\tilde{U}_\alpha \cap W_1) = h_\alpha(U_\alpha) \cap (\tilde{E}_\alpha^1 \times 0)$ is well defined when $\tilde{U}_\alpha \neq \emptyset$ and so (6.1) applies. Define $k_\alpha = (p_1 \cdot f_\alpha) \times p_2 : h_\alpha(\tilde{U}_\alpha) \to \tilde{E}^1_\alpha \times F_\alpha$. The derivative Dk_α at a point $(v,0) \in h_\alpha(\tilde{U}_\alpha \cap W_1)$ is given by the matrix $\begin{pmatrix} I & B \\ 0 & I \end{pmatrix}$ where $\|B\| \leq \|Df_\alpha(v,0)\|$. By the usual Size Estimate arguments we can define a function $\lambda: W_1 \to (0,1)$ with $\rho_1 > 1/\lambda$ on W_1, and an open set U_x containing x for $x \in W_1$ such that for some $\alpha = \alpha(x)$, $h_x = k_\alpha \cdot h_\alpha - c_{h_\alpha f x}$ is an $\mathfrak{M}^1$ isomorphism of U_x onto $B(0,\lambda(x))$ in $\tilde{E}^1_\alpha \times F_\alpha$. Let $\beta(x) = \beta(\alpha(x))$ and let $\tilde{G}_1$ consist of G_1 together with the charts $\{U_x, h_x\}$. We have the commutative diagram:

$$\begin{array}{ccc} \tilde{E}^1_\alpha \times F_\alpha \supset B(0,\lambda(x)) & \xrightarrow{\quad G_\alpha \quad} & \tilde{E}^2_\beta \times F_\alpha \\ p_1 = f_\alpha \Big\downarrow & \searrow^{p_2} & \Big\downarrow p_2 \\ \tilde{E}^1_\alpha & & F_\alpha \end{array}$$

with $G_\alpha[B(0,\lambda(x))] \subset \tilde{V}_\beta$, $\alpha = \alpha(x)$, $\beta = \beta(x)$, $x \in W_1$.

Now by an easy Metric Estimate we can choose $V_0 = V_0^{-1} \in \mathcal{U}_{X_1}$ such that $V_0 \cdot V_0[x] \subset h_x^{-1}(B(0,\lambda(x)/16))$ for all $x \in W_1$ and let $V_1 = V_0 \cap (f \times f)^{-1}(V_0)$ $V_1 \in \mathcal{U}_{X_1}$ by Lemma III.6.3.

By Theorem 3 and Proposition IV.3.10 $\{G_1 \in \mathfrak{M}^1(X_1,X_2)$: G_1 is properly transverse to W, $G_1^{-1}W \subset\subset V_1[W_1]$, $G \times G_1(X_1) \subset\subset V_2$ and $\|G_\alpha - G_{1\alpha}\|_{\mathcal{C}^1} < \lambda(x)/16$ for all $\alpha = \alpha(x)$ with $x \in W_1\}$ is a $\mathcal{C}^1$ neighborhood of G. Let G_1 be in this neighborhood.

If $y \in G_1^{-1}W$ and $x = fy$ then $(x_1,y) \in V_1$ for some $x_1 \in W_1$. So $(x,x_1) = (fy,fx_1)$ and $(x_1,y) \in V_0$. So $y \in V_0 \cdot V_0[x] \subset h_x^{-1}(B(0,\lambda(x)/16))$.

Since $\|p_2 - p_2 \cdot G_{1\alpha}\|_{\mathcal{C}^1} = \|p_2 \cdot (G_\alpha - G_{1\alpha})\|_{\mathcal{C}^1} < 1/2$ $(\alpha = \alpha(x))$, the function $K_x = p_1 \times p_2 \cdot G_{1\alpha}: B(0,\lambda(x)) \to \tilde{E}^1_\alpha \times F_\alpha$ is a local isomorphism with derivative and inverse given by the matrices:

$$DK_x = \begin{pmatrix} I & 0 \\ D_1 & D_2 \end{pmatrix} \qquad (DK_x)^{-1} = \begin{pmatrix} I & 0 \\ -D_2^{-1}D_1 & D_2^{-1} \end{pmatrix}$$

$D_i = D_i p_2 \circ G_{1\alpha} = p_2 \circ D_i G_{1\alpha}$ $(i = 1,2)$. Since $\|I - D_2\| < 1/4$ D_2 is an isomorphism with $\theta(D_2) < 4/3$. Since $\|D_1\| < 1/2$, $\theta(DK_x) < 2$. Hence, by the Size Estimate K_x has a unique inverse defined on $B(K_x(0), \lambda(x)/2)$ taking $K_x(0)$ to 0. This ball contains $B(0, \lambda(x)/4)$ since $\|0 - p_2 \circ G_{1\alpha}(0)\| = \|p_2 \circ G_\alpha(0) - p_2 \circ G_{1\alpha}(0)\| < \lambda(x)/8$ and
$K_x^{-1}(B(0,\lambda(x)/4)) \supset K_x^{-1}(B(K_x(0),\lambda(x)/8)) \supset B(0,\lambda(x)/16)$.

Thus, $s_x = K_x^{-1}$: $B(0,\lambda(x)/4) \cap (\tilde{E}^1_\alpha \times 0) \to \tilde{E}^1_\alpha \times F_\alpha$ is a section of $p_1 = f_\alpha$ with image in $h_x(G_1^{-1}W)$ and $y = h_x^{-1} s_x(0)$ is the unique point of $G_1^{-1}W$ with $fy = x$. Thus, $f|G_1^{-1}W$: $G_1^{-1}W \to W_1$ is a bijection and using the Inverse Function Theorem to estimate $\|s_x\|_{\mathfrak{M}^1}$ we see that $f|G_1^{-1}W$ is an $\mathfrak{M}_1$ isomorphism. Q.E.D.

The above proof can be adjusted to show that on the open set described by the theorem, the function $G \to (f|G^{-1}W)^{-1}$ is a continuous map to $\mathfrak{M}^1(X,X_1)$.

1. Grassmannian of a B-Space: For a B-space E, let $G(E)$ denote the Grassmannian of E: the set of closed split subspaces of E. If $E_0 \in G(E)$ with inclusion map i_0, we call a left inverse P for i_0 an E_0 projection, i.e. an E_0 projection is a projection with image E_0. Unless $E_0 = 0$, a case we will exclude until otherwise mentioned, $\theta(P) = \|P\| \geq \theta(i_0) = \theta(E_0)$. Letting $K(P)$ denote the kernel of P, we have an isomorphism $i_0 + i_K\colon E_0 \times K(P) \to E$ with inverse $P \times (I-P)$. Thus, $\theta(i_0 + i_K) \leq 1 + \|P\|$. Let $V_P^0 = \{E_1 \in G(E)\colon P|E_1\colon E_1 \to E_0 \text{ is an isomorphism}\}$. There is a bijective map $h_P\colon V_P^0 \to L(E_0;K(P))$ defined by $h_P(E_1) = T$ or $E_1 = E_T$ if $T = (I - P)\circ(P|E_1)^{-1}$, i.e. E_1 is the image of $\Gamma_T = i_0 + T\colon E_0 \to E$. Thus, $h_P(E_1) = T$ if E_1 is the graph of the map T. $\Gamma_T = (P|E_1)^{-1}$ is the graph map of T. Recall the homomorphism $Q\colon L(E_0;K(P)) \to Lis(E)$ defined by $Q_T = I + TP$ (cf. page 180). $\Gamma_T = Q_T \circ i_0$. Thus, we have:

$$\|T\| \leq (1 + \|P\|)\theta(P|E_1)$$

$$\max(\theta(P|E_1),\theta(\Gamma_T)) \leq \max(\|P\|,\|\Gamma_T\|) \leq \max(\|P\|, 1 + \|T\|) \tag{1.1}$$

$$\theta(Q_T) \leq 1 + \|TP\| \leq 1 + \|T\|\|P\|.$$

Note that if P_1 is an E_{T_1} projection, then $Q_{T_0 - T_1} P_1 Q_{T_1 - T_0} = Q_{T_0} Q_{T_1}^{-1} P_1 Q_{T_1} Q_{T_0}^{-1}$ is an E_{T_0} projection. Hence, by (1.1):

$$(1 + \|T_1 - T_0\|\|P\|)^{-2} \leq \theta(E_{T_1})/\theta(E_{T_0}) \leq (1 + \|T_1 - T_0\|\|P\|)^2. \tag{1.2}$$

In particular, $T \to \theta(E_T)$ is continuous on $L(E_0;K(P))$.

Now let P be an E_0 ($\in G(E)$) projection with kernel K and $\bar{P}$ be an $\bar{E}_0$ ($\in G(\bar{E})$) projection with kernel $\bar{K}$. Let $L \in L(E;\bar{E})$. With respect to the splittings induced by P and $\bar{P}$ we can write L as a matrix:

$$L = \begin{pmatrix} \bar{P}Li_0 & \bar{P}Li_K \\ (I-\bar{P})Li_0 & (I-\bar{P})Li_K \end{pmatrix} = \begin{pmatrix} A & B \\ C & D \end{pmatrix} .$$

To say that $L(E_{T_1}) \subset E_{\bar{T}_1}$, $T_1 \in L(E_0;K)$ and $\bar{T}_1 \in L(\bar{E}_0;\bar{K})$ says $L \cdot \Gamma_{T_1} = \Gamma_{\bar{T}_1} \cdot \bar{P} \cdot L \cdot \Gamma_{T_1} = \Gamma_{\bar{T}_1} \cdot (A + BT_1)$, i.e.

$$\begin{pmatrix} A & B \\ C & D \end{pmatrix} \begin{pmatrix} I \\ T_1 \end{pmatrix} = \begin{pmatrix} I \\ \bar{T}_1 \end{pmatrix} (A + BT_1)$$

(1.3) $$C + DT_1 = \bar{T}_1(A + BT_1)$$

(1.4) $$\bar{T}_1 A - C = (D - \bar{T}_1 B)T_1 .$$

1 <u>LEMMA</u>: Let $L \in L(E;\bar{E})$ with $L(E_T) \subset E_{\bar{T}}$ as above.

(a) Assume $L|E_T: E_T \to E_{\bar{T}}$ is an isomorphism. There exists $r > 0$ with $r^{-1} \leq O^*(\theta(L|E_T), \|P\|, \|\bar{P}\|, \|T\|, \|\bar{T}\|, \|L\|)$ such that on the r ball B in $L(E;\bar{E}) \times L(E_0;K)$ with center (L,T), the map $G: B \to L(\bar{E}_0;\bar{K})$ is well defined by $G(L_1,T_1) = \bar{T}_1$ if L_1 restricts to a linear isomorphism $L_1|E_{T_1}: E_{T_1} \to E_{\bar{T}_1}$. G is a C^∞ map and for any standard function space type $\mathfrak{M}$ we have:

$$\|G\|_{\mathfrak{M}}, \theta(L_1|E_{T_1}) \leq O^*(\theta(L|E_T), \|P\|, \|\bar{P}\|, \|T\|, \|\bar{T}\|, \|L\|) .$$

(b) For $q_{\bar{T}}: \bar{E} \to \bar{E}/E_{\bar{T}}$ the quotient map, assume $q_{\bar{T}} \cdot L \cdot i_K: K \to \bar{E}/E_{\bar{T}}$ is an isomorphism, i.e. L is transverse to $E_{\bar{T}}$ with $L^{-1}(E_{\bar{T}}) = E_T$. There exists $r > 0$ with $r^{-1} \leq O^*(\theta(q_{\bar{T}} \cdot L \cdot i_K), \|P\|, \|\bar{P}\|, \|T\|, \|\bar{T}\|, \|L\|)$ such that on the r ball B in $L(E;\bar{E}) \times L(\bar{E}_0;\bar{K})$ with center $(L,\bar{T})$, the map $\bar{G}: B \to L(\bar{E}_0;\bar{K})$ is well defined by $\bar{G}(L_1,\bar{T}_1) = T_1$ if L_1 is transverse to $E_{\bar{T}_1}$ with $L_1^{-1}(E_{\bar{T}_1}) = E_{T_1}$. $\bar{G}$ is a C^∞ map and for any standard function space type $\mathfrak{M}$ we have:

$$\|\bar{G}\|_{\mathfrak{M}}, \theta(q_{\bar{T}_1} \cdot L_1 \cdot i_K) \leq O^*(\theta(q_{\bar{T}} \cdot L \cdot i_K), \|P\|, \|\bar{P}\|, \|T\|, \|\bar{T}\|, \|L\|) .$$

PROOF: (a): Let $M = \max(\theta(L|E_T),\ldots,\|L\|)$ as above. $A + BT = \bar{P}L\Gamma_T = (\bar{P}|E_{\bar{T}})\cdot(L|E_T)\cdot(\Gamma_T)$ and so is an isomorphism with $\theta \leq O^*(M)$. $A_1 + B_1T_1 = \bar{P}L\Gamma_T + \bar{P}(L_1 - L)\Gamma_T + \bar{P}L_1(T_1 - T)$. Hence, for some r with $r^{-1} \leq O^*(M)$, $\|L_1 - L\|$ and $\|T_1 - T\| < r$ implies the latter two terms together have norm $\leq 1/2\ \theta(A + BT)$. Hence, $A_1 + B_1T_1$ is an isomorphism with $\theta(A_1 + B_1T_1) \leq 2\theta(A + BT) \leq O^*(M)$. Define $G(L_1,T_1) = (C_1 + D_1T_1)(A_1 + B_1T_1)^{-1}$. The parenthesized terms are polynomial in L_1 and T_1. That G is $\mathcal{C}^\infty$ and the estimate on $\mathfrak{M}$ norm now follow from Proposition II.3.3. That $L_1|E_{T_1} \to E_{\bar{T}_1}$ with $\bar{T}_1 = G(L_1,T_1)$ is a linear isomorphism follows from (1.3). Finally, $L_1|E_{T_1} = \Gamma_{\bar{T}_1}\cdot(A_1 + B_1T_1)\cdot P$, yielding the estimate of $\theta(L_1|E_{T_1})$. (b): Let $\bar{M} = \max(\theta(q_{\bar{T}}\cdot L\cdot i_K),\ldots,\|L\|)$ as above. The restriction $q_{\bar{T}}\colon \bar{K} \to \bar{E}/E_{\bar{T}}$ is an isomorphism whose inverse is given by factoring $(I - \bar{P})\cdot Q_{-\bar{T}} = (I - \bar{P}) - \bar{T}\bar{P}$ through $q_{\bar{T}}$. Hence, $\theta(q_{\bar{T}}\cdot L\cdot i_K)$ and $\theta((I - \bar{P})\cdot Q_{-\bar{T}}\cdot L\cdot i_K)$ have ratios bounded by $O^*(\bar{M})$. But $(I - \bar{P})\cdot Q_{-\bar{T}}\cdot L\cdot i_K = D - \bar{T}B$ which is consequently an isomorphism with $\theta(D - \bar{T}B) \leq O^*(M)$. Now proceed as in (a), defining $\bar{G}(L_1,\bar{T}_1) = (D_1 - \bar{T}_1B)^{-1}(\bar{T}_1A - C)$ and applying (1.4). Q.E.D.

Now define for P an E_0 projection:

$$D_P = \{T \in L(E_0;K(P)) : \|T\|\|P\| < 1\}$$

(1.5)

$$V_P = \{E_T : T \in D_P\} \subset V_P^0$$

Clearly, $h_P\colon V_P \to D_P$ is a bijection with $h_P(E_0) = 0$, the center of the ball D_P.

2 THEOREM: For any constant $\infty > K > 1$, let $\mathfrak{a}_K(G(E)) = \{V_P,h_P\}$ indexed by projections P with $\|P\| \leq K\theta(P(E))$ (when $K = 2$ we will drop the subscript). $(\mathfrak{a}_K(G(E)),\theta)$ is a semicomplete $\mathcal{C}^\infty$ adapted atlas on $G(E)$. In fact, for any standard function space type $\mathfrak{M}$, $\rho^{\mathfrak{M}}_{\mathfrak{a}_K} \leq O^*(K,\theta)$ and $1/\lambda_{\mathfrak{a}_K} \leq \theta$.

PROOF: If $V_P \cap V_{\bar{P}} \neq \emptyset$ then, with I the identity map of E, $h_{\bar{P}} h_P^{-1}$ is the restriction of $G(I)$ to $h_P(V_P \cap V_{\bar{P}}) = D_P \cap G(I)^{-1} D_{\bar{P}}$, i.e. $h_{\bar{P}} h_P^{-1}(T) = G(I,T)$ on an open set in $L(E_0;K(P))$, Since $\theta(I) = \theta(I: E_T \to E_{\bar{T}}) = 1$, $\|T\|, \|\bar{T}\| < 1$ and $\|P\|, \|\bar{P}\| \leq K\theta(E_0), K\theta(\bar{E}_0)$, we have $\|h_{\bar{P}} h_P^{-1}\|_{\mathfrak{M}} \leq O^*(K, \theta(E_0), \theta(\bar{E}_0))$ by Lemma 1. Now G_K is star bounded. In fact, by (1.2):

$$\sup \theta | V_P \leq 9 \inf \theta | V_P. \tag{1.6}$$

It follows that $\rho^{\mathfrak{M}}_{G_K} \leq O^*(K, \theta)$. Clearly, $\lambda_{G_K}(E_0) \geq \|P\|^{-1}$ for all E_0 projections P and so $\lambda_{G_K}(E_0) \geq \theta(E_0)^{-1}$. The result follows from the Regularity Lemma III.7.3 once we show that with the topology induced by G_K $G(E)$ is Hausdorff.

Let $L \in L(E;F)$ and define $M_L: G(E) \to [0,\infty)$ by $M_L(E_0) = \|L | E_0\|$. On $L(E_0;K(P))$, $M_L \circ h_P^{-1}(T_1) = M_L(E_{T_1}) = \|L | E_{T_1}\|$. Note that $\|L \circ Q_{T_1 - T_2} | E_{T_2}\| \leq M_L(E_{T_1})(1 + \|T_1 - T_2\| \|P\|)$ and since $Q_{T_1 - T_2} = I + (T_1 - T_2)P$, $\|L \circ Q_{T_1 - T_2} | E_{T_2}\| \geq M_L(E_{T_2}) - \|L\| \|T_1 - T_2\| \|P\|$. Thus, $|M_L(E_{T_1}) - M_L(E_{T_2})| \leq 2\|L\| \|P\| \|T_1 - T_2\|$. Hence, $M_L \circ h_P^{-1}$ is continuous and so M_L is continuous on $G(E)$. Now let $E_0 \not\supset \bar{E}_0$ be two distinct elements of $G(E)$. For P an E_0 projection, $(I - P) | \bar{E}_0$ is not 0 and so by the Hahn Banach theorem we can choose L_1 an element of the dual space of $K(P)$ with $L_1 \circ (I - P) | \bar{E}_0 \neq 0$. Let $L = L_1 \circ (I - P)$. $M_L(E_0) = 0$ and $M_L(\bar{E}_0) \neq 0$. Since the family of continuous functions $\{M_L: L \in L(E;R)\}$ distinguishes points $G(E)$ is Hausdorff. Q.E.D.

3 THEOREM: Let E and $\bar{E}$ be B-spaces. Define $O(E,\bar{E}) = \{(L, E_0) \in L(E;\bar{E}) \times G(E): L | E_0 \in Li(E_0;\bar{E})\}$. $O(E,\bar{E})$ is an open subset of the product $L(E;\bar{E}) \times G(E)$ and $\tilde{\theta}(L, E_0) = \max(\|L\|, \theta(L | E_0))$ induces a semi-complete refinement on it. $Li(E;\bar{E}) \times G(E)$ is open in $O(E,\bar{E})$ and the product structure (with the θ refinement structure on Li) is a semi-complete refinement of the restriction of the $\tilde{\theta}$ structure.

$G: O(E,\bar{E}) \to G(\bar{E})$ by $G(L,E_0) = G(L)(E_0) = G_{E_0}(L) = L(E_0)$ is a $\mathcal{C}^\infty$ map. For any standard function space type $\mathfrak{M}$, $G^*\theta \leq O^*(\tilde{\theta})$ and $\rho[G;(L(E;\bar{E}) \times \mathcal{G}(G(E)))|O,\mathcal{G}(G(\bar{E}))] \leq O^*(\tilde{\theta})$. $\pi_2 \times G: O(E,\bar{E}) \to G(E) \times G(\bar{E})$ is a $\mathcal{C}^\infty$ submersion. $\pi_1 \times G: Li(E;\bar{E}) \times G(E) \to Li(E;\bar{E}) \times G(\bar{E})$ is a metricly proper $\mathcal{C}^\infty$ embedding and $\pi_1 \times G: Lis(E;\bar{E}) \times G(E) \to Lis(E;\bar{E}) \times G(\bar{E})$ is a $\mathcal{C}^\infty$ isomorphism with respect to the product structures (θ refinement structures on Li and Lis).

<u>PROOF</u>: If $(L,E_0) \in O(E,\bar{E})$ and L_0 is the isomorphism of E_0 to $G(L,E_0) = L(E_0)$, we have:

$$(1.7)\qquad \begin{aligned} &\max(\theta(E_0),\theta(G(L,E_0)),\theta(L_0)) \leq O^*(\|L\|,\theta(L|E_0)) \\ &\max(\theta(E_0),\theta(L|E_0)) \leq O^*(\|L\|,\theta(G(L,E_0)),\theta(L_0)) \end{aligned}$$

Recalling that $\theta(L_0) = \bar{\theta}(L|E_0)$ these results follow from Proposition VI. 1.1 and the fact that if $\bar{P}$ is an $L(E_0)$ projection then $L_0^{-1}\circ\bar{P}\circ L$ is an E_0 projection. If $L \in Li(E;\bar{E})$ then

$$(1.8)\qquad \max(\theta(L),\theta(E_0)) \approx \max(\theta(L),\theta(L|E_0)) \approx \max(\theta(L),\theta(G(L,E_0)))$$

because if $\bar{P}$ is an $L(E_0)$ projection and P_1 is a left inverse for L or $L|E_0$, then $P_1\circ\bar{P}\circ L$ is an E_0 projection. Here $a \approx b$ means $a \leq O^*(b)$ and $b \leq O^*(a)$. (1.7) and Lemma 1 (a) imply that O is open, $\tilde{\theta}$ induces a semicomplete refinement and G is a $\mathcal{C}^\infty$ map with the required estimates. (1.8) implies that on $Li \times G(E)$, $\max(\theta(L),\theta(E_0))$ induces a refinement of $\tilde{\theta}$. This refinement structure is just the product.

For the submersion result let $(E_0,\bar{E}_0) \in G(E) \times G(\bar{E})$ with projections P and $\bar{P}$. Let $F_{P\bar{P}} = \{L_1 \in L(E;\bar{E}): L_1$ maps E_0 isomorphicly onto $\bar{E}_0\} = \{L_1: (I - \bar{P})L_1 i_0 = 0$ and $\bar{P}L_1 i_0 \in Lis(E_0;\bar{E}_0)\}$ which is open in the kernel of $L(i_0;I - \bar{P}): L(E;\bar{E}) \to L(E_0;K(\bar{P}))$. We have a commutative diagram:

$$\begin{array}{ccc} D_P \times D_{\bar{P}} \times F_{P\bar{P}} & \xrightarrow{\quad Q_{P\bar{P}} \quad} & O(E,\bar{E}) \\ \big\downarrow & & \big\downarrow \pi_2 \times G \\ D_P \times D_{\bar{P}} & \xrightarrow[\quad h_P^{-1} \times h_{\bar{P}}^{-1} \quad]{} & G(E) \times G(\bar{E}) \end{array}$$

with $Q_{P\bar{P}}(T,\bar{T},L_1) = (Q_{\bar{T}} \cdot L_1 \cdot Q_{-T}, E_T)$. The inverse of $Q_{P\bar{P}}$ on $(\pi_2 \times G)^{-1}(V_P \times V_{\bar{P}})$ is given by $(L_1,E_1) \to (T,\bar{T},Q_{-\bar{T}}L_1Q_T)$ with $T = h_P(E_1)$ and $\bar{T} = h_{\bar{P}}(L_1(E_1))$. Letting P and $\bar{P}$ vary over the indices of $\mathcal{G}_K(G(E))$ and $\mathcal{G}_K(G(\bar{E}))$, we have that $(\{(\pi_2 \times G)^{-1}(V_P \times V_{\bar{P}}), Q_{P\bar{P}}^{-1}\}, \mathcal{G}_K(G(E)) \times \mathcal{G}_K(G(\bar{E})))$ is an s admissible atlas pair for the submersion $\pi_2 \times G$.

$\pi_1 \times G$ is a $\mathcal{C}^\infty$ isomorphism on $\mathrm{Lis}(E;\bar{E}) \times G(E)$ because its inverse is given by the composition:

$$\mathrm{Lis}(E;\bar{E}) \times G(\bar{E}) \xrightarrow{\mathrm{inv} \times 1} \mathrm{Lis}(\bar{E};E) \times G(\bar{E}) \xrightarrow{\pi_1 \times G} \mathrm{Lis}(\bar{E};E) \times G(E) \xrightarrow{\mathrm{inv} \times 1} \mathrm{Lis}(E;\bar{E}) \times G(E).$$

If $L \in \mathrm{Li}(E;\bar{E})$ and S is a left inverse for L with $\theta(S) \leq 2\theta(L)$ then for $E_0 \in G(E)$, $S|L(E_0): L(E_0) \to E$ is a split injection with left inverse $\bar{P} \cdot L$ for $\bar{P}$ any $L(E_0)$ projection. Thus, $\theta(S|L(E_0)) \leq 2\max(\theta(L),\theta(L(E_0)))^2$. Let $O_S = \{E_1 \in G(\bar{E}): S|E_1 \in \mathrm{Li}(E_1;E)$ with $\theta(S|E_1) < 3\max(\theta(L),\theta(E_1))^2\}$. O_S is a full neighborhood of $G(L)(G(E))$ in $G(\bar{E})$. In fact, with $\bar{\mathcal{G}} = \mathcal{G}(G(\bar{E})), 1/\lambda_{\bar{\mathcal{G}}|O_S} \leq O^*(\theta(L),\theta)$ on $G(L)(G(E))$. Furthermore, $\{S\} \times O_S$ is open in $O(\bar{E},E) \cap \{S\} \times G(\bar{E})$ and so $G(S): O_S \to G(E)$ is well defined with $\rho^{m}(G(S);\bar{\mathcal{G}}|O_S,\mathcal{G}) \leq O^*(\theta(L),\theta)$. Next, let $B_L = B(L,(4\theta(L))^{-1})$. For $L_1 \in B_L$, $\|(L_1 - L)S\| < 1/2$ and so $Q_{L_1-L} = I + (L_1 - L)S \in \mathrm{Lis}(\bar{E})$ with $\theta < 2$. Note that since $S(L_1 - L)$ need not equal 0, $(Q_{L_1-L})^{-1}$ need not equal Q_{L-L_1}. Now with $G(Q^{-1})(\bar{E}_1,L_1)$ $= G((Q_{L_1-L})^{-1})(\bar{E}_1)$ we have the commutative diagram (because $L = (Q_{L_1-L})^{-1} \cdot L_1$):

$$\begin{array}{ccc} & & B_L \times G(\bar{E}) \\ & \nearrow^{\pi_1 \times G} & \big\downarrow \pi_1 \times G(Q^{-1}) \\ B_L \times G(E) & & B_L \times G(\bar{E}) \\ & & \cup \\ & \searrow_{\pi_1 \times G(L)} & B_L \times O_S \end{array}$$

$\pi_1 \times G(Q^{-1})$ is a $\mathcal{C}^\infty$ isomorphism with inverse $\pi_1 \times G(Q)$. Furthermore, $\rho^{\mathfrak{M}}(G(Q^\varepsilon)) \leq O^*(\theta(L),\pi_2{}^*\theta)$. Finally, $\pi_1 \times G(S): B_L \times O_S \to B_L \times G(E)$ is a left inverse for $\pi_1 \times G(L)$. $\pi_1 \times G$ is metricly proper on Li $\times\, G(E)$ by (1.8) and it now follows easily from Theorem VI.3.1 and inequality III.(5.1) that it is an embedding. Q.E.D.

The proof of Lemma 1 shows that if $(L_1,T_1) \in 1 \times h_P[O(E,\bar{E}) \cap (L(E;\bar{E}) \times V_P) \cap G^{-1}(V_{\bar{P}})]$ then $\bar{T}_1 = (1 \times h_{\bar{P}})\cdot G \cdot (1 \times h_P)^{-1}(L_1,T_1) = (C_1 + D_1T_1)(A_1 + B_1T_1)^{-1}$ where L_1 has matrix $\begin{pmatrix} A_1 & B_1 \\ C_1 & D_1 \end{pmatrix}$ with respect to the P and $\bar{P}$ splittings. $A_1 + B_1T_1$ is the isomorphism $(\bar{P}|E_{T_1})\cdot(L_1|E_{T_1})\cdot\Gamma_{T_1}$. Hence, by (1.6) and the estimate on $G^*\theta$, $\theta(A_1 + B_1T_1) \leq O^*(K,\tilde{\theta}(L_1,E_1))$ $(E_1 = h_P(T_1) = E_{T_1})$. Since inv is $\mathcal{L}ip^r$ (cf. p. 38) if U is open and bounded in $\mathrm{Li}(E;\bar{E})$ we have:

$$(1.9)\quad \|(1 \times h_{\bar{P}})\cdot G\cdot(1 \times h_P)^{-1}|1 \times h_P[(U \times V_P) \cap G^{-1}(V_{\bar{P}})]\|_{\mathcal{L}ip^r}$$
$$\leq O^*(\sup \theta|U, \inf \theta|V_P, K).$$

In particular, with U a neighborhood of the identity in $L(E;E)$ we see that $\mathcal{G}_K(G(E))$ is a convex, star bounded $\mathcal{L}ip^r$ atlas for all $r \geq 0$.

Lis(E) acts on G(E) by the map G of Theorem 3. Let $\mathrm{Lis}_I(E)$ be the component of the identity. Each V_P meets exactly one $\mathrm{Lis}_I(E)$ orbit in G(E). In fact, $Q^P: V_P \to \mathrm{Lis}_I(E)$ defined by $Q^P(E_1) = Q_T$ with $T = h_P(E_1)$ is a local section of the action, i.e. $G(Q^P(E_1),E_0) = E_1$. Hence, the

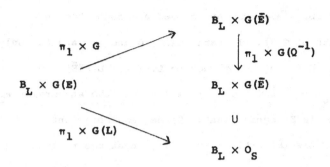

$Lis_I(E)$ orbits are open and closed and connected in $G(E)$, i.e. the components of $G(E)$. In particular, 0 and E are the only isolated points of $G(E)$. Defining the $\mathcal{C}^\infty$ action $Lis(E) \times Lis(\bar{E}) \times O(E,\bar{E}) \to O(E,\bar{E})$ by $(Q,\bar{Q},L,E_0) \to (\bar{Q}\cdot L\cdot Q^{-1}, Q(E_0))$ we see that the submersion $\pi_2 \times G$ is $Lis(E) \times Lis(\bar{E})$ equivariant. Since each component of O is a union of $Lis_I(E) \times Lis_I(\bar{E})$ orbits, it follows that the restriction of $\pi_2 \times G$ to any component of O is mapped onto the corresponding component of $G(E) \times G(\bar{E})$. Since the inclusion of $Li(E;\bar{E}) \times G(E)$ into O is equivariant the same is true for the restriction of $\pi_2 \times G$ to any component of $Li \times G(E)$. If $E_0 \in G(E)$ then the pull back of the submersion $\pi_2 \times G$ by the inclusion $G(\bar{E}) \xrightarrow{c_{E_0}\times 1} G(E) \times G(\bar{E})$ is essentially the map $G_{E_0}: Li(E;\bar{E}) \to G(\bar{E})$ and by Proposition VI.6.7 G_{E_0} is a $\mathcal{C}^\infty$ submersion. The most important example is G_E associating to L its image. Note that G_{E_0} is $Lis(\bar{E})$ equivariant and so maps any component of $Li(E;\bar{E})$ onto the corresponding component of $G(\bar{E})$.

It now follows from Corollary VI.1.6 that any component of $G(E)$ consisting of subspaces of finite dimension or finite codimension is semicompact.

Define the atlas $(\mathfrak{D}_K(\Gamma(E)), \mathfrak{G}_K(G(E))) = \{V_P, h_P, \varphi_P\}$ on the trivial bundle $\epsilon_E: G(E) \times E \to G(E)$ by

$$\varphi_P(E_1,v) = (T, Q_{-T}(v)) \qquad (T = h_P(E_1)). \tag{1.10}$$

4 <u>THEOREM</u>: $(\mathfrak{D}_K(\Gamma(E)), \mathfrak{G}_K(G(E)))$ is an s admissible $\mathcal{C}^\infty$ atlas on the trivial bundle ϵ_E. For any standard function space type, $\rho^{\mathfrak{M}}_{(\mathfrak{D}_K, \mathfrak{G}_K)} \le O^*(\theta, K)$. Let $\Gamma(E) = \{(E_1,v) \in G(E) \times E: v \in E_1\}$ and let γ_E denote the restriction of ϵ_E to $\Gamma(E)$, i.e. $\gamma_E(E_1,v) = E_1$. $(\mathfrak{D}_K(\Gamma(E)), \mathfrak{G}_K(G(E)))$ induces a locally split, semicomplete $\mathcal{C}^\infty$ subbundle atlas on γ_E. γ_E with the resulting $\mathcal{C}^\infty$ structure is called the canonical bundle on $G(E)$.

The atlas $(L(\mathfrak{D}_K(\Gamma(E)); \mathfrak{D}_K(\Gamma(E))), \mathfrak{G}_K(G(E)))$ on $L(\epsilon_E; \epsilon_E) = \epsilon_{L(E;E)}$ is an s admissible $\mathcal{C}^\infty$ atlas. Let $N(E) = \{(E_1,L) \in G(E) \times L(E;E): L(E_1) \subset E_1\}$ and let n_E denote the restriction of $\epsilon_{L(E;E)}$ to $N(E)$. $(L(\mathfrak{D}_K;\mathfrak{D}_K),\mathcal{G}_K)$ induces a locally split, semicomplete $\mathcal{C}^\infty$ subbundle atlas on n_E. n_E with the resulting $\mathcal{C}^\infty$ structure is called the abstract normal bundle of $G(E)$.

PROOF: The estimate is obvious by comparing $(\mathfrak{D}_K,\mathcal{G}_K)$ and $(\mathcal{G}_K \times E,\mathcal{G}_K)$. The subbundle results because $\varphi_P((\gamma_E)^{-1}V_P) = D_P \times E_0$ and $L(\varphi;\varphi)_P((n_E)^{-1}V_P)$ $= D_P \times L_{E_0}(E;E)$ where $L_{E_0}(E;E) = \{L \in L(E;E): L(E_0) \subset E_0\}$ $= \{L \in L(E;E): (I - P)\circ L\circ P = 0\}$. Note that $L \to (I - P)\circ L\circ P$ is a projection on $L(E;E)$. Q.E.D.

The motivation for the name of n_E comes from the identification of $L(E_0;K(P))$ with $L(E;E)/L_{E_0}(E;E)$ by the split short exact sequence $L_{E_0}(E;E) \to L(E;E) \leftrightarrows L(E_0;K(P))$ defined by $L \to (I - P)Li_0$ and $T \to i_K TP$ for $L \in L(E;E)$ and $T \in L(E_0;K(P))$.

For the rest of this section we will assume the spaces E_1,E_2, etc. are real Hilbert spaces.

Every closed subspace E_0 splits via the orthogonal projection, which has norm 1 (unless $E_0 = 0$). Thus, $\theta(E_0) = 1$. It follows that every injection and surjection between Hilbert spaces splits. In fact, if $A: E_1 \to E_2$ is a linear injection of Hilbert spaces then $\theta(A) = \bar{\theta}(A)$ as we can use the orthogonal projection to $A(E_1)$ followed by $\bar{A}^{-1}$ as a left inverse for A. Recall that if $A \in L(E_1;E_2)$ $(i = 1,2)$ then the adjoint $A^* \in L(E_2;E_1)$ is defined by $\langle Av_1,v_2\rangle_2 = \langle v_1,Av_2\rangle_1$ with $v_i \in E_i$ $(i = 1,2)$. Taking adjoints is an isometric isomorphism between $L(E_1;E_2)$ and $L(E_2;E_1)$ satisfying $(A^*)^* = A$ and $(AB)^* = B^*A^*$. Define $\mathrm{Sym}(E) =$ $\{A \in L(E;E): A = A^*\}$. Recall (eg. [14;Appendix]) that if $A \in \mathrm{Sym}(E)$ then $\|A\| = \sup\{|\langle Av,v\rangle|: \|v\| = 1\}$ and the spectrum of A lies in the interval $[\inf\{\langle Av,v\rangle: \|v\| = 1\}, \sup\{\langle Av,v\rangle: \|v\| = 1\}]$ with the endpoints in the spectrum. Define $\mathrm{Pos}(E) \subset \mathrm{Sym}(E)$ by: $A \in \mathrm{Sym}(E)$ lies in $\mathrm{Pos}(E)$

iff $A \in Lis(E)$ and $\langle Av,v\rangle > 0$ for all $v \neq 0$ iff $\inf\{\langle Av,v\rangle : \|v\| = 1\} > 0$. By the spectral mapping theorem these conditions are equivalent and

$$\theta(A) = \max(\sup\{\langle Av,v\rangle\}, \sup\{\langle Av,v\rangle^{-1}\}) \qquad A \in Pos(E) \tag{1.10}$$

where v varies with norm 1. Clearly, if $A \in Pos(E)$ then $A^{-1} \in Pos(E)$.

5 LEMMA: (a) A projection P in $L(E;E)$ is an orthogonal projection iff the following equivalent conditions hold: (1) $P \in Sym(E)$. (2) $K(P) = P(E)^{\perp}$. (3) $\|P\| \leq 1$.

(b) Taking adjoints defines a bijection between $Li(E_1;E_2)$ and $Ls(E_2;E_1)$ with $\theta(A^*) = \theta(A)$.

(c) The map $Li(E_1;E_2) \to L(E_2;E_2)$ associating to a linear injection the orthogonal projection onto its image is a C^∞ map (θ refined structure on the domain). If $A \in Li(E_1;E_2)$ then $A^*A \in Pos(E_1)$ with $\theta(A^*A) = \theta(A)^2$ and $A(A^*A)^{-1}A^*$ is the associated projection.

PROOF: (a): Standard, eg. see [27; pp. 83-84]. (b): B is a left (or right) inverse for A iff B^* is a right (resp. left) inverse for A^*. (c): $\inf\{\langle A^*Av,v\rangle : \|v\| = 1\} = m(A)^2$. Hence, by (1.10) $A^*A \in Pos(E_1)$ with $\theta(A^*A) = \bar{\theta}(A)^2 = \theta(A)^2$. $A(A^*A)^{-1}A^*$ is clearly in Sym and is idempotent, i.e. it is an orthogonal projection. $A^*A \in Lis(E_1)$ and $A^* \in Ls(E_2;E_1)$ and so the image of the projection is that of A. Q.E.D.

Now let $E_0 \in G(E)$ and P be the orthogonal E_0 projection. Let $B_P = B(P,1/2) \subset L(E;E)$. With respect to the splitting $E = E_0 \oplus E_0^{\perp}$, P is given by the matrix $\begin{pmatrix} I & 0 \\ 0 & 0 \end{pmatrix}$. If $L = \begin{pmatrix} A & B \\ C & D \end{pmatrix} \in B_P$ then $\|A - I\|, \|D\| < 1/2$. So $A \in Lis(E_0)$ with $\theta(A) \leq 2$. Define the isomorphisms:

$$J_L = \begin{pmatrix} A^{-1} & -A^{-1}B \\ 0 & I \end{pmatrix} \qquad J_L^{-1} = \begin{pmatrix} A & B \\ 0 & I \end{pmatrix}$$

then we have:

$$LJ_L = \begin{pmatrix} I & 0 \\ CA^{-1} & D - CA^{-1}B \end{pmatrix} .$$

Since J_L is an isomorphism with $J_L(E_0) = E_0$, $L(E_0) = L \circ J_L(E_0) =$ graph CA^{-1} and $L(E) = L \circ J_L(E)$ is the direct sum of the graph of CA^{-1} and the Image of $D - CA^{-1}B$ which is a not necessarily closed subspace of $E_0^{\perp}$.

6 LEMMA: If $L \in B_P$ and $L^2 = L$, i.e. L is a projection, then $D = CA^{-1}B$ and so the image of L is the graph of CA^{-1}.

PROOF: $L^2 = L$ implies $L(LJ_L) = LJ_L$ and so $B(D - CA^{-1}B) = 0$ and $D(D - CA^{-1}B) = D - CA^{-1}B$. Since $L^2 = L$, $D = CB + D^2$ and so if $Bv = 0$, $Dv = D^2v$ and since $\|D\| < 1$, $Dv = 0$. Thus, $D - CA^{-1}B = D(D - CA^{-1}B) = 0$. Q.E.D.

7 THEOREM: Let E be a real Hilbert space. The map $G(E) \to L(E;E)$ defined by associating to E_0 the orthogonal E_0 projection is a C^∞ embedding onto a semicompact, globally split C^∞ submanifold of $L(E;E)$.

PROOF: Let $n_E: N(E) \to G(E)$ be the abstract normal bundle of Theorem 4. Let $f: G(E) \to L(E;E)$ be the map described in the statement and define $F: N(E) \to L(E;E)$ by $F(E_0,L) = f(E_0) + L$. $|\|F(E_0,L)\| - \|L\|| \leq \|f(E_0)\| \leq 1$. Since $\theta = 1$ on $G(E)$ this means that the map F is metricly proper from the total space $N(E)$ to $L(E;E)$. Let $G_1(G(E))$ denote the atlas on $G(E)$ indexed by the orthogonal projections. If P is the orthogonal E_0 projection then $f \circ h_P^{-1}: D_P \to L(V;V)$ is defined by Lemma 5(c). It associates to T the orthogonal projection $\Gamma_T(\Gamma_T{}^*\Gamma_T)^{-1}\Gamma_T{}^*$ which has matrix (with respect to $E_0 \oplus E_0^{\perp}$):

$$P(T) = \begin{pmatrix} (I + T^*T)^{-1} & (I + T^*T)^{-1}T^* \\ T(I + T^*T)^{-1} & T(I + T^*T)^{-1}T^* \end{pmatrix} .$$

If $L \in L_{E_0}(E;E)$ then L has matrix $\begin{pmatrix} A & B \\ 0 & D \end{pmatrix}$ and $L(\varphi,\varphi)_P^{-1}(T,L) = (E_T, Q_T L Q_{-T})$ and $Q_T L Q_{-T}$ has the matrix:

$$M(T,L) = \begin{pmatrix} A - BT & B \\ TA - DT + TBT & D + TB \end{pmatrix}.$$

Thus, $F \circ L(\varphi,\varphi)_P^{-1}(T,L) = P(T) + M(T,L)$. Since $I + T^*T \in \mathrm{Pos}(E_0)$ with $\theta(I + T^*T) = 1 + \theta(T)^2 \leq 2$ it easily follows that F is a $\mathcal{C}^\infty$ map. Furthermore, identifying (T,L) with the matrix $\begin{pmatrix} A & B \\ T & D \end{pmatrix}$, the derivative $DF \circ L(\varphi,\varphi)_P^{-1}\ (0,0)$ is the identity map. So F is a local isomorphism about some full neighborhood of the zero section and f is a bijective $\mathcal{C}^\infty$ immersion. Now there exists $\varepsilon > 0$ independent of P such that $\tilde{D}_P = (f \circ h_P^{-1})^{-1} B_P$ contains the ε ball about 0. So with $\tilde{V}_P = h_P^{-1}(\tilde{D}_P)$, the subdivision $\{\tilde{V}_P, h_P\}$ of $G_1(G(E))$ is s admissible on $G(E)$. $\{B_P\} \cup \{L(E;E) - f(G(E))\}$ is a uniform open cover of $L(E;E)$ (since $f(G(E)) = \{P^2 = P\} \cap \{P^* = P\}$ is closed in $L(E;E)$). By Lemma 6, if $L = \begin{pmatrix} A & B \\ C & D \end{pmatrix} \in B_P \cap f(G(E))$ then L is the orthogonal projection onto the graph of $CA^{-1} = T$ and $\|T\| < 2 \cdot 2^{-1} = 1$. Hence, $B_P \cap f(G(E)) = f(\tilde{V}_P)$. Also, the reader can check that $\|T - T_1\| \leq K\|P(T) - P(T_1)\|$ for some constant K. Hence by Theorem VI.3.9(7) f is an embedding. By Corollary VI.5.3 for some full neighborhood O of the zero section in $N(E)$, $F|O\colon O \to L(E;E)$ is a $\mathcal{C}^\infty$ isomorphism onto an open set. Hence, ${}^{\tau}L(E;E) \cong \tau_{G(E)} \oplus n_E$ and so the embedding is split. Q.E.D.

Theorem 7 is the construction of the Grassmannian given by Palais [24].

8 <u>COROLLARY</u>: Let E be a real Hilbert space. The map $\perp\colon G(E) \to G(E)$ defined by $\perp(E_0) = E_0^\perp$ is a $\mathcal{C}^\infty$ isomorphism satisfying $\perp^2 = \perp$.

<u>PROOF</u>: Regarding $G(E)$ as a submanifold of $L(E;E)$, $\perp$ is the restriction of $T \to I - T$. Result by the Factoring Lemma. Q.E.D.

2. Grassmannian of a Bundle: Let $\pi: \mathbb{E} \to X$ be a metric bundle with admissible atlas $(\mathfrak{D},\mathfrak{a}) = \{U_\alpha, h_\alpha, \varphi_\alpha\}$, admissible Finsler $\|\ \|$ and admissible bound ρ. Define $g_\pi: G(\mathbb{E}) \to X$ by $G(\mathbb{E})_x = G(\mathbb{E}_x)$, i.e. $E_{0x} \in G(\mathbb{E})$ if E_{0x} is a closed split subspace of $\mathbb{E}_x$ in which case $g_\pi(E_{0x}) = x$. Define $G(\varphi_\alpha): g_\pi^{-1}(U_\alpha) \to h_\alpha(U_\alpha) \times G(F_\alpha)$ by $G(\varphi_\alpha)(E_{0x}) = (h_\alpha x, \varphi_{\alpha x}(E_{0x}))$. Now let αP vary over pairs with P a projection in $L(F_\alpha;F_\alpha)$ satisfying $\|P\| \leq 2\theta^\alpha(P(F_\alpha))$ (θ^α is the splitting bound defined on $G(F_\alpha)$) and let $G_{\alpha P} = G(\varphi_\alpha)^{-1}(h_\alpha(U_\alpha) \times V_P^\alpha)$ where V_P^α is the open set in $G(F_\alpha)$ defined by (1.5). Let $k_{\alpha P} = (1 \times h_P) \circ G(\varphi_\alpha): G_{\alpha P} \to E_\alpha \times L(P(F_\alpha);K(P))$. On $G(\mathbb{E})$ we can define θ directly via the Finsler $\|\ \|$ which makes $\mathbb{E}_x$ a B-space for all x and we can define θ^M and θ^m by $\theta^M(E_{0x}) = \sup\{\theta^\alpha(\varphi_{\alpha x}(E_{0x}))\}$. It is easy to prove that over the map g_π:

$$\theta \sim_\rho \theta^M \leq (g_\pi^* \rho_{(\mathfrak{D},\mathfrak{a})})^2 \theta^m. \tag{2.1}$$

In particular, the equivalence class of $\max(\theta, g_\pi^*\rho)$ is independent of the choices of admissible bound and Finsler and agrees with that of $\max(\theta^M, g_\pi^*\rho)$ and $\max(\theta^m, g_\pi^*\rho)$.

1 THEOREM: Let $\pi: \mathbb{E} \to X$ be an $\mathfrak{M}$ bundle with admissible Finsler, bound and atlas $(\mathfrak{D},\mathfrak{a})$ as above. Let $G(\mathfrak{D}) = \{G_{\alpha P}, k_{\alpha P}\}$. With θ defined by Finsler or atlas as above $\{(G(\mathfrak{D}), \max(\theta, g_\pi^*\rho))\}$ is a class of adapted $\mathfrak{M}$ atlases generating an $\mathfrak{M}$ metric structure on $G(\mathbb{E})$. g_π is then an $\mathfrak{M}$ submersion with admissible atlas pairs $(G(\mathfrak{D}),\mathfrak{a})$.

If $\mathfrak{M} \subset \mathcal{L}$ and π is semicomplete then $G(\mathbb{E})$ is semicomplete with $G(\mathfrak{D})$ s admissible and g_π is a semicomplete submersion with $(G(\mathfrak{D}),\mathfrak{a})$ s admissible (when $(\mathfrak{D},\mathfrak{a})$ is s admissible).

If $\mathfrak{a}$ is a convex (or star bounded) atlas then $G(\mathfrak{D})$ is convex (resp. star bounded). If $\mathfrak{M} \subset \mathcal{L}^r$, π is semicomplete and $(\mathfrak{D},\mathfrak{a})$ is a $\mathcal{L}ip^r$ atlas for π with $\mathfrak{a}$ star bounded then $G(\mathfrak{D})$ is a $\mathcal{L}ip^r$ atlas.

The trivial section $s_\pi: X \to G(\mathbb{E})$ defined by $s_\pi(x) = \mathbb{E}_x \in G(\mathbb{E}_x)$ is an

$\mathfrak{M}$ isomorphism onto an open and closed subset of $G(\mathbb{E})$. In particular, it is an $\mathfrak{M}$ section of g_π.

If $(\Phi,f)\colon \pi_1 \to \pi_2$ is an $\mathfrak{M}$ VB map with $(f^*\Phi,1)\colon \pi_1 \to f^*\pi_2$ a locally split $\mathfrak{M}$ VB injection then $(G(\Phi),f)\colon g\pi_1 \to g\pi_2$, defined by $G(\Phi)(E_{0x}) = \Phi(E_{0x}) \in G(\mathbb{E}_{2fx})$ with $E_{0x} \in G(\mathbb{E}_{1x})$, is an $\mathfrak{M}$ map of submersion. $G(\Phi,f) = G(\Phi)\cdot s_{\pi_1}\colon X_1 \to G(\mathbb{E}_2)$ is an $\mathfrak{M}$ map.

If π_1 is an $\mathfrak{M}$ bundle over X then $G\colon Li(\mathbb{E}_1;\mathbb{E}) \to G(\mathbb{E})$ defined by $G(T_x) = T_x(E_{0x}) \in G(\mathbb{E}_x)$, is an $\mathfrak{M}$ submersion. G is a semicomplete submersion if $\mathfrak{M} \subset \mathcal{L}$ and π_1,π are semicomplete.

If π is the trivial bundle ϵ_F then g_π can be identified at the atlas level with $p_1\colon X \times G(F) \to X$.

<u>PROOF</u>: Let $(\mathfrak{D}_1,\mathfrak{a}_1) = \{U_\alpha,h_\alpha,\varphi_\alpha\}$ and $(\mathfrak{D}_2,\mathfrak{a}_2) = \{V_\beta,g_\beta,\varphi_\beta\}$ be admissible atlases for π_1 and π_2. If $G_{\alpha P_1} \cap G(\Phi)^{-1}(G_{\beta P_2}) \neq \emptyset$ then $(k_{\beta P_2}\cdot G(\Phi)\cdot k_{\alpha P_1}^{-1})\cdot(1 \times h_{P_1})$ is the restriction of:

$$f_{\beta\alpha} \times [G\cdot(\Phi_{\beta\alpha} \times 1)]\colon h_\alpha(U_\alpha \cap f^{-1}V_\beta) \times G(F_\alpha^1) \qquad g_\beta(V_\beta) \times G(F_\beta^2)$$

composed with $1 \times h_{P_2}$. It follows from Theorem 1.3 that:

$$\text{(2.2)}\quad \max[G(\Phi)^*\theta_2^M,\rho^{\mathfrak{M}}(G(\Phi);G(\mathfrak{D}_1),G(\mathfrak{D}_2))] \leq O^*(\theta_1^M,g_{\pi_1}^*\rho^{\mathfrak{M}}_{(\Phi,f)},g_{\pi_1}^*\theta^M(f^*\Phi)).$$

Here θ^M is defined on $Li(\mathbb{E}_1;f^*\mathbb{E}_2)$ by the atlas pair $(\mathfrak{D}_1,f^*\mathfrak{D}_2;\mathfrak{a}_{1f})$, i.e. $\theta^M(f^*\Phi)(x) = \sup\{\theta(\Phi_{\beta\alpha}(h_\alpha x))\colon x \in U_\alpha \cap f^{-1}V_\beta\}$. Applying this result with (Φ,f) = identity, first with $(\mathfrak{D}_1,\mathfrak{a}_1) = (\mathfrak{D}_2,\mathfrak{a}_2)$ and then with $(\mathfrak{D}_1,\mathfrak{a}_1) \neq (\mathfrak{D}_2,\mathfrak{a}_2)$ we see first that $(G(\mathfrak{D}),\max(\theta,g_\pi^*\rho))$ is an adapted atlas and then that the metric structure is independent of the choice of $(\mathfrak{D},\mathfrak{a})$. Finally, we get that $G(\Phi)$ is $\mathfrak{M}$. That g_π is a submersion with admissible atlas pair $(G(\mathfrak{D}),\mathfrak{a})$ is clear. For the s_π result, recall that E is an isolated point of $G(E)$ with $V_P = \{E\}$ for $P = I_E$ and $D_P = \{0\} = L(E;\{0\})$. The principal part $h_{\alpha P}\cdot s_\pi\cdot h_\alpha^{-1} = 1 \times c_0\colon h_\alpha(U_\alpha) \to h_\alpha(U_\alpha) \times 0$. The semicompleteness results follow from the estimate on λ in Theorem 1.2.

Since $h_{\alpha P}(G_{\alpha P}) = h_\alpha(U_\alpha) \times D_P$, $G(\mathfrak{D})$ is convex iff $\mathfrak{G}$ is convex. If $\mathfrak{G}$ is star bounded then $G(\mathfrak{D})$ is star bounded by (1.6). That $G(\mathfrak{D})$ is a $\mathcal{L}ip^r$ atlas if $(\mathfrak{D},\mathfrak{G})$ is (and $\mathfrak{G}$ is star-bounded) follows from (1.9). If $(\mathfrak{D}_1,\mathfrak{D};\mathfrak{G}) = \{U_\alpha,h_\alpha,\varphi_\alpha^1,\varphi_\alpha\}$ is an admissible two-tuple for π_1 and π, then $G(\varphi_\alpha)\cdot G\cdot L(\varphi^1;\varphi)_\alpha^{-1} = 1 \times G_{F_\alpha} : h_\alpha(U_\alpha) \times Li(F_\alpha^1;F_\alpha) \to h_\alpha(U_\alpha) \times G(F_\alpha)$ and so G is a submersion by the proof of Theorem 1.3. Finally, for a trivial bundle it is easy to check that $G(\mathfrak{G} \times F) \to \mathfrak{G} \times \mathfrak{G}(G(F))$. Q.E.D.

2 <u>PROPOSITION</u>: Let $\pi: E \to X$ be a semicomplete $\mathcal{L}^r$ bundle and let $\pi_0: E_0 \to X_0$ be a $\mathcal{C}_u^r(d_0)$ (or $\mathcal{L}_a^r(d_0)$) bundle with respect to apm d_0 on X_0. Recall that for d_X an admissible metric on X we can regard π as an $\mathcal{L}^r(d_X)$ bundle. Let $(\Phi,f): \pi_0 \to \pi$ be a $\mathcal{C}_u^r(d_0,d_X)$ (or $\mathcal{L}_a^r(d_0,d_X)$) VB map with $(f^*\Phi,1): \pi_0 \to f^*\pi$ a $\mathcal{C}$ locally split VB injection. Regarding $G(E)$ as an $\mathcal{L}^r(d_G)$ manifold with respect to an admissible metric d_G on $G(E)$, $G(\Phi,f): X_0 \to G(E)$ is a $\mathcal{C}_u^r(d_0,d_G)$ (resp. $\mathcal{L}_a^r(d_0,d_G)$) map.

<u>PROOF</u>: Let $(\mathfrak{D},\mathfrak{G})$ be an s admissible $\mathcal{L}ip^r$ atlas for π with $\mathfrak{G}$ convex and star bounded and let $(\mathfrak{D}_0,\mathfrak{G}_0)$ be an admissible multiatlas for π_0 with $\mathfrak{G}_0$ star bounded. Adapt the proof of Theorem 1, using (1.9), to show that the transition map family is $\mathcal{C}_u$ (or $\mathcal{L}_a$). Q.E.D.

Now we define the atlas $(\Gamma(\mathfrak{D}),G(\mathfrak{D})) = \{G_{\alpha P},k_{\alpha P},\psi_{\alpha P}\}$ on $g_\pi{}^*\pi: g_\pi{}^*E \to G(E)$ with $\psi_{\alpha P}(E_{0x},v) = (h_\alpha x,T,Q_{-T}(p_2\varphi_\alpha v))$ where $v \in E_x$ and $(h_\alpha x,T) = k_{\alpha P}(E_{0x})$.

3 <u>THEOREM</u>: Let $\pi: E \to X$ be an $\mathfrak{M}$ bundle with admissible atlas $(\mathfrak{D},\mathfrak{G})$. $(\Gamma(\mathfrak{D}),G(\mathfrak{D}))$ is an admissible atlas on $g_\pi{}^*\pi$. If $\mathfrak{M} \subset \mathcal{L}$, π is semicomplete and $(\mathfrak{D},\mathfrak{G})$ is s admissible then $(\Gamma(\mathfrak{D}),G(\mathfrak{D}))$ is s admissible. If $\mathfrak{M} \subset \mathcal{L}^r$, π is semicomplete and $(\mathfrak{D},\mathfrak{G})$ is a $\mathcal{L}ip^r$ atlas with $\mathfrak{G}$ star bounded then $(\Gamma(\mathfrak{D}),G(\mathfrak{D}))$ is a $\mathcal{L}ip^r$ atlas. Let $\Gamma(E) = \{(E_{0x},v) \in g_\pi{}^*E: v \in E_{0x}\}$ and let γ_π denote the restriction of $g_\pi{}^*\pi$ to $\Gamma(E)$. $(\Gamma(\mathfrak{D}),G(\mathfrak{D}))$ induces a

locally split subbundle atlas on γ_π. γ_π is called the canonical subbundle of $g_\pi{}^*\pi$.

PROOF: $g_\pi : G(\mathfrak{D}) \to \mathfrak{G}$ is index preserving. The admissibility and $\mathscr{L}ip^r$ results follow from comparing $(\Gamma(\mathfrak{D}), G(\mathfrak{D}))$ and $(g_\pi{}^*\mathfrak{D}, G(\mathfrak{D}))$, i.e. look at the principal parts of the identity maps. The subbundle result is clear as in Theorem 1.4. Q.E.D.

4 COROLLARY: Let $\pi: E \to X$ be an $\mathfrak{M}$ metric bundle, $f: X_0 \to X$ be an $\mathfrak{M}$ metric map and $E_0 \subset f^*E$ with E_{0x} a closed split subspace of E_{fx} for all $x \in X_0$. Let $s_{E_0} : X_0 \to G(E)$ be defined by $s_{E_0}(x) = E_{0x} \in G(E_{fx})$. $f^*\pi|E_0$ is a locally split $\mathfrak{M}$ metric subbundle iff s_{E_0} is an $\mathfrak{M}$ map. If $\mathfrak{M} \subset \mathscr{L}$, π is semicomplete and $f^*\pi|E_0$ is a locally split $\mathfrak{M}$ subbundle then $f^*\pi|E_0$ is a semicomplete, locally split $\mathfrak{M}$ subbundle.

If π is a semicomplete $\mathscr{L}^r$ bundle, X_0 is a $\mathcal{C}_u^r(d_0)$ (or $\mathscr{L}^r(d_0)$) manifold with apm d_0 and f is a $\mathcal{C}_u^r(d_0,d_X)$ (resp. $\mathscr{L}_a^r(d_0,d_X)$) map then $f^*\pi|E_0$ is a $\mathcal{C}_u^r(d_0)$ (resp. $\mathscr{L}_a^r(d_0)$) subbundle of $f^*\pi$ iff s_{E_0} is a $\mathcal{C}_u^r(d_0,d_G)$ (resp. $\mathscr{L}_a^r(d_0,d_G)$) map.

PROOF: If $f^*\pi|E_0$ is a subbundle of $f^*\pi$ with inclusion map $(i,1)$, then $s_{E_0} = G(\Phi_f \cdot i, f)$ and the result follows from Theorem 1 and Proposition 2. Conversely, the identification of $s_{E_0}{}^*g_\pi{}^*\pi$ with $f^*\pi$ identifyies $s_{E_0}{}^*\Gamma(E)$ with E_0 and the converse follows from Theorem 3. Q.E.D.

5 COROLLARY: Let $f: X_0 \to X$ be a $\mathcal{C}^r$ immersion $(r \geq 1)$. If X_0, X and f are $\mathfrak{M}^r$ then $G(f)$, defined to be $G(Tf,f): X_0 \to G(TX)$ is an $\mathfrak{M}^{r-1}$ map. If X is a semicomplete $\mathscr{L}^r$ manifold, X_0 is a $\mathcal{C}_u^r(d_0)$ (or $\mathscr{L}^r(d_0)$) manifold with respect to apm d_0 on X_0 and f is a $\mathcal{C}_u^r(d_0,d_X)$ (resp. $\mathscr{L}_a^r(d_0,d_X)$) map then $G(f)$ is a $\mathcal{C}_u^{r-1}(d_0,d_G)$ (resp. $\mathscr{L}_a^{r-1}(d_0,d_G)$) map. $G(f)$ is metricly proper if f is. If $r \geq 2$ then $G(f)$ is a $\mathcal{C}^{r-1}$ immersion.

PROOF: Smoothness of $G(f)$ is clear from Corollary 4. Let ρ and ρ_1 be admissible bounds on X and $G(TX)$. $\rho_1 > g_\tau{}^*\rho$. If f is metricly proper then $f^*\rho > G(f)^*\rho_1 > G(f)^*g_\tau{}^*\rho = f^*\rho$. So $G(f)$ is metricly proper. If $r \geq 2$ then to show $G(f)$ is a $\mathcal{C}^{r-1}$ immersion we can reduce to the metricly proper case and apply Theorem VI.3.1. If P is a left inverse for $T_x f$ then $P \cdot T_{G(f)x} g_\tau$ is a left inverse for $T_x G(f)$. Q.E.D.

From now on we will refer to a $\mathcal{C}_u^r(d_0,d_X)$ (or $\mathcal{L}_a^r(d_0,d_X)$) map which is a $\mathcal{C}^1$ immersion as a $\mathcal{C}_u^r(d_0,d_X)$ (or $\mathcal{L}_a^r(d_0,d_X)$) immersion.

3. Smoothing Immersions: Let X be an $\mathfrak{M}^1$ manifold and $f\colon X_0 \to X$ a C^1 immersion. If $\mathcal{G} = \{U_\alpha, h_\alpha\}$ is an admissible $\mathfrak{M}^1$ atlas on X we define the multiatlas $f^*G(T\mathcal{G}) = \{U_{\alpha P}, h_{\alpha P}\}$ (P is a projection of E_α with $\|P\| \leq 2\theta(P(E_\alpha))$) by

$$(3.1) \qquad U_{\alpha P} = G(f)^{-1}G_{\alpha P} \qquad h_{\alpha P} = P \cdot h_\alpha \cdot f$$

where $G(f) = G(Tf,f)\colon X_0 \to G(TX)$ satisfies $g_\pi \cdot G(f) = f$. Hence, $U_{\alpha P} \subset f^{-1}U_\alpha$, i.e. $f\colon f^*G(T\mathcal{G}) \to \mathcal{G}$ is index preserving. Also, if $x \in U_{\alpha P}$ then $T_x(h_\alpha \cdot f)$ is a linear isomorphism of $T_x X$ onto the graph of a map $T_{\alpha P}(x) \in L(P(E_\alpha);K(P))$. In fact, $(h_\alpha fx, T_{\alpha P}(x)) = k_{\alpha P} \cdot G(f)(x)$. Hence, $T_x h_{\alpha P}$ is a linear isomorphism and so $f^*G(T\mathcal{G})$ is a C^1 multiatlas.

1 PROPOSITION: Let X be an $\mathfrak{M}^1$ manifold, X_0 be a C^1 manifold and $f\colon X_0 \to X$ be a C^1 immersion. Let ρ be an admissible bound on X and define θ on $G(TX)$ via admissible atlas or Finsler. Let $\mathcal{G}$ be an admissible $\mathfrak{M}^1$ atlas on X.

(a) If $f^*\rho > G(f)^*\theta$ then $(f^*G(T\mathcal{G}), f^*\rho)$ is a $\mathcal{C}^1$ adapted multiatlas inducing the unique $\mathcal{C}^1$ structure on X_0 with respect to which f is a metricly proper, $\mathcal{C}^1$ immersion. X_0 is regular if X is.

(b) If X_0 is an $\mathfrak{M}^1$ manifold and $f\colon X_0 \to X$ is an $\mathfrak{M}^1$ immersion then $f^*G(T\mathcal{G})$ is an admissible $\mathfrak{M}^1$ multiatlas on X_0.

(c) Let X be a semicomplete $\mathcal{L}^r$ manifold with admissible metric d_X, X_0 be a $\mathcal{C}_u^r(d_0)$ (or $\mathcal{L}^r(d_0)$) manifold with apm d_0 ($r \geq 1$). If f is a $\mathcal{C}_u^r(d_0,d_X)$ (resp. $\mathcal{L}^r(d_0,d_X)$) immersion then $f^*G(T\mathfrak{a})$ is an admissible $\mathcal{C}_u^r(d_0)$ (resp. $\mathcal{L}^r(d_0)$) multiatlas on X_0, provided $\mathfrak{a}$ is a convex, s admissible $\mathcal{L}ip^r$ atlas.

<u>PROOF</u>: Let $\mathfrak{a}_0 = f^*G(T\mathfrak{a})$ and d_G be an admissible metric on $G(TX)$. (a): $f^*\rho > f^*\rho_{\mathfrak{a}} \geq \rho^{\mathcal{C}}_{\mathfrak{a}_0}$. If $x \in U_{\alpha P} \cap U_{\beta\bar{P}}$ let U be a neighborhood of x in the intersection with $h_{\alpha P}|U$ injective. $D(h_{\beta\bar{P}}\circ(h_{\alpha P}|U)^{-1})(h_{\alpha P}x) = \Gamma_{\bar{T}}^{-1}\circ D(h_\beta h_\alpha^{-1})(h_\alpha fx)\circ\Gamma_T$ where $T = T_{\alpha P}(x)$ and $\bar{T} = T_{\beta\bar{P}}(x)$. Defining $\theta = \theta^M$ via $(T\mathfrak{a},\mathfrak{a})$, we have $f^*\rho > G(f)^*\theta^M$ and so $\rho(fx) > \sup\{\theta(\Gamma_T): T = T_{\alpha P}(x), x \in U_{\alpha P}\}$. Hence, $f^*\rho > \rho^{\mathcal{C}^1}_{\mathfrak{a}_0}$. So $(\mathfrak{a}_0,f^*\rho)$ generates a $\mathcal{C}^1$ structure on X_0. If $x \in U_{\alpha P}$ and $h_{\alpha P}|U$ is injective then $D(h_\alpha\circ f\circ(h_{\alpha P}|U)^{-1})(h_{\alpha P}x) = \Gamma_T$ ($T = T_{\alpha P}(x)$). Thus, f is a metricly proper $\mathcal{C}^1$ immersion by Theorem VI.3.1. Uniqueness and regularity as on pages 191 and 192. (b): We can assume f is metricly proper. Let $\mathfrak{a}_1 = \{V_\beta,g_\beta\}$ be an admissible $\mathfrak{M}^1$ atlas on X_0. The transition map $h_{\alpha P}\circ g_\beta^{-1}$: $g_\beta(V_\beta \cap U_{\alpha P}) \to P(E_\alpha)$ is P composed with the local representative for f, $f_{\alpha\beta}$: $g_\beta(V_\beta \cap U_{\alpha P}) \to E_\alpha$. Thus, $h_{\alpha P}\circ g_\beta^{-1}$ is an $\mathfrak{M}^1$ map and its norm near $g_\beta(x)$ is dominated by $\rho(fx)$. Also, by Corollary 2.5 $G(f)$ is an $\mathfrak{M}$ metric map and so $f^*\rho > G(f)^*\theta^M$. Hence, the derivative of $h_{\alpha P}\circ g_\beta^{-1}$ is an isomorphism at points of $g_\beta(V_\beta \cap U_{\alpha P})$ with θ at points near $g_\beta(x)$ dominated by $\rho(fx)$. So by the Inverse Function Theorem, the local inverse of $h_{\alpha P}\circ g_\beta^{-1}$ near $g_\beta x$ exists, is $\mathfrak{M}^1$ and has norm dominated by $\rho(fx)$. This shows that $\rho^{\mathfrak{M}^1}(1;\mathfrak{a}_1,\mathfrak{a}_0)$ and $\rho^{\mathfrak{M}^1}(1;\mathfrak{a}_0,\mathfrak{a}_1)$ are domainted by $f^*\rho$. The usual composition and gluing argument shows that $f^*\rho > \rho_{\mathfrak{a}_1\cup\mathfrak{a}_0}$. (c): By Corollary 2.5, $G(f)$ is $\mathcal{C}_u(d_0,d_G)$ and since $G(T\mathfrak{a})$ is s admissible on $G(TX)$ it follows that the cover $\{U_{\alpha P}\}$ is a $\mathfrak{N}_{d_0}$ uniform, open cover of X_0. Let $\mathfrak{a}_1$ be any admissible multiatlas on X_0 and compare $\mathfrak{a}_0$ and $\mathfrak{a}_1$ by estimating $\{j_1^r(h_{\alpha P}\circ g_\beta^{-1})\circ g_\beta\}$ using $\{P_\# j_1^r(h_\alpha\circ f\circ g_\beta^{-1})\circ g_\beta\}$ as in (b). The required uniform version of the

Inverse Function Theorem (Thm. II.3.4) is an easy adaptation of the original. Proposition III.6.7 is used for gluing. Q.E.D.

2 **COROLLARY**: Assume $\mathfrak{M} \subset \mathcal{L}$. Let $f: X_0 \to X_1$ be a metricly proper $\mathfrak{M}^1$ immersion with X_1 semicomplete. X_0 is semicomplete iff it is uniformly complete. In that case f is semicomplete.

PROOF: Let $\mathcal{G}$ be an s admissible atlas on X. X_0 is regular because X is. Since $G(f)$ is $\mathcal{L}$, it is $\mathcal{C}_u$ and so the cover $\{U_{\alpha P}\}$ is a uniform open cover of X_0. Result from Lemma V.1.2 and Lemma VI.3.2. Q.E.D.

ADDENDUM: If $f: X_0 \to X$ is a metricly proper $\mathcal{C}^1$ immersion with X a semicomplete $\mathcal{L}^1$ manifold and $G(f): X_0 \to G(TX)$ is a $\mathcal{C}_u$ map, then X_0 is a semicomplete $\mathcal{C}^1$ manifold iff it is uniformly complete. In that case, f is semicomplete. The proof is the same after a slight adjustment of Lemma VI.3.2.

That uniform completeness is needed is shown by looking at the inclusion of an open set. In general, any uniformly complete refinement of X_0 is semicomplete and f is a semicomplete immersion with respect to this refinement (though f is then no longer metricly proper).

3 **LEMMA**: Let $f: X_0 \to X$ be a C^1 immersion and $\mathcal{G}$ be a C^1 atlas on X. The VB map $(f^*Tf,1): \tau_{X_0} \to f^*\tau_X \cong G(f)^*g_\tau{}^*\tau_X$ induces an atlas identification between the multiatlases $(T(f^*G(T\mathcal{G})), f^*G(T\mathcal{G}))$ and the restriction of $(G(f)^*\Gamma(T\mathcal{G}), f^*G(T\mathcal{G}))$ to the subbundle $G(f)^*\gamma_\tau$.

PROOF: We compute the principal parts of the VB map $(G(f) \times Tf, G(f))$: $\tau_{X_0} \to g_\tau{}^*\tau_X$ regarded as a map of atlases $(T(f^*G(T\mathcal{G})), f^*G(T\mathcal{G})) \to (\Gamma(T\mathcal{G}), G(T\mathcal{G}))$. On $h_{\alpha P}(U_{\alpha P})$ we show that the principal part is the constant map to the inclusion in $L(P(E_\alpha); E_\alpha)$. Note that $G(f): f^*G(T\mathcal{G}) \to G(T\mathcal{G})$ is index preserving. If $x \in U_{\alpha P}$ and $k_{\alpha P}G(f)(x) = (h_\alpha fx, T_{\alpha P}(x))$ then the image of $T_x(h_\alpha \circ f)$ is the graph of $T = T_{\alpha P}(x)$ and if $h_{\alpha P}|U$ is injective, $D(h_\alpha f(h_{\alpha P}|U)^{-1})(h_{\alpha P}x)$ is the graph map $\Gamma_T\colon P(E_\alpha) \to E_\alpha$. To get the principal part with respect to $(\Gamma(TG), g(TG))$ we compose with Q_{-T} and get the inclusion. Restricting to $G(f)^*\nu_\tau$ the principal part is constantly the identity in $L(P(E_\alpha);P(E_\alpha))$. Q.E.D.

4 SMOOTHING LEMMA: (a) Let $f\colon X_0 \to X$ be a metricly proper $\mathfrak{M}^1$ immersion with X an $\mathfrak{M}^2$ manifold. f is a metricly proper, $\mathfrak{M}^2$ immersion (more precisely, X_0 has an $\mathfrak{M}^2$ metric structure, necessarily unique and necessarily compatible with the original $\mathfrak{M}^1$ structure, such that f is an $\mathfrak{M}^2$ immersion) iff $G(f)\colon X_0 \to G(TX)$ is an $\mathfrak{M}^1$ map.

(b) Assume $\mathcal{C}^1 \subset \mathfrak{M} \subset \mathcal{C}$. Let $f\colon X_0 \to X$ be a metricly proper, $\mathcal{C}^1$ immersion with X an $\mathfrak{M}^1$ manifold. f is a metricly proper $\mathfrak{M}^1$ immersion iff $G(f)$ is an $\mathfrak{M}$ map.

(c) Let X be a semicomplete $\mathcal{L}^{r+1}$ manifold ($r \geq 1$) and X_0 be a $\mathcal{C}^r_u(d_0)$ (or $\mathcal{L}^r(d_0)$) manifold with respect to apm d_0. Assume $f\colon X_0 \to X$ is a metricly proper $\mathcal{C}^r_u(d_0,d_X)$ (resp. $\mathcal{L}^r(d_0,d_X)$) immersion. X_0 is a $\mathcal{C}^{r+1}_u(d_0)$ (resp. $\mathcal{L}^{r+1}(d_0)$) manifold and f is a metricly proper $\mathcal{C}^{r+1}_u(d_0,d_X)$ (resp. $\mathcal{C}^{r+1}(d_0,d_X)$) immersion iff $G(f)$ is a $\mathcal{C}^r_u(d_0,d_G)$ (resp. $\mathcal{L}^r(d_0,d_G)$) map.

(d) Let X be a semicomplete $\mathcal{L}^1$ manifold and X_0 be a $\mathcal{C}^1$ manifold with apm d_0. Assume $f\colon X_0 \to X$ is a metricly proper, $\mathcal{C}^1$ immersion and with respect to d_0 f is a $\mathcal{C}_u$ (or $\mathcal{L}$) map. X_0 admits a compatible $\mathcal{C}^1_u(d_0)$ (resp. $\mathcal{L}^1(d_0)$) structure with f a $\mathcal{C}^1_u(d_0,d_X)$ (resp. $\mathcal{L}^1(d_0,d_X)$) metricly proper immersion iff $G(f)$ is, with respect to d_0, a $\mathcal{C}_u$ (resp. $\mathcal{L}$) map.

PROOF: (a): Let G be an admissible $\mathfrak{M}^2$ atlas on X. $(\Gamma(TG), G(TG))$ is an admissible $\mathfrak{M}^1$ atlas on $g_\tau{}^*\tau_X$ and so $(G(f)^*\Gamma(TG), f^*G(TG))$ is an admissible $\mathfrak{M}^1$ atlas on $G(f)^*g_\tau{}^*\tau_X \cong f^*\tau_X$ because $G(f)$ is $\mathfrak{M}^1$. By Lemma 3 there is an atlas identification between $(T(f^*G(TG)), f^*G(TG))$ and the subbundle multiatlas $(G(f)^*\Gamma(TG)|G(f)^*\Gamma(TX), f^*G(TG))$. Consequently, $((T(f^*G(TG)), f^*G(TG)), f^*\rho)$ is an $\mathfrak{M}^1$ adapted multiatlas on τ_{X_0} and (Tf,f)

is an $\mathfrak{M}^1$ map of adapted multiatlases. Thus, $(f^*G(TG), f^*\rho)$ is an adapted $\mathfrak{M}^2$ multiatlas and with respect to the induced metric structure f is an $\mathfrak{M}^2$ map. By Proposition 1(b) this $\mathfrak{M}^2$ structure on X_0 is compatible with the original $\mathfrak{M}^1$ structure and so f is an $\mathfrak{M}^2$ immersion. (b), (c) and (d) are completely similar. In (c) and (d) choose G a convex, star bounded s admissible $\mathcal{L}ip^r$ atlas on X $(r = 1 \text{ in (d)})$. Q.E.D.

Let $f: X_0 \to X$ be a locally split C^r immersion and X be an $\mathfrak{M}^r$ manifold $\mathcal{C}^1 \subset \mathfrak{M} \subset \mathcal{C}$. Proposition 1(a) implies that f is a $\mathcal{C}^1$ immersion iff $f^*\rho > G(f)^*\theta$. With respect to the $\mathcal{C}^1$ structure Lemma 4(b) implies that f is an $\mathfrak{M}^1$ immersion iff $G(f)$ is an $\mathfrak{M}$ map. Proceeding inductively using Lemma 4(a) we see that there is a unique $\mathfrak{M}^s$ structure on X_0 with $r \geq s \geq 1$ and regular if X is regular such that f is an $\mathfrak{M}^s$ immersion. s is characterized by the fact that $G(f)$ is not an $\mathfrak{M}^s$ map. Furthermore, if X is semicomplete and either $\mathfrak{M} \subset \mathcal{L}$ or $s > 1$, then X_0 is semicomplete iff it is uniformly complete. Given an apm d_0 on X_0, if X is a semicomplete $\mathcal{L}^r$ manifold and f is $\mathcal{C}_u$ (or $\mathcal{L}$) with respect to d_0 and d_X we can proceed similarly to find $r \geq s_1 \geq s_2 \geq 0$ such that f is a $\mathcal{C}_u^{s_1}(d_0,d_X)$ (or $\mathcal{L}^{s_2}(d_0,d_X)$) immersion.

There is a pitfall here which I should point out. Assume f is a C^2 immersion and X_0 is finite dimensional. Then f is a $\mathcal{C}^1$ immersion. Furthermore, $G(f): X_0 \to G(TX)$ is a C^1 immersion. Hence, $G(f)$ is a $\mathcal{C}^1$ immersion and so f is a $\mathcal{C}^2$ immersion. This need not be true. The error is that although $G(f)$ is a $\mathcal{C}^1$ immersion for a $\mathcal{C}^1$ structure on X_0 characterized by this fact, and $g_\tau \cdot G(f) = f$ is a $\mathcal{C}^1$ map with respect to this structure, f might <u>not</u> be a $\mathcal{C}^1$ immersion with respect to this new structure, i.e. the $G(f)$ induced $\mathcal{C}^1$ structure might not agree with the f induced $\mathcal{C}^1$ structure. They do agree iff $G(f)$ is $\mathcal{C}^1$ with respect to the f induced structure. For example, let $f: R \to R^3$ be a curve. Use the semicompact group structure on R^3 (cf. p. 90). A reparametrization of the curve is a homeomorphism of R, i.e. a single chart atlas on R. Clearly, parametrization by arclength gives the $\mathcal{C}^1$ structure induced by f and $G(f)$ is essentially the unit tangent map df/ds. The transition map between the f induced structure and the $G(f)$ induced structure is thus the curvature function. If the curvature is unbounded then f is not a $\mathcal{C}^2$ metricly proper immersion.

4. __Leaf Immersions__: Let X be an $\mathfrak{M}^r$ manifold. Let $g_\tau^1\colon G^1(TX) \to X$ be the Grassman submersion of τ_X and inductively define $g_\tau^i\colon G^i(TX) \to G^{i-1}(TX)$ to be the Grassman submersion of $\tau_{G^{i-1}}$ $(2 \le i \le r)$. $G^i(TX)$ is an $\mathfrak{M}^{r-i}$ manifold, semicomplete if $\mathfrak{M}^{r-i} \subset \mathcal{L}$ and X is semicomplete, and g_τ^i is an $\mathfrak{M}^{r-i}$ submersion. If $f\colon X_0 \to X$ is an $\mathfrak{M}^r$ immersion then we can define inductively $G^i(f) = G(G^{i-1}(f))\colon X_0 \to G^i(TX)$ an $\mathfrak{M}^{r-i}$ map, metricly proper if f is, and an immersion $i < r$. By convention, $G^0(TX) = X$ and $G^0(f) = f$.

1 __PROPOSITION__: Let X be a semicomplete $\mathcal{L}^r$ manifold $(r \ge 1)$ with admissible metric and bound d_X, ρ. Let d_i, ρ_i be admissible metric and bound on the semicomplete $\mathcal{L}^{r-i}$ manifold $G^i(TX)$ $(1 \le i \le r)$. Let X_0 be a regular $\mathcal{L}^r$ manifold and $f\colon X_0 \to X$ be a metricly proper, $\mathcal{L}^r$ immersion. Let d_{X_0} be an admissible metric on X_0 and d_0 be an apm on X_0. Define $d_f = f^*d_X$ and $d_{G^i(f)} = G^i(f)^*d_i$ $(1 \le i \le r)$. These are apm on X_0.

(a) There exists a (necessarily unique) $\mathcal{L}^i(d_0)$ structure on X_0 such that f is a metricly proper $\mathcal{L}^i(d_0,d_X)$ immersion iff $d_0 >_{f^*\rho} d_{G^i(f)}$ $(1 \le i \le r)$.

(b) There exists a (necessarily unique) $\mathcal{C}_u^i(d_0)$ structure on X_0 such that f is a metricly proper, $\mathcal{C}_u^i(d_0,d_X)$ immersion iff the map $(X_0,d_0,f^*\rho) \to (X_0,d_{G^i(f)},f^*\rho)$ is $\mathcal{C}_u$ $(1 \le i \le r)$.

__PROOF__: Since $d_i >_{\rho_i} g_\tau^{i*}d_{i+1}$, $f^*\rho > G^i(f)^*\rho_i$ and $G^{i-1}(f) = g_\tau^i \circ G^i(f)$:

(4.1) $$d_{X_0} >_{f*\rho} d_{G^r(f)} >_{f*\rho} \cdots >_{f*\rho} d_{G(f)} >_{f*\rho} d_f .$$

Since $G^i(f)$ is $\mathcal{L}$ for $1 \le i \le r$ these are apm by Lemma V.1.1. The uniqueness of the metric structures follows from Proposition 3.1. Corollary 2.5 and induction imply the necessity of the conditions on d_0 in order that the metric structures exist. We prove sufficiency by induction. Let $\mathcal{a}$ be an s admissible, convex, star bounded $\mathcal{L}ip^r$ atlas on X. The cover $\{G_{\alpha P}\}$ of G(TX) is d_1 uniform, open and so $\{U_{\alpha P}\}$ is a $d_{G(f)}$ uniform open cover of X and so by hypothesis on d_0 it is a d_0 uniform, open cover. G(f) is clearly a $d_{G(f)}$ $\mathcal{L}$ map and so it is d_0 $\mathcal{L}$ (or $\mathcal{C}_u$) if d_0 satisfies the hypothesis in (a) (resp. (b)). Hence, $f = g_\tau \circ G(f)$ is $d_0 - \mathcal{L}$ (or $\mathcal{C}_u$). The result for $i = 1$ then follows from the Smoothing Lemma 3.4 (d). We prove the result for $1 < i \le r$. Using (4.1) we apply the $i - 1$ result to f and G(f). We get two $\mathcal{L}^{i-1}(d_0)$ (or $\mathcal{C}_u^{i-1}(d_0)$) structures on X_0 characterized by the fact that f or G(f) is an $\mathcal{L}^{i-1}(d_0,d_X)$ or $\mathcal{L}^{i-1}(d_0,d_1)$ (resp. $\mathcal{C}_u^{i-1}(d_0,d_X)$ or $\mathcal{C}_u^{i-1}(d_0,d_1)$) immersion. With respect to the G(f) induced structure, $f = g_\tau \circ G(f)$ is an $\mathcal{L}^{i-1}(d_0,d_X)$ (resp. $\mathcal{C}_u^{i-1}(d_0,d_X)$) map. We avoid the pitfall of pages 247 and 248 by noting that by uniqueness the underlying $\mathcal{C}^{i-1}$ structures of the f and G(f) structures agree. Hence, f is an $\mathcal{C}^{i-1}$ immersion with respect to the G(f) structure and so by uniqueness the f and G(f) induced $\mathcal{L}^{i-1}(d_0)$ (resp. $\mathcal{C}_u^{i-1}(d_0)$) structures agree. The result for i now follows from the Smoothing Lemma 3.4 (c). Q.E.D.

2 <u>COROLLARY</u>: Let X, d_X,ρ,d_i,ρ_i $(1 \le i \le r)$ be as in Proposition 1. Assume X_0 is a regular $\mathcal{C}^r$ manifold with apm d_0 and $f: X_0 \to X$ is a metricly proper $\mathcal{C}^r$ immersion. There exists a (necessarily unique) $\mathcal{C}_u^r(d_0)$ structure on X_0 such that f is a metricly proper, $\mathcal{C}_u^r(d_0,d_X)$ immersion iff the map $(X_0,d_0,f*\rho) \to (X_0,d_{G^r(f)},f*\rho)$ is $\mathcal{C}_u$.

PROOF: Since $G^r(f)$ is only $\mathcal{C}$, (4.1) holds only up to $d_{G^{r-1}(f)}$. We still have $d_{G^r(f)} >_{f^*\rho} d_{G^{r-1}(f)}$ but it needn't be true a priori that $d_{X_0} >_{f^*\rho} d_{G^r(f)}$, i.e. $d_{G^r(f)}$ needn't be an apm. Necessity and uniqueness are just as in Proposition 1. The proof is by induction on r. $r = 1$ is just as in Proposition 1. For the inductive step we apply the inductive hypothesis to $G(f)$ and Proposition 1 to f to get $\mathcal{C}_u^{r-1}(d_0)$ structures induced by $G(f)$ and f. Then proceed to apply the Smoothing Lemma as in Proposition 1. Q.E.D.

Let X be a semicomplete $\mathcal{L}^r$ manifold as in Proposition 1 and Corollary 2. A map $f: X_0 \to X$ is called a <u>$\mathcal{C}_u^r$ leaf immersion</u> (or <u>$\mathcal{L}^r$ leaf immersion</u>) ($r \geq 1$) if the following equivalent conditions hold: (1) There exists a $\mathcal{C}^r$ (resp. $\mathcal{L}^r$) structure on X_0 (necessarily regular and unique) such that $f: X_0 \to X$ is a metricly proper $\mathcal{C}^r$ (resp. $\mathcal{L}^r$) immersion and the map $(X_0, d_f, f^*\rho) \to (X_0, d_{G^r(f)}, f^*\rho)$ is $\mathcal{C}_u$ (resp. $\mathcal{L}$). Note that the inverse is necessarily $\mathcal{L}$. (2) There exists $\mathcal{C}_u^r(d_f)$ (resp. $\mathcal{L}^r(d_f)$) structure on X_0 (necessarily unique) such that f is a metricly proper $\mathcal{C}_u^r(d_f, f_X)$ (resp. $\mathcal{L}^r(d_f, d_X)$) immersion. Equivalence follows from Proposition 1 and Corollary 2. Note that by Corollary 3.2 and its Addendum the $\mathcal{C}^r$ (resp. $\mathcal{L}^r$) manifold X_0 is semicomplete iff X_0 is $\mathfrak{U}_{X_0}$ uniformly complete.

3 <u>**LEMMA**</u>: Let $g: X_1 \to X_2$ be an $\mathcal{L}^r$ map of semicomplete $\mathcal{L}^r$ manifolds ($r \geq 1$) with admissible metrics and bounds $d_{X_1}, d_{X_2}, \rho_1, \rho_2$. Let $f_i: X_0 \to X_i$ with $f_2 = g \cdot f_1$ and let $d_{f_i} = f_i^* d_{X_i}$ ($i = 1,2$). Assume f_2 is a $\mathcal{C}_u^r$ (or $\mathcal{L}^r$) leaf immersion and f_1 is a $\mathcal{C}_u^r(d_{f_2}, d_{X_1})$ (or $\mathcal{L}^r(d_{f_2}, d_{X_1})$) map. Then $f_1^*\rho_1 \sim f_2^*\rho_2$, $(X_0, d_{f_2}, f_2^*\rho_2) \to (X_0, d_{f_1}, f_1^*\rho_1)$ is a $\mathcal{C}_u$ (resp. $\mathcal{L}$) map with an $\mathcal{L}$ inverse and f_1 is a $\mathcal{C}_u^r$ (resp. $\mathcal{L}^r$) leaf immersion with the same apm metric structure on the domain as for f_2.

<u>**PROOF**</u>: $g: (X_1, d_{X_1}, \rho_1) \to (X_2, d_{X_2}, \rho_2)$ is an $\mathcal{L}$ map. Pulling back by f_1,

$(X_0,d_{f_1},f_1^*\rho_1) \to (X_0,d_{f_2},f_2^*\rho_2)$ is an $\mathcal{L}$ map. By hypothesis, f_1: $(X_0,d_{f_2},f_2^*\rho_2) \to (X_1,d_{X_1},\rho_1)$ is a $\mathcal{C}_u$ (resp. $\mathcal{L}$) map. Pulling back yields the regular pseudometric space result. That f_1 is a metricly proper, $\mathcal{C}^r$ immersion follows just as in the proof of Corollary 2.5. Since f_1 is a $\mathcal{C}_u^r(d_{f_2},d_{X_1})$ (resp. $\mathcal{L}^r(d_{f_2},d_{X_1})$) map it is a $\mathcal{C}_u^r(d_{f_1},d_{X_1})$ (resp. $\mathcal{L}^r(d_{f_1},d_{X_1})$) map by the pseudometric space result. Hence, f_1 is a leaf immersion. Q.E.D.

4 <u>COROLLARY</u>: If f is a $\mathcal{C}_u^r$ (or $\mathcal{L}^r$) leaf immersion with $r \geq 2$ then $G(f)$ is a $\mathcal{C}_u^{r-1}$ (resp. $\mathcal{L}^{r-1}$) leaf immersion.

<u>PROOF</u>: Apply Lemma 3 to g_τ. Q.E.D.

5 <u>PROPOSITION</u>: Let X be a semicomplete $\mathcal{L}^1$ manifold. Let f: $X_0 \to X$ be a C^1 immersion. The following are equivalent: (1) f is a $\mathcal{C}_u^1$ (or $\mathcal{L}^1$) leaf immersion. (2) The identity $(X_0,d_f,f^*\rho) \to (X_0,d_{G(f)},f^*\rho_1)$ is a $\mathcal{C}_u$ (resp. $\mathcal{L}$) map of regular pseudometric spaces. (3) $G(f)\circ f^{-1}$: $(f(X_0),d_X,\rho) \to (G(TX),d_G,\rho_1)$ is a well defined $\mathcal{C}_u$ (resp. $\mathcal{L}$) map of regular pseudometric spaces.

<u>PROOF</u>: That (1) implies (2) and (2) is equivalent to (3) are clear. That (2) implies (1) follows from Proposition 3.1 (a) and (1) of the leaf immersion definition. Q.E.D.

6 <u>COROLLARY</u>: Let X be a semicomplete $\mathcal{L}^r$ manifold. Let f: $X_0 \to X$ be a C^r immersion ($r \geq 1$). The following are equivalent: (1) $f,G(f),\ldots,G^{r-1}(f)$ are $\mathcal{C}_u^1$ (resp. $\mathcal{L}^1$) leaf immersions. (2) The identity maps among $(X_0,d_{G^i(f)},f^*\rho_i)$ $(0 \leq i \leq r$ with $G^0(f) = f)$ are all $\mathcal{C}_u$ (resp. $\mathcal{L}$) maps of regular pseudometric spaces. If X_0 is a finite dimensional space and the image $f(X_0)$ is a compact subset of X then these conditions imply that f is a $\mathcal{C}_u^r$ (resp. $\mathcal{L}^r$) leaf immersion.

PROOF: The equivalence of (1) and (2) follows from Proposition 5 applied to $f, G(f), \ldots, G^{r-1}(f)$. Clearly, if f is a $\mathcal{C}_u^r$ leaf immersion (resp. $\mathcal{L}^r$ leaf immersion) then these conditions hold. The converse is not true in general. Under the compactness hypothesis it is true. We do the $\mathcal{C}_u$ case. The result for $r = 1$ is trivial. For $r > 1$, the image of $G(f)$ is compact in $G(TX)$ because $G(f)(X_0) = G(f) \circ f^{-1}(f(X_0))$ and we can use (3) in Proposition 5. So by inductive hypothesis there are $\mathcal{C}_u^{r-1}(d_f)$ and $\mathcal{C}_u^{r-1}(d_{G(f)})$ structures induced on X_0 such that f and $G(f)$ are $\mathcal{C}_u^{r-1}(d_f, d_X)$ and $\mathcal{C}_u^{r-1}(d_{G(f)}, d_1)$ immersions respectively. By hypothesis d_f and $d_{G(f)}$ are uniformly equivalent and we can regard both as $\mathcal{C}_u^{r-1}(d_f)$ structures. As usual $f = g_\tau \circ G(f)$ is a $\mathcal{C}_u^{r-1}(d_f, d_X)$ map with respect to the $G(f)$ induced structure. To apply the Smoothing Lemma, it suffices by uniqueness to show that f is a $\mathcal{C}^1$ immersion with respect to the $G(f)$ induced structure. By Proposition 5 (3) the images of Tf and $TG(f)$ are continuous subbundles of $\tau_X | f(X_0)$ and $\tau_{G(TX)} | G(f)(X_0)$. $g_\tau : G(f)(X_0) \to f(X_0)$ is a homeomorphism and because f is a C^2 immersion (Tg_τ, g_τ) is a continuous VB isomorphism of these subbundles. Let $\| \|_X$ and $\| \|_G$ be admissible continuous Finslers on τ_X and $\tau_{G(TX)}$. Let S be the unit sphere bundle of the image of $TG(f)$. S is compact. Since $Tg_\tau * \| \|_X$ is never 0 on S there exists $K > 1$ such that $K^{-1} \leq Tg_\tau * \| \|_X | S \leq K$, i.e. the Finslers $\| \|_G$ and $Tg_\tau * \| \|_X$ are equivalent on Image of $TG(f)$. So we can define $\theta(T_x G(f))$ by using $Tg_\tau * \| \|_X$ and the $G(f)$ induced structure on X_0 and get $\rho(fx) > \theta(T_x G(f)) = \theta(T_x f)$ where the latter is defined using $\| \|_X$ and the $G(f)$ structure on X_0. So f is a $\mathcal{C}^1$ immersion of the $G(f)$ structure. Q.E.D.

The leaf immersion idea comes from foliations. Let X be an $\mathfrak{M}^r$ manifold $(r \geq 2)$ and π be an $\mathfrak{M}^{r-1}$ locally split subbundle of τ_X. π is called <u>involutive</u> if whenever s_1 and s_2 are C^1 sections of π_U for U open in X the Lie product of the vectorfields $[s_1, s_2]$, which is a priori a section of τ_U, is actually a section of π_U. An atlas $\mathcal{G} = \{U_\alpha, h_\alpha\}$ is called a foliation atlas for π if the atlas $(T\mathcal{G}, \mathcal{G})$ induces a subbundle atlas on π, i.e. $E^1_\alpha \times E^2_\alpha = E_\alpha$ and $Th_\alpha(\pi^{-1}U_\alpha) = h_\alpha(U_\alpha) \times E^1_\alpha \times 0$ for all α. Clearly, if π admits a C^1 foliation atlas, it is involutive. The converse is:

7 FROBENIUS THEOREM: Let X be an $\mathfrak{M}^r$ manifold $(r \geq 2)$ and π be an involutive $\mathfrak{M}^{r-1}$ locally split subbundle of τ_X. There exist admissible $\mathfrak{M}^{r-1}$ atlases on X which are foliation atlases for π. If X is semi-complete and $\mathfrak{M}^{r-1} \subset \mathcal{L}$ there exist s admissible $\mathfrak{M}^{r-1}$ foliation atlases.

PROOF: The proof in Lang [14; Chap. VI] of the Frobenius Theorem is easily adapted using the differential equations estimates Proposition II. 3.9 to prove the metric version above. In the semicomplete case we note that by Corollary 2.4 π is a semicomplete subbundle. Q.E.D.

If $\mathcal{G} = \{U_\alpha, h_\alpha\}$ is a foliation atlas for π then define $\mathcal{G}_\pi = \{U_{\alpha v}, h_{\alpha v}\}$ indexed by pairs with $v \in p_2 \circ h_\alpha(U_\alpha)$.

$$U_{\alpha v} = h_\alpha^{-1}(E^1_\alpha \times \{v\})$$

$$h_{\alpha v} = p_1 \circ h_\alpha \quad .$$

Since it is clear that $p_2 \circ h_{\beta\alpha} | h_\alpha(U_\alpha) \cap (E^1_\alpha \times \{v\})$ is locally constant (as its derivative is zero), it follows that $\mathcal{G}_\pi$ is an $\mathfrak{M}^{r-1}$ atlas on the set underlying X. Denote by X_π the $\mathfrak{M}^{r-1}$ manifold generated by the class $\{(\mathcal{G}_\pi, \rho)\}$ with ρ an admissible bound on X. The "identity" map $i_\pi : X_\pi \to X$ is an $\mathfrak{M}^{r-1}$ bijective immersion. X_π is called the leaf space of π. In particular, the continuity of i_π shows that X_π is Hausdorff. Since $G(i_\pi) = s_\pi \circ i_\pi$ is $\mathfrak{M}^{r-1}$ it follows from the Smoothing Lemma that X_π has a unique $\mathfrak{M}^r$ structure such that i_π is a metricly proper $\mathfrak{M}^r$ immersion. X_π is semicomplete if X is and $\mathfrak{M}^{r-2} \subset \mathcal{L}$.

If X is a semicomplete $\mathcal{L}^r$ manifold $(r \geq 2)$ and π is an involutive

$\mathcal{L}^{r-1}$ subbundle of τ_X with convex, star bounded s admissible $\mathcal{L}ip^{r-1}$ foliation atlas $\mathcal{G} = \{U_\alpha, h_\alpha\}$ (cf. Prop. V.2.2) then $\tilde{\mathcal{G}}_\pi = \{U_\alpha, p_1 \circ h_\alpha\}$ is an $\mathcal{L}^{r-1}(d_{i_\pi})$ multiatlas on X_π and clearly $i_\pi : \tilde{\mathcal{G}}_\pi \to \mathcal{G}$ is an index preserving $\mathcal{L}^{r-1}(d_{i_\pi}, d_X)$ map. Hence, i_π is an $\mathcal{L}^{r-1}$ leaf immersion. Since $G(i_\pi) = s_\pi \circ i_\pi$ is an $\mathcal{L}^{r-1}(d_{i_\pi}, d_1)$ map it follows from the Smoothing Lemma that i_π is an $\mathcal{L}^r$ leaf immersion.

Thus the inclusion of the space of leaves of a foliation is a -very- special case of a leaf immersion. On the other hand, in defining leaf immersions, Hirsch, Pugh and Shub proved the following theorem [11; Thm. 6.2] which shows that leaf immersions admit atlases with some of the properties of foliation atlases.

8 <u>PLAQUATION THEOREM</u>: Assume $\mathfrak{M} \subset \mathcal{L}$. Let X be a semicomplete $\mathfrak{M}^r$ manifold and $f: X_0 \to X$ be a semicomplete $\mathcal{C}_u^r$ leaf immersion $(r \geq 1)$. There exist s admissible $\mathfrak{M}^r$ atlases $\mathcal{G} = \{U_\alpha^3, h_\alpha\}$, which are $\mathcal{L}ip^r$ and satisfy condition $\mathcal{L}$ and the following:

For all α, either $U_\alpha^3 \cap \overline{f(X_0)} = \emptyset$, in which case we define $U_\alpha^i = U_\alpha^3$ $i = 1,2,3$, or $E_\alpha = E_\alpha^0 \times \bar{E}_\alpha$ and $h_\alpha(U_\alpha^3) = B(0,\epsilon_\alpha) \times \bar{B}(0,6\epsilon_\alpha)$ with B and $\bar{B}$ balls in E_α^0 and $\bar{E}_\alpha$, in which case we define $U_\alpha^i = h_\alpha^{-1}(B(0,\epsilon_\alpha) \times \bar{B}(0,2i\epsilon_\alpha))$ $i = 1,2,3$. $0 < \epsilon_\alpha < 1$ with $\sup \rho|U_\alpha^3 \sim \inf \rho|U_\alpha^3 \sim \epsilon_\alpha^{-1}$ as α varies with $U_\alpha^3 \cap \overline{f(X_0)} \neq \emptyset$. Also, $\cup\{h_\alpha^{-1}(B(0,\frac{1}{2}\epsilon_\alpha) \times \bar{B}(0,\frac{1}{2}\epsilon_\alpha)) : U_\alpha^3 \cap \overline{f(X_0)} \neq \emptyset\} \supset \overline{f(X_0)}$.

Define $V_\alpha^3 = f^{-1}U_\alpha^3$ and let $\{V_{\alpha\lambda} : \lambda \in \Lambda_\alpha^3\}$ be an indexing of the components of V_α^3. For $i = 1,2$ let $\Lambda_\alpha^i = \{\lambda : V_{\alpha\lambda} \cap f^{-1}U_\alpha^i \neq \emptyset\}$ and $V_\alpha^i = \cup\{V_{\alpha\lambda} : \lambda \in \Lambda_\alpha^i\}$.

For $i = 1,2$ and $\lambda \in \Lambda_\alpha^i$, $h_\alpha f(V_{\alpha\lambda})$ is the graph of a $\mathcal{C}^r$ map $g_{\alpha\lambda} : B(0,\epsilon_\alpha) \to \bar{B}(0,2(i+1)\epsilon_\alpha)$ defining $g_\alpha : B(0,\epsilon_\alpha) \times \Lambda_\alpha^2 \to B(0,\epsilon_\alpha) \times \bar{B}(0,6\epsilon_\alpha)$ by $g_\alpha(v,\lambda) = (v, g_{\alpha\lambda}(v))$. $\|Dg_{\alpha\lambda}\|_0 \leq 1$ and $\sup \rho|U_\alpha^3 > \|D^{r_1} g_{\alpha\lambda}\|_0$ uniformly in λ for $1 < r_1 \leq r$.

For all $K,n \geq 1$ there exist $K_1, n_1 \geq 1$ such that for

$v_1, v_2 \in B(0,\epsilon_\alpha), \lambda_1, \lambda_2 \in \Lambda_\alpha^2$:

$$(4.2)\qquad \begin{aligned} &\|(v_1, g_{\alpha\lambda_1}(v_1)) - (v_2, g_{\alpha\lambda_2}(v_2))\| < (K_1(\sup \rho|U_\alpha^3)^{n_1})^{-1} \quad \text{implies} \\ &\|j^r g_{\alpha\lambda_1}(v_1) - j^r g_{\alpha\lambda_2}(v_2)\| < (K(\sup \rho|U_\alpha^3)^{n})^{-1}. \end{aligned}$$

If f is an $\mathcal{L}^r$ leaf immersion, we can improve the (4.2) condition to: There exist constants $K, n \geq 1$ such that for $v_1, v_2 \in B(0,\epsilon_\alpha)$, $\lambda_1, \lambda_2 \in \Lambda_\alpha^2$:

$$(4.3)\qquad \|j^r g_{\alpha\lambda_1}(v_1) - j^r g_{\alpha\lambda_2}(v_2)\| \leq K \sup(\rho|U_\alpha^3)^n \|(v_1, g_{\alpha\lambda_1}(v_1)) - (v_2, g_{\alpha\lambda_2}(v_2))\|.$$

An atlas satisfying the above conditions is called a <u>plaquation atlas</u> for f. If $f: X_0 \to X$ is a continuous map with X a semicomplete $\mathcal{L}^r$ manifold and X_0 a topological manifold and if X admits a plaquation atlas as above, then f is a $\mathcal{C}_u^r$ leaf immersion with respect to a unique metric structure on X_0. If (4.3) holds then f is an $\mathcal{L}^r$ leaf immersion.

<u>PROOF</u>: Begin with $\mathcal{G} = \{U_\alpha, h_\alpha\}$ an s admissible, star bounded $\mathfrak{M}^r$ atlas on X which is $\mathcal{L}ip^r$ and satisfies condition $\mathcal{L}$. Assume ρ is an admissible bound with $\rho > 1/\lambda_{\mathcal{G}}$. For each $x \in \overline{f(X_0)}$ choose $\alpha = \alpha(x)$ such that $B(h_\alpha(x), 1/\rho(x)) \subset h_\alpha(U_\alpha)$. Recall from Proposition 5 the $\mathcal{C}_u$ section $G(f) \cdot f^{-1}$ of $G(\tau_X)|f(x_0)$. By uniform continuity it extends to a $\mathcal{C}_u$ section, also denoted $G(f) \cdot f^{-1}$, of $G(\tau_X)|\overline{f(X_0)}$. Choose $P = P_x$ a projection of E_α onto $E^0 = E_x^0 = T_x h_\alpha(G(f) \cdot f^{-1}(x))$ with $\|P\| \leq 2\theta(E^0)$. Let $K = K_x$ be the kernel of P and define $h_x: U_x \to E^0 \times K$ by $h_x(y) = (P \times I - P)(h_\alpha(y) - h_\alpha(x))$. By uniform continuity of $G(f) \cdot f^{-1}$ we can define $\epsilon: \overline{f(X_0)} \to (0,1)$ so that $\rho \sim 1/\epsilon$ on $\overline{f(X_0)}$, $B(0, 6\epsilon(x)) \subset h_x(U_\alpha)$ and $\overline{f(X_0)} \cap h_x^{-1}(B(0,6\epsilon(x))) \subset (G(f) \cdot f^{-1})^{-1} U_{\alpha,P}$. Finally, let $U_x^3 = h_x^{-1}(B(0,\epsilon(x)) \times \bar{B}(0,6\epsilon(x)))$ where the latter balls lie in E^0 and K. Let $\mathcal{G}_1$ consist of the charts $\{U_x, h_x\}$, indexed by $x \in \overline{f(X_0)}$ and the charts $\{U_\alpha^3, h_\alpha\}$ where $U_\alpha^3 = U_\alpha - \overline{f(X_0)}$ for α

in the index set of $\mathcal{G}$. $\mathcal{G}_1$ is clearly a star bounded, s admissible $\mathfrak{M}^r$ atlas on X and since h_x differs from $h_{\alpha(x)}$ by an affine map with bound estimated by $\rho(x)$, $\mathcal{G}_1$ also $\mathcal{L}ip^r$ and satisfies condition $\mathcal{L}$. I claim that $\mathcal{G}_1$ is a plaquation atlas for f.

First, we note that $f^*\mathcal{G}_1$ defined to be $\{V_x^3, P_x \cdot h_x \cdot f\}$ (indexed by $x \in \overline{f(X_0)}$) is a $\mathcal{U}_{d_f}$ uniform refinement of $f^*\mathcal{G}(T\mathcal{G}_1)$. So by Proposition 3.1, $f^*\mathcal{G}_1$ is an admissible $\mathcal{C}_u^r(d_f)$ multiatlas on X_0. We define $d = d_{\mathcal{G}_1}$, $d_f = f^*d$ and $d_0 = d_{\mathcal{G}_0}$ where $\mathcal{G}_0$ is any atlas subdivision of $f^*\mathcal{G}_1$.

<u>CLAIM</u>: Let $y \in f^{-1}U_x^i$ (i = 1 or 2) i.e. $h_x f(y) \in B(\epsilon(x)) \times \bar{B}(2i\epsilon(x))$ and $v \in B(\epsilon(x))$. There exists a unique continuous path z_t in V_x^i with $z_0 = y$ and $P_x \cdot h_x \cdot f(z_t) = v_t = tv + (1-t)P_2 \cdot h_x \cdot f(y)$.

In the statement of the Claim and for the rest of the proof we have suppressed mention of the center, 0, of the balls. What we have to show is that v_t lifts continuously to a path z_t in $f^{-1}U_x^3$ with $\bar{P} \cdot h_x \cdot f(z_t) \subset \bar{B}(2(i+1)\epsilon(x))$. In fact, we will show that $\bar{P} \cdot h_x \cdot f(z_t)$ lies in the closed ball $\bar{B}[r]$ where $r = \|\bar{P} \cdot h_x \cdot f(y)\| + \|P \cdot h_x \cdot f(y) - v\| < 2(i+1)\epsilon(x)$. $P \cdot h_x \cdot f\colon f^{-1}U_x^3 \to B(\epsilon(x))$ is a local homeomorphism. Thus, we can choose $t_1 \leq 1$ to be the maximum t_1 such v_t lifts to a unique path z_t $0 \leq t < t_1$ with $z_0 = y$. Assume for a moment that $\bar{P} \cdot h_x \cdot f(z_t) \in \bar{B}[r]$ for $0 \leq t < t_1$. Then for $t_0, \bar{t}_0 < t_1$, $d_0(z_{t_0}, z_{\bar{t}_0}) \leq \|P \cdot h_x \cdot f(z_{t_0}) - P \cdot h_x \cdot f(z_{\bar{t}_0})\| = |t_0 - \bar{t}_0| \|v - v_0\|$. Thus, as $t_0 \to t_1$, z_{t_0} is d_0 Cauchy in the $f^*\rho$ bounded set $f^{-1}U_x^3$. So z_{t_0} converges to $z_{t_1} \in X_0$ by uniform completeness of X_0. By Lemma III.2.1, $h_x^{-1}([v_0, v] \times \bar{B}[r])$ is closed in X. Hence, $f(z_{t_1})$ lies in this set and so in U_x^3. This extends z_t to t_1 and so beyond t_1. Thus, by maximality $t_1 = 1$.

We complete the proof of the Claim by showing that for $\bar{P} \cdot h_x \cdot f(z_t) \in B[r]$ for $0 \leq t < t_1$. First, note that on the compact set $z_{[0,t]}$, $P \cdot h_x \cdot f$ is injective. Hence, on some open set $U_t \subset B(\epsilon(x))$ about the segment $[v_0, v_t]$ we can define the principal part of f with respect to the charts $\{f^{-1}U_x^3, P\circ h_x\circ f\}$ and $\{U_x^3, h_x\}$. This function $h_x\circ f\circ(P\circ h_x\circ f)^{-1}\colon U_t\to B(\epsilon(x))\times\bar{B}(6\epsilon(x))$ maps v_s to $h_x\circ f(z_s)$ for $0\le s\le t$. Furthermore, it is the graph of a function $g_t\colon U_t\to\bar{B}(6\epsilon(x))$ whose derivative is the local representative of the Grassman map $G(f)\circ f^{-1}$ on $f(P\circ h_x\circ f^{-1})(U_t)$. Now $G(f)\circ f^{-1}$ maps $U_x^3\cap f(X_0)$ into $U_{\alpha(x)P}$ and so $\|Dg_t\|_0\le 1$. Hence, $\|\bar{P}\circ h_x\circ f(z_t) - \bar{P}\circ h_x\circ f(y)\| = \|g_t(v_t) - g(v_0)\| \le \int_0^t \|D_{v_s}g_t(v - v_0)\|ds \le \|v - v_0\|$. Since $v_0 = P\circ h_x\circ f(y)$, $\|\bar{P}\circ h_x\circ f(z_t)\| \le \|v - v_0\| + \|\bar{P}\circ h_x\circ f(y)\| = r$. This proves the Claim.

The Claim and Lemma II.3.5 now imply that for $i = 1,2$ and $\lambda\in\Lambda_x^i$, $P\circ h_x\circ f\colon V_{x\lambda}\to B(\epsilon(x))$ is a bijection and so the principal part of $f|V_{x\lambda}$ with respect to $\{f^{-1}U_x^3, P\circ h_x\circ f\}$ and $\{U_x^3, h_x\}$ is well defined and is the graph of a map $g_{x\lambda}\colon B(\epsilon(x))\to\bar{B}(2(i+1)\epsilon(x))$. $g_{x\lambda}$ is obtained by gluing together the maps g_t of the proof of the Claim and so, as in that proof, $\|Dg_{x\lambda}\|_0\le 1$. Since $(f^*\mathcal{G}_1, f^*\rho)$ is an admissible adapted $\mathcal{C}^r$ multiatlas on X_0, $(\mathcal{G}_1,\rho)$ is a star bounded admissible $\mathfrak{M}^r$ atlas on X and f is a $\mathcal{C}^r$ map, it follows that $\sup\rho|U_x^3 > \|D^{r_1}g_{x\lambda}\|_0$ uniformly in λ for $1 < r_1\le r$.

Since $\mathcal{G}_1$ is a star bounded, $\mathcal{L}ip^r$ atlas satisfying condition $\mathcal{L}$ it is an admissible atlas for the $\mathcal{L}^r(d)$ structure induced by the admissible apm d, cf. Theorem V.2.4. Since f is a $\mathcal{C}_u^r$ (or $\mathcal{L}^r$) leaf immersion, $f^*\mathcal{G}_1$ is an admissible $\mathcal{C}_u^r(d_f)$ (resp. $\mathcal{L}^r(d_f)$) atlas on X_0, cf. Proposition 3.1(c). Thus, for $1\le r_1\le r$, the family

$\{D^{r_1}(h_x\circ f\circ(P\circ h_x\circ f)^{-1})\circ P\circ h_x\circ f\colon (V_x^3, d_f, f^*\rho)\to L_s^{r_1}(E^0; E^0\times K)\}$ is a $\mathcal{C}_u$ (resp. $\mathcal{L}$) family of maps. Clearly, this holds iff the family

$\{D^{r_1}(\bar{P}\circ h_x\circ f\circ(P\circ h_x\circ f)^{-1})\circ P\circ h_x\circ f\colon (V_x^3, d_f, f^*\rho)\to L_s^{r_1}(E^0;K)\}$ is a $\mathcal{C}_u$ (resp. $\mathcal{L}$) family. This implies the same result with the domains restricted to $(V_x^2, d_f, f^*\rho)$. Note that $D^{r_1}(\bar{P}\circ h_x\circ f\circ(P\circ h_x\circ f|V_{x\lambda})^{-1}) = D^{r_1}(g_{x\lambda})$ for $\lambda\in\Lambda_x^2$. Next note that if $y_1\in V_{x\lambda_1}$ and $y_2\in V_{x\lambda_2}$ then by condition $\mathcal{L}$ on $\mathcal{G}_1$ we can replace the distance $d(fy_1, fy_2)$ by $\|h_x f y_1 - h_x f y_2\| = \|(v_1, g_{x\lambda_1}(v_1)) - (v_2, g_{x\lambda_2}(v_2))\|$ where $v_i = P\circ h_x\circ f(y_i)$ $(i = 1,2)$. (4.2) (resp. (4.3)) follows. This proves that $\mathcal{G}_1$ is a plaquation atlas.

For the converse note that this argument is reversible and from (4.2) we get that the family $\{D^{r_1}(h_\alpha \cdot f \cdot (P_\alpha \cdot h_\alpha \cdot f)^{-1}) \cdot P_\alpha \cdot h_\alpha \cdot f\colon (V_\alpha^2, d_f, f^*\beta) \to L_s^{r_1}(E_\alpha^0; E_\alpha^0 \times \bar{E}_\alpha)\}$ is $\mathcal{C}_u$ and $\{f^{-1}U_\alpha^2\}$ is a d_f uniform open subdivision of $\{V_\alpha^2\}$, where P_α is the projection of E_α onto E_α^0. Thus, $\{f^{-1}U_\alpha^2, P_\alpha \cdot h_\alpha \cdot f\}$ is a $\mathcal{C}_u^r(d_f)$ atlas and f is a $\mathcal{C}_u^r(d_f, d)$ map. Similarly, for the $\mathcal{L}^r$ case. Q.E.D.

If $\mathcal{G}$ is a plaquation atlas then for each index α

$$V_\alpha^3 = f^{-1}U_\alpha^3 \supset V_\alpha^2 \supset f^{-1}U_\alpha^2 \supset V_\alpha^1 \supset f^{-1}U_\alpha^1. \tag{4.4}$$

The only inclusion that is not a priori obvious is $f^{-1}U_\alpha^2 \supset V_\alpha^1$ and it follows because $g_{\alpha\lambda}$ maps into $\bar{B}(0, 4\epsilon_\alpha)$ for $\lambda \in \Lambda_\alpha^1$. We also define $\Lambda_\alpha^{3/2} = \{\lambda\colon V_{\alpha\lambda} \cap (h_\alpha \cdot f)^{-1}(0 \times B(0, 4\epsilon_\alpha)) \neq \emptyset\}$ and $V_\alpha^{3/2} = \cup\{V_{\alpha\lambda}\colon \lambda \in \Lambda_\alpha^{3/2}\}$. Since $V_\alpha^1 \subset f^{-1}U_\alpha^2$ it follows that:

$$V_\alpha^2 \supset V_\alpha^{3/2} \supset V_\alpha^1. \tag{4.5}$$

Letting P_α denote the projection of E_α onto E_α^0 it follows that for $i = 1, 3/2, 2, 3$, $f^*\mathcal{G}^i = \{V_\alpha^i, P_\alpha \cdot h_\alpha \cdot f\}$ is an s admissible $\mathcal{C}^r$ multiatlas on X_0. For $i = 3$ it is a $\mathcal{C}_u^r(d_f)$ multiatlas while for $i = 1, 3/2, 2$ this need not be true only because V_α^i needn't be d_f open. However $\{f^{-1}U_\alpha^1\}$ is a d_f open, uniform cover of X_0 refining $\{V_\alpha^i\}$. On the other hand, for $i = 1, 3/2, 2$ the principal parts of f with respect to $f^*\mathcal{G}^i$ and $\mathcal{G}$ have the nice special form given by the aggregate plaque maps g_α.

If f is injective then $g_\alpha | 0 \times \Lambda_\alpha^2$ defines a bijection of Λ_α^2 onto a subset $\tilde{\Lambda}_\alpha^2$ of $\bar{B}(0, 6\epsilon_\alpha)$. Let $\tilde{\Lambda}_\alpha^i = g_\alpha(0 \times \Lambda_\alpha^i)$ for $i = 1, 3/2$. By definition, $\tilde{\Lambda}_\alpha^{3/2} = \tilde{\Lambda}_\alpha^2 \cap \bar{B}(0, 4\epsilon_\alpha)$ and

$$0 \times \tilde{\Lambda}_\alpha^{3/2} = h_\alpha(f(X_0) \cap U_\alpha^3) \cap (0 \times \bar{B}(0, 4\epsilon_\alpha)). \tag{4.6}$$

We now apply the plaquation atlas to factorization questions for leaf immersions. We require the following topological lemma whose nice proof was shown to me by Eli Goodman.

9 LEMMA: Let X be an arcwise connected topological space. If $\{F_n\}$ is a sequence of pairwise disjoint closed sets covering X then $F_n = X$ for some n and the other sets are empty.

PROOF: By arcwise connectedness we are reduced to the case when $X = [0,1]$. Consider that case and assume that the sequence $F_0, F_1, \ldots$ is a non-trivial cover arranged so that $0,1 \in F_0 \cup F_1$. $X - F_0 \cup F_1$ is a countable disjoint union of open intervals. Choose I_1 to be such an interval and let $A_0 = F_0 \cup F_1$. Let F_{k_1} contains the midpoint of I_1 and $A_1 = \cup_{i=0}^{k_1} F_i$. $A_1 \cap I_1$ is compact since the endpoints of I_1 lie in A_0 and $A_0 \cap I_1 = \emptyset$. So we can define I_2 the left-most open interval of $I_1 - A_1$. Let F_{k_2} contain the midpoint of I_2, $A_2 = \cup_{i=0}^{k_2} F_i$ and I_3 to be the right-most open interval of $I_2 - A_2$. Continuing inductively, alternating left and right choices we get a decreasing sequence of open intervals I_n with $\bar{I}_{n+2} \subset I_n$ and $I_n \cap F_i = \emptyset$ for $i \leq n$. Since the diameters go to zero, $\cap I_n = \cap \bar{I}_n$ is a point which cannot lie in any F_n. Contradiction. Q.E.D.

10 PROPOSITION: Let X_2 be a semicomplete $\mathfrak{M}^r \cap \mathcal{L}^1$ manifold $(r \geq 1)$ and X_1 be a semicomplete, separable $\mathfrak{M}^r$ manifold. Let $f: X_1 \to X_2$ be a metricly proper, injective $\mathfrak{M}^r$ immersion which is a $\mathcal{C}_u^1$ leaf immersion. Let Y be an $\mathfrak{M}^r$ manifold and $g_i: Y \to X_i$ $(i = 1,2)$ with $f \cdot g_1 = g_2$. If g_2 is $\mathfrak{M}^r$ then g_1 is.

PROOF: By the Factoring Lemma VI.3.13 it suffices to show that g_1 is continuous. Hence, we need only verify that f satisfies the hypotheses of Lemma VI.3.14. Let $\mathcal{G} = \{U_\alpha^3, h_\alpha\}$ be a plaquation atlas for f. For $x \in X_1$ choose α with $fx \in U_\alpha^1$. Because X_1 is separable and locally connected an open set has only countable many components and so Λ_α^3 is countable. For each $\lambda \in \Lambda_\alpha^2$, the graph of $g_{\alpha\lambda}$ is closed in $B(0,\epsilon_\alpha) \times \bar{E}_\alpha$ and so its inverse image under h_α, denoted F_λ, is closed in U_α^3. Since f is injective the countable family $\{F_\lambda : \lambda \in \Lambda_\alpha^2\}$ is pairwise disjoint. By (4.4), $fx \in F_{\lambda_0}$, i.e. $x \in V_{\alpha\lambda_0}$, for some $\lambda_0 \in \Lambda_\alpha^1$ and $F_{\lambda_0} \subset U_\alpha^2$. Clearly, $V_{\alpha\lambda_0}$ is path connected and $f: V_{\alpha\lambda_0} \to F_{\lambda_0}$ is a homeomorphism. Finally, let F be the path component of fx in $U_\alpha^2 \cap f(X_0)$. By (4.4) $\{F_\lambda \cap F: \lambda \in \Lambda_\alpha^2\}$ is a countable, pairwise disjoint, closed cover of F. Since $F_{\lambda_0} \subset F$, Lemma 9 implies $F_{\lambda_0} = F$. Q.E.D.

In practice, Proposition 10 is applied to cases like the inclusion of a single leaf of a foliation. The extreme opposite of a leaf immersion of a separable manifold is a bijective leaf immersion. For example, the inclusion of the entire leaf space of a foliation into the ambient manifold is of this type. Clearly, if $f: X_1 \to X_2$ is a bijective leaf immersion then letting $Y = X_2$ and g_2 be the identity, g_2 factors through f iff f is an isomorphism. For such immersions factorization theorems require appropriate conditions on the tangent maps.

11 DEFINITION: Let $f: X_1 \to X_2$ be an injective, C^1 locally split immersion. f satisfies the Tangent Factoring Property if whenever $g: Y \to X_2$ is a C^1 map with $g(Y) \subset f(X_1)$ and $Tg(TY) \subset Tf(TX_1)$ then the map $g_1: Y \to X_1$ with $f \circ g_1 = g$ is continuous (and hence C^1).

12 LEMMA: An injective C^1 immersion $f: X_1 \to X_2$ satisfies the Tangent Factoring Property in general if it satisfies it for maps $g: Y \to X_2$ with Y a bounded open interval in $\mathbb{R}$.

PROOF: Since the Tangent Factoring Property is a local property we can reduce to the following situation: (U_1,h_1) and (U_2,h_2) are charts for X_1 and X_2 with $h_2(U_2)$ the product of open balls $B_1 \times \bar{B}$, $h_1(U_1) = B_1$ and $h_2 f h_1^{-1}: B_1 \to B_1 \times \bar{B}$ is the inclusion. Y is an open ball with $g(Y) \subset U_2$.

All the balls are centered at 0 and $h_2g(0) = (0,0)$. Since $B_1 \times 0$ is closed in $B_1 \times \bar{B}$, U_1 is open and closed in $f^{-1}U_2$. Applying the Tangent Factoring Property to g composed with the rays from 0 in Y we see that $g(Y) \subset f(U_1)$. If $P_1: B_1 \times \bar{B} \to B_1$ is the projection then $P_1h_2f: U_1 \to B_1$ is a homeomorphism. If $g_1: Y \to U_1 \subset X_1$ is the set map factorization of g then $(P_1h_2f)g_1 = P_1h_2g$ which is continuous. Hence, g_1 is continuous. Q.E.D.

13 **PROPOSITION:** Let X_2 be a semicomplete $\mathcal{L}^1$ manifold. If $f: X_1 \to X_2$ is an injective, semicomplete $\mathcal{L}^1$ leaf immersion then f satisfies the Tangent Factoring Property.

PROOF: Let $\mathcal{G} = \{U_\alpha^3, h_\alpha\}$ be a plaquation atlas for f. Applying Lemma 12 and reducing to the local case we need only show: If $a_t: J \to X_2$ is a C^1 path with image in U_α^1 for some α, $a_t \in f(X_1)$ with $a_t' \in Tf(TX_1)$ for $t \in J$ and $b_t: J \to X_1$ is defined by $f \circ b_t = a_t$, then b_t is continuous. We can assume $0 \in J$. For $t \in J$ let $\lambda_t \in \Lambda_\alpha^1$ be defined by $b_t \in V_{\alpha\lambda_t}$. Since $P_\alpha \circ h_\alpha \circ f: V_{\alpha\lambda_0} \to B(0,\epsilon_\alpha)$ is a homeomorphism and $P_\alpha \circ h_\alpha \circ f(b_t) = P_\alpha \circ h_\alpha(a_t)$, it suffices to show that $\lambda_t = \lambda_0$ for $t \in J$, i.e. if $v_t = P_\alpha \circ h_\alpha(a_t)$ we prove that $h_\alpha(a_t)$ (which equals $(v_t, g_{\alpha\lambda_t}(v_t))$) is equal to $(v_t, g_{\alpha\lambda_0}(v_t))$. The key fact is that $(h_\alpha \circ a)_t' \in T(h_\alpha \circ f)(T_{b_t}X_1)$ which is the graph of $D_{v_t}g_{\alpha\lambda_t}$, i.e. $(h_\alpha \circ a)_t' = (v_t', D_{v_t}g_{\alpha\lambda_t}(v_t'))$. Now let $\delta_t = (v_t, g_{\alpha\lambda_0}(v_t)) - (v_t, g_{\alpha\lambda_t}(v_t))$. $\delta_t' = (v_t', D_{v_t}g_{\alpha\lambda_0}(v_t')) - (v_t', D_{v_t}g_{\alpha\lambda_t}(v_t'))$. Noting that the first coordinates cancel, we can apply (4.3) and get $\|\delta_t'\| \le K_\alpha\|\delta_t\|\|v_t'\|$ for all $t \in J$. On J_1 any compact subinterval containing $0_t \|v_t'\|$ is bounded by some constant M_{J_1}. Now by the usual differential inequality argument (see for example [8;Sec. I.6]) $\|\delta_t\| \le \|\delta_0\| \exp(K_\alpha M_{J_1}|t|)$ on J_1. Since $\delta_0 = 0$, $\delta_t = 0$ on J_1 and hence on J. Q.E.D.

14 **PROPOSITION:** Let π be a C^{r-1} locally split, involutive subbundle of the tangent bundle of a C^r manifold X $(r \ge 2)$. The inclusion of the leaf

space $i_\pi : X_\pi \to X$ satisfies the Tangent Factoring Property.

PROOF: If $r \geq 3$ then we can put a semicomplete $\mathcal{L}^2$ structure on X such that π is a locally split $\mathcal{L}^1$ subbundle of τ_X. Then i_π is a $\mathcal{L}^1$ leaf immersion and Proposition 13 applies. However, the direct proof is easy and applies when $r = 2$: If a_t is a C^1 path lying in a chart of a foliation atlas for π and a_t' lies in the subbundle then $(\bar{P}\cdot h_\alpha \cdot a)_t' = 0$ and so $\bar{P}\cdot h_\alpha \cdot a_t$ is constant. The proof is then completed as in Lemma 12 or Proposition 13. Q.E.D.

A C^1 immersion $f: X_0 \to X$ is said to be of finite códimension if $\dim T_{fx}X/T_x f(T_x X_0) < \infty$ for all $x \in X_0$.

15 LAMINATION THEOREM: Let X be a semicomplete $\mathcal{L}^1$ manifold and $f: X_0 \to X$ be an injective $\mathcal{C}^1$ leaf immersion of finite codimension. Assume f has a closed image, i.e. $f(X_0)$ is closed in X. Let $\mathcal{Q} = \{U_\alpha^3, h_\alpha\}$ be a plaquation atlas for f and J be a bounded open interval. The following are equivalent:

(a) If $a_t: J \to X$ is a continuous path which is the pointwise limit of continuous paths factoring continuously through f, then a_t factors continuously through f.

(b) If $a_t: J \to X$ is a continuous path which is the uniform limit of continuous paths factoring continuously through f then a_t factors continuously through f.

(c) If $a_t: J \to X$ is a C^1 path and $\{b_t^n: J \to X_0\}$ is a sequence of C^1 paths with $\{f\cdot b_t^n\}$ and $\{(f\cdot b_t^n)' = Tf\cdot(b_t^n)'\}$ converging uniformly to a_t and a_t' respectively then a_t factors continuously through f and hence to a C^1 path $b_t: J \to X_0$.

(d) For each α with $U_\alpha^3 \cap f(X_0) \neq \emptyset$ there is an open set $U_\alpha^{3/2}$ with $U_\alpha^3 \supset U_\alpha^{3/2} \supset U_\alpha^1$ and $g_\alpha: B(0,\epsilon_\alpha) \times \tilde{\Lambda}_\alpha^{3/2} \to B(0,\epsilon_\alpha) \times \bar{B}(0,6\epsilon_\alpha)$ is a homeomorphism onto $h_\alpha(f(X_0) \cap U_\alpha^{3/2})$.

If f is an $\mathcal{L}^1$ leaf immersion then these conditions hold and in (d) both g_α and g_α^{-1} are Lipschitz with $L(g_\alpha)$ and $L(g_\alpha^{-1}) \leq \exp[K \sup(\phi|U_\alpha^3)^n]$ for some constants K and n.

__PROOF:__ Note first that if any of (a) - (c) hold for some J then they hold for any other J by reparametrizing linearly. (d) says that, locally, $f(X_0)$ is topologically just the product of $B(0,\epsilon_\alpha)$ and a closed subset of $\bar{B}(0,4\epsilon_\alpha)$ (cf. (4.6)). To prove (d) $\Rightarrow$ (a): Let $\{a_t^n\}$ be a sequence of paths factoring through f with $a_t^n \to a_t$ as $n \to \infty$. As usual, the result is local and we can assume the images lie in U_α^1 for some α. Note that $a_t \in f(X_0)$ because the latter is closed. For each n there exists $\lambda_n \in \tilde{\Lambda}_\alpha^1$ such that $a_t^n \in f(V_{\alpha\lambda_n})$ for all t, i.e. if $v_t^n = P_\alpha \cdot h_\alpha(a_t^n)$ and $v_t = P_\alpha \cdot h_\alpha(a_t)$ then $h_\alpha(a_t^n) = g_\alpha(v_t^n, \lambda_n)$ for all t, i.e. $\bar{P}_\alpha \cdot g_\alpha^{-1} \cdot h_\alpha(a_t^n)$ is independent of t. Hence, $\lambda = \operatorname{Lim}_n \bar{P}_\alpha \cdot g_\alpha^{-1} \cdot h_\alpha(a_t^n) = P_\alpha \cdot g_\alpha^{-1} \cdot h_\alpha(a_t)$ is independent of t, i.e. $a_t = g_\alpha(v_t, \lambda)$. Thus, a_t factors to a continuous path in $V_{\alpha\lambda}$.

Clearly, (a) implies (b) implies (c). We now show that (c) implies (d). The main step is to show that if $(v_n,\lambda_n),(v,\lambda) \in B(0,\epsilon_\alpha) \times \tilde{\Lambda}_\alpha^2$ then $\{g_\alpha(v_n,\lambda_n)\}$ converges to $g_\alpha(v,\lambda)$ iff $\{(v_n,\lambda_n)\}$ converges to (v,λ). In either case, $\{v_n\}$ converges v and so we must show that $\{g_{\alpha\lambda_n}(v_n)\}$ converges to $\{g_{\alpha\lambda}(v)\}$ iff $\{\lambda_n\}$ converges to λ. Finally, since $\|g_{\alpha\lambda_n}\|_L \leq 1$ we have $\|g_{\alpha\lambda_n}(v_n) - g_{\alpha\lambda}(v)\| \leq \|g_{\alpha\lambda_n}(v_n) - g_{\alpha\lambda_n}(v)\| + \|g_{\alpha\lambda}(v) - g_{\alpha\lambda_n}(v)\| \leq \|v_n - v\| + \|g_{\alpha\lambda_n}(v) - g_{\alpha\lambda}(v)\|$. So with $v \in B(0,\epsilon_\alpha)$ we have to show that $\{\lambda_n\}$ converges to λ iff $\{g_{\alpha\lambda_n}(v)\}$ converges to $g_{\alpha\lambda}(v)$. Note that, by definition of $\tilde{\Lambda}_\alpha^2$, $g_{\alpha\lambda}(0) = \lambda$ and $g_{\alpha\lambda_n}(0) = \lambda_n$. With $t \in J$, an open interval containing $[0,1]$, let $v_t = tv$. Define $a_t^n = h_\alpha^{-1}(v_t, g_{\alpha\lambda_n}(v_t))$ and $z_t^n = g_{\alpha\lambda_n}(v_t)$. Since f has finite codimension the closed ball $\bar{B}[0,5\epsilon_\alpha]$ is compact and with J_1 an open interval containing $[0,1]$ with closure in J,Ascoli's Theorem implies that $\{z_t^{n_k}\}$ converges uniformly to z_t $(t \in J_1)$ for some subsequence. By (4.2) the sequence of derivatives $\{(z_t^{n_k})' = D_{v_t} g_{\alpha\lambda_{n_k}}(v)\}$ is a uniformly Cauchy sequence and so z_t is differentiable on J_1 and $(z_t^{n_k})'$ converges uniformly to z_t'. Thus, with $\bar{a}_t = h_\alpha^{-1}(v_t, z_t)$ the hypotheses of (c) apply and so $\bar{a}_t$ factors continuously through f. Hence, $\bar{a}_t \in f(V_{\alpha\bar{\lambda}})$ for some $\bar{\lambda}$ and $z_t = g_{\alpha\bar{\lambda}}(v_t)$. By hypothesis either $g_{\alpha\lambda}(0) = \text{Lim}\, g_{\alpha\lambda_n}(0) = z_0 = g_{\alpha\bar{\lambda}}(0)$ or $g_{\alpha\lambda}(v) = \text{Lim}\, g_{\alpha\lambda_n}(v) = z_1 = g_{\alpha\bar{\lambda}}(v)$ and so by injectivity $\lambda = \bar{\lambda}$ and $z_t = g_{\alpha\lambda}(v_t)$. Since this holds for every convergent subsequence of $\{z_t^n\}$ it follows that $\{z_t^n\}$ converges to $g_{\alpha\lambda}(v_t)$. Hence, both $\{\lambda_n = z_0^n\}$ converges to $z_0 = \lambda$ and $\{g_{\alpha\lambda_n}(v) = z_1^n\}$ converges to $z_1 = g_{\alpha\lambda}(v)$. This proves the equivalence.

Now we show that $g_\alpha(B(0,\epsilon_\alpha) \times \tilde{\Lambda}_\alpha^{3/2})$ is open in $h_\alpha(f(X_0) \cap U_\alpha^3)$. This is because $g_\alpha(B(0,\epsilon_\alpha) \times \tilde{\Lambda}_\alpha^{3/2}) \subset B(0,\epsilon_\alpha) \times \bar{B}(0,5\epsilon_\alpha)$ and by the above arguments, $\bar{P}_\alpha \cdot g_\alpha^{-1}: (B(0,\epsilon_\alpha) \times \bar{B}(0,5\epsilon_\alpha)) \cap h_\alpha(f(X_0) \cap U_\alpha^3) \to \bar{B}(0,6\epsilon_\alpha)$ is continuous. Hence, $g_\alpha(B(0,\epsilon_\alpha) \times \tilde{\Lambda}_\alpha^{3/2})$ which is the inverse under this map of $\bar{B}(0,4\epsilon_\alpha)$ is open in $h_\alpha(f(X_0) \cap U_\alpha^3)$. Let V be an open set in $B(0,\epsilon_\alpha) \times \bar{B}(0,6\epsilon_\alpha)$ with $V \cap h_\alpha(f(X_0) \cap U_\alpha^3) = g_\alpha(B(0,\epsilon_\alpha) \times \tilde{\Lambda}_\alpha^{3/2})$. By (4.4) and (4.5) this equation still holds if we replace V by $V \cup (B(0,\epsilon_\alpha) \times \bar{B}(0,2\epsilon_\alpha))$. Let $U_\alpha^{3/2} = h_\alpha^{-1}(V \cup (B(0,\epsilon_\alpha) \times \bar{B}(0,2\epsilon_\alpha)))$. This proves (d).

Finally, if f is an $\mathcal{L}^1$ leaf immersion then the proof of Proposition 13 applied to the ray tv shows that $\|g_{\alpha\lambda_1}(v) - g_{\alpha\lambda_2}(v)\| \le \|\lambda_1 - \lambda_2\| \exp[K_\alpha\|v\|]$ and $\|\lambda_1 - \lambda_2\| \le \|g_{\alpha\lambda_1}(v) - g_{\alpha\lambda_2}(v)\| \exp[K_\alpha\|v\|]$. Since $L(g_{\alpha\lambda}) \le 1$ for all λ, we get $L(g_\alpha)$ and $L(g_\alpha^{-1}) \le 1 + \exp[K_\alpha]$ with $K_\alpha = K_1(\sup \rho|U_\alpha^3)^{n_1}$ for constants K_1 and n_1. By increasing the constants we can absorb the 1. Q.E.D.

A $\mathcal{C}_u^r$ (or $\mathcal{L}^r$) leaf immersion satisfying the condition of Theorem 15 is called a <u>$\mathcal{C}_u^r$ lamination</u> (resp. an $\mathcal{L}^r$ lamination). Note that if $f(X_0)$ is closed and f is a $\mathcal{C}_u^1$ leaf immersion then $Tf(TX_0)$ is the total space of a subbundle of $\tau_X|f(X_0)$ and so is closed in TX. Thus, if a_t is a path satisfying the hypotheses of (c) then $a_t \in f(X_0)$ and $a_t' \in Tf(TX_0)$. Thus, an injective $\mathcal{C}^r$ immersion with closed image and finite codimension is a $\mathcal{C}^r$ lamination if it satisfies the Tangent Factoring Property. The above theorem shows that a lamination is what Ruelle and Sullivan call a partial foliation [25; especially the Remark on page 319].

Combining the proof of Theorem III.8.2 with the usual proofs about maximal intervals of existence (cf. [7; Sec. II.3]) it is easy to show that if X is a semicompact $\mathcal{C}^\infty$ manifold and ξ is a $\mathcal{C}^0$ vectorfield on X (i.e. a continuous vectorfield which is bounded with respect to an admissible Finsler on τ_X) then the maximal interval of existence of any solution path is the whole real line R. If ξ is nonvanishing, X_0 is a disjoint union of copies of R and $f: X_0 \to X$ is some solution path on each copy of R then f is a $\mathcal{C}_u^1$ leaf immersion by Proposition 5. If ξ is an $\mathscr{L}$ vectorfield, then f is an $\mathscr{L}^1$ leaf immersion. Assume that the set of solutions has been so chosen that f is bijective, i.e. each point of X lies on exactly one of the chosen solutions. The Tangent Factoring Property for f is then exactly the local uniqueness property for solutions of ξ. In particular, Proposition 13 generalizes the uniqueness Theorem for Lipschitz vectorfields and the proofs are clearly related. The gap between the Tangent Factoring Property and the weaker property (c) of Theorem 15 is illustrated by the following example. While the example is given on R^3 and is unbounded it can easily be compactified by using a bump function about the origin and regarding the vectorfield as on a fundamental domain of the torus.

Define the vectorfield ξ on R^3 by:

$$\begin{aligned} \dot{x} &= 1 \\ \dot{y} &= |y|^{3/2} / (y^2 + z^2)^{1/2} \qquad (4.7) \\ \dot{z} &= 0 \end{aligned}$$

Thus, for each value of z we have a vectorfield on the plane parallel to the xy plane with x the time variable. On the open set $\{z \neq 0\}$, ξ is C^∞ and so solutions are unique. When $z = 0$, ξ is the classic example $\frac{dy}{dx} = |y|^{1/2}$. On the plane, $\{z = 0\}$ let X_0 be determined by the set of solutions intersecting the line $\{y = 0\}$ in exactly one point. This set of solutions yields a C^1 lamination $f_0 \colon X_0 \to R^2$ of the plane but the Tangent Factoring Property fails because the path $x = t$, $y = 0$, $z = 0$ doesn't factor continuously through f. Define X_1 to be the union of X_0 and a copy of R for each solution in the set $\{z \neq 0\}$. f_1 extends to $f_1 \colon X_1 \to R^3$ a bijective C^1 leaf immersion. f_1 is not a C^1 lamination because the previously mentioned path is the limit of the sequence of paths $x = t$, $y = 0$, $z = 1/n$.

CHAPTER VIII: FINITE DIMENSIONAL MANIFOLDS

1. Manifolds of Finite Type: A $\mathcal{C}^1$ manifold X is called k dimensional, i.e. $\dim X = k$ if $\dim T_xX = k$ for all $x \in X$ or, equivalently, if $\dim E_\alpha = k$ for all α whenever $\{U_\alpha, h_\alpha\}$ is an atlas on X. By Proposition VI.1.5 if $\mathfrak{A} = \{U_\alpha, h_\alpha\}$ is an admissible atlas for a k dimensional manifold we can choose an isomorphism $Q_\alpha: E_\alpha \to R^k$ with $\theta(Q_\alpha) \leq 2^k$, and replace h_α by $Q_\alpha \cdot h_\alpha$. Thus, if $\dim X = k$ we need only deal with R^k atlases on X i.e. atlases with $E_\alpha = R^k$ for all α. Recall that R^k bears the product, i.e. the sup, norm. Similarly if $\pi: E \to X$ is a k dimensional bundle, i.e. $\dim E_x = k$ for all x, we need only use atlases $\{U_\alpha, h_\alpha, \varphi_\alpha\}$ with $F_\alpha = R^k$ for all α.

1 LEMMA: Let X be a connected, paracompact Hausdorff space. If X is locally separable then X is separable.

PROOF: By paracompactness the neighborhoods of the diagonal form a uniformity for X and every open cover is uniform with respect to this uniformity (cf. [13; pp. 155-157]). Thus, we can choose V an open neighborhood of the diagonal with $V = V^{-1}$ and $V \cdot V[x]$ separable for all $x \in X$. If A_0 is a countable dense subset of A in X then $V[A] \subset V \cdot V[A_0]$ and the latter is a countable union of separable subsets. Hence, $V[A]$ is separable. Thus, $\cup V^n[x_0]$ for any $x_0 \in X$ is open, closed, separable and nonempty. By connectedness, X is separable. Q.E.D.

2 COROLLARY: A regular $\mathcal{C}^1$ manifold X is separable iff T_xX is separable for all $x \in X$ and X has a countable number of components. In particular, a regular $\mathcal{C}^1$ finite dimensional manifold is separable iff it has a countable number of components.

PROOF: Since the components are open and closed this follows from Lemma 1. Q.E.D.

3 COROLLARY: Let X be a regular $\mathcal{C}^1$ manifold with T_xX separable for all $x \in X$. Any uniform open cover of X has a locally finite uniform refinement.

PROOF: We will assume X is separable. By Corollary 2 this is true of the components of X and we leave the details of the generalization to the reader. If $\{x_n\}$ is a dense sequence in X and $V \in \mathfrak{U}_X$ then $\{V[x]: x \in X\}$ refines $\{V \cdot V[x_n]: n = 1,2,\ldots\}$. Hence, any uniform open cover has a countable, uniform open refinement. Let $\{U_n\}$ be a uniform, countable open cover of X and let $W = W^{-1} \in \mathfrak{U}_X$ induce a $W \cdot W$ trimming $\{\tilde{U}_n\}$ of $\{U_n\}$ (cf. p. 133). We use Dieudonne's scalloping process [14; p. 35]. Let $V_n = U_n - \cup\{\tilde{U}_i: i < n\}$. If $x \in X$ and $W[x] \cap \tilde{U}_n \neq \emptyset$ then $W[x] \subset W \cdot W[\tilde{U}_n] \subset U_n$. Hence, if n is the smallest integer such that $W[x] \subset U_n$ then $W[x] \subset V_n$. Hence, $\{V_n\}$ is a uniform open cover. If $W[x] \subset \tilde{U}_i$ then $W[x] \cap V_n = \emptyset$ for $n \geq i$. Hence, $\{V_n\}$ is locally finite. Q.E.D.

If $\mathfrak{U} = \{U_\alpha\}$ is an open cover of X we define: $N_{\mathfrak{U}}, \bar{N}_{\mathfrak{U}}: X \to [1,\infty]$ and $n_{\mathfrak{U}} \in [1,\infty]$ by $N_{\mathfrak{U}}(x) = \#\{\alpha: x \in U_\alpha\}$, $\bar{N}_{\mathfrak{U}}(x) = \inf \#\{\alpha: U_\alpha \cap U \neq \emptyset\}$ (infinum taken over all open sets U containing x) and $n_{\mathfrak{U}} = \inf_V \sup_x \#\{\alpha: U_\alpha \cap V[x] \neq \emptyset\}$ $(V \in \mathfrak{U}_X)$. Clearly, $n_{\mathfrak{U}} \geq \bar{N}_{\mathfrak{U}} \geq N_{\mathfrak{U}}$. $\mathfrak{U}$ is called point finite if $N_{\mathfrak{U}} < \infty$, of <u>finite type</u> if $\sup N_{\mathfrak{U}} < \infty$ and locally finite if $\bar{N}_{\mathfrak{U}} < \infty$. A point finite cover needn't be locally finite, but if $\tilde{\mathfrak{U}} = \{\tilde{U}_\alpha\}$ is a W trimming of $\mathfrak{U}$, a uniform open cover of X with $W = W^{-1} \in \mathfrak{U}_X$ then:

$$N_{\mathfrak{U}}(x) \geq \#\{\alpha: W[x] \cap \tilde{U}_\alpha\} \geq \bar{N}_{\tilde{\mathfrak{U}}}(x). \tag{1.1}$$

In particular, if $\mathfrak{U}$ is of finite type then $\sup N_{\mathfrak{U}} \geq n_{\tilde{\mathfrak{U}}}$.

If X is a regular metric manifold with admissible metric d and $\mathfrak{U} = \{U_\alpha\}$ is a uniform open cover of finite type we define the Munkres construction [16; Lemma 2.7]. Define $d_\alpha: X \to [0,1]$ by $d_\alpha(x) = \min(1, d(x, X - U_\alpha))$. For $\{\alpha_1,\ldots,\alpha_j\}$ any set of j distinct indices define $U_{\{\alpha_1,\ldots,\alpha_j\}} = \{x: \min(d_{\alpha_1},\ldots,d_{\alpha_j})(x) > \max\{d_\alpha(x): \alpha \neq \alpha_1,\ldots,\alpha_j\}\}$. Let $\mathfrak{U}_j = \{U_{\{\alpha_1,\ldots,\alpha_j\}}\}$ $j = 1,2,\ldots N = \sup N_{\mathfrak{U}}$ and $\mathfrak{U}' = \cup_{j=1}^{N} \mathfrak{U}_j$. Since $U_{\{\alpha_1,\ldots,\alpha_j\}} \subset U_{\alpha_1} \cap \cdots \cap U_{\alpha_j}$, $\mathfrak{U}'$ is a refinement of $\mathfrak{U}$. If $x \in X$ then $d_\alpha(x) = 0$ for all except $N_{\mathfrak{U}}(x) \leq N$ values of α and $\max_\alpha d_\alpha(x) = \delta_{\mathfrak{U}}(d)(x)$. So for some choice of j and $\{\alpha_1,\ldots,\alpha_j\}$ $\min(d_{\alpha_1}(x),\ldots,d_{\alpha_j}(x)) - \max\{d_\alpha(x): \alpha \neq \alpha_1,\ldots,\alpha_j\} \geq \delta_{\mathfrak{U}}(d)(x)/N$. Hence, the ball $B^d(x, \delta_{\mathfrak{U}}(d)(x)/2N)$ lies in $U_{\{\alpha_1,\ldots,\alpha_j\}}$, i.e.

$$\delta_{\mathfrak{U}'}(d) \geq \delta_{\mathfrak{U}}(d)/2N, \tag{1.2}$$

and $\mathfrak{U}'$ is a uniform open cover. Clearly, the elements of $\mathfrak{U}_j$ are pairwise disjoint. Since $x \notin U_{\{\alpha_1,\ldots,\alpha_j\}}$ if $j > N_{\mathfrak{U}}(x)$ we have

$$N_{\mathfrak{U}'} \leq N_{\mathfrak{U}}. \tag{1.3}$$

Thus, $\mathfrak{U}'$ is also of finite type. Note that if W.W refines $\mathfrak{U}'$ and $\tilde{\mathfrak{U}}' = \cup \tilde{\mathfrak{U}}_j$ is the associated trimming then for each j, $W[x]$ intersects at most one element of $\tilde{\mathfrak{U}}_j$.

If $\mathfrak{U}_1 = \{U_\alpha\}$ and $\mathfrak{U}_2 = \{V_\beta\}$ are uniform open covers then $\mathfrak{U}_1 \cap \mathfrak{U}_2 = \{U_\alpha \cap V_\beta\}$ is and:

$$N_{\mathfrak{U}_1 \cap \mathfrak{U}_2} = N_{\mathfrak{U}_1} \cdot N_{\mathfrak{U}_2}, \quad \bar{N}_{\mathfrak{U}_1 \cap \mathfrak{U}_2} = \bar{N}_{\mathfrak{U}_1} \cdot \bar{N}_{\mathfrak{U}_2}, \quad n_{\mathfrak{U}_1 \cap \mathfrak{U}_2} \leq n_{\mathfrak{U}_1} \cdot n_{\mathfrak{U}_2}. \tag{1.4}$$

If $f: X_1 \to X_2$ is a $\mathcal{C}_u$ map of regular metric manifolds and $\mathfrak{U} = \{U_\alpha\}$ is a uniform open cover of X_2 then $f^{-1}\mathfrak{U} = \{f^{-1}U_\alpha\}$ is a uniform open cover of X_1 and:

$$N_{f^{-1}\mathfrak{U}} = f^* N_{\mathfrak{U}}, \quad \bar{N}_{f^{-1}\mathfrak{U}} \leq f^* \bar{N}_{\mathfrak{U}}, \quad n_{f^{-1}\mathfrak{U}} \leq n_{\mathfrak{U}}. \tag{1.5}$$

4 <u>THEOREM</u>: Assume $\mathfrak{M}$ is a standard function space type, $\mathfrak{M} = \mathcal{C}_u^r$ $(r \geq 1)$ or $\mathfrak{M} = \mathcal{C}^\infty$. Let X be a finite dimensional semicomplete $\mathfrak{M}$ manifold. The following conditions are equivalent and if they hold X is called an <u>$\mathfrak{M}$ manifold of finite type</u>. Let $k = \dim X$ and $\mathfrak{s}$ be an admissible bound on X.

(a) Every uniform open cover of X has a uniform open refinement of finite type.

(b) There exists $\mathcal{C} = \{U_\alpha, h_\alpha\}$ an s admissible $\mathfrak{M}$ atlas on X with $\{U_\alpha\}$ of finite type.

(c) There exists $\mathcal{C} = \{U_\alpha, h_\alpha\}$ an s admissible $\mathcal{L}$ multiatlas on X with $\{U_\alpha\}$ of finite type.

(d) For every uniform open cover $\mathfrak{U}$ of X there exists an admissible $\mathfrak{M}$ atlas $\mathcal{C} = \{U_\alpha^3, h_\alpha\}$ such that (1) $h_\alpha(U_\alpha^3) = B(0, 3\epsilon_\alpha) \subset R^k$. (2) $\sup \mathfrak{s}|U_\alpha^3 \sim \inf \mathfrak{s}|U_\alpha^3 \sim \epsilon_\alpha^{-1}$. (3) $\{U_\alpha^1\}$ is a uniform open cover of X where $U_\alpha^1 = h_\alpha^{-1}(B(0,\epsilon_\alpha))$. (4) $\sup \bar{N}_{\mathcal{C}} < \infty$. (5) $\mathcal{C}$ refines $\mathfrak{U}$. In particular, $\mathcal{C}$ is star bounded, s admissible and convex.

An atlas $\mathcal{C}$ satisfying conditions (1) - (4) of (d) will be called a <u>square atlas</u> for X.

<u>PROOF</u>: (d) $\to$ (a) $\to$ (b) $\to$ (c) is clear. Beginning with an s admissible $\mathfrak{M}$ multiatlas of finite type and an open cover $\mathfrak{U} = \{V_\beta\}$ we will construct an $\mathfrak{M}$ atlas satisfying (d). This shows (b) $\to$ (d) and when $\mathfrak{M} = \mathcal{L}$ (c) $\to$ (d). Since (d) $\to$ (a) and (a) depends only on the underlying regular pseudometric space structure the proof will be completed.

Let $\mathcal{C} = \{U_\alpha, h_\alpha\}$ be an s admissible R^k multiatlas of finite type. We can trim and so assume by (1.1) that $n_{\mathcal{C}} < \infty$.

Let d be the metric associated to any atlas subdivision of $\mathcal{C}$ and assume that $\mathfrak{s}$ satisfies:

(1.6) $$\mathfrak{s} \geq 2 \max(\mathfrak{s}_{\mathcal{C}}, 1/\delta_{\mathcal{C}\cap\mathfrak{U}}(d), 1/\kappa(d)).$$

We begin with a subdivision to get a star bounded cover. Let $U_{\alpha n} = U_\alpha \cap \{2^n < \rho^d < 2^{n+3}\}$ $n = 0,1,\ldots$ By (1.4) and (1.5) $\bar{N}_{\{U_{\alpha n}\}} \leq 4\bar{N}_{\mathfrak{a}}$. If $x_1 \in B^d(x,(2\rho^d(x))^{-1})$ then the ratios of $\rho^d(x)$ and $\rho^d(x_1)$ lie in the interval $(2^{-1},2)$ since $1/\rho^d$ is d-Lipschitz with constant ≤ 1. Since $\rho^d(x) \in [2^{n+1},2^{n+2}]$ for some $n = 0,1,\ldots$ it follows that $\{B^d(x,(2\rho^d(x))^{-1})\}$ refines $\{\{2^n < \rho^d < 2^{n+3}\}\}$. Hence, for each $x \in X$ we can choose α, n and β such that

(1.7) $$B^d(x,2^{-n-4}) \subset U_{\alpha n} \cap V_\beta.$$

The square atlas is built using the integer lattice Z^k in R^k. Define $\mathfrak{A}_i^k = \{B(z,i): z \in Z^k\}$. If $v \in R^k$ there exists $z \in Z^k$ with $\|v - z\| \leq 1/2$ and so $B(v,2^{-1}) \subset B(z,1)$. Hence, $\delta_{\mathfrak{A}_1^k} \geq 2^{-1}$. $\bar{N}_{\mathfrak{A}_3^1} \leq 7$ and since $\mathfrak{A}_3^k = p_1^{-1}\mathfrak{A}_3^1 \cap\ldots\cap p_k^{-1}\mathfrak{A}_3^1$ where $p_1,\ldots,p_k$ are the projections of R^k to R it follows from (1.4) and (1.5) that $\bar{N}_{\mathfrak{A}_3^k} \leq 7^k$. Hence, if $\epsilon > 0$ and $\epsilon\mathfrak{A}_i^k = \{B(\epsilon z,\epsilon i): z \in Z^k\}$:

(1.8) $$\delta_{\epsilon\mathfrak{A}_1^k} \geq \epsilon/2 \text{ and } \bar{N}_{\epsilon\mathfrak{A}_1^k} < \bar{N}_{\epsilon\mathfrak{A}_3^k} \leq 7^k.$$

Now let $\epsilon_n = 2^{-n-9}$. For $z \in Z^k$ let $U_{\alpha nzj}^4$ denote a component of $h_\alpha^{-1}(B(\epsilon_n z,4\epsilon_n))$ mapping bijectively under h_α onto $B(\epsilon_n z,4\epsilon_n) \subset h_\alpha(U_{\alpha n})$ in R^k. Let J be the index set of quadruples (α,n,z,j) so defined. For each (α,n,z,j) in J and $i = 1,2,3$ let $U_{\alpha nzj}^i$ denote the component of $h_\alpha^{-1}(B(\epsilon_n z,i\epsilon_n))$ contained in $U_{\alpha nzj}^4$. We prove: (1) $U_{\alpha nzj}^4 \subset U_{\alpha n}$. (2) $2^9\rho^d(x) \geq \epsilon_n^{-1} \geq 2^6\rho^d(x)$ for $x \in U_{\alpha nzj}^4$. (3) $h_\alpha: U_{\alpha nzj}^i \to B(\epsilon_n z,i\epsilon_n)$ is a homeomorphism. (4) $\{U_{\alpha nzj}^4\}$ refines $\mathfrak{A}$. (5) Letting $\mathfrak{a}^i = \{U_{\alpha nzj}^i,h_\alpha\}$, $\mathfrak{a}^i$ is an s admissible atlas on X with $1/\lambda_{\mathfrak{a}^i} \leq 2^{10}\rho^d$ $i = 1,\ldots,4$. In particular, $\{U_{\alpha nzj}^1\}$ is a uniform open cover of X. (6) $\bar{N}_{\mathfrak{a}^3} \leq 4\cdot 7^k\bar{N}_{\mathfrak{a}}$. This will prove (d) with the square atlas defined by $\{U_{\alpha nzj}^3,h_\alpha - c_{\epsilon_n z}\}$.

(1) and (3) hold by definition of the quadruples in J and (2)

follows from (1). If $x_1, x_2 \in U^4_{\alpha nzj}$ then $d(x_1,x_2) \le \|h_\alpha x_1 - h_\alpha x_2\|$ and so with $h_\alpha x = \epsilon_n z$, (4) follows from (1.7). To prove (5), let $x \in X$ and choose (α,n) such that $B^d(x,2^{-n-4}) \subset U_{\alpha n}$, cf. (1.7). By the proof of Lemma V.1.2 there is an open set U in $B^d(x,2^{-n-4})$ containing x which h_α maps homeomorphicly onto $B(h_\alpha x, 2^{-n-4})$. $B(h_\alpha x, \epsilon_n/2) \subset B(\epsilon_n z, \epsilon_n) \subset B(\epsilon_n z, 4\epsilon_n) \subset B(h_\alpha x, 5\epsilon_n) \subset B(h_\alpha x, 2^{-n-4})$. Thus, for the triple (α,n,z) there is a quadruple (α,n,z,j) defined with $x \in U^1_{\alpha nzj}$ and $\lambda_{\mathcal{G}1}(x) \ge \epsilon_n/2 \ge (2^{10}\rho^d(x))^{-1}$ by (2).

Finally, we prove (6). For each triple (α,n,z) define $C_{\alpha nz} = \text{closure } \cup_j U^3_{\alpha nzj}$. By Theorem III.2.2 applied to $\mathcal{G}^4$ the inverse of the $d_{\mathcal{G}^4}$ distance from $U^3_{\alpha nzj}$ to $X - U^4_{\alpha nzj}$ is $\le O^*(\epsilon_n^{-1})$. Hence, $C_{\alpha nz} \subset \cup_j U^4_{\alpha nzj}$ and for each $x \in C_{\alpha nz}$ there is a unique component $U^4_{\alpha nzj}(x)$ containing x. Let $x \in X$. Choose U_0 a neighborhood of x intersecting at most $4\bar{N}_{\mathcal{G}}(x)$ members of $\{U_{\alpha n}\}$. Since $h_\alpha(U_{\alpha n})$ is bounded $U_0 \cap U^4_{\alpha nzj} \ne \emptyset$ for only finitely many triples (α,n,z). For each such triple, define $G_{\alpha nz} = X - C_{\alpha nz}$ if $x \notin C_{\alpha nz}$. If $x \in C_{\alpha nz}$, choose a ball $B_{\alpha n}$ about $h_\alpha x$ intersecting $\epsilon_n \mathfrak{A}^k_3$ in at most 7^k members and let $G_{\alpha nz} = U^4_{\alpha nzj}(x) \cap h_\alpha^{-1}(B_{\alpha n})$. Define $U_1 = \cap\{G_{\alpha nz}\colon U^4_{\alpha nzj} \cap U_0 \ne \emptyset \text{ for some } (\alpha,n,z,j) \in J\} \cap U_0$. $U_1 \cap U^3_{\alpha nzj} \ne \emptyset$ implies $U_0 \cap U^4_{\alpha nzj} \ne \emptyset$ and $x \in C_{\alpha nz}$. Hence, for $U_1 \cap U^3_{\alpha nzj} \ne \emptyset$ to hold there are at most $4\bar{N}_{\mathcal{G}}(x)$ choices for (α,z) then at most 7^k choices for z and then at most 1 choice for j. This proves (6). Q.E.D.

5 <u>COROLLARY</u>: Let X be an $\mathfrak{M}$ manifold of finite type with $\mathfrak{M}$ a standard function space type, $\mathfrak{M} = \mathcal{C}^r_u$ $(r \ge 1)$ or $\mathfrak{M} = \mathcal{C}^\infty$. Given any s admissible R^k atlas we can obtain by refining and translating a square atlas $\mathcal{G}$ with $n_{\mathcal{G}} < \infty$.

<u>PROOF</u>: Given $\mathcal{G}_0 = \{U_\alpha, h_\alpha\}$ we can refine to get a star bounded atlas satisfying condition $\mathcal{L}$. Refine to get the cover of finite type and trim

to get $n_{G_0} < \infty$. Thus, we assume that G_0 is star bounded, satisfies condition $\mathcal{L}$ and has $n_{G_0} < \infty$. Let $d = d_{G_0}$ and assume $\rho \geq 2/\delta_{G_0}(d)$. Then we can choose ϵ_α such that (1) $\epsilon_\alpha^{-1} \geq 5 \sup \rho|U_\alpha$. (2) $\inf \rho|\text{Star}(U_\alpha) > \epsilon_\alpha^{-1}$. (3) If d_α is the norm metric on R^k pulled back by h_α to U_α then $\epsilon_\alpha d_\alpha \leq d \leq \epsilon_\alpha^{-1} d_\alpha$ on U_α. As in the proof of Theorem 4 let J consist of all pairs (α,z) with $B(\epsilon_\alpha z, 4\epsilon_\alpha) \subset h_\alpha(U_\alpha)$ and define $U_{\alpha z}^i = h_\alpha^{-1}(B(\epsilon_{\alpha z}, i\epsilon_\alpha))$ $i = 1,\ldots,4$. For each α, $(\alpha,z) \in J$ for only finitely many z. For $x \in X$ choose α such that $B^d(x,\rho(x)^{-1}) \subset U_\alpha$ and so $B(h_\alpha x, 5\epsilon_\alpha) \subset h_\alpha(U_\alpha)$. Choose z with $\|h_\alpha x - \epsilon_\alpha z\| < \epsilon_\alpha/2$. Then $B(h_\alpha x, \epsilon_\alpha/2) \subset B(\epsilon_\alpha z, \epsilon_\alpha) \subset B(\epsilon_\alpha z, 4\epsilon_\alpha) \subset B(h_\alpha x, 5\epsilon_\alpha) \subset h_\alpha(U_\alpha)$. By (3) $B^d(x,\epsilon_\alpha^2/2) \subset U_{\alpha z}^1$ and $(\alpha,z) \in J$.

Choose $n \geq 1$ such that (4) $\rho^n|\text{Star}(U_\alpha) \geq \epsilon_\alpha^{-2}$ and (5) $B^d(x,\rho(x)^{-n})$ intersects at most n_{G_0} members of $\{U_\alpha\}$. For each α define $C_\alpha = \text{closure} \cup_z U_{\alpha z}^3$. By (3) we have: (6) $d(C_\alpha, X - U_\alpha) \geq \epsilon_\alpha^2$. Let $x \in X$. There are at most n_{G_0} indices and with $B^d(x,\rho(x)^{-n}) \cap C_\alpha \neq \emptyset$. Since $\{B^d(x,\rho(x)^{-1})\}$ refines $\{U_\alpha\}$, $x \in \text{Star}(U_\alpha)$ for each such α and so by (4) $\rho(x)^{-n} \leq \epsilon_\alpha^2$ for each such α and so by (5) $B^d(x,\rho(x)^{-n}) \subset U_\alpha$ and $h_\alpha(B^d(x,\rho(x)^{-n})) \subset B(h_\alpha x, \epsilon_\alpha)$ intersects at most 7^k members of $\epsilon_\alpha \mathfrak{A}_3^k$. Thus, $B^d(x,\rho(x)^{-n}) \cap U_{\alpha n}^3 \neq \emptyset$ for at most $7^k n_{G_0}$ pairs (α,n), i.e. $n_{G^3} \leq 7^k n_{G_0}$. Translate to get the square atlas. Q.E.D.

<u>ADDENDUM</u>: If we begin with $G_0 = \{U_\alpha, h_\alpha\}$ and ϵ_α satisfying (1), (2), (3) above then for any uniform open cover $\mathfrak{A}$ on X we can define $\epsilon_\alpha(\mathfrak{A}) = \min(\epsilon_\alpha, 4^{-1}\inf \delta_{\mathfrak{A}}(d)|U_\alpha)$ and perform the above construction, obtaining a different index set $J_{\mathfrak{A}}$ and a square atlas $G(\mathfrak{A})$ refining $\mathfrak{A}$ with $\rho_{G(\mathfrak{A}) \cup G_0} \leq O^*(\rho_{G_0})$ and $n_{G(\mathfrak{A})} \leq 7^k n_{G_0}$.

In particular, Corollary 5 implies that if $\pi: E \to X$ is a semicomplete bundle over a manifold of finite type, we can use the transfer of atlas construction to get admissible atlases $(\mathfrak{F}, G)$ for π with G a square atlas on X.

6 PROPOSITION: (a) Let X be a metric manifold of finite type. Any refinement of X is a metric manifold of finite type. If U is open in X then the canonical refinement structure on U is of finite type.

(b) If X_1 and X_2 are metric manifolds of finite type then $X_1 \times X_2$ is a metric manifold of finite type.

(c) Let $\pi: \mathbb{E} \to X$ be a finite dimensional, semicomplete bundle with X of finite type. The total space $\mathbb{E}$ and the Grassmanian $G(\mathbb{E})$ are manifolds of finite type.

PROOF: (a): If $\mathcal{G}$ is s admissible for X and of finite type then it is s admissible for any refinement. $\mathcal{G}|U$ is s admissible for the canonical refinement structure on U. (b): If $\mathcal{G}_i$ is s admissible and of finite type on X_i $(i = 1,2)$ then $\mathcal{G}_1 \times \mathcal{G}_2$ is s admissible and, by (1.4) and (1.5), of finite type on $X_1 \times X_2$. (c): If $(\mathfrak{D},\mathcal{G}) = \{U_\alpha,h_\alpha,\varphi_\alpha\}$ is an s admissible atlas for π with $\mathcal{G}$ of finite type on X then $\mathfrak{D}$ is an s admissible atlas on $\mathbb{E}$ of finite type by (1.5). Let $\ell = \dim \mathbb{E}_x$ for all x in X. We can assume $F_\alpha = R^\ell$ for all α. By putting the Hilbert norm on R^ℓ and applying Theorem VII.1.7 we see that $G(R^\ell)$ is a closed, bounded set in $L(R^\ell;R^\ell)$ and so is compact. We can choose $P_1,\ldots,P_K$ such that $\{V_{P_i} : i = 1,\ldots,K\}$ is a covering of $G(R^\ell)$ with a nonzero Lebesgue number. The refinement $\{G_{\alpha P_i}, k_{\alpha P_i}\}$ of $G(\mathfrak{D})$ is s admissible by the proof of Theorem VII.2.1 and denoting this atlas $G_K(\mathfrak{D})$ we have $N_{G_K(\mathfrak{D})} \leq K(g_\pi * \bar{N}_{\mathcal{G}})$ and $N_{G_K(\mathfrak{D})} \leq K(g_\pi * N_{\mathcal{G}})$ So $G_K(\mathfrak{D})$ is of finite type. Q.E.D.

7 THEOREM: Assume X is an $\mathfrak{M}$ manifold of finite type with $\mathfrak{M} \subset \mathcal{L}^1$. Let X_0 be a regular, uniformly complete finite dimensional $\mathfrak{M}$ manifold. If there exists a $\mathcal{C}^1_u$ immersion $f: X_0 \to X$ then X_0 is of finite type.

PROOF: That X_0 is semicomplete follows from Corollary VII.2.5, Corollary VLL.2.2 and its Addendum. Let $\mathcal{G}$ be s admissible $\mathcal{L}^1$ atlas of finite type on X and let $G_K(T\mathcal{G})$ be the associated atlas of finite type on

$G(TX)$, cf. proof of Proposition 6 (c). $G(f)$ is $\mathcal{C}_u$ and so the open cover $\{U_{\alpha P_i}\} = G(f)^{-1}\{G_{\alpha P_i}\}$ is a uniform open cover of finite type on X_0. By Proposition VII.3.1 (b) $f^*G_K(T\mathcal{G})$ is an admissible $\mathcal{C}^1$ multiatlas on X_0. Since it is of finite type, X_0 is of finite type by Theorem 4(c). Q.E.D.

As the reader is probably aware it is a standard dimension theory result that any open cover $\mathfrak{U}$ of a k dimensional space has an open refinement $\mathfrak{U}_1$ with $N_{\mathfrak{U}_1} \leq k + 1$. I have been unable to prove that if $\mathfrak{U}$ is a uniform open cover of a semicomplete finite dimensional manifold then $\mathfrak{U}$ has a uniform open refinement $\mathfrak{U}_1$ with $N_{\mathfrak{U}_1}$ bounded. I conjecture that it is true, i.e. that any finite dimensional, semicomplete manifold is a manifold of finite type. Lacking a proof of this, Theorem 7 suffices as a test that a manifold be of finite type in most applications. We will soon see that Theorem 7 has a strong converse.

Choose, once and for all, a C^∞ bump function $\varphi: R \to [0,1]$ with $\varphi(x) = 1$ $(|x| \leq 1)$ and $\varphi(x) = 0$ $(|x| \geq 2)$. For $0 < \epsilon \leq 1$ let $\varphi_\epsilon(x) = \varphi(\epsilon^{-1}x)$. So $\varphi_\epsilon(x) = 1$ $(|x| \leq \epsilon)$ and $\varphi(x) = 0$ $(|x| \geq 2\epsilon)$. Clearly, for any standard function space type $\|\varphi_\epsilon\|_{\mathfrak{M}} \leq O^*(\epsilon^{-1}, \|\varphi\|_{\mathfrak{M}})$ where by $\|\varphi\|_{\mathfrak{M}}$ we mean $\|\varphi|U\|_{\mathfrak{M}}$ for any bounded open set U containing $[-2,2]$. On R^k define $\varphi_\epsilon^k(x_1,\ldots,x_k) = \varphi_\epsilon(x_1)\cdot\ldots\cdot\varphi_\epsilon(x_k)$. φ_ϵ^k is C^∞ with $\varphi_\epsilon^k(x) = 1$ $(\|x\| \leq \epsilon)$ and $\varphi_\epsilon^k(x) = 0$ $(\|x\| \geq 2\epsilon)$. For each k, $\|\varphi_\epsilon^k\|_{\mathfrak{M}} \leq O^*(\epsilon^{-1}, \|\varphi\|_{\mathfrak{M}})$. Now let $\mathcal{G} = \{U_\alpha^3, h_\alpha\}$ be a square atlas on a manifold of finite type with $k = \dim X$. Define $\bar{\varphi}_\alpha = \varphi_{\epsilon_\alpha}^k \cdot h_\alpha$ on U_α^3 and $= 0$ on $X - U_\alpha^2$. Since $\{U_\alpha^1\}$ covers X and $\{U_\alpha^3\}$ is locally finite, $\bar{\varphi} = \Sigma_\alpha \bar{\varphi}_\alpha: X \to R$ satisfies $1 \leq \bar{\varphi} \leq N_{\mathcal{G}}$. Clearly, $\rho^{\mathfrak{M}}(\bar{\varphi}_\alpha; \mathcal{G}, R) \leq O^*(\rho_{\mathcal{G}}, \epsilon_\alpha^{-1}, \|\varphi\|_{\mathfrak{M}})$ with the constants depending on $\mathfrak{M}$ and $\dim X$. Since each point x has a neighborhood intersecting $\bar{N}_{\mathcal{G}}(x)$ members of $\{U_\alpha^3\}$ we have $\rho^{\mathfrak{M}}(\bar{\varphi}; \mathcal{G}, R) \leq O^*(\rho_{\mathcal{G}}, \bar{N}_{\mathcal{G}}, \epsilon^{-1}, \|\varphi\|_{\mathfrak{M}})$. Thus, we can define $\varphi_\alpha = \bar{\varphi}_\alpha/\bar{\varphi}$ and get a family $\mathcal{P}_{\mathcal{G}} = \{\varphi_\alpha\}$ associated to $\mathcal{G}$ satisfying:

(1) $\varphi_\alpha: X \to [0,1]$ with $\{\varphi_\alpha > 0\} \subset U_\alpha^2$. (2) $\Sigma_\alpha \varphi_\alpha = 1$.

(3) $\rho^{\mathfrak{M}}(\varphi_\alpha; \mathcal{G}, R) \leq O^*(\rho_{\mathcal{G}}, \bar{N}_{\mathcal{G}}, \epsilon^{-1}, \|\varphi\|_{\mathfrak{M}})$ with constants depending on $\mathfrak{M}$ and

dim X where $\varepsilon(x) = \inf\{\varepsilon_\alpha : x \in U_\alpha^2\}$.

8 PROPOSITION: Assume X is an $\mathfrak{M}$ manifold of finite type ($\dim X = k$) with $\mathfrak{M}$ a standard function space type. Let $\mathcal{G}$ be an s admissible R^k atlas, d and ρ be admissible metric and bound on X. There exist constants K and n and a transverse bounded $V \in \mathfrak{U}_X$ such that for every uniform open cover $\mathfrak{U} = \{V_\beta\}$ of finite type there is defined $\mathcal{P}(\mathfrak{U}) = \{\varphi_\beta\}$ satisfying:

(a) $\varphi_\beta : X \to R$ with $\varphi_\beta \geq 0$ and $\Sigma\varphi_\beta = 1$.

(b) Support $\varphi_\beta = \text{closure}\{\varphi_\beta > 0\} \subset V_\beta$.

(c) $\rho^{\mathfrak{M}}(\varphi_\beta,\mathcal{G},R)(x) \leq K \max(\rho(x),(\inf \delta_{\mathfrak{U}}(d)|V[x])^{-1})^n$.

PROOF: Refine $\mathcal{G}$ as in the proof of Corollary 5 to get a star bounded s admissible atlas $\mathcal{G}_0 = \{U_\alpha, h_\alpha\}$ with $n_{\mathcal{G}_0} < \infty$ and satisfying condition $\mathcal{L}$. Let $V = \cup_\alpha U_\alpha \times U_\alpha$. Define $\varepsilon_\alpha(\mathfrak{U})$ as in the Addendum to Corollary 5 and construct the square atlas $\{U_{\alpha z}^3, h_\alpha - c_z\} = \mathcal{G}(\mathfrak{U})$ described in the Addendum. Let $\alpha z \to \beta(\alpha z)$ be an index map with $U_{\alpha z}^3 \subset V_{\beta(\alpha z)}$ and define $\varphi_\beta = \Sigma\{\varphi_{\alpha z} : \beta(\alpha z) = \beta\}$ (or $\varphi_\beta = 0$ if $\beta(\alpha z)$ never equals β). (a) and (b) are clear from (1) and (2) for the partition $\mathcal{P}(\mathcal{G}(\mathfrak{U}))$. (c) follows from (3) for $\mathcal{P}(\mathcal{G}(\mathfrak{U}))$ and the estimates of the Addendum since $\rho(x) > \max(\rho_{\mathcal{G}_0}(x), 1/\delta_{\mathcal{G}_0}(d_{\mathcal{G}_0})(x), \sup\{\varepsilon_\alpha^{-1} : x \in U_\alpha\})$ and $\max(\rho, 1/\delta_{\mathfrak{U}}(d)) > 1/\delta_{\mathfrak{U}}(d_{\mathcal{G}_0})$ with constants independent of $\mathfrak{U}$. Q.E.D.

ADDENDUM: If X is a $\mathcal{C}^\infty$ manifold of finite type and $\mathcal{G}$ is an s admissible $\mathcal{C}^\infty$ atlas then $\mathcal{P}(\mathfrak{U})$ consists of $\mathcal{C}^\infty$ functions with (c) holding for all standard function space types $\mathfrak{M}$. If X is a $\mathcal{C}_u^r$ manifold of finite type ($r \geq 1$) then the family $\{j^r(\varphi_\beta \cdot h_\alpha^{-1}) \cdot h_\alpha : (U_\alpha, d, \rho) \to J^r(R^k, R)\}$ is a $\mathcal{C}_u$ family.

9 THEOREM: Assume X is an $\mathfrak{M}^1$ manifold of finite type with $\mathfrak{M}$ a standard function space type, $\mathfrak{M} = \mathcal{C}_u^r$ ($r > 0$) or $\mathfrak{M} = \mathcal{C}^\infty$.

(a) There exists a metricly proper $\mathfrak{M}^1$ immersion of X into R^K (K large). If X is separable the immersion can be chosen to be a homeomorphism onto its image.

(b) There exists a metricly proper $\mathfrak{M}^1$ embedding of X into a Hilbert space. If X is separable then the range can be chosen separable.

PROOF: Begin with a square atlas $\mathcal{G}_0$ with $n_{\mathcal{G}_0} < \infty$. For (b) let $\mathcal{G} = \mathcal{G}_0$. For (a) apply the Munkres construction, trim and call the result $\mathcal{G}$. In either case, we have a star bounded s admissible R^k atlas $\mathcal{G} = \{U_\alpha, h_\alpha\}$ with $V \in \mathfrak{U}_X$ such that with $x \in X, V[x] \cap U_\alpha \neq \emptyset$ for at most N choices of α. For (a) we have a decomposition of the index set of $\mathcal{G}$ into $J_1, \ldots, J_N$ such that with $x \in X, V[x] \cap U_\alpha \neq \emptyset$ for at most one α in each J_i. Let $\{\varphi_\alpha\}$ be a refining partition of unity as in Proposition 8. Let ρ be an admissible bound and choose ρ_α between $\sup \rho|U_\alpha$ and $2 \sup \rho|U_\alpha$ for each α. Define $f_\alpha\colon X \to R^{k+2}$ ($k = \dim X$) by $f_\alpha(x) = (\varphi_\alpha(x)h_\alpha(x), \varphi_\alpha(x)\rho_\alpha, \varphi_\alpha(x))$ $x \in U_\alpha$ and $= 0$ $x \notin U_\alpha$. By Proposition 8 there exist constants K_1, n_1 such that $K_1\rho^{n_1} \geq \rho^{\mathfrak{M}^1}(f_\alpha;\mathcal{G}, R^{k+2})$ for all α if $\mathfrak{M}$ is a standard function space type. If $\mathfrak{M} = \mathcal{C}_u^r$ then the family $\{j^{r+1}(f_\alpha \circ h_\alpha^{-1})h_\alpha\}$ is $\mathcal{C}_u$.

(a): Let $f_i = \Sigma\{f_\alpha\colon \alpha \in J_i\}\colon X \to R^{k+2}$ $i = 1,\ldots,N$ and $f = f_1 \times \ldots \times f_N\colon X \to R^K$ $K = N(k+2)$. Since $f_i|V[x] = f_\alpha|V[x]$ for some $\alpha \in J_i$, f is an $\mathfrak{M}^1$ map. Since $\Sigma_\alpha \varphi_\alpha(x) = 1$ and $\varphi_\alpha(x) > 0$ for at most N α's there exists α_x with $\varphi_{\alpha_x} \geq N^{-1}$. Hence, $\|f(x)\| \geq \rho_{\alpha_x}/N \geq \rho(x)/N$. Thus, f is metricly proper. If this $\alpha_x \in J_i$ then $T_x f_i = (\varphi_\alpha(x)T_x h_\alpha + T_x\varphi_\alpha \cdot h_\alpha, \rho_\alpha T_x\varphi_\alpha, T_x\varphi_\alpha)$. Hence $P_x(u,t_1,t_2) = (T_x h_\alpha)^{-1}(\varphi_\alpha(x)^{-1}[u - (T_x\varphi_\alpha(t_2))h_\alpha x])$ is a left inverse for $T_x f_i$ and so $P_x \circ \pi_i$ is a left inverse for $T_x f$. It follows that f is an $\mathfrak{M}^1$ immersion. If X is separable then $\mathcal{G}$ is countable and the ρ_α's can be chosen with $\inf_\beta\{|\rho_\alpha - \rho_\beta|\colon \beta \neq \alpha\} > 0$ for each α. Let $x_0 \in X$ and $\{x_n\colon n = 1,2,\ldots\}$ be a sequence in X with $f(x_n)$ converging to $f(x_0)$. If $f_i(x_0) \neq 0$ we can assume $f_i(x_n) \neq 0$ for all n. Let $\alpha_n = \alpha_{x_n} \in J_i$ $n = 0,1,\ldots$ Then $(\varphi_{\alpha_n}(x_n)\rho_{\alpha_n}, \varphi_{\alpha_n}(x_n)) \to (\varphi_{\alpha_0}(x_0)\rho_{\alpha_0}, \varphi_{\alpha_0}(x_0)) \neq 0$ and so $\rho_{\alpha_n} = \rho_{\alpha_0}$ for n large enough. Hence, x_n is eventually in U_{α_0} and $h_{\alpha_0}(x_n) \to h_{\alpha_0}(x_0)$. Thus, $x_n \to x_0$.

(b): Let E^{k+2} be R^{k+2} with the Hilbert space norm. The "identity" $Q\colon R^{k+2} \to E^{k+2}$ satisfies $\|Q^{-1}\| \leq 1$ and $\|Q\| \leq (k+2)^{1/2}$. If $\{E_\alpha\}$ is a family of Hilbert spaces then let $\Sigma_\alpha E_\alpha$ denote the completion of the weak sum of the family, i.e. of $\{\{V_\alpha\}\colon v_\alpha = 0$ for all but finitely many α's$\}$, with respect to the norm $(\Sigma_\alpha \|v_\alpha\|^2)^{1/2}$. Letting $g_\alpha = Q\cdot f_\alpha$, define $\Sigma g_\alpha = g\colon X \to \Sigma_\alpha E_\alpha$ where each E_α is a copy of E^{k+2}. Note that g maps X into the weak sum. In fact $g_\alpha(x) \neq 0$ at most N α's. Thus $g|V[x]$ factors through an inclusion of E^K, $K = N\,(k+2)$ and so g is an $\mathfrak{M}^1$ map. It is a metricly proper $\mathfrak{M}^1$ immersion just as in (a). Let $\mathfrak{a}_1 = \{U_\alpha, Q\cdot h_\alpha\}$ and $d = d_{\mathfrak{a}_1}$. Since $\mathfrak{a}$ is a square atlas $\mathfrak{a}_1$ is convex. Assume $\|g(x_1) - g(x_2)\| < 1/2N$. Let α be chosen with $\varphi_\alpha(x_1) \geq 1/N$. Since $\|g(x_1) - g(x_2)\| \geq |\varphi_\alpha(x_1) - \varphi_\alpha(x_2)|$, $\varphi_\alpha(x_2) > 1/2N$. Hence, $x_1, x_2 \in U_\alpha$ and by convexity: $d(x_1,x_2) < \|h_\alpha x_1 - h_\alpha x_2\| \leq N\|\varphi_\alpha(x_1)h_\alpha x_1 - \varphi_\alpha(x_1)h_\alpha x_2\| \leq N[\|\varphi_\alpha(x_1)h_\alpha x_1 - \varphi_\alpha(x_2)h_2 x_2\| + |\varphi_\alpha(x_1) - \varphi_\alpha(x_2)|\,\|h_\alpha x_2\|] \leq 4kN\|g(x_1) - g(x_2)\|$, where we have written h_α for $Q\cdot h_\alpha$ throughout and we have used $\|h_\alpha x_2\| \leq 3k$. Thus, $4kN\|g(x_1) - g(x_2)\| \geq \min(d(x_1,x_2),1)$ and so g is an embedding. If X is separable then $g(X)$ is separable and so the closed subspace generated by $g(X)$ is separable. In fact, since $\mathfrak{a}$ is countable $\Sigma_\alpha E_\alpha$ is separable. Q.E.D.

10 PROPOSITION: (a) Let $\pi_i\colon E_i \to X$ be semicomplete metric bundles over a metric manifold of finite type $(i = 1,2)$. If $(\phi,1)\colon \pi_1 \to \pi_2$ is a locally split metric VB injection or surjection then $(\phi,1)$ splits.

(b) Every semicomplete, finite dimensional metric bundle over a metric manifold of finite type is (isomorphic to) a split subbundle of a trivial bundle ϵ_{R^K} (K large).

PROOF: (a): Choose an atlas two-tuple $(\mathfrak{D}_1,\mathfrak{D}_2;\mathfrak{a})$ for $(\xi,1)$ with $\mathfrak{a}$ a square atlas. Glue the obvious splittings over each chart of $\mathfrak{a}$ together by the partition of unity associated with $\mathfrak{a}$. (b): This is the bundle analogue of Theorem 9. The reader can adapt Lang's proof [14; Prop. III.9]. Q.E.D.

2. Approximation and Smoothing: Let $A \subset U$ in a regular metric manifold X with A closed and U open. $A \subset\subset U$ iff $\{U, X - A\}$ is a uniform open cover of X. In fact, if d is an admissible metric on X, then:

$$(2.1) \qquad \delta_{\{U,X-A\}}(d)(x) = \max(\delta^U(d)(x), \min[d(x,A),1]).$$

Note that for any x, $\delta^U(d)(x) + \min(d(x,A),1) \geq \min(d(A,X-U),1)$. Hence, from (2.1) we get:

$$(2.2) \qquad 2\delta_{\{U,X-A\}}(d) \geq \min(d(A,X - U),1) = \inf \delta^U(d)|A$$
$$= \inf \delta_{\{U,X-A\}}(d)|A \geq \inf \delta_{\{U,X-A\}}(d).$$

Note that if A is bounded and $A \subset\subset U$, then $\inf \delta^U(d)|A > 0$ and so $\inf \delta_{\{U,X-A\}}(d) \geq (\inf \delta^U(d)|A)/2 > 0$.

1 PROPOSITION: Assume $\mathfrak{M}_1 \subset \mathfrak{M}$ and $\mathfrak{M}_1 \subset \mathscr{L}$ with $\mathfrak{M}$ and $\mathfrak{M}_1$ standard function space types. Let X be an $\mathfrak{M}_1$ manifold of finite type and $\pi: \mathbb{E} \to X$ be an $\mathfrak{M}$ bundle. If $A \subset\subset U$ in X with A closed and U open there is a continuous linear map of topological groups $L: \mathfrak{M}(\pi_U) \to \mathfrak{M}(\pi)$ with $L(s)|A = s|A$ and $L(s)|X - U = 0$.

PROOF: Let $(\mathfrak{D}_0,\mathfrak{a}_0)$ be an atlas for π and $\mathfrak{a}$ be a square atlas for X. Let d and ρ be admissible metric and bound for X. By Proposition 8 for every pair $A \subset\subset U$ we can construct an $\mathfrak{M}_1$ partition of unity $\{\varphi_U,\varphi_{X-A}\}$ refining $\{U,X-A\}$. Let $L(s) = \varphi_U \cdot s$ on U and $= 0$ on $X - U$. It follows from Proposition 8 that there exists a transverse bounded

$V \in \mathfrak{U}_X$ and constants K, n depending on $(\mathfrak{D}_0, \mathcal{G}_0)$, $\mathcal{G}$, d and ρ but not on A or U such that $\tilde{\rho}(L(s);\mathcal{G}_0,\mathfrak{D}_0)(x) \leq$
$K \max(\rho(x), (\inf \delta_{\{U,X-A\}}(d)|V[x]|)^{-1})^n \tilde{\rho}(s;\mathcal{G}_0,\mathfrak{D})(x)$ for x in U.
$\tilde{\rho}(L(s);\mathcal{G}_0,\mathfrak{D}_0)(x) = 0$ for $x \in X - U$. Q.E.D.

2 <u>THEOREM</u>: Assume $\mathfrak{M}$ is a standard function space type with $\mathfrak{M} \subset \mathcal{L}^r$ $(r \geq 0)$ or $\mathfrak{M} = \mathcal{C}_u^t$ $(t > r \geq 0)$ or $\mathfrak{M} = \mathcal{C}^\infty$. Let $\pi\colon E \to X$ be a semicomplete $\mathfrak{M}$ bundle over an $\mathfrak{M}$ manifold of finite type. The image of the inclusion $\mathfrak{M}(\pi) \to \mathcal{C}^r(\pi)$ has image dense in the closed subspace $\mathcal{C}_u^r(\pi)$.

<u>PROOF</u>: We follow De Rham [4]. Choose $(\mathfrak{D},\mathcal{G})$ = an admissible $\mathfrak{M}$ and $\mathcal{L}ip^r$ atlas for π with $\mathcal{G} = \{U_\alpha^3, h_\alpha\}$ a square atlas is on X satisfying $\epsilon_\alpha \leq 6^{-1}$ for all α and $(\inf\{\epsilon_\alpha\colon x \in U_\alpha^3\})^{-1}$ an admissible bound on X. Let $\{\varphi_\alpha\}$ be the associated partition of unity. Assume $N = \sup \bar{N}_{\mathcal{G}}$ and $k = \dim X$.

Let $s \in \mathcal{C}(\pi)$ and let $\{s_\alpha\colon B(0,3\epsilon_\alpha) \to F_\alpha\}$ be the principal parts of s. Define $\tilde{s}_\alpha = (\varphi_\alpha \cdot h_\alpha^{-1}) s_\alpha$. It is the α principal part of the section $\varphi_\alpha s$. The support of $\tilde{s}_\alpha$ is in $B(0,2\epsilon_\alpha)$. If $s \in \mathcal{C}_u^r(\pi)$ then the family $\{j^r s_\alpha\colon (B(0,3\epsilon_\alpha),\|\ \|,\epsilon_\alpha^{-1}) \to J^r(R^k;F_\alpha)\}$ is a $\mathcal{C}_u$ family by condition $\mathcal{L}$. By the uniform $\mathcal{L}^r$ estimates on $\{\varphi_\alpha\}$ the family $\{j^r\tilde{s}_\alpha\colon B(0,3\epsilon_\alpha) \to J^r(R^k;F_\alpha)\}$ is also a $\mathcal{C}_u$ family.

Beginning with the original bump function φ, define $\psi\colon R \to R$ by $\psi = \varphi_{4^{-1}}/\int_{-\infty}^{\infty} \varphi_{4^{-1}}(t)dt$. Let $\psi_\epsilon(t) = \epsilon^{-1}\psi(\epsilon^{-1}t)$ for $0 < \epsilon \leq 1$ and $\psi_\epsilon^k(t_1,\ldots,t_k) = \psi_\epsilon(t_1)\cdot\ldots\cdot\psi_\epsilon(t_k) = \epsilon^{-k}\psi(\epsilon^{-1}t_1)\cdot\ldots\cdot\psi(\epsilon^{-1}t_k)$. Clearly, ψ_ϵ^k is C^∞ with $\int_{R^k} \psi_\epsilon^k = 1$ and with $\operatorname{supp} \psi_\epsilon^k \subset B(0,2^{-1}\epsilon)$ and for any standard function space type $\|\psi_\epsilon^k\| \leq O^*(\epsilon^{-1},\|\psi\|)$ with constants depending on k and the function space type.

For each choice of a family $\{\delta_\alpha\}$ with $\delta_\alpha \leq \epsilon_\alpha$ and $\epsilon_\alpha^{-1} > \delta_\alpha^{-1}$ in α, define $\tilde{u}_\alpha(x) = \int_{R^k} \psi_{\delta_\alpha}^k(x-y)\tilde{s}_\alpha(y)dy = \int_{R^k} \psi_{\delta_\alpha}^k(y)\tilde{s}_\alpha(x-y)dy$.

Note that $\psi^k_{\delta_\alpha}(x-y)\tilde{s}_\alpha(y) = 0$ unless $\|y\| < 2\epsilon_\alpha$ and $\|x-y\| < \delta_\alpha/2$ and $\psi^k_{\delta_\alpha}(y)\tilde{s}_\alpha(x-y) = 0$ unless $\|y\| < \delta_\alpha/2$ and $\|x-y\| < 2\epsilon_\alpha$. The integral is thus well defined and the support of $\tilde{u}_\alpha \subset B(0,5\epsilon_\alpha/2)$. Define $J_{\{\delta_\alpha\}}(s) = \Sigma_\alpha \tilde{u}_\alpha \cdot h_\alpha$. When $\mathfrak{M}$ is standard or $\mathfrak{M} = \mathcal{C}^r_u$, I claim that the linear map $J_{\{\delta_\alpha\}}: \mathcal{C}(\pi) \to \mathfrak{M}(\pi)$ is a continuous map of topological groups. Letting $I_\alpha = (-3\epsilon_\alpha, 3\epsilon_\alpha)$, and using uniform continuity of $\tilde{s}_\alpha$ on I^k_α the proof of Proposition II.1.6(a) shows that the map $\mathfrak{M}(I^k_\alpha \times I^k_\alpha, R) \to \mathfrak{M}(I^k_\alpha, F_\alpha)$ defined by $q(x,y) \to \int_{I^k_\alpha} q(x,y)\tilde{s}_\alpha(y)dy$ is a continuous linear map of norm $\leq (6\epsilon_\alpha)^k \|\tilde{s}_\alpha\|_0 \leq \|s_\alpha\|_0$. Hence, $\|\tilde{u}_\alpha\|_{\mathfrak{M}} \leq 0^*(\delta^{-1}_\alpha, \|\psi\|_{\mathfrak{M}})\|s_\alpha\|_0$ with constants depending on $\mathfrak{M}$ and k. Letting $V = \cup_\alpha U^3_\alpha \times U^3_\alpha$ and using $N = \sup N_{\mathcal{G}}$ it follows that $\tilde{\rho}^{\mathfrak{M}}(J_{\{\delta_\alpha\}}(s);\mathcal{G},\mathfrak{D})(x) \leq 0^*(N, \sup \rho_{(\mathfrak{D},\mathcal{G})}|V[x], (\inf\{\delta_\alpha : x \in U^3_\alpha\})^{-1}, \|\psi\|_{\mathfrak{M}}) \sup \tilde{\rho}^{\mathcal{C}}(s;\mathcal{G},\mathfrak{D})|V[x]$. Since V is transverse bounded, $\epsilon^{-1}_\alpha > \delta^{-1}_\alpha$ and $(\inf\{\epsilon_\alpha : x \in U^3_\alpha\})^{-1}$ is an admissible bound on X, it follows that $J_{\{\delta_\alpha\}}(s)$ is in $\mathfrak{M}(\pi)$ and $J_{\{\delta_\alpha\}}$ is continuous. If $\mathfrak{M} = \mathcal{C}^t_u$, $J_{\{\delta_\alpha\}}$ is continuous into $\mathcal{C}^t(\pi)$. That the image actually lies in $\mathcal{C}^t_u(\pi)$ follows from the uniform $\mathcal{L}^t$ estimates on $\{\tilde{u}_\alpha\}$. If $\mathfrak{M} = \mathcal{C}^\infty$, $J_{\{\delta_\alpha\}}$ maps continuously into $\mathcal{C}^r(\pi)$ for all r and so into $\mathcal{C}^\infty$.

To complete the proof, we show that if $s \in \mathcal{C}^r_u(\pi)$ then $J_{\{\delta_\alpha\}}(s)$ approaches s in $\mathcal{C}^r(\pi)$ as the δ_α's approach 0 in the right way. $J_{\{\delta_\alpha\}}(s) - s = \Sigma_\alpha[(\tilde{u}_\alpha \cdot h_\alpha) - \varphi_\alpha s]$. So $N_{\mathcal{G}} \leq N$ implies $\tilde{\rho}^{\mathcal{C}^r}(J_{\{\delta_\alpha\}}(s)-s)(x) \leq N \max_\alpha \tilde{\rho}^{\mathcal{C}^r}(\tilde{u}_\alpha \cdot h_\alpha - \varphi_\alpha s)(x)$. $\tilde{\rho}^{\mathcal{C}^r}(\tilde{u}_\alpha \cdot h_\alpha - \varphi_\alpha s)(x) = 0$ if $x \notin U^3_\alpha$ and $\leq 0^*(\sup \rho_{(\mathfrak{D},\mathcal{G})}|U^3_\alpha)\|\tilde{u}_\alpha - \tilde{s}_\alpha\|_{\mathcal{C}^r}$ if $x \in U^3_\alpha$. It suffices given $n,K \geq 1$ to show we can choose $\{\delta_\alpha\}$ so that $\|\tilde{u}_\alpha - \tilde{s}_\alpha\|_{\mathcal{C}^r} < \epsilon^n_\alpha/K$. But $j^r(\tilde{u}_\alpha - \tilde{s}_\alpha)(x) = \int_{R^k}\psi^k_{\delta_\alpha}(y)[j^r(\tilde{s}_\alpha)(x-y) - j^r(\tilde{s}_\alpha)(x)]dy$. Because the family $\{j^r(\tilde{s}_\alpha)\}$ is $\mathcal{C}_u$ there exist constants $K_1, n_1 \geq 1$ such that $\|x_1 - x_2\| < \epsilon^{n_1}_\alpha/K_1$ implies $\|j^r(\tilde{s}_\alpha)(x_1) - j^r(\tilde{s}_\alpha)(x_2)\| < \epsilon^n_\alpha/K$. Let $\delta_\alpha = \epsilon^{n_1}_\alpha/K_1$. We get $\|\tilde{u}_\alpha - \tilde{s}_\alpha\|_{\mathcal{C}^r} < \epsilon^n_\alpha/K$ as required. Q.E.D.

If X is bounded and hence is semicompact, we have a sequence of operators $J_n = J_{\{\phi^n_\alpha\}}$ $n = 1,\ldots$ mapping $\mathcal{C}(\pi)$ into $\mathcal{C}^\infty(\pi)$ and regarded as operators in $L(\mathcal{C}^r_u(\pi);\mathcal{C}^r_u(\pi))$ the sequence converges strongly to the identity.

In general, the above theorem gives a nice interpretation of the function space type $\mathcal{C}^r_u$: it is the "closure" of $\mathcal{J}^r$ in the $\mathcal{C}^r$ topology.

3 <u>THEOREM</u>: If X is a semicomplete $\mathcal{C}^\infty$ manifold admitting a metricly proper $\mathcal{C}^\infty$ embedding into a Hilbert space, or more generally a globally split metricly proper $\mathcal{C}^\infty$ embedding into a B-space E, then X admits $\mathcal{C}^\infty$ exponentials. Examples are any $\mathcal{C}^\infty$ manifold of finite type or the Grassmann manifold of a B-space isomorphic to a Hilbert space.

Assume $\mathfrak{M}$ is a standard function space type with $\mathfrak{M} \subset \mathcal{J}^r$ $(r \geq 0)$ or $\mathfrak{M} = \mathcal{C}^t_u$ $(t > r \geq 0)$ or $\mathfrak{M} = \mathcal{C}^\infty$. If X is an above and X_0 is an $\mathfrak{M}$ manifold of finite type then the image of the inclusion $\mathfrak{M}(X_0,X) \to \mathcal{C}^r(X_0,X)$ is dense in the closed subset $\mathcal{C}^r_u(X_0,X)$.

<u>PROOF</u>: Regard X as a globally split $\mathcal{C}^\infty$ submanifold of E i.e. $X \subset E$ and there exists a $\mathcal{C}^\infty$ VB retraction $(\Phi,1_X)\colon \epsilon_E \to \tau_X$ where the trivial bundle ϵ_E is the restriction of τ_E to X. In the Hilbert space case this is the map $s_\tau\colon X \to G(E) \to L(E;E)$ where the latter map is the inclusion of Theorem VII.1.7. Let $e_0\colon E \times E \to E \times E$ be the inverse linear exponential map on E, i.e. $e_0(v_1,v_2) = (v_1,v_2 - v_1)$. Define e to be the composition:

$$X \times X \xrightarrow{\text{inc}} X \times E \xrightarrow{e_0} X \times E \xrightarrow{\Phi} TX.$$

Clearly, $e \circ \Delta_X = 0_X$ and $\tau_X \circ e = p_1$. If $(x,x) \in \Delta_X$ then $T_{(x,x)}(e)(\dot{x}_1,\dot{x}_2)$ $= T_{(x,0)}\Phi(\dot{x}_1,\dot{x}_2 - \dot{x}_1) = (\dot{x}_1,\Phi_x(\dot{x}_2 - \dot{x}_1)) = (\dot{x}_1,\dot{x}_2 - \dot{x}_1)$ where $T_{0(x)}TX$ is identified with $T_xX \times T_xX$ and $\Phi_x|T_xX$ is the identity. Hence, by Corollary 5.3 e restricted to some full neighborhood of Δ_X is the inverse of a normalized $\mathcal{C}^\infty$ exponential on X. The examples come from Theorem 1.9 (b) and Theorem VII.1.7.

Now by the Tubular Neighborhood Theorem VI.5.4, which in this case is an easy variation of the above arguments, there exists U a full neighborhood of X in E and a $\mathcal{C}^\infty$ retraction $\pi: U \to X$. Recall that $\mathcal{C}^r(X_1,U) = \{f: f(X_1) \subset\subset U\}$ is open in $\mathcal{C}^r(X_1,E)$. Identifying $\mathcal{C}^r(X_1,E)$ with the section space $\mathcal{C}^r(\epsilon_E^1)$, where ϵ_E^1 is the trivial bundle over X_1, Theorem 2 implies that $\mathfrak{M}(X_1,U)$ is dense in $\mathcal{C}_u^r(X_1,U)$. $\pi_*: \mathcal{C}_u^r(X_1,U) \to \mathcal{C}_u^r(X_1,X)$ is a continuous map and so if $f \in \mathcal{C}_u^r(X_1,X)$ with open neighborhood N, then there exists $g \in \mathfrak{M}(X_1,U) \cap \pi_*^{-1}N$ and so $\pi\cdot g \in \mathfrak{M}(X_1,X) \cap N$. Q.E.D.

<u>ADDENDUM</u>: If X is an above and $\pi: \mathbb{E} \to X$ is a $\mathcal{C}^\infty$ bundle admitting a $\mathcal{C}^\infty$ VB injection into a trivial bundle ϵ_F with F a Hilbert space, or more generally, a globally split $\mathcal{C}^\infty$ VB injection into any trivial bundle ϵ_F, then π admits $\mathcal{C}^\infty$ bundle exponentials by an argument analogous to the one above. Examples are any semicomplete, finite dimensional $\mathcal{C}^\infty$ bundle over a $\mathcal{C}^\infty$ manifold of finite type and the canonical bundle γ_E over the Grassmanian of a B-space E isomorphic to a Hilbert space (cf. Prop. 1.10(b)).

In applying these approximation results to splittings of bundles the following is crucial.

4 <u>LEMMA</u>: Assume $\mathfrak{M}$ is a standard function space type with $\mathfrak{M} \subset \mathcal{L}$. Let $\pi: \mathbb{E} \to X$ be a semicomplete $\mathfrak{M}$ bundle with $\mathfrak{M}$ subbundles $\pi_i: \mathbb{E}_i \to X$ $(i = 1,2)$ such that the sum of the inclusions $(i_1 + i_2, 1_X): \pi_1 \times \pi_2 \to \pi$ is a $\mathcal{C}$ VB isomorphism (and hence an $\mathfrak{M}$ VB isomorphism cf. Prop. III.5.5 (b)). Let $\| \ \|_i$ $(i = 1,2)$ be admissible Finslers on π_i $(i = 1,2)$ and let ρ be an admissible bound on X. If s_1 is the section of g_π associated to π_1 then for any choice of constants K,n: $\{E_{0x} \in G(\mathbb{E}): E_{0x}$ is the graph of a map $T_x: \mathbb{E}_{1x} \to \mathbb{E}_{2x}$ with $\|T_x\| < (K\rho(x)^n)^{-1}\}$ is a full neighborhood of $s_1(X)$ in $G(\mathbb{E})$. Such sets are a basis for the set of full neighborhoods of $s_1(X)$ in $G(\mathbb{E})$.

PROOF: Recall the semicomplete $\mathfrak{M}$ submersion $G: Li(\mathbb{E}_1;\mathbb{E}) \to G(\mathbb{E})$ (cf. Thm.VII.2.1). Regarding i_1 as a section of $L(\pi_1;\pi_2)|Li(\mathbb{E}_1;\mathbb{E})$ we have $G\cdot i_1 = s_1$. Given $K_1, n_1 \geq 1$ the set $\{\tilde{T}_x \in L(\mathbb{E}_1;\mathbb{E}_2): \|\tilde{T}_x - i_{1x}\| < (2K_\rho(x)^n)^{-1}\}$ (defined using the Finsler on π making $i_1 + i_2$ an isometry) is a vertically full neighborhood of $i_1(X)$ in $L(\mathbb{E}_1;\mathbb{E}_2)$ and so by Proposition III.7.15 it is a full neighborhood of $i_1(X)$ rel $L(\mathbb{E}_1;\mathbb{E}_2)$ and so rel $Li(\mathbb{E}_1;\mathbb{E}_2)$. Now $G\cdot i_1 = s_1$ implies that $i_1(X)$ is vertically bounded with respect to the submersion G and so by Proposition VI.4.8 the image under G of the above set is a full neighborhood of $s_1(X)$. If $(P_i, 1_X): \pi \to \pi_i$ are the projections $(i = 1,2)$ then $G(\tilde{T}_x)$ is the graph of $(P_2\tilde{T}_x)\cdot(P_1\tilde{T}_x)^{-1}$ which has norm $\leq (K_\rho(x)^n)^{-1}$, as $\theta(P_1T_x) \leq 2$. That such sets form a basis for the full neighborhoods of $s_1(X)$ comes from the converse, i.e. if $T_x: \mathbb{E}_{1x} \to \mathbb{E}_{2x}$ has norm $\leq (2K_\rho(x)^n)^{-1}$ then $i_{1x} + i_{2x}\cdot T_x: \mathbb{E}_{1x} \to \mathbb{E}_x$ differs from i_{1x} by $i_{2x}\cdot T_x$ which has norm $< (2K_\rho(x)^n)^{-1}$. Q.E.D.

5 SMOOTHING THEOREM: Assume that $\mathfrak{M}$ is a standard function space type with $\mathfrak{M} \subset \mathscr{L}$ or $\mathfrak{M} = \mathcal{C}^r_u$ $(\infty > r \geq 0)$.

(a) Let X be an $\mathfrak{M}^1$ manifold of finite type. There exists a semicomplete $\mathcal{C}^\infty$ structure on X compatible with the original $\mathfrak{M}^1$ structure, and hence of finite type. Moreover, any two such $\mathcal{C}^\infty$ structures are $\mathcal{C}^\infty$ isomorphic and if $\mathfrak{M} \subset \mathcal{C}^r_u$ $(r \geq 0)$ then the isomorphism can be chosen $\mathcal{C}^{r+1}$ close to the identity map.

(b) Let X be a $\mathcal{C}^\infty$ manifold of finite type and $\pi: \mathbb{E} \to X$ a semicomplete, finite dimensional $\mathfrak{M}$ bundle over X. There exists a semicomplete $\mathcal{C}^\infty$ structure on π with base space the $\mathcal{C}^\infty$ manifold X and compatible with the original $\mathfrak{M}$ structure. Moreover, any two such $\mathcal{C}^\infty$ structures are $\mathcal{C}^\infty$ VB isomorphic over the identity on X and if $\mathfrak{M} \subset \mathcal{C}^r_u$ $(r \geq 0)$ then the isomorphism can be chosen $\mathcal{C}^r$ close to the identity map.

PROOF: We follow Palais [24]. Let $j: X \to R^k$ be a metricly proper $\mathfrak{M}^1$

immersion (cf. Thm. 1.9). Identifying $j^*\tau_R K$ with the trivial bundle $\epsilon_R K$ over X, the $\mathfrak{M}$ VB injection $(j^*Tj, 1): \tau_X \to \epsilon_R K$ splits by Proposition 1.10(a). Thus, there is an $\mathfrak{M}$ map $X \to G(R^K)$ representing a complementary bundle for τ_X in $\epsilon_R K$. By Theorem 3 and Lemma 4 we can $\mathcal{C}$ approximate this map by an $\mathfrak{M}^1$ map $s_\pi: X \to G(R^K)$ and so get an $\mathfrak{M}^1$ subbundle $\pi: E \to X$ of $\epsilon_R K$ such that $(j^*Tj + i_\pi): \tau_X \oplus \pi \to \epsilon_R K$ is an $\mathfrak{M}$ VB isomorphism. Furthermore the proof of Corollary VII.2.4 identifies π with the pull back of the canonical bundle $s_\pi^*\gamma_R K$. Thus, there is an $\mathfrak{M}^1$ VB map $(S_\pi, s_\pi): \pi \to \gamma_R K$ with $(s_\pi^* S_\pi, 1)$ an $\mathfrak{M}^1$ VB isomorphism. It clearly follows that $S_\pi: E \to \Gamma(R^K)$ is properly transverse to $G(R^K)$ with $X = S_\pi^{-1}(G(R^K))$, identifying base spaces with zero section images.

Now $J_0: X \times R^K \to R^K$ defined by $J(x,v) = f(x) + v$ is an $\mathfrak{M}^1$ map with $J_0|X = j$, (identifying $X = X \times 0$). The restriction $J = J_0|E$ has tangent map along the zero section given by the $\mathfrak{M}$ VB isomorphism $j^*Tj + i_\pi$. Hence, by Corollary VI.5.3 there exists an open set O with $X \subset\subset O$ rel E such that J is a metricly proper local isomorphism of O onto an open subset of R^K. This is just a reproof of the Tubular Neighborhood Theorem. Put the canonical refinement structure on O with admissible bound ρ_O. $(\{O,J\}, \rho_O)$ is an admissible $\mathfrak{M}^1$ adapted multiatlas for the refinement structure. It is also an adapted $\mathcal{C}^\infty$ atlas inducing a semicomplete $\mathcal{C}^\infty$ structure on O. The original $\mathfrak{M}^1$ structure is compatible with the $\mathcal{C}^\infty$ structure and so the latter is of finite type.

Since R^K is isomorphic to a Hilbert space, $\Gamma(R^K)$ is a $\mathcal{C}^\infty$ submanifold of $L(R^K; R^K) \times R^K$ which is isomorphic to a Hilbert space. O is a $\mathcal{C}^\infty$ manifold of finite type with $\pi|O: O \to X$ an $\mathfrak{M}^1$ submersion. $S_\pi: O \to \Gamma(R^K)$ is an $\mathfrak{M}^1$ map properly transverse to $G(R^K)$ with $\pi|S_\pi^{-1}(G(R^K))$ an $\mathfrak{M}^1$ isomorphism onto X. By Theorem 3 we can $\mathcal{C}^1$ approximate S_π by a $\mathcal{C}^\infty$ map $S: O \to \Gamma(R^K)$. By Theorem VI.6.10 we can choose S $\mathcal{C}^1$ close enough that S is properly transverse to $G(R^K)$ and $\pi: S^{-1}(G(R^K)) \to X$ is an $\mathfrak{M}^1$ isomorphism. Since $S^{-1}(G(R^K))$ is a semicomplete $\mathcal{C}^\infty$ submanifold of O, this gives a compatible $\mathcal{C}^\infty$ structure on X.

If X has two $\mathcal{C}^\infty$ structure of finite type with the same $\mathfrak{M}^1$ structure ($\mathfrak{M} \subset \mathcal{C}^r_u$) then the identity map is an $\mathfrak{M}^1$ isomorphism which by Theorem 3 can be $\mathcal{C}^{r+1}$ approximated by a $\mathcal{C}^\infty$ map. Since the set of $\mathfrak{M}^1$ isomorphisms is $\mathcal{C}^1$ open in $\mathfrak{M}^1(X,X)$ this $\mathcal{C}^\infty$ map can be chosen to be an $\mathfrak{M}^1$ and hence $\mathcal{C}^\infty$ isomorphism.

(b): By Proposition 1.10 we can regard π as a split $\mathfrak{M}$ subbundle of a trivial bundle $\epsilon_R K$, which clearly has a $\mathcal{C}^\infty$ structure. If $\pi \oplus \pi_1 \cong \epsilon_R K$ is an $\mathfrak{M}$ VB isomorphism we can, by Theorem 3 $\mathcal{C}$ approximate s_π and s_{π_1} by $\mathcal{C}^\infty$ maps to get a $\mathcal{C}^\infty$ splitting $\tilde{\pi} \oplus \tilde{\pi}_1 \cong \epsilon_R K$. By Lemma 4 we can choose the approximations so that the $\mathfrak{M}$ map $\tilde{\pi} \to \pi$ is a $\mathcal{C}$ VB isomorphism and hence an $\mathfrak{M}$ VB isomorphism. This gives π a compatible $\mathcal{C}^\infty$ bundle structure. Given two compatible $\mathcal{C}^\infty$ bundle structures, the identity is an $\mathfrak{M}$ section of $L(\pi;\pi)$ is can be $\mathcal{C}^r$ approximated by a $\mathcal{C}^\infty$ section. By Lemma III.7.10 applied to $Lis(\mathbb{E};\mathbb{E})$ we can choose this section to represent an $\mathfrak{M}$ VB isomorphism and hence a $\mathcal{C}^\infty$ VB isomorphism. Q.E.D.

From now on when we say <u>metric manifold of finite type</u> or <u>bundle of finite type</u> we will assume $\mathcal{C}^\infty$ and finite dimensionality.

6 **PROPOSITION**: Assume $(\mathfrak{M}_1,\mathfrak{M},\mathfrak{M}_2)$ is a standard triple. Let X_0 be a bounded $\mathfrak{M}$ manifold and X_1,X_2 semicomplete $\mathfrak{M}_1^2$ manifolds admitting $\mathfrak{M}_1^1$ exponentials. Let $F: X_1 \to X_2$ be an $\mathfrak{M}_1^1$ map and consider the $\mathfrak{M}_2^1$ map $F_*: \mathfrak{M}(X_0,X_1) \to \mathfrak{M}(X_0,X_2)$. If F is a globally split immersion (or submersion) then F_* is a globally split immersion (resp. submersion).

PROOF: Let $(\phi,1): F^*\tau_{X_2} \to \tau_{X_1}$ be an $\mathfrak{M}_1$ VB map which is a left (or right) inverse for $(F^*TF,1)$. $(\phi_*,1): \mathfrak{M}(X_0,F^*\tau_{X_2}) \to \mathfrak{M}(X_0,\tau_{X_1})$ is a left (resp. right) inverse for $((F^*TF)_*,1)$ which is identified with $((F_*)^*(TF)_*,1)$ and hence with $((F_*)^*T(F_*),1)$ by the pull back identification of page 141 (5) and the tangent bundle identification of Theorem IV.5.1. Q.E.D.

ADDENDUM: If $(\mathfrak{M}_1,\mathfrak{M},\mathfrak{M}_2) = (\mathcal{L}^{r+s},\mathcal{C}^r,\mathcal{L}^s)$ and X_0 is a semicompact $\mathcal{L}^r$ manifold then $F_*: \mathcal{C}_u^r(X_0,X_1) \to \mathcal{C}_u^r(X_0,X_2)$ is a globally split immersion (or submersion) by the same argument.

Note that if X_1 and X_2 are of finite type and F is an $\mathfrak{M}_1^1$ submersion (or immersion) then it is globally split by Proposition 1.10 (a). X_1 and X_2 admit $\mathcal{C}^\infty$ exponentials by Theorem 3.

7 **THEOREM:** Let X_0, X_1 and X_2 be manifolds of finite type and $F: X_1 \to X_2$ be a $\mathcal{C}^\infty$ submersion. Consider the continuous map $F_*: \mathcal{C}_u^r(X_0,X_1) \to \mathcal{C}_u^r(X_0,X_2)$ and $f \in \mathcal{C}^\infty(X_0,X_2)$. $\mathcal{C}^\infty(X_0,X_1) \cap (F_*)^{-1}(f)$ is dense in $(F_*)^{-1}(f)$.

PROOF: This theorem is a slight, but important, refinement of Theorem 3. As its proof is similar we will only sketch it. Let $j: X_1 \to E$ be a $\mathcal{C}^\infty$ embedding of X_1 into a Hilbert space and let $\epsilon_E^i: X_i \times E \to X_i$ $(i = 1,2)$ be the trivial bundles. We identify ϵ_E^1 with $F^*\epsilon_E^2$ and $(F \times j)^*\epsilon_E^2 * \epsilon_E^2$. $(F \times j,1): F \to \epsilon_E^2$ is a $\mathcal{C}^\infty$ embedding of the submersion F into the bundle ϵ_E^2. Thus, the vertical tangent bundle of F, τ_F (cf. page 222) is a $\mathcal{C}^\infty$ subbundle of ϵ_E^1 and its complementary normal bundle $\pi: \mathbb{E} \to X_1$ is a $\mathcal{C}^\infty$ subbundle of ϵ_E^1. Let $J_0: X_1 \times E \to X_2 \times E$ be defined by $J_0(x,v) = (Fx, j(x) + v)$ and restrict as in Theorem 3 to some full open neighborhood O of the zero section of π so that $J_0|O = J: O \to X_2 \times E$ is a $\mathcal{C}^\infty$ isomorphism onto an open set (use Prop. VI.6.9 (c)). Let $O_2 = J(O)$ and $\pi_1 = \pi \circ J^{-1}$ on O_2. We have the commutative diagram:

$$\begin{array}{ccc} X_1 & \underset{F\times j}{\overset{\pi_1}{\rightleftarrows}} & O_2 \subset X_2 \times E \\ & {\scriptstyle F}\searrow \quad \swarrow{\scriptstyle \epsilon_E^2} & \\ & X_2 & \end{array}$$

If $f \in \mathcal{C}^\infty(X_0,X_2)$ and $g \in \mathcal{C}_u^r(X_0,X_1)$ with $F \circ g = f$ then $(F \times j) \circ g$ can be regarded as a $\mathcal{C}_u^r$ section of $f^*\epsilon_E^2$ and so by Theorem 2 it can be $\mathcal{C}^r$

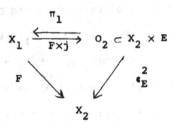

approximated by a $\mathcal{C}^\infty$ section which can be regarded as a map $f \times g_2: X_0 \to X_1 \times E$. It can be chosen so that $(f \times g_2)(X_0) \subset\subset O_2$. Then $g_1 = \pi_1 \circ (f \times g_2): X_0 \to X_1$ is a $\mathcal{C}^\infty$ map with $F \circ g_1 = f$ and since $\pi_1 \circ (F \times j) \circ g = g$, g_1 $\mathcal{C}^r$ approximates g. Q.E.D.

In practice this result is most commonly applied with $\pi: E \to X_2$ a bundle of finite type, $X_1 = G(E)$ and $F = g_\pi$. With $X_0 = X_2$ and $f = 1_{X_2}$ this result yields $\mathcal{C}^r$ approximation of $\mathcal{C}^r_u$ subbundles of π by $\mathcal{C}^\infty$ subbundles of π.

8 <u>PROPOSITION</u>: Assume $(\mathfrak{M}_1, \mathfrak{M}, \mathfrak{M}_2)$ is a standard triple. Let $F: X_1 \to X_2$ be a surjective $\mathfrak{M}$ submersion of semicompact manifolds of finite type. Assume X is a semicomplete $\mathfrak{M}_1^2$ manifold admitting $\mathfrak{M}_1^1$ exponentials.

(a) If $\pi: E \to X_2$ is an $\mathfrak{M}$ bundle then $F^*: \mathfrak{M}(\pi) \to \mathfrak{M}(F^*\pi)$ is a split injection.

(b) $F^*: \mathfrak{M}(X_2, X) \to \mathfrak{M}(X_1, X)$ is a metricly proper, globally split $\mathfrak{M}_2^1$ embedding.

(c) If $\pi: E \to X$ is a semicomplete $\mathfrak{M}_1$ bundle admitting fiber exponentials then $(F^*, F^*): \mathfrak{M}(X_2, \pi) \to \mathfrak{M}(X_1, \pi)$ is an $\mathfrak{M}_2$ VB map and $((F^*)^*F^*, 1): \mathfrak{M}(X_2, \pi) \to (F^*)^*\mathfrak{M}(X_1, \pi)$ is a split $\mathfrak{M}_2$ VB injection.

<u>PROOF</u>: We sketch the arguments. (a) is similar to (c). For (c) let $\mathcal{G}_1 = \{U_\alpha, h_\alpha\}$ and $\mathcal{G}_2 = \{V_\alpha, g_\alpha\}$ with $h_\alpha(U_\alpha) = g_\alpha(V_\alpha) \times O_\alpha$ with O_α open and $F_\alpha = p_1$. Choose $v_\alpha \in O_\alpha$ and define $i_\alpha = h_\alpha^{-1} \circ (1 \times c_{v_\alpha}) \circ g_\alpha: V_\alpha \to X_1$ for all α. Let $\{\varphi_\alpha\}$ be a partition of unity refining $\mathcal{G}_2$. For $(f_2, s) \in (F^*)^*\mathfrak{M}(X_1, E)$, i.e. $f_2: X_2 \to X$ and $s: X_1 \to E$ with $\pi \circ s = f_2 \circ F$, let $\Phi(f_2, s) = \Sigma_\alpha \varphi_\alpha (i_\alpha^* s)$. $(\Phi, 1): (F^*)^*\mathfrak{M}(X_1, \pi) \to \mathfrak{M}(X_2, \pi)$ is a VB map and it is a left inverse for $((F^*)^*F^*, 1)$. That $(\Phi, 1)$ is an $\mathfrak{M}_2$ VB map follows from uniform estimates for $(i_\alpha^*, i_\alpha^*): \mathfrak{M}(X_1, \pi) \to \mathfrak{M}(V_\alpha, \pi)$.

For (b), note that (c) implies that F^* is a globally split $\mathfrak{M}_2^1$ immersion since $(T(F^*), F^*)$ is identified with (F^*, F^*). That F^* is an embedding follows by choosing an exponential on X and noting that the following diagram commutes where $f_1 = f_2 \circ F$:

$$\begin{array}{ccc} V_{f_2} \cap (F^*)^{-1} V_{f_1} & \xrightarrow{\ H_{f_2}\ } & \mathfrak{M}(f_2{}^*\tau_X) \\ {\scriptstyle F^*}\big\downarrow & & \big\downarrow{\scriptstyle F^*} \\ V_{f_1} & \xrightarrow[\ H_{f_1}\ ]{} & \mathfrak{M}(f_1{}^*\tau_X) \end{array}$$

Thus, $F^*(V_{f_2} \cap (F^*)^{-1} V_{f_1}) = H_{f_1}^{-1}(F^*\mathfrak{M}(f_2{}^*\tau_X))$ by Lemma VI.4.9. Result by Theorem VI.3.9. Q.E.D.

9 __PROPOSITION__: Assume $(\mathfrak{M}_1, \mathfrak{M}, \mathfrak{M}_2)$ is a standard triple. Let X_1, X_2 and X be manifolds of finite type with X_1 and X_2 semicompact. Let $F: X_1 \to X_2$ be an $\mathfrak{M}$ embedding.

(a) If $\pi: E \to X_2$ is an $\mathfrak{M}$ bundle then $F^*: \mathfrak{M}(\pi) \to \mathfrak{M}(F^*\pi)$ is a split surjection.

(b) $F^*: \mathfrak{M}(X_2, X) \to \mathfrak{M}(X_1, X)$ is a $\mathcal{C}^\infty$ submersion (not necessarily surjective).

(c) If $\pi: E \to X$ is a semicomplete $\mathfrak{M}_1$ bundle admitting fiber exponentials then $F^*: \mathfrak{M}(X_2, \pi) \to \mathfrak{M}(X_1, \pi)$ is an $\mathfrak{M}_2$ VB map and $((F^*)^*F^*, 1): \mathfrak{M}(X_2, \pi) \to (F^*)^*\mathfrak{M}(X_1, \pi)$ is a split $\mathfrak{M}_2$ VB surjection. If π is $\mathcal{C}^\infty$ and the splitting can be chosen to be $\mathcal{C}^\infty$.

__PROOF__: (a) is like (c) and (c) implies (b). To prove (c), let $e: D \to V$ be a $\mathcal{C}^\infty$ exponential for X and $T: (\pi \times 1)_V \to (1 \times \pi)_V$ be an $\mathfrak{M}_1$ VB isomorphism. By the Tubular Neighborhood Theorem we can choose U open in X_2 and $R: U \to X_1$ a $\mathcal{C}^\infty$ map with $F(X_1) \subset\subset U$ and $R \circ F = 1_{X_1}$. Let i_U be the inclusion of U into X_2. Choose $\mathcal{C}^\infty$ functions $\varphi_V: X \times X \to [0,1]$ and $\varphi_U: X_2 \to [0,1]$ with $\varphi_V | \Delta_X$ and $\varphi_U | F(X_1) \equiv 1$ and $\varphi_V | X \times X - V$ and $\varphi_U | X_2 - U \equiv 0$. Define $(\tilde{T}, 1): \pi \times 1 \to 1 \times \pi$ by $\tilde{T}(v, y_2) = \varphi_V(y_1, y_2) T(v, y_2)$ if $(\pi v, y_2) = (y_1, y_2) \in V$ and by $\tilde{T}(v, y_2) = 0$ otherwise. For

$(f_2,s) \in (F^*)^*\mathfrak{M}(X_1,E)$, i.e. $f_2\colon X_2 \to X$ and $s\colon X_1 \to E$ with $\pi\circ s = f_1 = f_2\circ F$ we use the diagram:

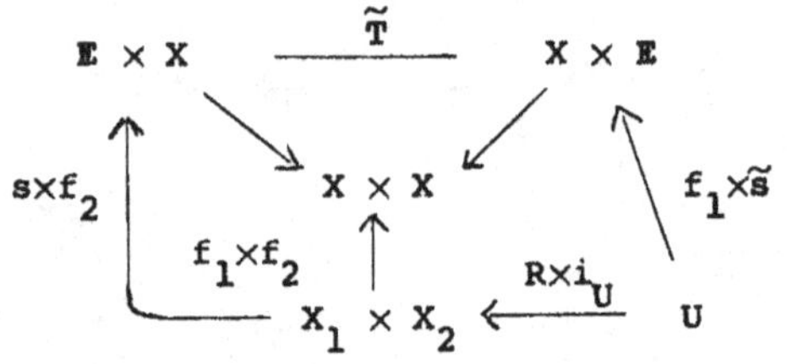

to define $\tilde{s}\colon U \to E$ with $\pi\circ\tilde{s} = f_2|U$, i.e. $\tilde{s} = p_2\circ\tilde{T}\circ(s\times f_2)\circ(R\times i_U)$. $\Phi(f_2,s) = \varphi_U\tilde{s}\colon X_2 \to E$ defines a right inverse, $(\Phi,1)\colon (F^*)^*\mathfrak{M}(X_1,E) \to \mathfrak{M}(X_2,E)$, for $((F^*)^*F^*,1)$. By looking at the atlases on $\mathfrak{M}(X_1,\pi)$ and $\mathfrak{M}(X_2,\pi)$ defined by e and T it is easy to show that $(\Phi,1)$ is an $\mathfrak{M}_2$ VB map. Q.E.D.

We close by combining the smoothing results of this section with those of Section VII.3.

10 <u>THEOREM</u>: Assume that $\mathfrak{M}$ is a standard function space type with $\mathfrak{M} \subset \mathcal{L}^r$ $(r \geq 0)$. Let X_0 be a manifold of finite type and X be a semicomplete $\mathfrak{M}^2$ manifold.

(a) Let $f\colon X_0 \to X$ be a metricly proper $\mathfrak{M}^1$ immersion. The following conditions are equivalent:

(1) $G(f)\colon X_0 \to G(TX)$ is an $\mathfrak{M}^1$ map.

(2) There exists $g\colon X_0 \to X_0$ an $\mathfrak{M}^1$ isomorphism such that $f\circ g\colon X_0 \to X$ is a metricly proper $\mathfrak{M}^2$ immersion.

(3) There exist $g\colon X_0 \to X_0$ $\mathfrak{M}^1$ isomorphisms arbitrarily $\mathcal{C}^{r+1}$ close to the identity such that $f\circ g\colon X_0 \to X$ is a metricly proper $\mathfrak{M}^2$ immersion.

(b) Let $f\colon X_0 \to X$ be a $\mathcal{C}_u^{r+1}$ (or $\mathcal{L}^{r+1}$) leaf immersion. The following conditions are equivalent:

(1) With d and d_G admissible metrics on X and $G(TX)$ and

$d_f = f^*d$, $G(f): X_0 \to G(TX)$ is a $\mathcal{C}_u^{r+1}(d_f, d_G)$ map (resp. an $\mathcal{L}^{r+1}(d_f, d_G)$ map).

(2) There exists $g: X_0 \to X_0$ a $\mathcal{C}^{r+1}$ isomorphism (resp. an $\mathcal{L}^{r+1}$ isomorphism) such that $f \circ g: X_0 \to X$ is a $\mathcal{C}_u^{r+2}$ leaf immersion (resp. an $\mathcal{L}^{r+2}$ leaf immersion).

(3) There exist $g: X_0 \to X_0$ $\mathcal{C}^{r+1}$ isomorphisms arbitrarily $\mathcal{C}^{r+1}$ close to the identity such that $f \circ g: X_0 \to X$ is a $\mathcal{C}_u^{r+2}$ leaf immersion (resp. an $\mathcal{L}^{r+2}$ leaf immersion).

PROOF: In each case (3) implies (2) is obvious. That (2) implies (1) follows from the fact that $G(f \circ g) = G(f) \circ g$ and hence $G(f) = G(f \circ g) \circ g^{-1}$ when g is a C^1 automorphism. In (b) note that $g: (X_0, d_{f \circ g}, (f \circ g)^*\rho) \to (X_0, d_f, f^*\rho)$ and its inverse are isometries of regular pseudometric spaces. Now if (1) holds in (a), then by the Smoothing Lemma VII.3.4, f is an $\mathfrak{M}^2$ metricly proper immersion with respect to a unique semicomplete $\mathfrak{M}^2$ structure on X_0 which agrees with the original $\mathfrak{M}^1$ structure and consequently is of finite type. By Theorem 5 (a) there is a compatible $\mathcal{C}^\infty$ structure of finite type on X_0 with respect to which f is an $\mathfrak{M}^2$ immersion, and there exist maps $g: X_0 \to X_0$ which are $\mathcal{C}^\infty$ isomorphisms from the original $\mathcal{C}^\infty$ structures to the new one which are $\mathcal{C}^{r+1}$ close to the identity. Hence, $f \circ g$ is an $\mathfrak{M}^2$ immersion of the original structure. Since the two $\mathcal{C}^\infty$ structures agree up to $\mathfrak{M}^1$, g is an $\mathfrak{M}^1$ automorphism of the original structure. Thus, (1) implies (3) in (a). The proof in (b) is the same with the additional remark that since f is a $\mathcal{C}_u^{r+2}$ or $\mathcal{L}^{r+2}$ leaf immersion from the new structure and g is a $\mathcal{C}^\infty$ isomorphism between structures, $f \circ g$ is a $\mathcal{C}_u^{r+2}$ or $\mathcal{L}^{r+2}$ leaf immersion because $G^{r+2}(f \circ g) = G^{r+2}(f) \circ g$ and so $g^*d_f = d_{f \circ g}$ and $g^*d_{G^{r+2}(f)} = d_{G^{r+2}(f \circ g)}$. Q.E.D.

3. $\mathcal{C}_u$ and $\mathcal{L}$ Finslers: For applications in the next section we need to understand uniform continuity of Finslers. To this end note that if E is a B-space with norm $\| \ \|$ then the space of norms $(\mathcal{N}(E), \mathrm{dist}, \rho_{\| \ \|})$

(cf. pages 55-56) is a regular metric space. In fact, by Proposition III.3.2(a) we have, since $\exp[1/2] < 2$:

$$(3.1) \qquad (\rho_{\|\ \|})^{\text{dist}} < 2\, \rho_{\|\ \|}.$$

This implies that if $\{E_\alpha\}$ is a family of B-spaces then $\{\mathcal{R}(E_\alpha)\}$ is a regular family of regular metric spaces.

Let X be a regular metric manifold with apm d and $\pi: \mathbb{E} \to X$ be a $\mathcal{C}_u(d)$ (or $\mathcal{L}(d)$) bundle. A Finsler $\|\ \|$ on π is called a $\mathcal{C}_u(d)$ (resp. $\mathcal{L}(d)$) Finsler for π if for $(\mathfrak{D},\mathcal{G}) = \{U_\alpha, h_\alpha, \varphi_\alpha\}$ an admissible $\mathcal{C}_u(d)$ (resp. $\mathcal{L}(d)$) multiatlas on π and ρ a bound on X the family of principal parts for the Finsler $\{n_\alpha: U_\alpha \to \mathcal{R}(F_\alpha)\}$ defined by $(\mathfrak{D},\mathcal{G})$ (cf. page 58) is a $\mathcal{C}_u$ (resp. $\mathcal{L}$) family of maps where $\{U_\alpha\}$ is the regular family obtained by restriction from (X,d,ρ) and $\{\mathcal{R}(F_\alpha)\}$ is the regular family defined as above. Since $\rho|U_\alpha > (n_\alpha)^*\rho_{\|\ \|_\alpha}$ uniformly in α, $\|\ \|$ is an admissible Finsler on π. To prove independence of the choice of admissible multiatlas note first that we can intersect covers and apply Proposition III.6.7 and so assume $(\mathfrak{D}_1,\mathcal{G}_1) = \{U_\alpha, h^1_\alpha, \varphi^1_\alpha\}$. Then the principal part n^1_α of $\|\ \|$ with respect to $(\mathfrak{D}_1,\mathcal{G}_1)$ is the composition:

$$U_\alpha \xrightarrow{I_\alpha \times n_\alpha} \operatorname{Lis}(F^1_\alpha;F_\alpha) \times \mathcal{R}(F_\alpha) \longrightarrow \mathcal{R}(F^1_\alpha)$$

where I_α is the principal part of the identity map: $(\mathfrak{D}_1,\mathcal{G}_1) \to (\mathfrak{D},\mathcal{G})$ and the second map is the pullback map of Proposition III.3.2(d). Since $\{I_\alpha\}$ and $\{n_\alpha\}$ are $\mathcal{C}_u$ (resp. $\mathcal{L}$) families the composition family $\{n^1_\alpha\}$ is $\mathcal{C}_u$ (resp. $\mathcal{L}$) by the estimates of the proof of Proposition III.3.2(d).

1 PROPOSITION: (a) Let X be a regular metric manifold with apm d and F be a B-space. If X admits a $\mathcal{C}_u(d)$ (or $\mathcal{L}(d)$) metric structure then ϵ_F does and the trivial Finsler is $\mathcal{C}_u(d)$ (resp. $\mathcal{L}(d)$).

(b) If π is a $\mathcal{C}_u(d)$ subbundle of a $\mathcal{C}_u(d)$ bundle π_0 and $\|\ \|$ is a $\mathcal{C}_u(d)$ Finsler on π_0 then $\|\ \|$ restricts to a $\mathcal{C}_u(d)$ Finsler on π.

Similarly, for $\mathcal{L}$(d) Finslers.

(c) If π_1 and π_2 are $\mathcal{C}_u$(d) bundles over X with $\mathcal{C}_u$(d) Finslers then the associated product Finsler on $\pi_1 \oplus \pi_2$ and the associated operator Finsler on $L^p_\epsilon(\pi_1;\pi_2)$ are $\mathcal{C}_u$(d) Finslers. Similarly for $\mathcal{L}$(d) Finslers.

(d) Let $\pi: E \to X$ be a $\mathcal{C}_u$(d) bundle, X_1 be a $\mathcal{C}_u(d_1)$ manifold and $f: X_1 \to X$ be a $\mathcal{C}_u(d_1,d)$ map. If $\| \; \|$ is a $\mathcal{C}_u$(d) Finsler on π then $\Phi_f^*\| \; \|$ is a $\mathcal{C}_u(d_1)$ Finsler on $f^*\pi$. Similarly for $\mathcal{L}$(d) Finslers.

<u>PROOF</u>: (a) is obvious. (b): If $(\mathfrak{I},\mathfrak{a})$ is an admissible multiatlas inducing the subbundle then the principal part n_α is the composition $U_\alpha \xrightarrow{n^0_\alpha} \mathcal{P}(F^0_\alpha) \xrightarrow{i^*_\alpha} \mathcal{P}(F_\alpha)$ where $i_\alpha: F_\alpha \to F^0_\alpha$ is the inclusion. Note that $(i_\alpha^*)^*\text{dist} \le \text{dist}$ and $(i_\alpha^*)^*\rho_{\| \; \|_\alpha} \le \rho_{\| \; \|_\alpha}$. (c) follows easily from the proof of Proposition III.3.2 (b) and (c). For (d) choose atlases so that f is index preserving and note that $n^1_\alpha = n_\beta \circ f_\alpha$ where $\beta = \beta(\alpha)$ and n_β, n^1_α are principal parts for $\| \; \|$ and $\Phi_f^*\| \; \|$, respectively. Q.E.D.

The most important special case occurs when X is a semicomplete metric manifold and d is an admissible metric. In this case we will use the term $\mathcal{C}_u$ or $\mathcal{L}$ Finsler.

2 <u>PROPOSITION</u>: Let $\pi: E \to X$ be a semicomplete $\mathcal{L}$ bundle.

(a) If X is a manifold of finite type then π admits $\mathcal{L}$ Finslers.

(b) If $\| \; \|$ is a $\mathcal{C}_u$ (or $\mathcal{L}$) Finsler on π then $\| \; \|: E \to R$ is a $\mathcal{C}_u$ (resp. $\mathcal{L}$) map.

<u>PROOF</u>: (a): Let $(\mathfrak{I},\mathfrak{a})$ be an s admissible, $\mathcal{L}$ip atlas for π with $\mathfrak{a}$ a square atlas satisfying $n_{\mathfrak{a}} < \infty$. Let $V \in \mathfrak{U}_X$ with $\#\{\alpha: U_\alpha \cap V[x] \ne \emptyset\} \le n_{\mathfrak{a}}$ for all $x \in X$. With $\mathcal{P}$ the partition of unity associated with $\mathfrak{a}$, I claim that the Finsler $\| \; \|^{\mathcal{P}}$ defined by $\mathcal{P}$ and $(\mathfrak{I},\mathfrak{a})$ is an $\mathcal{L}$ Finsler.

In fact, by inequality III.(3.3) the family of principal parts $\{n_\alpha|U_\alpha \cap V[x]\}$ is $\mathcal{L}$. So $\{n_\alpha\}$ is $\mathcal{L}$ by Proposition III.6.7. (b): Let $(\mathfrak{F},\mathfrak{G})$ be an s admissible $\mathcal{L}$ip atlas for π. $\| \|\circ\varphi_\alpha^{-1}: h_\alpha(U_\alpha) \times F_\alpha \to R$ is given by the composition:

$$h_\alpha(U_\alpha) \times F_\beta \xrightarrow{n_\alpha \times 1} \eta(F_\alpha) \times F_\alpha \longrightarrow R$$

where the second map is the evaluation map of Proposition III.3.2(e). The result follows from III.(3.2). Q.E.D.

In general, if π is an $\mathcal{L}$ subbundle of a trivial bundle then π admits $\mathcal{L}$ Finslers. Now let $(\mathfrak{M}_1,\mathfrak{M},\mathfrak{M}_2)$ be a standard triple, π_0 be a bounded $\mathfrak{M}$ bundle and π be a semicomplete $\mathfrak{M}_1$ bundle admitting fiber exponentials. If π is a globally split $\mathfrak{M}_1$ subbundle of a trivial bundle ϵ_F then $\mathfrak{M}\mathcal{L}(\pi_0;\pi)$ is a globally split $\mathfrak{M}_2$ subbundle of $\epsilon_{\mathfrak{M}(L(\pi_0;\epsilon_F^0))}$ (cf. page 141). Consequently, $\mathfrak{M}\mathcal{L}(\pi_0;\pi)$ as well as π admit $\mathcal{L}$ Finslers. In particular, if X admits an $\mathcal{L}^1$ immersion into a B-space F then τ_X admits $\mathcal{L}$ Finslers. Let X_0 be a bounded $\mathfrak{M}$ manifold and X be a semicomplete $\mathfrak{M}_1^2$ manifold admitting $\mathfrak{M}_1^1$ exponentials. If $f: X \to F$ is a globally split $\mathfrak{M}_1^1$ immersion into a B-space then $f_*: \mathfrak{M}(X_0,X) \to \mathfrak{M}(X_0,F)$ is a globally split $\mathfrak{M}_2^1$ immersion into a B-space. Thus, $\tau_{\mathfrak{M}(X_0,X)}$ as well as τ_X admit $\mathcal{L}$ Finslers.

4. <u>Semicompact Structures</u>: Every smooth manifold or bundle admits semicompact structures as we now show.

1 <u>LEMMA</u>: Let $\mathfrak{M} = \mathcal{C}$, $\mathcal{L}$ or $\mathfrak{B}^1$ and let $\pi: E \to X$ be a semicomplete $\mathfrak{M}^r$ bundle $(1 \leq r < \infty)$. If X has s admissible atlases $\mathfrak{G}$ with $\rho_{\mathfrak{G}}^{\mathcal{C}^1}$ bounded then X has s admissible atlases $\mathfrak{G}$ with $\rho_{\mathfrak{G}}^{\mathfrak{M}^r}$ and $1/\lambda_{\mathfrak{G}}$ bounded. If, in addition, π has s admissible atlases $(\mathfrak{D},\mathfrak{G})$ with $\rho_{(\mathfrak{D},\mathfrak{G})}^{\mathcal{C}}$ bounded then π has atlases $(\mathfrak{D},\mathfrak{G})$ with $\rho_{(\mathfrak{D},\mathfrak{G})}^{\mathfrak{M}^r}$ and $1/\lambda_{\mathfrak{G}}$ bounded. If X is of finite type then $\mathfrak{G}$ can

be chosen to be an R^k atlas of finite type.

<u>PROOF</u>: Begin with $\mathfrak{a}_1 = \{U_\alpha, h_\alpha\}$ s admissible and with $\rho^{\mathcal{C}^1}_{\mathfrak{a}_1}$ bounded by K. Choose ρ an admissible bound with $\rho \geq \max(1/\delta_{\mathfrak{a}_1}(d), \rho^{\mathfrak{M}^r}_{\mathfrak{a}_1})$ where $d = d_{\mathfrak{a}_1}$. Let $\epsilon(x) = 1/2\rho^d(x)$. There exists $\alpha = \alpha(x)$ such that $B^d(x,\epsilon(x)) \subset U_\alpha$ and so $B(h_\alpha x, \epsilon(x)) \subset h_\alpha(B^d(x,\epsilon(x)))$. Let $U_x = h_\alpha^{-1}(B^d(h_\alpha x, \epsilon(x)))$ and $h_x = h_\alpha - c_{h_\alpha x}$. $\mathfrak{a}_2 = \{U_x, h_x\}$ is s admissible. In fact, $K \geq \rho^{\mathcal{C}^1}_{\mathfrak{a}_1} \geq \rho^{\mathcal{C}^1}_{\mathfrak{a}_2}$, $\rho \geq \rho^{\mathfrak{M}^r}_{\mathfrak{a}_1} \geq \rho^{\mathfrak{M}^r}_{\mathfrak{a}_2}$ and $\lambda_{\mathfrak{a}_2} \geq \epsilon$. Note that if $y \in U_x$ then $2\rho^d(x) \geq \rho^d(y) \geq 2^{-1}\rho^d(x)$ and so $4\rho^d(y) \geq \epsilon(x)^{-1} \geq \rho^d(y)$. In particular, if $U_x \cap U_y \neq \emptyset$ then $16 \geq \epsilon(y)/\epsilon(x) \geq 16^{-1}$. Let $\mathfrak{a} = \{U_x, g_x\}$ with $g_x = \epsilon(x)^{-1}h_x$. Clearly, $\lambda_{\mathfrak{a}} = 1$ and $\rho^{\mathcal{C}}_{\mathfrak{a}} \leq 1$. By Corollary III.1.3 $\mathfrak{a}$ is s admissible. If $U_x \cap U_y \neq \emptyset$ then $\|D^p g_{xy}\| = \epsilon(y)^p \epsilon(x)^{-1}\|D^p h_{xy}\|$ and $\|D^p g_{xy}\|_L = \epsilon(y)^{p+1}\epsilon(x)\|D^p h_{xy}\|_L$ for $p = 0,1,\ldots$ Since $1/\epsilon \geq \rho^{\mathfrak{M}^1}_{\mathfrak{a}_2}$ it follows that $16K \geq \rho^{\mathfrak{M}^r}_{\mathfrak{a}}$.

In the bundle case, begin with $(\mathfrak{D}_1, \mathfrak{a}_1)$ such that $\rho^{\mathcal{C}}_{(\mathfrak{D}_1,\mathfrak{a}_1)}$ and $\rho^{\mathcal{C}^1}_{(\mathfrak{D}_1,\mathfrak{a}_1)}$ are bounded by K and choose $\rho \geq \rho^{\mathfrak{M}^r}_{(\mathfrak{D}_1,\mathfrak{a}_1)}$, too. At each step use the transfer of atlas construction. If $(\mathfrak{D}_2, \mathfrak{a}_2) = \{U_x, h_x, \varphi_x\}$ and $(\mathfrak{D}, \mathfrak{a}) = \{U_x, g_x, \psi_x\}$ then $\|D^p \psi_{xy}\|_0 = \epsilon(y)^p \|D^p \varphi_{xy}\|_0$ and $\|D^p \psi_{xy}\|_L = \epsilon(y)^{p+1}\|D^p \varphi_{xy}\|_L$ for $p = 0,1,\ldots$ So $16K \geq \rho^{\mathfrak{M}^r}_{(\mathfrak{D},\mathfrak{a})}$.

In the case of a manifold of finite type we can begin with $\mathfrak{a}_1$ an R^k atlas of finite type by refining and renorming. In the process we may multiply $\rho^{\mathfrak{M}^r}$ and $\rho^{\mathcal{C}^1}$ be a factor $\leq O^*(2^k)$. Apply the proof of Theorem 1.4 to get a square atlas $\mathfrak{a}_2 = \{U^3_\gamma, h_\gamma\}$ with $K_1 \min\{1/\epsilon_\gamma : x \in U^3_\gamma\} \geq \max(\rho^{\mathfrak{M}^r}_{\mathfrak{a}_2}(x), \max\{1/\epsilon_\gamma : x \in U^3_\gamma\}$ for some constant K_1. If $\mathfrak{a} = \{U^3_\gamma, \epsilon^{-1}_\gamma h_\gamma\}$ then $\lambda_{\mathfrak{a}} = 1$ and $\max(3, K_1^2 K) \geq \rho^{\mathfrak{M}^r}_{\mathfrak{a}}$. Q.E.D

2 <u>LEMMA</u>: Let $\mathfrak{M} \subset \mathcal{L}$ and $\pi: E \to X$ be an $\mathfrak{M}$ bundle. π has admissible atlases $(\mathfrak{D}, \mathfrak{a})$ with $\rho^{\mathcal{C}}_{(\mathfrak{D},\mathfrak{a})}$ bounded. If π is semicomplete and admits

$\mathcal{C}_u$ Finslers then π has s admissible atlases $(\mathfrak{D},\mathfrak{G})$ with $\rho^{\mathcal{C}}_{(\mathfrak{D},\mathfrak{G})}$ bounded. Let X be an $\mathfrak{M}^1$ manifold. X has atlases with $\rho^{\mathcal{C}^1}_{\mathfrak{G}}$ bounded. If X is semicomplete and τ_X admits $\mathcal{C}_u$ Finslers then X has s admissible atlases with $\rho^{\mathcal{C}^1}_{\mathfrak{G}}$ bounded.

__PROOF:__ We can always refine and translate charts in an atlas to get $\rho^{\mathcal{C}}_{\mathfrak{G}}$ bounded. In the process $\rho^{\mathfrak{M}}_{\mathfrak{G}}$ is increased, at most, by a factor of 2. Furthermore, in the semicomplete case we can so transform an s admissible atlas without losing s admissibility. Let $(\mathfrak{D},\mathfrak{G}) = \{U_\alpha, h_\alpha, \varphi_\alpha\}$ be an admissible atlas and $\| \ \|$ a continuous admissible Finsler for π with $\mathfrak{G}$ star bounded and $\rho^{\mathcal{C}}_{\mathfrak{G}}$ bounded. In the semicomplete case choose $(\mathfrak{D},\mathfrak{G})$ s admissible and $\| \ \|$ $\mathcal{C}_u$. Using continuity of $\| \ \|$ we can refine $(\mathfrak{D},\mathfrak{G})$ so that $\mathrm{dist}(n_\alpha(x), n_\alpha(y)) < 1$ whenever $x,y \in U_\alpha$. Using uniform continuity we can do this without losing s admissibility in the semicomplete case. Now for each α choose $x = x(\alpha) \in U_\alpha$ and renorm F_α so that $\varphi_{\alpha x}: E_x \to F_\alpha$ is an isometry. Since $\mathfrak{G}$ is star bounded this adjustment yields an admissible atlas. With this final atlas, equation III(3.1) implies $\theta(\varphi_{\alpha\beta}(x)) \le \exp(\mathrm{dist}[n_\alpha(x), n_\alpha(x(\alpha))] + \mathrm{dist}[n_\beta(x), n_\beta(x(\beta))]) \le e^2$. So $\rho^{\mathcal{C}}_{(\mathfrak{D},\mathfrak{G})}$ is bounded. In the manifold case, we begin with $\mathfrak{G}$ on X and $\| \ \|$ on τ_X and refine so that with respect to $(T\mathfrak{G},\mathfrak{G})$ $\mathrm{dist}(n_\alpha(x), n_\alpha(y)) < 1$ whenever $x,y \in U_\alpha$. Renorm E_α so that $T_{x(\alpha)}h_\alpha: T_xX \to E_\alpha$ is an isometry. Refine and translate to get $\rho^{\mathcal{C}}_{\mathfrak{G}}$ bounded. As above $\rho^{\mathcal{C}}_{(T\mathfrak{G},\mathfrak{G})} = \rho^{\mathcal{C}^1}_{\mathfrak{G}}$ is bounded. Q.E.D.

3 __THEOREM:__ Let $\mathfrak{M} = \mathcal{C}$, $\mathcal{L}$ or $\mathfrak{B}^1$ and $\infty > r \ge 2$. Any semicomplete $\mathfrak{M}^r$ structure on a manifold X with respect to which τ_X admits $\mathcal{C}_u$ Finslers is a refinement of a semicompact $\mathfrak{M}^r$ structure. Any semicomplete $\mathfrak{M}^r$ structure on a bundle $\pi: E \to X$ with respect to which π and τ_X admit $\mathcal{C}_u$ Finslers is a refinement of a semicompact $\mathfrak{M}^r$ structure. In either case if the structure on X is that of an $\mathfrak{M}^r$ manifold of finite type then the $\mathcal{C}_u$

Finslers always exist and the semicompact structures can be chosen of finite type.

PROOF: Immediate from Lemmas 1 and 2. If X is of finite type $\mathcal{C}_u$ Finslers exist on τ_X and π by Proposition 3.2(a). Q.E.D.

4 THEOREM: (a) Let $\infty > r \geq 2$. Any paracompact C^r manifold admits semicompact $\mathcal{C}^r$ structures. Any C^r bundle over a paracompact C^r manifold admits semicompact $\mathcal{C}^r$ structures.

(b) If a C^∞ manifold admits any $\mathcal{C}^\infty$ adapted atlases then it admits semicompact $\mathcal{C}^\infty$ structures. If a C^∞ bundle admits any $\mathcal{C}^\infty$ adapted atlases then it admits semicompact $\mathcal{C}^\infty$ structures.

(c) Any finite dimensional, paracompact C^∞ manifold admits semicompact $\mathcal{C}^\infty$ structures of finite type. Any C^∞ bundle over a finite dimensional, paracompact C^∞ manifold admits semicompact $\mathcal{C}^\infty$ structures with the base space structure of finite type.

PROOF: We prove the space results as the bundle proofs are similar.

(a): Begin with a $\mathcal{C}^r$ atlas $\mathfrak{A}$ (see page 47) on X. $(\mathfrak{A},\rho_{\mathfrak{A}})$ generates a $\mathcal{C}^r$ structure on X within which we can choose by Lemma 2 a $\mathcal{C}^r$ atlas $\mathfrak{A}_1$ with $\rho^{\mathcal{C}^1}_{\mathfrak{A}_1}$ bounded. $\max(\rho_{\mathfrak{A}\cup\mathfrak{A}_1},1/\lambda_{\mathfrak{A}_1})^{d_{\mathfrak{A}_1}}$ generates a semicomplete refinement to which Lemma 1 applies. In it we can choose $\mathfrak{A}_2$ a $\mathcal{C}^r$ atlas with $\rho^{\mathcal{C}^r}_{\mathfrak{A}_2}$ and $1/\lambda_{\mathfrak{A}_2}$ bounded. $\mathfrak{A}_2$ generates a semicompact $\mathcal{C}^r$ structure.

(b): If $(\mathfrak{A},\rho)$ is a $\mathcal{C}^\infty$ adapted atlas on X then as in Lemma 2 we can adjust $\mathfrak{A}$ to get $\rho^{\mathcal{C}^1}_{\mathfrak{A}}$ bounded. Then let $d = d_{\mathfrak{A}}$ and define $\rho_1(x) = \max(\rho^d(x),\rho^{\mathcal{C}^n}_{\mathfrak{A}}(x))$ if $n \leq \rho^d(x) < n+1$ and let $\rho_2 = \rho_1^d$. Then $\rho_2 \geq \rho^{\mathcal{C}^n}_{\mathfrak{A}}$ on $\{\rho^d \geq n\}$ and so since $\rho > \rho^{\mathcal{C}^n}_{\mathfrak{A}}$ there exist constants $\{K_n\}$ such that $K_n\rho_2 \geq \rho^{\mathcal{C}^n}_{\mathfrak{A}}$ for $n = 2,3,\ldots$. Applying the construction of Lemma 1 to $(\mathfrak{A},\max(\rho_2,1/\lambda_{\mathfrak{A}}))$ we get an atlas $\mathfrak{A}_1$ such that $1/\lambda_{\mathfrak{A}_1}$ and $\rho^{\mathcal{C}^n}_{\mathfrak{A}_1}$ are bounded for $n = 1,2,\ldots$ $\mathfrak{A}_1$ generates a semicompact $\mathcal{C}^\infty$ structure.

(c): Begin with a locally finite C^∞ atlas and shrink as on page 47 to obtain an atlas $\mathfrak{a}$ which is $\mathcal{C}^r$ for each r. Adjust the atlas as in Lemma 2 to get $\rho_{\mathfrak{a}}^{\mathcal{C}^1}$ bounded. Now refine to get $N_{\mathfrak{a}} \leq k+1$ where k is the dimension of X. This is possible by a classic result in dimension theory [28; Thm V.1]. By considering the components separately we can assume X is connected and hence σ compact [13; pp. 214-215]. Let $\{A_n\}$ be an sequence of compact subsets with union X such that $A_n \subset$ Interior A_{n+1}. Define $\rho_1(x) = \rho_{\mathfrak{a}}^{\mathcal{C}^n}(x)$ for $x \in A_{n+1} - A_n$ and let $\rho_2 = \rho_1^{d_{\mathfrak{a}}}$. Complete the proof as in (b). Q.E.D.

5. <u>Compact Sets and Compact Manifolds:</u> Since any open cover of a compact space has a finite subcover and a nonzero Lebesgue number, it follows that a compact C^∞ manifold X has a unique $\mathcal{C}^\infty$ structure. This structure is semicompact and of finite type. Any metric d with topology that of X is uniformly equivalent to an admissible metric. If $\pi: E \to X$ is a C^∞ bundle on X then π has a unique $\mathcal{C}^\infty$ structure. This structure makes π a semicompact $\mathcal{C}^\infty$ bundle and any continuous Finsler $\| \ \|$ on π yielding the correct topology on the fibers of π is an admissible Finsler for π. Any C^r map on X, section of π or VB map on π is $\mathcal{C}_u^r$.

1 <u>LEMMA</u>: Let X be a semicomplete metric manifold.

(a) Let $\mathfrak{a}$ be an s admissible atlas and $d = d_{\mathfrak{a}}$. The closed ball $B^d[x,r]$ is of finite $\mathfrak{U}_X$ diameter ($\infty > r > 0$) iff it is a bounded subset of X.

(b) Assume X is finite dimensional. A closed subset A of X is of finite $\mathfrak{U}_X$ diameter iff it is compact and lies in a component of X.

<u>PROOF</u>: (a): Any subset of finite $\mathfrak{U}_X$ diameter is bounded. Conversely, if $B^d[x,r]$ is bounded then $B = B^d(x;r+\delta)$ is bounded for some $\delta > 0$ and on B the uniformity $\mathfrak{U}_X$ is generated by sets of the form $V = \{d < \epsilon\}$ with $\epsilon > 0$.

Let N be an integer greater than $(r+\delta)/\epsilon$. If y is in $B^d[x,r]$ there is an $\mathfrak{a}$ chain connecting x and y with length less than $r + \delta$. So the associated p.l. path lies on B and we can choose points $x = x_1,\dots,x_N = y$ on the path with $d(x_i,x_{i+1}) < \epsilon$. Hence, $y \in V^N[x]$.
(b): A compact set lying in $\cup_n V^n[x]$ lies in some $V^N[x]$. Hence, a compact set in a component of X has $\mathfrak{U}_X$ finite diameter. For the converse, we can choose $V \in \mathfrak{U}_X$ with $V[x]$ having compact closure for all $x \in X$. Then choose W a closed element of $\mathfrak{U}_X$ with $W = W^{-1}$ and $W \cdot W \subset V$. If A is compact then $W[A]$ has compact closure because if $W[x_1],\dots,W[x_n]$ $(x_1,\dots,x_n \in A)$ cover A then $V[x_1] \cup \dots \cup V[x_n] \supset W[A]$. So by induction $W^N[x]$ has compact closure for all N and x. Thus, if A is closed and has finite $\mathfrak{U}_X$ diameter, it is compact. Q.E.D.

2 <u>COROLLARY</u>: Let X be a connected, semicompact finite dimensional metric manifold with s admissible atlas $\mathfrak{a}$. A closed subset A of X is compact iff it is of finite $\mathfrak{U}_X$ diameter iff it is of finite $d_{\mathfrak{a}}$ diameter.

<u>PROOF</u>: The first equivalence from Lemma 1 (b) and the second from Lemma 1 (a). Q.E.D.

3 <u>COROLLARY</u>: Let X be a separable, semicomplete metric manifold of finite type. Let R be given the semicompact $\mathcal{C}^\infty$ group metric structure. There exist $\mathcal{C}^\infty$ functions $\varphi: X \to R$ which are topologically proper, i.e. $\varphi^{-1}(K)$ is compact in X if K is compact in R.

<u>PROOF</u>: First note that we need only construct an $\mathcal{L}$ map $\varphi_1: X \to R$ which is topologically proper. For then by Theorem 2.3 there exists $\varphi: X \to R$ a $\mathcal{C}^\infty$ map with $|\varphi_1(x) - \varphi(x)| < 1$ for all x in X. $\{|\varphi| \leq N\}$ is then a closed subset of $\{|\varphi_1| \leq N + 1\}$ which is compact. So φ is topologically proper. To find φ_1 we need only consider the $\mathcal{C}^2$ structure on X which by Theorem 4.2 is a refinement of a semicompact $\mathcal{C}^2$ structure. Let $\mathfrak{a}$ be

an s admissible atlas for the semicompact structure and let $\{X_1, X_2, \ldots\}$ be a listing of the components of X. Choose $x_n \in X_n$ and define $\varphi_1(x) = d_G(x_n, x) + n$ for $x \in X_n$. φ_1 is d_G Lipschitz and so is an $\mathcal{L}$ map. By Lemma 1 $\{\varphi_1 \leq N\}$ is compact for all $N < \infty$, i.e. φ_1 is topologically proper. Q.E.D.

City College of New York

Bibliography:

1. Abraham, R. "Lectures of Smale on Differential Topology" Columbia University Lecture Notes, 1962.

2. Abraham, R. and Robbins, J. "Transversal Mappings and Flows" Benjamin, New York, 1967.

3. Davis, Dean and Singer. "Complemented Subspaces and Λ Systems in Banach Spaces" Israel J. of Math., (6), 1968 pp. 303-309.

4. De Rham, Georges. "Varietes Differentiables" Hermann, Paris, 1960.

5. Dieudonne, J. "Foundations of Modern Analysis" Academic Press, New York, 1960.

6. Eliason, H. "Geometry of Manifolds of Maps" J. of Diff. Geom. (1), 1967, pp. 169-194.

7. Franks, J. "Notes on Manifolds of C^r Mappings" Northwestern University Lecture Notes, 1971.

8. Hale, J. "Ordinary Differential Equations" Wiley-Interscience, New York, 1969.

9. Hartman, P. "Ordinary Differential Equations" John Wiley and Sons, New York, 1964.

10. Hirsch, M. and Pugh, C. "Stable Manifolds and Hyperbolic Sets" Proc. Symp. Pure Math. Vol. 14, AMS, Providence, R.I., 1970, pp. 133-164.

11. Hirsch, Pugh and Shub. "Invariant Manifolds" Bull. AMS (76), 1970, pp. 1015-1019.

12. Hirsch, Pugh and Shub. "Invariant Manifolds" preprint.

13. Kelley, J. "General Topology" Van Nostrand, Princeton, N.J., 1955.

14. Lang, S. "Differential Manifolds" Addison-Wesley, Reading, Mass. 1972.

15. Michael, E. "Continuous Selections I" Ann. of Math., (63), 1956, pp. 361-382.

16. Munkres, J. "Elementary Differential Topology" Princeton University

Press, Princeton, N.J., 1961.

17. Murray, F.J. "On Complementary Manifolds and Projections in L_p and ℓ_p" Trans. AMS, (41), 1937, pp. 139-152.

18. Narasimhan, R. "Several Complex Variables" The University of Chicago Press, Chicago, Ill. 1971.

19. Nelson, E. "Topics in Dynamics I: Flows" Princeton University Press, Princeton, N. J., 1969.

20. Palais, R. "Lectures on the Differential Topology of Infinite Dimensional Manifolds" Lecture Notes, Brandeis University, 1965.

21. Palais, R. "Foundations of Global Analysis" Benjamin, New York, 1968.

22. Palais, R. "Lusternik-Schnirelman Theory on Banach Manifolds" Topology, (5), 1966, pp. 115-132.

23. Palais, R. "Critical Point Theory and the Minimax Principle" Proc. Symp. Pure Math. Vol. 15, AMS, Providence, R.I., 1970, pp. 185-212.

24. Palais, R. "Equivariant, Real Algebraic Differential Topology, Part 1: Smoothness Categories and Nash Manifolds", preprint.

25. Ruelle and Sullivan "Currents, Flows and Diffeomorphisms" Topology, (14), 1975, pp. 319-328.

26. Sondow, J. "When is a Manifold a Leaf of Some Foliation?" Bull. AMS, (81), 1975, pp.

27. Yosida, K. "Functional Analysis" Springer-Verlag, Berlin, 1965.

28. Hurewicz and Wallman "Dimension Theory" Princeton University Press, Princeton, N.J., 1941.

29. Eels, J. "A Setting for Global Analysis" Bull. AMS, (72), 1966, pp. 751-807.

30. Treves, F. "Basic Linear Partial Differential Equations" Academic Press, New York, 1975.

Index

Vol. 489: J. Bair and R. Fourneau, Etude Géométrique des Espaces Vectoriels. Une Introduction. VII, 185 pages. 1975.

Vol. 490: The Geometry of Metric and Linear Spaces. Proceedings 1974. Edited by L. M. Kelly. X, 244 pages. 1975.

Vol. 491: K. A. Broughan, Invariants for Real-Generated Uniform Topological and Algebraic Categories. X, 197 pages. 1975.

Vol. 492: Infinitary Logic: In Memoriam Carol Karp. Edited by D. W. Kueker. VI, 206 pages. 1975.

Vol. 493: F. W. Kamber and P. Tondeur, Foliated Bundles and Characteristic Classes. XIII, 208 pages. 1975.

Vol. 494: A Cornea and G. Licea. Order and Potential Resolvent Families of Kernels. IV, 154 pages. 1975.

Vol. 495: A. Kerber, Representations of Permutation Groups II. V, 175 pages. 1975.

Vol. 496: L. H. Hodgkin and V. P. Snaith, Topics in K-Theory. Two Independent Contributions. III, 294 pages. 1975.

Vol. 497: Analyse Harmonique sur les Groupes de Lie. Proceedings 1973–75. Edité par P. Eymard et al. VI, 710 pages. 1975.

Vol. 498: Model Theory and Algebra. A Memorial Tribute to Abraham Robinson. Edited by D. H. Saracino and V. B. Weispfenning. X, 463 pages. 1975.

Vol. 499: Logic Conference, Kiel 1974. Proceedings. Edited by G. H. Müller, A. Oberschelp, and K. Potthoff. V, 651 pages 1975.

Vol. 500: Proof Theory Symposion, Kiel 1974. Proceedings. Edited by J. Diller and G. H. Müller. VIII, 383 pages. 1975.

Vol. 501: Spline Functions, Karlsruhe 1975. Proceedings. Edited by K. Böhmer, G. Meinardus, and W. Schempp. VI, 421 pages. 1976.

Vol. 502: János Galambos, Representations of Real Numbers by Infinite Series. VI, 146 pages. 1976.

Vol. 503: Applications of Methods of Functional Analysis to Problems in Mechanics. Proceedings 1975. Edited by P. Germain and B. Nayroles. XIX, 531 pages. 1976.

Vol. 504: S. Lang and H. F. Trotter, Frobenius Distributions in GL_2-Extensions. III, 274 pages. 1976.

Vol. 505: Advances in Complex Function Theory. Proceedings 1973/74. Edited by W. E. Kirwan and L. Zalcman. VIII, 203 pages. 1976.

Vol. 506: Numerical Analysis, Dundee 1975. Proceedings. Edited by G. A. Watson. X, 201 pages. 1976.

Vol. 507: M. C. Reed, Abstract Non-Linear Wave Equations. VI, 128 pages. 1976.

Vol. 508: E. Seneta, Regularly Varying Functions. V, 112 pages. 1976.

Vol. 509: D. E. Blair, Contact Manifolds in Riemannian Geometry. VI, 146 pages. 1976.

Vol. 510: V. Poènaru, Singularités C^∞ en Présence de Symétrie. V, 174 pages. 1976.

Vol. 511: Séminaire de Probabilités X. Proceedings 1974/75. Edité par P. A. Meyer. VI, 593 pages. 1976.

Vol. 512: Spaces of Analytic Functions, Kristiansand, Norway 1975. Proceedings. Edited by O. B. Bekken, B. K. Øksendal, and A. Stray. VIII, 204 pages. 1976.

Vol. 513: R. B. Warfield, Jr. Nilpotent Groups. VIII, 115 pages. 1976.

Vol. 514: Séminaire Bourbaki vol. 1974/75. Exposés 453 – 470. IV, 276 pages. 1976.

Vol. 515: Bäcklund Transformations. Nashville, Tennessee 1974. Proceedings. Edited by R. M. Miura. VIII, 295 pages. 1976.

Vol. 516: M. L. Silverstein, Boundary Theory for Symmetric Markov Processes. XVI, 314 pages. 1976.

Vol. 517: S. Glasner, Proximal Flows. VIII, 153 pages. 1976.

Vol. 518: Séminaire de Théorie du Potentiel, Proceedings Paris 1972–1974. Edité par F. Hirsch et G. Mokobodzki. VI, 275 pages. 1976.

Vol. 519: J. Schmets, Espaces de Fonctions Continues. XII, 150 pages. 1976.

Vol. 520: R. H. Farrell, Techniques of Multivariate Calculation. X, 337 pages. 1976.

Vol. 521: G. Cherlin, Model Theoretic Algebra – Selected Topics. IV, 234 pages. 1976.

Vol. 522: C. O. Bloom and N. D. Kazarinoff, Short Wave Radiation Problems in Inhomogeneous Media: Asymptotic Solutions. V. 104 pages. 1976.

Vol. 523: S. A. Albeverio and R. J. Høegh-Krohn, Mathematical Theory of Feynman Path Integrals. IV, 139 pages. 1976.

Vol. 524: Séminaire Pierre Lelong (Analyse) Année 1974/75. Edité par P. Lelong. V, 222 pages. 1976.

Vol. 525: Structural Stability, the Theory of Catastrophes, and Applications in the Sciences. Proceedings 1975. Edited by P. Hilton. VI, 408 pages. 1976.

Vol. 526: Probability in Banach Spaces. Proceedings 1975. Edited by A. Beck. VI, 290 pages. 1976.

Vol. 527: M. Denker, Ch. Grillenberger, and K. Sigmund, Ergodic Theory on Compact Spaces. IV, 360 pages. 1976.

Vol. 528: J. E. Humphreys, Ordinary and Modular Representations of Chevalley Groups. III, 127 pages. 1976.

Vol. 529: J. Grandell, Doubly Stochastic Poisson Processes. X, 234 pages. 1976.

Vol. 530: S. S. Gelbart, Weil's Representation and the Spectrum of the Metaplectic Group. VII, 140 pages. 1976.

Vol. 531: Y.-C. Wong, The Topology of Uniform Convergence on Order-Bounded Sets. VI, 163 pages. 1976.

Vol. 532: Théorie Ergodique. Proceedings 1973/1974. Edité par J.-P. Conze and M. S. Keane. VIII, 227 pages. 1976.

Vol. 533: F. R. Cohen, T. J. Lada, and J. P. May, The Homology of Iterated Loop Spaces. IX, 490 pages. 1976.

Vol. 534: C. Preston, Random Fields. V, 200 pages. 1976.

Vol. 535: Singularités d'Applications Differentiables. Plans-sur-Bex. 1975. Edité par O. Burlet et F. Ronga. V, 253 pages. 1976.

Vol. 536: W. M. Schmidt, Equations over Finite Fields. An Elementary Approach. IX, 267 pages. 1976.

Vol. 537: Set Theory and Hierarchy Theory. Bierutowice, Poland 1975. A Memorial Tribute to Andrzej Mostowski. Edited by W. Marek, M. Srebrny and A. Zarach. XIII, 345 pages. 1976.

Vol. 538: G. Fischer, Complex Analytic Geometry. VII, 201 pages. 1976.

Vol. 539: A. Badrikian, J. F. C. Kingman et J. Kuelbs, Ecole d'Eté de Probabilités de Saint Flour V-1975. Edité par P.-L. Hennequin. IX, 314 pages. 1976.

Vol. 540: Categorical Topology, Proceedings 1975. Edited by E. Binz and H. Herrlich. XV, 719 pages. 1976.

Vol. 541: Measure Theory, Oberwolfach 1975. Proceedings. Edited by A. Bellow and D. Kölzow. XIV, 430 pages. 1976.

Vol. 542: D. A. Edwards and H. M. Hastings, Čech and Steenrod Homotopy Theories with Applications to Geometric Topology. VII, 296 pages. 1976.

Vol. 543: Nonlinear Operators and the Calculus of Variations, Bruxelles 1975. Edited by J. P. Gossez, E. J. Lami Dozo, J. Mawhin, and L. Waelbroeck, VII, 237 pages. 1976.

Vol. 544: Robert P. Langlands, On the Functional Equations Satisfied by Eisenstein Series. VII, 337 pages. 1976.

Vol. 545: Noncommutative Ring Theory. Kent State 1975. Edited by J. H. Cozzens and F. L. Sandomierski. V, 212 pages. 1976.

Vol. 546: K. Mahler, Lectures on Transcendental Numbers. Edited and Completed by B. Diviš and W. J. Le Veque. XXI, 254 pages. 1976.

Vol. 547: A. Mukherjea and N. A. Tserpes, Measures on Topological Semigroups: Convolution Products and Random Walks. V, 197 pages. 1976.

Vol. 548: D. A. Hejhal, The Selberg Trace Formula for PSL (2, ℝ). Volume I. VI, 516 pages. 1976.

Vol. 549: Brauer Groups, Evanston 1975. Proceedings. Edited by D. Zelinsky. V, 187 pages. 1976.

Vol. 550: Proceedings of the Third Japan – USSR Symposium on Probability Theory. Edited by G. Maruyama and J. V. Prokhorov. VI, 722 pages. 1976.

Vol. 551: Algebraic K-Theory, Evanston 1976. Proceedings. Edited by M. R. Stein. XI, 409 pages. 1976.

Vol. 552: C. G. Gibson, K. Wirthmüller, A. A. du Plessis and E. J. N. Looijenga. Topological Stability of Smooth Mappings. V, 155 pages. 1976.

Vol. 553: M. Petrich, Categories of Algebraic Systems. Vector and Projective Spaces, Semigroups, Rings and Lattices. VIII, 217 pages. 1976.

Vol. 554: J. D. H. Smith, Mal'cev Varieties. VIII, 158 pages. 1976.

Vol. 555: M. Ishida, The Genus Fields of Algebraic Number Fields. VII, 116 pages. 1976.

Vol. 556: Approximation Theory. Bonn 1976. Proceedings. Edited by R. Schaback and K. Scherer. VII, 466 pages. 1976.

Vol. 557: W. Iberkleid and T. Petrie, Smooth S^1 Manifolds. III, 163 pages. 1976.

Vol. 558: B. Weisfeiler, On Construction and Identification of Graphs. XIV, 237 pages. 1976.

Vol. 559: J.-P. Caubet, Le Mouvement Brownien Relativiste. IX, 212 pages. 1976.

Vol. 560: Combinatorial Mathematics, IV, Proceedings 1975. Edited by L. R. A. Casse and W. D. Wallis. VII, 249 pages. 1976.

Vol. 561: Function Theoretic Methods for Partial Differential Equations. Darmstadt 1976. Proceedings. Edited by V. E. Meister, N. Weck and W. L. Wendland. XVIII, 520 pages. 1976.

Vol. 562: R. W. Goodman, Nilpotent Lie Groups: Structure and Applications to Analysis. X, 210 pages. 1976.

Vol. 563: Séminaire de Théorie du Potentiel. Paris, No. 2. Proceedings 1975–1976. Edited by F. Hirsch and G. Mokobodzki. VI, 292 pages. 1976.

Vol. 564: Ordinary and Partial Differential Equations, Dundee 1976. Proceedings. Edited by W. N. Everitt and B. D. Sleeman. XVIII, 551 pages. 1976.

Vol. 565: Turbulence and Navier Stokes Equations. Proceedings 1975. Edited by R. Temam. IX, 194 pages. 1976.

Vol. 566: Empirical Distributions and Processes. Oberwolfach 1976. Proceedings. Edited by P. Gaenssler and P. Révész. VII, 146 pages. 1976.

Vol. 567: Séminaire Bourbaki vol. 1975/76. Exposés 471–488. IV, 303 pages. 1977.

Vol. 568: R. E. Gaines and J. L. Mawhin, Coincidence Degree, and Nonlinear Differential Equations. V, 262 pages. 1977.

Vol. 569: Cohomologie Etale SGA 4½. Séminaire de Géométrie Algébrique du Bois-Marie. Edité par P. Deligne. V, 312 pages. 1977.

Vol. 570: Differential Geometrical Methods in Mathematical Physics, Bonn 1975. Proceedings. Edited by K. Bleuler and A. Reetz. VIII, 576 pages. 1977.

Vol. 571: Constructive Theory of Functions of Several Variables, Oberwolfach 1976. Proceedings. Edited by W. Schempp and K. Zeller. VI. 290 pages. 1977

Vol. 572: Sparse Matrix Techniques, Copenhagen 1976. Edited by V. A. Barker. V, 184 pages. 1977.

Vol. 573: Group Theory, Canberra 1975. Proceedings. Edited by R. A. Bryce, J. Cossey and M. F. Newman. VII, 146 pages. 1977.

Vol. 574: J. Moldestad, Computations in Higher Types. IV, 203 pages. 1977.

Vol. 575: K-Theory and Operator Algebras, Athens, Georgia 1975. Edited by B. B. Morrel and I. M. Singer. VI, 191 pages. 1977.

Vol. 576: V. S. Varadarajan, Harmonic Analysis on Real Reductive Groups. VI, 521 pages. 1977.

Vol. 577: J. P. May, E_∞ Ring Spaces and E_∞ Ring Spectra. IV, 268 pages. 1977.

Vol. 578: Séminaire Pierre Lelong (Analyse) Année 1975/76. Edité par P. Lelong. VI, 327 pages. 1977.

Vol. 579: Combinatoire et Représentation du Groupe Symétrique, Strasbourg 1976. Proceedings 1976. Edité par D. Foata. IV, 339 pages. 1977.

Vol. 580: C. Castaing and M. Valadier, Convex Analysis and Measurable Multifunctions. VIII, 278 pages. 1977.

Vol. 581: Séminaire de Probabilités XI, Université de Strasbourg. Proceedings 1975/1976. Edité par C. Dellacherie, P. A. Meyer et M. Weil. VI, 574 pages. 1977.

Vol. 582: J. M. G. Fell, Induced Representations and Banach *-Algebraic Bundles. IV, 349 pages. 1977.

Vol. 583: W. Hirsch, C. C. Pugh and M. Shub, Invariant Manifolds. IV, 149 pages. 1977.

Vol. 584: C. Brezinski, Accélération de la Convergence en Analyse Numérique. IV, 313 pages. 1977.

Vol. 585: T. A. Springer, Invariant Theory. VI, 112 pages. 1977.

Vol. 586: Séminaire d'Algèbre Paul Dubreil, Paris 1975–1976 (29ème Année). Edited by M. P. Malliavin. VI, 188 pages. 1977.

Vol. 587: Non-Commutative Harmonic Analysis. Proceedings 1976. Edited by J. Carmona and M. Vergne. IV, 240 pages. 1977.

Vol. 588: P. Molino, Théorie des G-Structures: Le Problème d'Equivalence. VI, 163 pages. 1977.

Vol. 589: Cohomologie l-adique et Fonctions L. Séminaire de Géométrie Algébrique du Bois-Marie 1965–66, SGA 5. Edité par L. Illusie. XII, 484 pages. 1977.

Vol. 590: H. Matsumoto, Analyse Harmonique dans les Systèmes de Tits Bornologiques de Type Affine. IV, 219 pages. 1977.

Vol. 591: G. A. Anderson, Surgery with Coefficients. VIII, 157 pages. 1977.

Vol. 592: D. Voigt, Induzierte Darstellungen in der Theorie der endlichen, algebraischen Gruppen. V, 413 Seiten. 1977.

Vol. 593: K. Barbey and H. König, Abstract Analytic Function Theory and Hardy Algebras. VIII, 260 pages. 1977.

Vol. 594: Singular Perturbations and Boundary Layer Theory, Lyon 1976. Edited by C. M. Brauner, B. Gay, and J. Mathieu. VIII, 539 pages. 1977.

Vol. 595: W. Hazod, Stetige Faltungshalbgruppen von Wahrscheinlichkeitsmaßen und erzeugende Distributionen. XIII, 157 Seiten. 1977.

Vol. 596: K. Deimling, Ordinary Differential Equations in Banach Spaces. VI, 137 pages. 1977.

Vol. 597: Geometry and Topology, Rio de Janeiro, July 1976. Proceedings. Edited by J. Palis and M. do Carmo. VI, 866 pages. 1977.

Vol. 598: J. Hoffmann-Jørgensen, T. M. Liggett et J. Neveu, Ecole d'Eté de Probabilités de Saint-Flour VI – 1976. Edité par P.-L. Hennequin. XII, 447 pages. 1977.

Vol. 599: Complex Analysis, Kentucky 1976. Proceedings. Edited by J. D. Buckholtz and T. J. Suffridge. X, 159 pages. 1977.

Vol. 600: W. Stoll, Value Distribution on Parabolic Spaces. VIII, 216 pages. 1977.

Vol. 601: Modular Functions of one Variable V, Bonn 1976. Proceedings. Edited by J.-P. Serre and D. B. Zagier. VI, 294 pages. 1977.

Vol. 602: J. P. Brezin, Harmonic Analysis on Compact Solvmanifolds. VIII, 179 pages. 1977.

Vol. 603: B. Moishezon, Complex Surfaces and Connected Sums of Complex Projective Planes. IV, 234 pages. 1977.

Vol. 604: Banach Spaces of Analytic Functions, Kent, Ohio 1976. Proceedings. Edited by J. Baker, C. Cleaver and Joseph Diestel. VI, 141 pages. 1977.

Vol. 605: Sario et al., Classification Theory of Riemannian Manifolds. XX, 498 pages. 1977.

Vol. 606: Mathematical Aspects of Finite Element Methods. Proceedings 1975. Edited by I. Galligani and E. Magenes. VI, 362 pages. 1977.

Vol. 607: M. Métivier, Reelle und Vektorwertige Quasimartingale und die Theorie der Stochastischen Integration. X, 310 Seiten. 1977.

Vol. 608: Bigard et al., Groupes et Anneaux Réticulés. XIV, 334 pages. 1977.